Advances in Intellig

Volume 906

Series Editor

Janusz Kacprzyk, Systems Research Institute, Polish Academy of Sciences, Warsaw, Poland

Advisory Editors

Nikhil R. Pal, Indian Statistical Institute, Kolkata, India
Rafael Bello Perez, Faculty of Mathematics, Physics and Computing, Universidad Central de Las Villas, Santa Clara, Cuba
Emilio S. Corchado, University of Salamanca, Salamanca, Spain
Hani Hagras, School of Computer Science and Electronic Engineering, University of Essex, Colchester, UK
László T. Kóczy, Department of Automation, Széchenyi István University, Gyor, Hungary
Vladik Kreinovich, Department of Computer Science, University of Texas at El Paso, El Paso, TX, USA
Chin-Teng Lin, Department of Electrical Engineering, National Chiao Tung University, Hsinchu, Taiwan
Jie Lu, Faculty of Engineering and Information Technology, University of Technology Sydney, Sydney, NSW, Australia
Patricia Melin, Graduate Program of Computer Science, Tijuana Institute of Technology, Tijuana, Mexico
Nadia Nedjah, Department of Electronics Engineering, University of Rio de Janeiro, Rio de Janeiro, Brazil
Ngoc Thanh Nguyen, Faculty of Computer Science and Management, Wrocław University of Technology, Wrocław, Poland
Jun Wang, Department of Mechanical and Automation Engineering, The Chinese University of Hong Kong, Shatin, Hong Kong

The series "Advances in Intelligent Systems and Computing" contains publications on theory, applications, and design methods of Intelligent Systems and Intelligent Computing. Virtually all disciplines such as engineering, natural sciences, computer and information science, ICT, economics, business, e-commerce, environment, healthcare, life science are covered. The list of topics spans all the areas of modern intelligent systems and computing such as: computational intelligence, soft computing including neural networks, fuzzy systems, evolutionary computing and the fusion of these paradigms, social intelligence, ambient intelligence, computational neuroscience, artificial life, virtual worlds and society, cognitive science and systems, Perception and Vision, DNA and immune based systems, self-organizing and adaptive systems, e-Learning and teaching, human-centered and human-centric computing, recommender systems, intelligent control, robotics and mechatronics including human-machine teaming, knowledge-based paradigms, learning paradigms, machine ethics, intelligent data analysis, knowledge management, intelligent agents, intelligent decision making and support, intelligent network security, trust management, interactive entertainment, Web intelligence and multimedia.

The publications within "Advances in Intelligent Systems and Computing" are primarily proceedings of important conferences, symposia and congresses. They cover significant recent developments in the field, both of a foundational and applicable character. An important characteristic feature of the series is the short publication time and world-wide distribution. This permits a rapid and broad dissemination of research results.

** **Indexing: The books of this series are submitted to ISI Proceedings, EI-Compendex, DBLP, SCOPUS, Google Scholar and Springerlink** **

More information about this series at http://www.springer.com/series/11156

N. R. Shetty · L. M. Patnaik · H. C. Nagaraj ·
Prasad Naik Hamsavath · N. Nalini
Editors

Emerging Research in Computing, Information, Communication and Applications

ERCICA 2018, Volume 2

Editors
N. R. Shetty
Central University of Karnataka
Kalaburagi, Karnataka, India

L. M. Patnaik
National Institute of Advanced Studies
Bangalore, Karnataka, India

H. C. Nagaraj
Nitte Meenakshi Institute of Technology
Bangalore, Karnataka, India

Prasad Naik Hamsavath
Nitte Meenakshi Institute of Technology
Bangalore, Karnataka, India

N. Nalini
Nitte Meenakshi Institute of Technology
Bangalore, Karnataka, India

ISSN 2194-5357   ISSN 2194-5365 (electronic)
Advances in Intelligent Systems and Computing
ISBN 978-981-13-6000-8   ISBN 978-981-13-6001-5 (eBook)
https://doi.org/10.1007/978-981-13-6001-5

Library of Congress Control Number: 2018966829

© Springer Nature Singapore Pte Ltd. 2019
This work is subject to copyright. All rights are reserved by the Publisher, whether the whole or part of the material is concerned, specifically the rights of translation, reprinting, reuse of illustrations, recitation, broadcasting, reproduction on microfilms or in any other physical way, and transmission or information storage and retrieval, electronic adaptation, computer software, or by similar or dissimilar methodology now known or hereafter developed.
The use of general descriptive names, registered names, trademarks, service marks, etc. in this publication does not imply, even in the absence of a specific statement, that such names are exempt from the relevant protective laws and regulations and therefore free for general use.
The publisher, the authors and the editors are safe to assume that the advice and information in this book are believed to be true and accurate at the date of publication. Neither the publisher nor the authors or the editors give a warranty, expressed or implied, with respect to the material contained herein or for any errors or omissions that may have been made. The publisher remains neutral with regard to jurisdictional claims in published maps and institutional affiliations.

This Springer imprint is published by the registered company Springer Nature Singapore Pte Ltd.
The registered company address is: 152 Beach Road, #21-01/04 Gateway East, Singapore 189721, Singapore

Organizing Committee

ERCICA 2018

The Fifth International Conference on "Emerging Research in Computing, Information, Communication and Applications," ERCICA 2018, was held during July 27–28, 2018, at the Nitte Meenakshi Institute of Technology (NMIT), Bangalore, and organized by the Departments of CSE and MCA, NMIT.

Chief Patrons

Dr. N. V. Hegde, President, Nitte Education Trust, Mangalore, India.
Dr. N. R. Shetty, Chancellor, Central University of Karnataka, Kalaburagi, and Advisor, Nitte Education Trust, Mangalore, India.

Conference Chair

Dr. H. C. Nagaraj, Principal, NMIT, Bangalore, India.

Program Chairs

Dr. Prasad Naik Hamsavath, HOD, MCA, NMIT, Bangalore, India.
Dr. N. Nalini, Professor, CSE, NMIT, Bangalore, India.

Publisher

Springer

Advisory Chairs

Dr. K. Sudha Rao, Advisor, Admin and Management, NMIT, Bangalore, India.
Mr. Rohit Punja, Administrator, NET, Mangalore, India.
Dr. Jharna Majumdar, Dean (R&D), NMIT, Bangalore, India.
Mr. K. A. Ranganatha Setty, Dean (Academic), NMIT, Bangalore, India.

Advisory Committee

Dr. L. M. Patnaik, INSA Senior Scientist, NIAS, Bangalore, India.
Dr. B. S. Sonde, Former Vice Chancellor, Goa University, Goa, India.
Dr. D. K. Subramanian, Former Dean and Professor, IISc, Bangalore, India.
Dr. K. D. Nayak, Former OS & CC, R&D (MED & MIST), DRDO, India.
Dr. Kalidas Shetty, Founding Director of Global Institute of Food Security and International Agriculture (GIFSIA), North Dakota State University, Fargo, USA.
Dr. Kendall E. Nygard, Professor of Computer Science and Operations Research, North Dakota State University, Fargo, USA.
Dr. Sathish Udpa, Dean and Professor, Michigan State University, Michigan, USA.
Dr. K. N. Bhat, Visiting Professor, Center for Nano Science and Engineering-CeNSE, IISc, Bangalore, India.
Dr. K. R. Venugopal, Principal, UVCE, Bangalore, India.
Dr. C. P. Ravikumar, Director, Technical Talent Development at Texas Instruments, Bangalore, India.
Dr. Navakanta Bhat, Chairperson, Center for Nano Science and Engineering-CeNSE, IISc, Bangalore, India.
Dr. Anand Nayyar, Professor, Researcher and Scientist in Graduate School, Duy Tan University, Da Nang, Vietnam.

Program Committee

Dr. Savitri Bevinakoppa, Professional Development and Scholarship Coordinator, School of IT and Engineering, Melbourne Institute of Technology (MIT), Australia.
Dr. P. Ramprasad, Professor, Department of CSE and IT, Manipal University, Dubai.
Dr. Ohta Tsuyoshi, Department of Computer Sciences, Shizuoka University, Japan.

Organizing Committee

Dr. Sonajharia Minz, Professor, School of Computer and Systems Sciences, Jawaharlal Nehru University, New Delhi, India.
Dr. Sanjay Kumar Dhurandher, Professor and Head, Department of Information Technology, Netaji Subhas Institute of Technology, New Delhi, India. Networks, University Putra Malaysia, Malaysia.
Dr. Ramesh R. Galigekere, Professor, Department of Biomedical Engineering, Manipal Institute of Technology, Manipal, Karnataka, India.
Dr. K. G. Srinivasa, CBP Government Engineering College, New Delhi, India.
Dr. S. Ramanarayana Reddy, HOD, Department of CSE, IGDTU for Women, Kashmere Gate, Delhi, India.
Dr. K. Raghavendra, Professor, School of Computing, National University of Singapore, Singapore.
Dr. Abhijit Lele, Principal Consultant, Robert Bosch, Bangalore, India.
Dr. Bappaditya Mandal, Faculty of Science, Engineering and Computing, Kingston University, London.

Organizing Co-chairs

Dr. M. N. Thippeswamy, Professor and Head, Department of CSE, NMIT, Bangalore, India.
Dr. H. A. Sanjay, Professor and Head, Department of ISE, NMIT, Bangalore, India.
Dr. S. Sandya, Professor and Head, Department of ECE, NMIT, Bangalore, India.
Dr. H. M. Ravikumar, Professor and Head, Department of EEE, NMIT, Bangalore, India.

Preface

The Fifth International Conference on "Emerging Research in Computing, Information, Communication and Applications," ERCICA 2018, is an annual event organized at the Nitte Meenakshi Institute of Technology (NMIT), Yelahanka, Bangalore, India.

ERCICA aims to provide an interdisciplinary forum for discussion among researchers, engineers and scientists to promote research and exchange of knowledge in computing, information, communication and related applications. This conference will provide a platform for networking of academicians, engineers and scientists and also will enthuse the participants to undertake high-end research in the above thrust areas.

ERCICA 2018 received more than 400 papers from all over the world, viz. from China, UK, Africa, Saudi Arabia and India. The ERCICA Technical Review Committee has followed all necessary steps to screen more than 400 papers by going through six rounds of quality checks on each paper before selection for presentation/publication in Springer proceedings.

The acceptance ratio is only 1:3.

Bangalore, India
July 2018

Prasad Naik Hamsavath
N. Nalini

Acknowledgements

First of all, we would like to thank Prof. N. R. Shetty who has always been the guiding force behind this event's success. It was his dream that we have striven to make a reality. Our thanks to Prof. L. M. Patnaik, who has monitored the whole activity of the conference from the beginning till its successful end.

Our special thanks to Springer and especially the editorial staff who were patient, meticulous and friendly with their constructive criticism on the quality of papers and outright rejection at times without compromising the quality of the papers as they are always known for publishing the best international papers.

We would like to express our gratitude to all the review committee members of all the themes of computing, information, communication and applications and the best-paper-award review committee members.

Finally, we would like to express our heartfelt gratitude and warmest thanks to the ERCICA 2018 organizing committee members for their hard work and outstanding efforts. We know how much time and energy this assignment demanded, and we deeply appreciate all the efforts to make it a grand success.

Our special thanks to all the authors who have contributed to publishing their research work in this conference and participated to make this conference a grand success. Thanks to everyone who have directly or indirectly contributed to the success of this conference ERCICA 2018.

Regards
Program Chairs
ERCICA 2018

About the Conference

ERCICA 2018

The Fifth International Conference on "Emerging Research in Computing, Information, Communication and Applications," ERCICA 2018, is an annual event jointly organized by the Departments of CSE and MCA during July 27–28, 2018, at the Nitte Meenakshi Institute of Technology (NMIT), Yelahanka, Bangalore, India.

ERCICA 2018 is organized under the patronage of Prof. N. R. Shetty, Advisor, Nitte Education Trust. Dr. L. M. Patnaik, Technical Advisor, NMIT, and Dr. H. C. Nagaraj, Principal, served as Conference Chairs, and Program Chairs of the conference were Dr. Prasad Naik Hamsavath, Professor and Head, MCA, and Dr. N. Nalini, Professor, CSE, NMIT, Bangalore, Karnataka.

ERCICA aims to provide an interdisciplinary forum for discussion among researchers, engineers and scientists to promote research and exchange of knowledge in computing, information, communication and related applications. This conference will provide a platform for networking of academicians, engineers and scientists and also will enthuse the participants to undertake high-end research in the above thrust areas.

For ERCICA 2019, authors are invited to submit the manuscripts of their original and unpublished research contributions to ercica.chair@gmail.com (ERCICA Web site: http://nmit.ac.in/ercica/ercica.html). All the submitted papers will go through a peer review process, and the corresponding authors will be notified about the outcome of the review process. There will be six rounds of quality checks on each paper before selection for presentation/publication. Authors of the selected papers may present their papers during the conference.

Theme Editors

Computing

Dr. N. Nalini
Dr. Thippeswamy M. N.

Information

Dr. Sanjay H. A.
Dr. Shakti Mishra

Communication

Dr. H. C. Nagaraj
Dr. Raghunandan S.
Prof. Sankar Dasiga

Application

Dr. Prasad Naik Hamsavath
Prof. Sitaram Yaji

Contents

Use of Blockchain for Smart T-Shirt Design Ownership 1
Ashley Alexsius D'Souza and Okstynn Rodrigues

A Feasibility Study and Simulation of 450 kW Grid Connected
Solar PV System at NMIT, Bangalore 7
B. Smitha, N. Samanvita and H. M. Ravikumar

Prediction of a Dam's Hazard Level 19
Urna Kundu, Srabanti Ghosh and Satyakama Paul

Elemental Racing 29
Lingala Siva Karthik Reddy, Karthik Koka, Amiya Kumar Dash
and Manjusha Pandey

A Survey on Existing Convolutional Neural Networks and Waste
Management Techniques and an Approach to Solve Waste
Classification Problem Using Neural Networks 45
Tejashwini Hiremath and S. Rajarajeswari

L1-Regulated Feature Selection in Microarray Cancer Data
and Classification Using Random Forest Tree 65
B. H. Shekar and Guesh Dagnew

Automated Delineation of Costophrenic Recesses on Chest
Radiographs 89
Prashant A. Athavale and P. S. Puttaswamy

Using Location-Based Service for Interaction 103
K. M. Deepika, Piyush Chaterjee, Sourav Kishor Singh, Writtek Dey
and Yatharth Kundra

Modeling Implementation of Big Data Analytics in Oil
and Gas Industries in India 113
Dilip Motwani and G. T. Thampi

Practical Market Indicators for Algorithmic Stock Market Trading: Machine Learning Techniques and Grid Strategy 121
Ajithkumar Sreekumar, Prabhasa Kalkur and Mohammed Moiz

A Review on Feature Selection Algorithms 133
Savina Colaco, Sujit Kumar, Amrita Tamang and Vinai George Biju

A Review on Ensembles-Based Approach to Overcome Class Imbalance Problem 155
Sujit Kumar, J. N. Madhuri and Mausumi Goswami

"College Explorer" An Authentication Guaranteed Information Display and Management System 173
Sonali Majumdar, K. M. Monika Patel, Arushi Gupta and M. N. Thippeswamy

Activity-Based Music Classifier: A Supervised Machine Learning Approach for Curating Activity-Based Playlists 185
B. P. Aniruddha Achar, N. D. Aiyappa, B. Akshaj, M. N. Thippeswamy and N. Pillay

New Password Embedding Technique Using Elliptic Curve Over Finite Field 199
D. Sravana Kumar, C. H. Suneetha and P. Sirisha

Performance of Wind Energy Conversion System During Fault Condition and Power Quality Improvement of Grid-Connected WECS by FACTS (UPFC) 211
Sudeep Shetty, H. L. Suresh, M. Sharanappa and C. H. Venkat Ramesh

Comprehensive Survey on Hadoop Security 227
Maria Martis, Namratha V. Pai, R. S. Pragathi, S. Rakshatha and Sunanda Dixit

Descriptive Data Analysis of Real Estate Using Cube Technology 237
Gursimran Kaur and Harkiran Kaur

Temporal Information Retrieval and Its Application: A Survey 251
Rakshita Bansal, Monika Rani, Harish Kumar and Sakshi Kaushal

A Survey on Multi-resolution Methods for De-noising Medical Images 263
G. Bharath, A. E. Manjunath and K. S. Swarnalatha

Performance Analysis of Vedic Multiplier with Different Square Root BK Adders 273
Ranjith B. Gowda, R. M. Banakar and Basavaprasad

Analysis of Traffic Characteristics of Skype Video Calls Over the Internet 287
Gulshan Kumar and N. G. Goudru

Smart Tourist Guide (Touristo) 299
M. R. Sowmya, Shashi Prakash, Shubham K. Singh, Sushent Maloo and Sachindra Yadav

Thefted Vehicle Identification System and Smart Ambulance System in VANETs 313
S. R. Nagaraja, N. Nalini, B. A. Mohan and Afroz Pasha

Performance Study of OpenMP and Hybrid Programming Models on CPU–GPU Cluster 323
B. N. Chandrashekhar and H. A. Sanjay

Signature Analysis for Forgery Detection 339
Dinesh Rao Adithya, V. L. Anagha, M. R. Niharika, N. Srilakshmi and Shastry K. Aditya

Optimal Sensor Deployment and Battery Life Enhancement Strategies to Employ Smart Irrigation Solutions for Indian Agricultural Sector 351
M. K. Ajay, H. A. Sanjay, Sai Jeevan, T. K. Harshitha, M. Farhana mobin and K. Aditya Shastry

Smart Waste Monitoring Using Wireless Sensor Networks 363
T. V. Chandan, R. Chaitra Kumari, Renu Tekam and B. V. Shruti

Digital Filter Technique Used in Signal Processing for Analysing of ECG Signal 371
A. E. Manjunath, M. V. Vijay Kumar and K. S. Swarnalatha

GeoFencing-Based Accident Avoidance Notification for Road Safety 379
Bhavyashree Nayak, Priyanka S. Mugali, B. Raksha Rao, Saloni Sindhava, D. N. Disha and K. S. Swarnalatha

An Innovative IoT- and Middleware-Based Architecture for Real-Time Patient Health Monitoring 387
B. A. Mohan and H. Sarojadevi

Travelling Salesman Problem: An Empirical Comparison Between ACO, PSO, ABC, FA and GA 397
Kinjal Chaudhari and Ankit Thakkar

Twitter Data Sentiment Analysis on a Malayalam Dataset Using Rule-Based Approach 407
Deepa Mary Mathews and Sajimon Abraham

An IoT-Based Smart Water Microgrid and Smart Water Tank Management System 417
Shubham Kumar, Sushmita Yadav, H. M. Yashaswini and Sanket Salvi

Smart Sensing for Vehicular Approach 433
Mukesh Chandra Sah, Chandan Kumar Sah, Shuhaib Akhter Ansari, Anjit Subedi, A. C. Ramachandra and P. Ushashree

Android Malware Detection Techniques 449
Shreya Khemani, Darshil Jain and Gaurav Prasad

A Comparative Study of Machine Learning Techniques for Emotion Recognition 459
Rhea Sharma, Harshit Rajvaidya, Preksha Pareek and Ankit Thakkar

IoT-Based Smart Parking System 465
G. Abhijith, H. A. Sanjay, Aditya Rajeev, Chidanandan, Rajath and Mohan Murthy

Smart Agricultural Monitoring System Using Internet of Things 473
H. V. Asha, K. Kavya, S. Keerthana, G. Kruthika and R. Pavithra

Detecting Healthiness of Leaves Using Texture Features 483
Srishti Shetty, Zulaikha Lateef, Sparsha Pole, Vidyadevi G. Biradar, S. Brunda and H. A. Sanjay

A Survey on Different Network Intrusion Detection Systems and CounterMeasure 497
Divya Rajput and Ankit Thakkar

Compressed Sensing for Image Compression: Survey of Algorithms 507
S. K. Gunasheela and H. S. Prasantha

Intrusion Detection System Using Random Forest on the NSL-KDD Dataset 519
Prashil Negandhi, Yash Trivedi and Ramchandra Mangrulkar

Dual-Mode Wide Band Microstrip Bandpass Filter with Tunable Bandwidth and Controlled Center Frequency for C-Band Applications 533
Shobha I. Hugar, Vaishali Mungurwadi and J. S. Baligar

ADHYAYAN—An Innovative Interest Finder and Career Guidance Application 541
Akshay Talke, Virendra Patil, Sanyam Raj, Rohit Kr. Singh, Ameya Jawalgekar and Anand Bhosale

Implementation of Cure Clustering Algorithm for Video Summarization and Healthcare Applications in Big Data 553
Jharna Majumdar, Sumant Udandakar and B. G. Mamatha Bai

Redundancy Management of On-board Computer in Nanosatellites 565
Shubham, Vikash Kumar, Vishal Pandey, K. Arun Kumar and S. Sandya

A Fault Tolerant Architecture for Software Defined Network 571
Bini Y. Baby, B. Karunakara Rai, N. Karthik, Akshith Chandra, R. Dheeraj and S. RaviShankar

Optimal Thresholding in Direct Binary Search Visual Cryptography for Enhanced Bank Locker System 589
Sandhya Anne Thomas and Saylee Gharge

Comparison Between the DDFS Implementation Using the Look-up Table Method and the CORDIC Method 601
Anish K. Navalgund, V. Akshara, Ravali Jadhav, Shashank Shankar and S. Sandya

Adding Intelligence to a Car 609
Komal Suresh, Svati S. Murthy, Usha Nanthini, Shilpa Mondal and P. Raji

Thermal Care and Saline Level Monitoring System for Neonatal Using IoT 619
Huma Kousar Sangreskop

Home Security System Using GSM 627
P. Mahalakshmi, Raunak Singhania, Debabrata Shil and A. Sharmila

Automatic Toll Tax Collection Using GSM 635
P. Mahalakshmi, Viraj Pradip Puntambekar, Aayushi Jain and Raunak Singhania

Facial Expression Recognition by Considering Nonuniform Local Binary Patterns 645
K. Srinivasa Reddy, E. Sunil Reddy and N. Baswanth

About the Editors

Prof. N. R. Shetty is the Chancellor of Central University of Karnataka, Kalaburagi, and Chairman of the Review Commission for the State Private University Karnataka. He is currently serving as an advisor to the Nitte Meenakshi Institute of Technology (NMIT), Bangalore. He is also founder Vice-President of the International Federation of Engineering Education Societies (IFEES), Washington DC, USA. He served as Vice Chancellor of Bangalore University for two terms and President of the ISTE, New Delhi for three terms. He was also a member of the AICTE's Executive Committee and Chairman of its South West Region Committee.

Prof. L. M. Patnaik obtained his Ph.D. in Real-Time Systems in 1978, and his D.Sc. in Computer Systems and Architectures in 1989, both from the Indian Institute of Science, Bangalore. From 2008 to 2011, he was Vice Chancellor of the Defense Institute of Advanced Technology, Deemed University, Pune. Currently he is an Honorary Professor with the Department of Electronic Systems Engineering, Indian Institute of Science, Bangalore, and INSA Senior Scientist and Adjunct Professor with the National Institute of Advanced Studies, Bangalore.

Dr. H. C. Nagaraj completed his B.E. in Electronics & Communication from the University of Mysore in 1981, his M.E. in Communication Systems from P.S.G College of Technology, Coimbatore in 1984. He was awarded Ph.D. (Biomedical Signal Processing and Instrumentation) from Indian Institute of Technology Madras, Chennai in 2000. Dr. Nagaraj has teaching experience spanning more than 35 years. He was the Chairman, BOS of IT/BM/ML of Visvesvaraya Technological University, Belagavi for 2010–13 and Member, Academic Senate of VTU for 06 years w.e.f. April 2010. Further extended for a period of three years w.e.f 02-06-2016. He has the credit of publishing more than 40 technical papers and has also published a book-"VLSI Circuits", Star-Tech Education, Bangalore in 2006. He has won the Best Student Paper Award at the 5th National Conference of Biomechanics held at I.I.T. Madras, Chennai in 1996 and Best Paper Award at the Karnataka State Level Seminar on "Introduction of Flexible System in Technical

Education" under Visveswaraiah Technological University in 1999 at P.E.S. Institute of Technology, Bangalore. Presently, he is the Dean, Faculty of Engineering, Visvesvaraya Technological University Belgaum, for three years from 2016 to 2019 and Member of the Court, Pondicherry University. He is the Member of Karnataka State Innovation Council, Government of Karnataka and Member of NAAC (UGC) Peer Team to assess the institutions for Accreditation. He has also visited as an Expert Member of the UGC, New Delhi for inspecting the colleges seeking Autonomous Status.

Dr. Prasad Naik Hamsavath is a Professor and Head of the Department of Master of Computer Applications at Nitte Meenakshi Institute of Technology, Bangalore. He completed his Ph.D. at Jawaharlal Nehru University, New Delhi, India. Dr. Prasad N H has more than 12 years of experience in different roles in both public and private sector enterprises, including the Ministry of Human Resource and Development, New Delhi, Government of India. He received the prestigious "Dr. Abdul Kalam Life Time Achievement Award" and also received a "Young Faculty" award at the 2nd Academic Brilliance Awards.

Dr. N. Nalini is a Professor at the Department of Computer Science and Engineering at Nitte Meenakshi Institute of Technology, Bangalore. She received her MS from BITS, Pilani in 1999 and her Ph.D. from Visvesvaraya Technological University in 2007. She has more than 21 years of teaching and 14 years of research experience. She has written numerous international publications, and **Received "Bharath Jyoti Award" by India International Friendship Society, New Delhi on 2012**, from Dr. Bhishma Narain Singh, former Governor of Tamilnadu and Assam. She received the **"Dr. Abdul Kalam Life time achievement National Award" for excellence in Teaching, Research Publications, and Administration by International Institute for Social and Economic Reforms, IISER, Bangalore on 29th Dec 2014**. She is also the recipient of "Distinguished Professor" award by TechNext India 2017 in association with Computer Society of India-CSI, Mumbai Chapter and "Best Professor in Computer Science & Engineering "award by **26th Business School Affaire & Dewang Mehta National Education Awards (Regional Round) on 5th September 2018, at Bangalore**. She is a lifetime member of the ISTE, CSI, ACEEE and IIFS.

Use of Blockchain for Smart T-Shirt Design Ownership

Ashley Alexsius D'Souza and Okstynn Rodrigues

Abstract This study aims to provide a reliable solution for the T-shirt designer to gain ownership of his design using the blockchain technology. As blockchain is a decentralized technology, it will provide authentication for the ownership of the artwork among various non-trusting members. In this research paper, we have explained how to resolve this issue for the designer as well as the customer. Using this method, a novel distributed application can be created.

Keywords Bitcoin · Blockchain · Smart contract · Supply chain

1 Introduction

On every online shopping site, it is a common behavior to download designs and use them as your own. If the original designer does not get any credit for his design, he is at loss and his business can be in trouble. The ability to manage the T-shirt design ownership in an easy manner and streamline the process for maximum efficiency can be achieved using blockchain. In this paper, we suggest how the owner of a design can gain ownership of his design using the blockchain technology. In Sect. 1, we introduce blockchain. In Sect. 2, we introduce smart contracts. In Sect. 3, we present our problem statement. In Sect. 4, we discuss how this problem can be solved using blockchain. In Sect. 5, we discuss how our solution will benefit the designers and customers. In Sect. 6, we give the future scope. We conclude in Sect. 7.

A. A. D'Souza · O. Rodrigues (✉)
Padre Conceicao College of Engineering, Verna-Goa, India
e-mail: mecta2k7@gmail.com

A. A. D'Souza
e-mail: ashalexsius@gmail.com

© Springer Nature Singapore Pte Ltd. 2019
N. R. Shetty et al. (eds.), *Emerging Research in Computing, Information, Communication and Applications*, Advances in Intelligent Systems and Computing 906,
https://doi.org/10.1007/978-981-13-6001-5_1

1.1 What Is Blockchain

A blockchain is a digital ledger which is distributed in nature [1]. All the transactions carried out are recorded in a series of blocks. Multiple copies are made which are spread over multiple nodes (computers). Each block is made up of a time-stamped batch of transactions to be included in the ledger. Each blockchain block is uniquely identified by a cryptographic signature.

All the blocks in the blockchain are back-linked to each other, so they refer to the cryptographic signature of the previous block in the chain and that chain can be traced all the way back to the very first block that it created at the start.

Blockchain is a distributed public general ledger thus making it a perfect match for supply chain management.

1.2 History of Blockchain

Initially, blockchain is the original source code for Bitcoin, the first major blockchain innovation. Bitcoin is a digital currency, invented in October 2008. The source code was made open source in January 2009. Bitcoin really took off in 2013, as more and more Web sites started accepting this virtual currency [1]. Satoshi Nakamoto [2], the inventor of bitcoin (still his real identity remains unknown), coined the initial bitcoins. The interest of banks, businesses, and governmental organization in bitcoin has rapidly increased over years.

1.3 Advantages of Blockchain

Blockchain is a distributed database, which has multiple copies across multiple computers forming a peer-to-peer network, wherein there is no single, centralized database or server [3]. Rather the blockchain database exists across a network of machines which is decentralized in nature. The transactions on the blockchain are digitally signed, using the public key cryptography technique. This technique consists of two keys, public and private. The public key is used to sign and encrypt a message that is being sent. The recipient will use its private key to decrypt the message. If any other person other than the recipient tries to decrypt the message, it will be difficult for him to do so.

2 Smart Contracts in Blockchain

A smart contract is a computer program code that is capable of self-executing contracts with the terms of an agreement between the buyer and seller across a blockchain network [4]. The transactions are transparent, traceable and irreversible. The agreements are carried out among anonymous parties without the need of a central authority or legal system. The agreements enforce themselves [5].

Smart contracts function as "Multi-Signature" accounts wherein funds are only spent if a certain percentage of people agree. They help manage agreements between users. Utility to other contracts is provided as well. They also store information about an application such as membership records or domain registrations. Smart contracts are autonomous and automatic [6].

3 Current Problem Statement

The designer designs a logo. He is the owner of his design. Another person uses the design and prints T-shirts with that design. The owner of that design does not receive any credit for his design. The royalty amount is, therefore, not credited to his account. He is at loss not only from a monetary point of view but also because the design which was designed by the owner is used by some other person for business purpose.

4 Solution Using Blockchain

Using blockchain, we will provide solution not only for the designer of the design but also for the customer.

4.1 Solution for the Designer

Using blockchain, we will provide individual owner with the ownership to his own design. The blockchain technology ensures that all the participants in a decentralized network share the identical view of the real world [7]. The owner can maintain his account where his identity is authenticated. The owner can claim his own design which cannot be copied by any other designer. If such a case arises, the original owner of the design can file a claim request and the owner can either sue the copier or ask for a royalty. Figure 1 shows the flow diagram for the designer.

4.2 Solution for the Customer

The customer will not be able to differentiate between an original design and a fake design. As market conditions change, he can verify autonomous and instantaneous

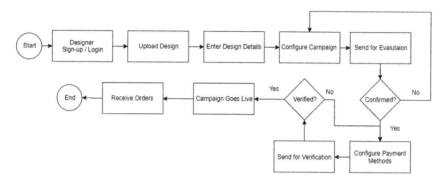

Fig. 1 Flow diagram for the designer

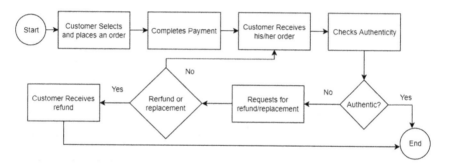

Fig. 2 Flow diagram for the customer

transactions across nodes [8]. So, using blockchain we are providing a solution to the customer wherein he will be able to differentiate between an original design and a fake design. Using identity verification, the customer can verify the origin of the design as this system is verifiable. He can, therefore, ensure for himself that he is purchasing an original product. Figure 2 shows the flow diagram for the customer.

5 Benefits

With the help of our blockchain technology, both the stakeholders will benefit the designer as well as the customer.

5.1 Benefits for Designer

The designer is the original designer of that design and he should get credit for his design. We will provide him with credit for his design using the blockchain technology. The blockchain application will track the origin of the design and the owner of the design will receive a notification. Every time a sale takes place a royalty amount is credited to his account. He will be able to track the number of sales his design has made. He can see each and every delivery being made at multiple locations [9]. Each transaction is automatically recorded in the blockchain ledger therefore making it possible for him to see every delivery made at different locations.

5.2 Benefits for Customer

The customer will now be able to differentiate between an original design and a fake design. He will not be cheated for the amount he pays to purchase the product with that design as he knows it is a genuine product and not a fake product. The customer will get value for the amount he pays for the product. The customer can monitor the transaction as well as see what is happening at every step [10]. This will give the customer satisfaction which in turn will make him a happy customer.

6 Future Scope

We plan to incorporate and resolve the issue of copyright for the original designer. Here the original owner of the design can copyright his design. If anybody else copies his design, he can file a copyright case against that person.

7 Conclusion

Using blockchain, the original owner of the design will benefit from our technology. He will be satisfied because his design has received recognition. He will receive the royalty amount he deserves. Using our blockchain technology, the customer will be happy because he has received value for the amount he has paid. He will be satisfied that he has not been sold a fake product but an original product. Therefore, our system is secure and reliable.

References

1. The blockchain for education, an introduction. http://hackeducation.com/2016/04/07/blockchain-education-guide.
2. Bitcoin: A peer-to–peer electronic cash system. https://bitcoin.org/bitcoin.pdf.
3. Blockchain technology set to revolutionize the supplychain. https://due.com/blog/blockchain-technology-supply-chain.
4. Smartcontracts. https://www.investopedia.com/terms/s/smart-contracts.asp.
5. How do ethereum smart contracts work? https://www.coindesk.com/information/ethereum-smart-contracts-work/.
6. How blockchain revolutionizes supply chain management. http://www.digitalistmag.com/finance/2017/08/23/how-the-blockchain-revolutionizes-supply-chain-management-05306209.
7. Greenspan, G. *(2015). MultiChain private blockchain—white paper.* [Online]. Available: http://www.multichain.com/white-paper/.
8. Lacey, S. (2016). *The energy blockchain: How bitcoin could be a catalyst for the distributed grid.* [Online]. Available: http://www.greentechmedia.com/articles/read/the-energy-blockchain-could-bitcoin-be-a-catalyst-forthe-distributed-grid.
9. Regulation of blockchain revolution: Current statue and future trends. http://www.iamwire.com/2017/09/regulation-of-blockchain-revolution-current-status-and-future-trends/167370.
10. Blockchain in the supply chain: Too much hype. https://www.forbes.com/sites/stevebanker/2017/09/01/blockchain-in-the-supply-chain-too-much-hype/#56382abe198c.

A Feasibility Study and Simulation of 450 kW Grid Connected Solar PV System at NMIT, Bangalore

B. Smitha, N. Samanvita and H. M. Ravikumar

Abstract The consumption of energy can be reduced by efficiently using the available resources and effectively energy bill is reduced by considering photovoltaic system, which is most promising nowadays. In this paper, a feasibility study and simulation model on MATLAB/SIMULINK of 450-kW grid-connected solar PV system is considered for NMIT campus. The energy consumption at the campus is studied, and the number of billed units in kWh is considered for the last two years. The modelling of PV array, their integration with MPPT in SIMULINK environment are described. The deployment of available energy resources along with the incoming PV system is studied for effective usage of electricity. The simulation results are shown, the performance of the incoming PV system and its feasibility is described as obtained.

Keywords Fossil fuel · Solar PV system · MPPT · SIMULINK

1 Introduction

The fast-expanding economy and growing economic activities demanding for quality and quantity in energy sources. India is producing 66% of energy by using fossil fuels [1]. Coal, natural gas and oil are fossil fuels which consisting of hydrocarbons, produces carbon dioxide and other poisonous gases when they are burnt. These gases are the main reason for global warming. By using energy sources which gives clean energy or free from carbon dioxide or renewable energy sources, we can reduce our dependence upon the fossil fuels. As the sun is the major source of renewable energy, the generation of power from the solar PV system is fast developing in India.

B. Smitha (✉) · N. Samanvita · H. M. Ravikumar
Department of EEE, NMIT, Bangalore 560064, India
e-mail: smitha.b@nmit.ac.in; smithamnt@gmail.com

N. Samanvita
e-mail: samanvitha@nmit.ac.in

H. M. Ravikumar
e-mail: hmrgama@gmail.com

Nitte Meenakshi Institute of Technology (NMIT) an autonomous institute is spread in 23 acre of land in Bengaluru. The college is well equipped with laboratories and workshops, full-fledged central computation facility, a good library and other facilities. The hostels for Boys and Girls, Stationary Shop, Bakery, Canteen, Xerox shop, College Buses, Bank ATMS, Staff quarters, Temple, Open Air Theatre, Auditorium, Coffee Shop, Maggy station are available in the college campus.

The population density at NMIT is increasing yearwise and demands for increase in energy consumption. The college has consumed an average of 127,223 units per month in the academic year 2016 and 2017 and spent an average of Rs. 1,233,946 per month to the BESCOM. In order to meet this demand, the institute has planned to generate the power by installing the rooftop solar PV system to meet its energy demands.

In this paper, the feasibility study of the existing power system is discussed. The simulation model of 450 kW rooftop solar power plant using MATLAB/SIMULINK R17a is studied.

2 Related Work

Investigation at the Mu'tah campus was consuming 96 MWh/annum [2]. The authors had designed solar on-grid PV system of capacity of 56.7 kW, which produces the electricity of 97.02 MWh/annum to the grid, and they have evaluated the cost of the plant installation and payback period [2].

The feasibility study of grid-connected PV system at BVUCOEP campus, which consumes 49,000 units per month from the sanctioned load of 187 kWp, had estimated the payback period of the solar PV installation [3].

At MMUT, Gorakhpur the grid-connected solar PV system capacity of 100 kW has been modelled and simulated on MATLAB, by using MPPT technology to detect the peak power [4].

The simulation model of 100-kW grid-connected PV system based on the mathematical model developed at BRCM college with different aspects such as module temperature and shading done and found the simulated results which were very close to practical result [5].

3 Existing Power Supply Arrangement at NMIT

3.1 Bengaluru Electricity Supply Company Limited (BESCOM), Bangalore

BESCOM is supplying 11 kV to NMIT, which will be step down 415 V using a high-tension (HT) 500 kVA transformer. The college has taken the contract demand

Table 1 The monthly energy consumed in 2016 at NMIT

Month	No of billed units in kWh (2016)	Energy bill amount in Rs.
JAN	100,100	871,351
FEB	119,980	1,039,287
MAR	135,880	1,186,888
APR	149,360	1,387,372
MAY	108,920	1,011,537
JUN	86,860	695,269
JUL	84,180	784,741
AUG	90,520	843,646
SEPT	120,800	1,129,259
OCT	124,940	1,155,058
NOV	131,420	1,214,877
DEC	108,820	1,025,142

of 450 kVA from BESCOM. The billing is calculated based on 75% of the contract demand or recorded demand, whichever is higher. The demand charges are charged at Rs. 230/kVA of billing demand billed in BESCOM bill. The college has DG backup, which operates during the power outages.

3.2 Emergency Power Supply

The college campus has two DG sets of 500 kVA and 320 kVA capacity, which have been installed to provide the backup to the college. These DG sets have been connected to the whole campus. Depending on the load, change over switch that operate between 500 kVA and 320 kVA DG set. Currently, the college is using 500-kVA DG during the working hours and 320-kVA DG is used during the evening time and holidays.

3.3 Energy Consumption in the College

In this subsection, the monthly energy consumption in the academic years 2016 and 2017 is briefly discussed and corresponding monthly bills paid is presented in Tables 1 and 2 and corresponding energy consumed is shown in Fig. 1. Figure 2 shows the bill amount paid during 2016 and 2017.

Table 2 Monthly energy consumed in 2017

Month	No. of billed units in kWh (2017)	Energy bill amount in Rs.
JAN	109,440	1,016,551
FEB	110,300	1,025,826
MAR	153,020	1,417,800.86
APR	162,322	1,591,372
MAY	124,920	1,138,259
JUN	102,860	997,742
JUL	105,180	1,051,847
AUG	131,880	1,320,707
SEPT	143,580	1,436,865
OCT	133,840	1,338,440
NOV	125,400	1,234,806
DEC	123,936	1,237,141

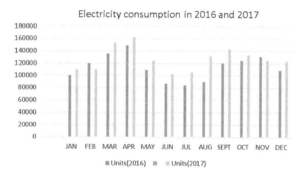

Fig. 1 Graphical representation of electricity consumption (Units) in 2016 and 2017

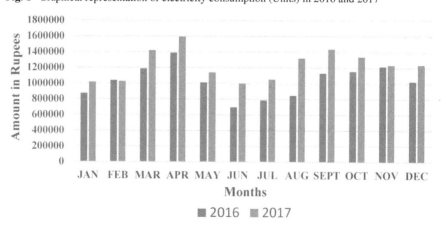

Fig. 2 Bill amount paid during 2016 and 2017

Table 3 Solar energy generation

Month	Solar radiation (kWh/m^2/day)	AC energy (kWh)
JAN	5.25	54,143
FEB	5.59	52,113
MAR	6.08	61,745
APR	5.53	54,930
MAY	4.86	50,313
JUN	4.83	49,076
JUL	4.58	47,990
AUG	4.24	44,606
SEPT	4.64	46,920
OCT	4.57	47,826
NOV	5.18	51.578
DEC	5.2	53,729

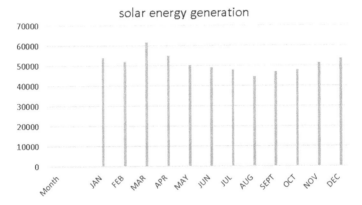

Fig. 3 Graphical representation of solar energy generation

The college is planned to install the solar rooftop PV system of 450-kWp capacity. The expected radiation and energy generation from PV Watts calculator by NREL [6] at NMIT for capacity of 450 kW shown in Table 3. The graphical representation of solar energy is shown in Fig. 3.

The monthly average consumption of energy (units) is 127,223.166 kWh.
The total amount paid to BESCOM, by considering actual energy consumed and 75% of the contract demand or recorded demand, whichever higher is Rs. 1,153,714.91.
The monthly average energy generation from solar system is 51,247.41 kWh.
The amount for 51,247.41 kWh is Rs. 279,298.42 @ Rs. 5.45/kWh.
If we reduce the BESCOM contract demand from 450 kVA to 300 kVA, after solar system installation, the amount for 225 kVA (75% of contract demand) is Rs. 51,750 per month has to pay.

The monthly average of total units going to be consumed from the BESCOM after the solar installation is 75,976 kWh.

The amount for 75,976 kWh is Rs. 641,997 @ Rs. 8.45/Month.

Therefore, the total amount paid to BESCOM after installation of solar system is Rs. 948.296/Month.

The amount saving by the solar installation will be Rs. 205,419/Month (Approximately).

4 Simulation Model

In this section, the dynamic and comprehensive simulation model is presented for grid-associated photovoltaic production system [7, 8]. The created framework comprises of one 450 kW photovoltaic exhibit, DC-to-DC converter and insulated-gate bipolar transistor (IGBT) inverter using the pulse-wave modulation (PWM) method with a confined transformer, intended for accomplishing the most extreme power tip, utility meter and a metering arrangement. The block diagram of Grid–Tied solar electric system is depicted in Fig. 4.

4.1 Modelling of Photovoltaic Cell

The SIMULINK model of photovoltaic solar cell is depicted in Fig. 5. A solar cell is the essential portion of photovoltaic module [9]. A sunlight-based cell is made of semiconductor silicon P–N intersection. The doping is obtained by adding the impurities on it. PV cluster changes over solar radiations into power, and in this way creating vitality out of the sustainable power source asset and the sun-oriented cell is its fundamental unit. The current versus power attributes and current versus voltage attributes of a sun-powered cell is for the most part subject to the sun-oriented

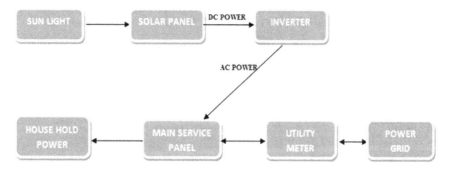

Fig. 4 Grid–Tied solar electric system

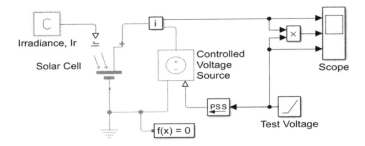

Fig. 5 Solar photovoltaic cell

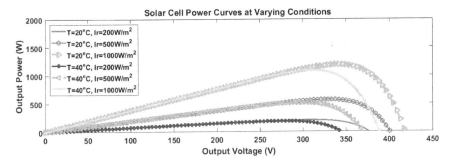

Fig. 6 Solar cell power at varying conditions

radiations. Power varieties can be expected reasons like change in climate conditions or because of load varieties. Figure 6 demonstrates the variation of power concerning the temperature and Sun illumination.

Photovoltaic cells are delicate to heat. The bandwidth of semiconductor PN junction is reduced by increasing the temperature. As the temperature builds, the open-circuit voltage reduces, subsequently reducing the fill factor lastly reducing the efficiency of a solar cell. It is prescribed to work at 25 °C. The power yield for various working temperatures is appeared in Fig. 6.

4.2 Irradiation Level at NMIT Campus in a Year

The average Sun irradiation received at NMIT is 5049 W/m^2/year (refer to Table 3). Figure 7 depicts the insulation levels of 12 months.

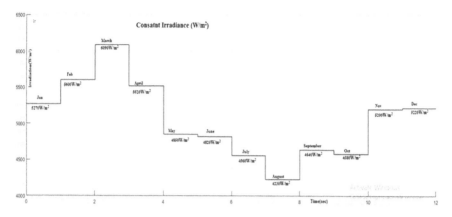

Fig. 7 Insulation level

4.3 SIMULINK Module Description

In this work, 450 kW photovoltaic arrays are integrated with a DC-to-DC boost converter and a 3-Ø voltage-source converter. The working principle of MPPT controller (refer to Fig. 8) follows the 'Incremental Conductance and Integral Regulator strategy' [10].

The 450-kW grid-connected solar PV array model is made with the following components:

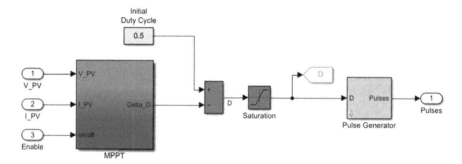

Fig. 8 Boost converter control (MPPT)

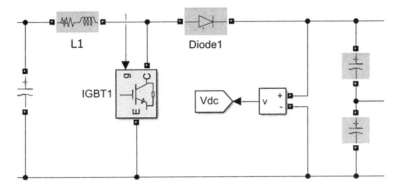

Fig. 9 DC-to-DC boost converter

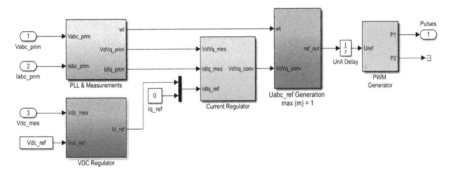

Fig. 10 Voltage-source converter main controller

Module	Description
Photovoltaic cluster	This cluster delivers a peak power of 450 kW at an average sun irradiation of 5049 W/m^2
DC-DC boost converter	In order to increase the PV natural voltage to 500 V DC, boost converter with frequency of 5 KHz is utilized
3-level 3-Ø VSC	The role of voltage-source converter is to change the 500-V DC link voltage to 260-V AC by maintaining the unity power factor
Variable capacitor bank (10 K)	Capacitor filter bank removes the harmonics released by VSC

Figure 9 shows the DC-to-DC boost converter. A 450-kW photovoltaic cluster utilizes 330 Sun Power modules. This cluster comprises of 66 strings of five series (66 × 5 × 4.5 × 305.2 W = 450 kW). Figure 10 depicts the voltage-source converter main controller. PV exhibit is a point-by-point model of 450-kW cluster associated with a 25-kV grid and after at last to the principle network by means of

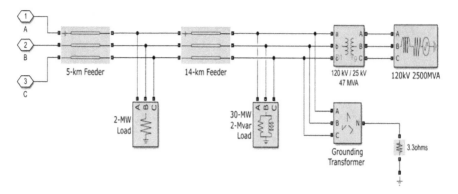

Fig. 11 Simulation of utility grid

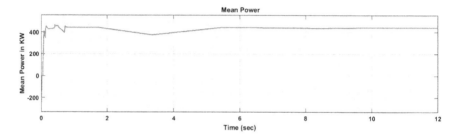

Fig. 12 Output signal P$_{mean}$(kW), Vdc boost and duty cycle

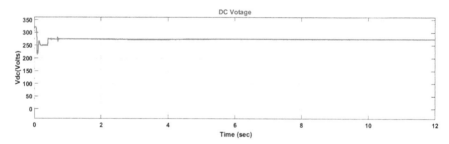

Fig. 13 Output of Vdc boost

converting DC-to-DC and a 3-level 3-Ø-insulated-gate bipolar transistor or semiconductor inverter using the pulse-wave modulation strategy and connected transformer. The pulse width of PWM is switched to obtain the regulated controlled voltage and besides to decrease the consonant substance. Figure 11 shows the utility grid, which is a very basic level distribution and transmission system. The average power in kW, output of boost converter (Vdc) and duty cycle are depicted in Figs. 12, 13 and 14, respectively. Table 4 shows the simulation parameters used.

With the association of the photovoltaic exhibit with the network, it will be extremely useful for individuals to use the sunlight-based vitality which is created

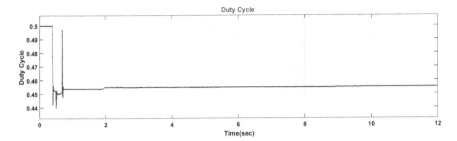

Fig. 14 Duty cycle

Table 4 Solar module specifications list

No. of photovoltaic cells configured in series	96
Voltage (open-circuited)	Voc = 64.2 V
Current (short-circuited)	Isc = 5.96 A
At peak power, the values of current and voltages	Imp = 5.58 A and Vmp = 54.7 V

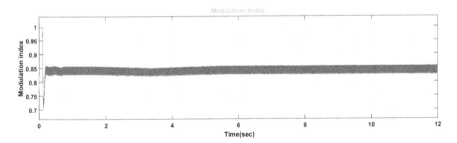

Fig. 15 Modulation index

at exceptionally far and henceforth individuals will not be denied of this vitality regardless of how far they are from the plant gave that PV cluster is associated with grid. The modulation index is shown in Fig. 15.

5 Conclusion

In this paper, the feasibility study on energy consumption at NMIT campus is briefly discussed. The paper also presents the modelling of 450-kW PV power plant. The simulation is done in the MATLAB–SIMULINK R17a.

The installation of solar power plant of capacity 450 kW at NMIT campus expected to generates 51,247.41 kWh per month and possible to save approximately Rs. 200,000 per month in the energy bill. We can improve the PF by installing the capacitor bank, and it is possible to save diesel fuel using solar batteries. We can also supply the energy to the BESCOM, whenever it is not used by the college. The implementation of solar PV project can solve the problems in energy crisis but also helps in reducing the pollution.

References

1. Naganagowda, H. Solar power hand book. National Training Centre for Solar Technology KPCL, Govt of Karnataka Bengaluru.
2. Mohammad, I., Al-Najideen & Alrwashdeh, S. (2017). Design of a solar photovoltaic system to cover the electricity demand for the faculty of Engineering —Mutah University in Jordan. *Resource-Efficient Technologies, 3*,440–445.
3. Jha, A., Baba, A., Jagga, S., Sharan, S., & Kulkarni, S. U. (2016 April). Feasibility study of grid connected rooftop solar PV setup in BVUCOEP campus. *International Journal of Engineering Science and Computing*, 6(4), 4811–4813. ISSN- 2321 3361, .
4. Jaiswal, P., Srivastava, S. K., & Sahay, K. B. (2016). Modeling & simulation of proposed 100 KW solar PV array power plant for MMMUT Gorakhpur. In *International Conference on Emerging Trends in Electrical Electronics & Sustainable Energy Systems (ICETEESES)*, Sultanpur, (pp. 261–266). https://doi.org/10.1109/iceteeses.2016.7581391.
5. Sharma, N. (2016, June). Modeling, Simulation and Analysis of 100 kw grid connected PV system using MATLAB/SIMULINK. *International Journal of Advanced Research in Computer and Communication Engineering*, 5(6).
6. NREL's PVwatts Calculator. https://sam.nrel.gov. Accessed on 12 March 2018.
7. Smart Grid. *IEEE, Conference drive smart grids*. Eetimes.com (2009-03-19). Retrieved on 2011-05-14.
8. Demo model of 100kW Grid Connected PV array. https://in.mathworks.com/help/physmod/sps/examples/detailed-model-of-a-100-kw-grid-connected-pv-array.html. Accessed on 10 January 2018.
9. Tripathi, A., & Sahay, K. B. (2016). Modeling & simulation of proposed grid connected 10 MW Solar PV array power plant at Lucknow. In *2016 IEEE 1st International Conference on Power Electronics, Intelligent Control and Energy Systems (ICPEICES)*, Delhi, (pp. 1–5). https://doi.org/10.1109/icpeices.2016.7853236.
10. de Brito, M. A. G., Sampaio, L. P., Luigi, G., e Melo, G. A., & Canesin, C. A. (2011). Comparative analysis of MPPT techniques for PV applications. In *International Conference on Clean Electrical Power (ICCEP)*, Ischia, 2011, (pp. 99–104). https://doi.org/10.1109/iccep.2011.6036361.

Prediction of a Dam's Hazard Level

A Case Study from South Africa

Urna Kundu, Srabanti Ghosh and Satyakama Paul

Abstract South Africa has a vast infrastructure of dams. Since the country receives very little rainfall, these dams assume prime importance in storing water and sustaining agriculture, industry, household, etc. Thus prediction of their multiple hazard levels (in this case, three) is of prime importance. In addition, South Africa lacks skilled personnels to classify these dam's hazards. Under such a framework, this work is an application of single and ensemble decision trees in a multi-class supervised learning framework to predict the hazard level of a dam. The result obtained is highly promising and at is above 94%. With the implementation of the algorithm, we expect to address the problem of paucity of skilled personnels.

Keywords Dam hazard-level prediction · Multiclass classification · Imbalanced classes · Decision trees · C5.0 · Tree bagging · Random forest · t-SNE · South Africa

1 Introduction

South Africa (SA) has a extensive infrastructure of dams.[1] Publications [2] from the Water and Sanitation Department of SA show that in 2016, there were 5226 registered dams in the country. Investment in such huge infrastructure is necessary as the country receives one of the least rainfall in the world [3] and with abundant sunshine, it is able

---

[1] As per the National Water Act (1998) [1] of SA, a dam is defined as an existing or proposed structure that can be used for containing, storing, or impounding water.

U. Kundu · S. Ghosh
WNS, Plot 8A, RMZ Centennial, Whitefield, Bengaluru 560048, Karnataka, India
e-mail: urnakundu@gmail.com

S. Ghosh
e-mail: srabanti.ghosh.research@gmail.com

S. Paul (✉)
Oracle, Prestige Tech Park, Marathahalli Ring Road, Bengaluru 560103, Karnataka, India
e-mail: satyakama.paul@gmail.com

© Springer Nature Singapore Pte Ltd. 2019
N. R. Shetty et al. (eds.), *Emerging Research in Computing, Information, Communication and Applications*, Advances in Intelligent Systems and Computing 906,
https://doi.org/10.1007/978-981-13-6001-5_3

to hold very little water in its ground. Thus these dams are required for conserving water that can be used for industry, agriculture, and domestic purposes.

Also as the larger dams are more than 30 years old, infrastructural integrity of the dams has to be ensured through adherence to a long list of safety regulations [4]. Safety regulation checks are done by approved professional persons (APPs). These APPs are mostly professionally certified engineers, technologists, and technicians. One of the primary functions of these individuals is to classify the hazard potential[2] of the dams into low, significant, and high.

Given the present shortage of technical skill in SA, these APPS are very few in number. One estimate [4] shows that currently there are less than 100 APPs in the country. With such low number of APPs, safety inspection of dams is a challenge. By another statistics [4], due to the less number of APPs, in 2014–2015, only 58% of the targeted number of dams could be inspected.

This paper addresses the problem of low APPs by using single and ensemble decision tree algorithms to predict the hazard potential of the dams. Our work feeds the basic characteristics of the dams (such as its wall height, crest length, and surface area) as inputs into the model. Using simple and ensemble decision trees, the models can predict dam hazard to a very high degree of accuracy, over 93%. We believe that the research can help in addressing the skill shortage problem in a dam's hazard prediction and overall in dam safety enforcement.

2 Literature Review

Much research in the area of dam hazard prediction are the models that consider features which are measured over a considerable time period. Since our data are cross-sectional in nature, review of researches based upon time series models is out of our scope. Some of the notable guidelines and reports in the area of dam hazard prediction/dam safety management are by the International Commission on Large Dams [5], the Australian National Committee on Large Dams [6], and the Canadian guidelines [7]. While these reports provide detailed instructions on various aspects of dam safety, their objective is to provide generic guidelines rather than dealing with a specific aspect of predicting a dam's hazard in a supervised machine learning framework.

To the best knowledge of the authors, not much work has been carried out in the supervised learning framework to create models that predict the hazard potential of dams. However in their work [8] Danso-Amoako et al. uses a single hidden layer, artificial neural network, with back propagation of error to predict a dam's risk (as a continuous value feature) with 40 features and 5000 data points.

---

[2]Hazard potential of a dam is the same as it's hazard level, and from here on they are used interchangeably to mean the same.

Table 1 List of original features

S. no.	Feature	S. no.	Feature
1	no.of.dam (Ca)	21	wall.type (Co)
2	warms.dam.id (Ca)	22	wall.height (Co)
3	name.of.dam (Ca)	23	crest.length..m. (Co)
4	water.management.area (Ca)	24	spillway.type (Ca)
5	quaternary.drainage.area (Ca)	25	capacity..1000.cub.m. (Co)
6	latitude.deg (Co)	26	surface.area..ha. (Co)
7	lat.min (Co)	27	catchment.area..sq.km. (Co)
8	lat.sec (Co)	28	purpose(Ca)
9	longitude.deg (Co)	29	owner.name (Ca)
10	long.min (Co)	30	designer (Ca)
11	long.sec (Co)	31	contractor (Ca)
12	town.nearest (Ca)	32	registration.date (Co)
13	distance.from.town (Co)	33	size (Ca)
14	name.of.farm (Ca)	34	hazard.potential (Ca)
15	municipal.district (Ca)	35	category (Ca)
16	province.code (Ca)	36	classification.date (Co)
17	region.code (Ca)	37	sector (Ca)
18	completion.date (Co)	38	date.last.dsi (Co)
19	completion.date.raised (Co)	39	number.last.dsi (Co)
20	river.or.watercourse (Ca)	40	target.date (Co)

All dates are year values and hence assumed continuous

3 Problem Statement

The original dataset is obtained from the Dept. of Water and Sanitation, SA [2]. The dataset consists of 5226 data instances (rows) and 40 features (columns). Table 1 shows the list of original features. They are a mixture of continuous (*Co*) and categorical (*Ca*) types of data. Parenthesis beside each feature shows its type. The hazard potential (feature 34) of a dam can be of three levels—high, significant, and low. The objective of this work is to correctly predict the hazard potential of a dam as a function of the rest of the relevant features.

4 Data Preprocessing and Experimental Set-Up

Data preprocessing consists of a number of steps. At first, we identify and remove 16 features that are names and hence have no statistical significance. They do not explain any common intrinsic characteristics of the dams. Such features are numbered 1, 2, 3, 5, 12, 14, 15, 20, 29, 30, 31, 32, 36, 38, 39, and 40 in Table 1. Second, we discard all those features that have 30% or more missing values from the feature space. In the process we drop two features— *completion.date.raised* and *catchment.area..sq.km*. Third, we discard categorical features that have too many levels. Features *spillway.type* and *purpose* that 302 and 247 levels, respectively, and hence are discarded. Thus we are left with a feature space of 20 features (=40−16−2−2) and 5161 rows. Next we remove all those rows of data (from the feature space) for which at least one feature value (column data) is missing. This leaves our dataset with 4535 rows. It may also be noted that the latitude and longitude features (feature 6–11) are in degrees, minutes, and seconds. Thus each of them are represented by three columns of data. In order to reduce computational time and complexity without loss of information, we transform latitude and longitude from their respective degree, minutes, and seconds to decimal format—hence shrinking 6 columns of information to 2 columns. Hence, our final feature set consists of 4535 rows and 16 (=20−6+2) columns.

In continuation of the problem statement mentioned in Sect. 3, our response is *hazard.potential*, and the predictors are *water.management.area*, *distance.from.town*, *province.code*, *region.code*, *completion.date*, *wall.type*, *wall.height*, *crest.length.m*, *capacity.1000.cub.m.*, *surface.area.ha.*, *size category*, *sector*, *latitude.dec*, and *longitude.dec*.

5 Exploratory Data Analysis

This section is an EDA. Here we consider single feature(s) and their combinations to find certain interesting patterns in the data. The topmost panel of Fig. 1 shows that in the total population, 61.98% dams have low hazard potential followed by significant 31.20% and high 6.81%. Thus, we are dealing with a imbalanced multiclass classification problem. The middle panel shows that large size dams have higher level of hazard than the medium or smaller dams. Lastly, the bottom panel shows that Western Cape has more dams than the rest of the provinces and also proportionately more high hazard dams than the rest of SA.

Figure 2 is the density distribution of the continuous predictors that denote the physical characteristics of the dams, with respect to each level of *hazard.potential*. In the two right-hand panels, the distributions for various levels of hazards overlap one another for *crest.length* and *capacity*. Similarly in the left-hand panels, the density distributions for each level of hazard considerably overlap each other for *surface.area.ha.* and *wall.height*. This means that the physical characteristics of the dams are not adequate features in separating one hazard potential from another.

Lastly, we visualize the spatial distribution of the hazard potential of 4535[3] dams across 15 predictors, using t distributed Stochastic Neighborhood Embedding (t-SNE) algorithm. While there are several techniques[4] that can visualize the nonlinear

---

[3] After preprocessing, we have 4535 rows of data.

[4] Shannon map, Isomap, Curvilinear component analysis, Locally linear embedding, etc.

Prediction of a Dam's Hazard Level 23

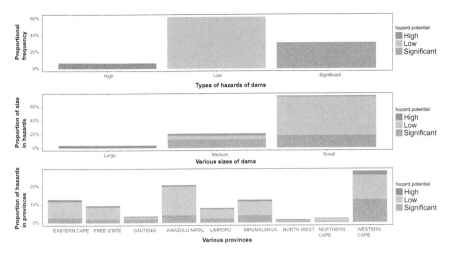

Fig. 1 Bar plot showing various features of the dams

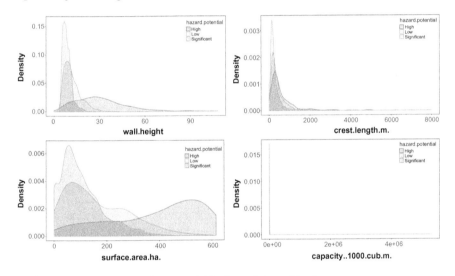

Fig. 2 Density plot of the continuous variables for each hazard level

relationship between high dimensional data and lower dimensional spaces, these techniques are unable to retain both the global and local proximity relationships between the data points on a single map [9]. Hence, we choose t-SNE to visualize how various hazard levels of the dams are distributed. In a realistic estimate, we assume the perplexity parameter of t-SNE to take 3 values - 5, 10, and 15. Perplexity is a measure of the number of neighbors that exist in the vicinity of a data point. We rationally assume that for a particular dam, the number of dams in its vicinity cannot exceed 15. Figure 3 shows the spatial distribution of each dam in terms of its

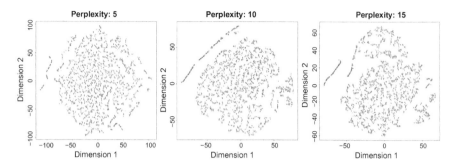

Fig. 3 t-SNE visualization for various perplexity values (High-Red, Significant-Green, Low-Blue)

hazard level. In the central part of all the three panels, we can find a considerable overlap between the three hazard levels. Thus, our research problem is a nonlinear multi-class classification problem, where one class cannot be separated from others through linear separating hyperplanes.

6 Theoretical Considerations

6.1 Feature Subset Selection

In our classification problem, the predictor space consists of 15 features. Our objective is to find (and remove) if there are any predictors that are redundant, i.e., their removal do not affect the overall classification accuracy. There are four major advantages of doing the exercise: One, it reduces the computational time and complexity of the final model; two, with reduction of the irrelevant features—the model is better interpretable; three, curse of dimensionality can be avoided; and four: reduce over fitting [10].

In our work, we compare the results of two feature subset selection algorithms to select a common set of features. The first algorithm we use is Boruta. Boruta is a wrapper around the random forest classifier, where inside the random forest the algorithm compares each feature against a group of "shadow"[5] features. During comparison, each feature's Z score is compared against the shadow feature which has maximum Z score (among the shadow features). For any feature whose Z score is bigger than the maximum Z score of the shadow feature, its accepted as an important feature. Correspondingly, if the feature's Z score is less than the maximum Z score of a shadow feature, it is considered unimportant [11]. The second feature subset selection algorithm used is Recursive Feature Elimination (RFE). Let us assume a n-fold cross validation setting. In each fold, a RFE does the following: (1) it trains a

---

[5] A shadow feature is created by shuffling the values of an original feature.

classifier on 1 to m predictors by creating m models and recursively eliminates those predictors that do not contribute to the classification accuracy on the test set. (2) Step 1 is repeated for n times, and then the average classification accuracy is shown [12].

6.2 Decision Trees

In this sub subsection, we will discuss our choice of model and their theoretical basis. Decision trees are supervised algorithms suitable for both classification and regression. It is an inverted-tree-like structure where the root nodes are at the top and leaves at the bottom. Depending on feature importance, split takes place on the feature which has got the highest importance, followed by the other features whose importance is less compared to the former. Each split on the feature is carried with the objective to create more homogeneous classes, and the splitting criterion stops when we achieve purer classes. The nodes in which all the purer classes are stored are called leaves from where no further split takes place. Here, we use three decision tree algorithms: C5.0, Tree bagging, and Random Forest.

C5.0 algorithm is a modified form of C4.5 algorithm and is more memory efficient, faster and creates simpler trees. C5.0 permits models which are rule based in nature and helps in evaluating variable importance and involves pessimistic pruning which evaluates every subtree specifically to determine whether the entire tree should be simplified or not. The algorithm prunes the model by utilizing independent conditional statements which further helps in evaluating specific rules. Mathematically, the process of evaluation and elimination is formulated as:

$$p(x_k|x_1, x_2, \ldots, x_N), p(x_k|x_2, x_2, \ldots, x_N), \ldots, p(x_k|x_1, x_2, \ldots, x_N)$$

where p = error rate and $x_1, x_2, \ldots, x_k$ are the independent conditional statements. The entire process involves removal of the worst rule by comparing the error rate to the pessimistic error rate that has already been predetermined. The recalculation of the pessimistic error is performed iteratively on further smaller trees, so long all the conditions have reached the threshold of baseline error rate or gets totally removed.

Random Forest is a bootstrap aggregating (bagging) technique which creates trees in parallel in order to reduce the model complexity. It takes random samples (with replacement) and random number of features,[6] aggregate the learning of each (weak learning) decision tree together and using the principle of voting strategy outputs the final result in classification settings. By doing so, it reduces the model complexity by bringing the variance down without compromising on bias. Tree bagging is similar to Random forest with the only difference being that in Treebag models, all predictors (and not a random subset of predictors) are passed into the model.

---

[6]In general, in the case of regression we take one-third of total number of predictors; and in classification, square root of the total number of predictors.

Table 2 Feature importance ranking

Feature importance using Boruta	MFI	Feature importance using RFE	Accuracy
category	137.357	*category*	0.9435
capacity	28.760	*capacity*	0.9418
surface.area	22.466	*wall.height*	0.9416
wall.height	21.704	*surface.area*	0.9413
sector	21.662	*size*	0.9411
size	21.144	*sector*	0.9409
latitude.dec	13.953	*latitude.dec*	0.9407
longitude.dec	13.900	*longitude.dec*	0.9398
crest.length	12.113	*province.code*	0.9391
distance.from.town	10.996	*region.code*	0.9327
region.code	10.966	*distance.from.town*	0.9312
province.code	10.706	*water.management.area*	0.9310
completion.date	8.731	*crest.length*	0.9310
wall.type	8.720	*completion.date*	0.9288
water.management.area	8.511		

MFI Mean of feature importance

7 Results and Discussions

In this section, we discuss the results of the carried experiments. To reiterate, the objective of our work is to predict the *hazard.potential* based on the 15 predictors (as referred to in Sect. 4). For benefits enumerated in Sect. 6.1, we first carry out feature subset selection to identify whether there are any redundant features. Table 2 shows the results of feature importance on 15 predictors (and 4535 data points) carried through two algorithms—Boruta and RFE.

While Boruta confirms all features as important, RFE confirms 14 as important—leaving *wall.type* as the only non-important feature. It might also be noted that Boruta predicts *wall.type* as the second least important feature. In addition, though not exactly in the same order, yet both algorithms show similar ranking of feature importance for the first eight predictors. Thus to keep parity in the suggested list of important feature, we consider all features except *wall.type* for our final modeling exercise.

Next, we randomly shuffle the dataset (to remove any sampling bias) and split it into training set and test set in 3:1 ratio. Thus, the training set consists of 3403 data points and the test set remaining 1132 points. Also, in order to increase the generalization performance of our models, we use a 10-fold cross validation with 5 repeats. Table 3 shows the results from three implemented decision trees—single C5.0, ensemble - Tree bagging and Random forest on the test set.

Table 3 Results of various performance metrics on the test data

Performance	Algorithms		
	C5.0	Tree bagging	Random forest
Overall accuracy	0.934	0.944	0.94
Cohen's kappa	0.871	0.893	0.884
No. information rate	0.620	0.620	0.620
p value[Acc > NIR]	<2e−16	<2.2e−16	<2.2e−16
Sensitivity			
High	0.857	0.870	0.857
Significant	0.898	0.938	0.935
Low	0.960	0.956	0.952
Specificity			
High	0.993	0.992	0.993
Significant	0.951	0.947	0.942
Low	0.930	0.967	0.963

As expected, the ensemble Tree bagging and Random forest algorithms combine the single weak learner decision trees to provide a strong learner that in turn give better overall predictive accuracy than the single C5.0 tree. The overall predictive accuracy of the three models is above 93%. In addition, since our classes are highly imbalanced (refer the top panel of Fig. 1), we do not solely depend on overall accuracy but further consider the Cohen's kappa metrics for the three models. A high value of kappa (above 0.87 for all the three models) show that there is a very high level of agreement between the accuracies predicted by the models and that due to models based on random chances. Also the very low p value shows that the models are statistically significant at 95% confidence interval. Upon consideration of values of overall accuracy, kappa, sensitivity[7] and specificity,[8] we see that tree bagging model is best suited for prediction of a dam's hazard potential.

8 Conclusions

This research uses a multi-class-supervised machine learning approach (using single and ensemble decision trees) to predict the hazard potential of SA dams. Our results from EDA show complex interactions between the multiple predictors and the response and there is no one hyperplane that separates the three levels of the dams from one another. Under such a framework, we use single and ensemble decision trees to

---

[7]Sensitivity refers to the metric that evaluates when there is an actual event and how often does the classification model predict it as an actual event.
[8]Specificity evaluates when there is an actual non-event and how often does the classification model predict it as a non-event.

predict the three levels of hazards of the dams. The results are highly promising at 94% for the best model. Once deployed in a server with interactive user interfaces (to input new values of the predictors), we expect the model to do highly accurate predictions and hence decrease dependence on APPs who have to physically visit dam sites and classify its hazard level. As a future direction in research, we wish to implement Gradient Boosting Algorithm and Stacked Ensemble and expect further improvement in predictive accuracy.

References

1. National Water Act, Act No 36 of 1998, http://www.dwa.gov.za/Documents/Legislature/nw_act/NWA.pdf
2. Dam Sefety Office, Dept. of Water and Sanitation - Republic of South Africa, http://www.dwaf.gov.za/DSO/Publications.aspx
3. The World Bank, Average precipitation in depth (mm per year), World Bank, http://data.worldbank.org/indicator/AG.LND.PRCP.MM?end=2014&start=2010&year_high_desc=true
4. Dam safty - ensuring the interity of SA's 5000+ registered dams, South African Water Research Commission. http://www.wrc.org.za/Lists/Knowledge%20Hub%20Items/Attachments/11496/WW_Nov15_dam%20safety.pdf
5. ICOLD Bulletin on Dam Safety Management. http://www.sgmconsulting.com.au/images/Dam_Safety_Management_Bulletin_ICOLD_2007.pdf
6. Guidelines. https://www.ancold.org.au/?page_id=334
7. Hartford, D. N. D., & Baecher, G. B. (2004). *Risk and Uncertainty in Dam Safety*. Inst of Civil Engineers Pub. ISBN-13:978-0727736390
8. Danso-Amoako, E., Scholz, M., Kalimeris, N., Yang, Q., & Shao, J. (2012). Predicting dam failure risk for sustainable flood retention basins: A generic case study for the wider Greater Manchester area. In *Computers, Environment and Urban Systems* (Vol. 36, pp. 423–433).
9. Maaten, L., & Hilton, G. (2008). Visualizing Data using t-SNE. *Journal of Machine Learning Research*, 9, 2579–2605.
10. Kohavi, R., & John, G. H. (1997). Wrappers for feature subset selection. In *Applied Intelligence* (pp. 273–324).
11. Kursa, M. B., & Rudnicki, W. R. (2010). Feature selection with the boruta package. *Journal of Statistical Software*, 36(11), 1–13.
12. Guyon, I., Weston, J., & Barnhill, S. (2002). Gene selection for cancer classification using support vector machines. In *Machine Learning* (Vol. 46, pp. 389–422).

Elemental Racing

Lingala Siva Karthik Reddy, Karthik Koka, Amiya Kumar Dash and Manjusha Pandey

Abstract This paper presents a 3D motion sensor racing game for Android/iOS devices called *Elemental Racing,* in which players can race a car against opponent racers connected through local area network. There are many popular racing games for mobile phones like Asphalt 8, Real Racing 3, and NFS et al. These have dominated the Android racing game market with their extensive graphics and detail, but they are lacking when it comes to newer gameplay mechanisms, the objective of this game is to provide a fun yet intelligent experience to the player. To guarantee these objectives, we have consolidated multiple strategic upgrading mechanisms in the game which makes use of the concepts of elemental powers like *Fire, Water, Wind, Lighting, and Earth.*

Keywords 3D android motion sensor racing game · Online/offline multiplayer · Strategic combative gameplay · Fire · Water · Wind · Lightning and Earth

1 Introduction

Video games give people the chance to experience a wide variety of things, more closely than any other media. To some people, they are just a way to pass time, but to some, they provide an escape from reality. Video games allow people to explore and express themselves. Video games have the ability to make people's fantasies come true. Scientifically, it has been proved that after playing video games people show an increase in their cognitive function, strategic planning skills, and deductive reasoning skills. They have also shown an increase in gray matter in areas allied with memory and motor functioning of hands. Many games are also being used as a means of education.

L. Siva Karthik Reddy (✉) · K. Koka
Department of IT, KIIT Deemed to be University, Bhubaneswar 751024, India
e-mail: l.sivakarthikreddy@gmail.com

A. K. Dash · M. Pandey
Department of CSE, KIIT Deemed to be University, Bhubaneswar 751024, India

© Springer Nature Singapore Pte Ltd. 2019
N. R. Shetty et al. (eds.), *Emerging Research in Computing, Information, Communication and Applications*, Advances in Intelligent Systems and Computing 906,
https://doi.org/10.1007/978-981-13-6001-5_4

The 3D racing games genre has witnessed an evolution from classics such as Super Mario Kart and Road Rash to contemporary PC, Console, Android, and IOS games such as Asphalt, Need for speed, and Forza. Games like Asphalt 8, Real Racing 3, and NFS et al. all have intuitive controls and high graphical detail which makes the racing environment attractive but are lacking when it comes to making players think strategically.

Elemental Racing is a 3D racing game with multiple facilities, and the core values of the game are to offer players an exciting, challenging, thought-provoking, enjoyable, and entertaining experience through his/her communication with the game. Unlike all the other Android racing games, here players will think strategically and use the "elemental system" accordingly to win the race. It is not just a racing game with good graphics but it also provides facilities like online/offline multiplayer gameplay, car upgrading system, "elemental system," and community interactions through social media so that player can continuously engage and retain to the game.

Elemental system is a keyword used in the gameplay, now what does it mean? Like I said we have implemented the powerup system with upgrades. In this game, we used the term elemental system as the terminology to define the powerups so that players can play strategically by thinking how to defend themselves and Counterattack on others to win the race.

2 Related Work

Author/year	Paper title	Proposal	Advantages	Tools/software
Marvin et al. 2016	Development of a cognitive vehicle system for simulation of driving behavior [1]	A car racing simulator where the players race a car against AI bots in a 3D atmosphere by effectively using the waypoint and trigger system for AI cars and conditional monitoring system for the efficient calculation of the nonlinear relationship between the input and the output vectors so that the steering and breaking output levels are enhanced [1]	The objective of this game is to provide a challenging experience to the player while competing with the AI bots neck to neck in a racing combat	Unity, MonoDevelop, C# Programming language
Yoppy et al. 2017	Path-finding car racing game using dynamic path-finding algorithm and algorithm A* [2]	The method used by non-player character for path-finding in this game is AI*A algorithm to find the shortest path	The arrangement of both methods can be executed well in racing car games with unhindered track conditions as well as track conditions with static obstacles	Heuristic algorithm, games, automobiles, algorithm design and analysis, search problems, and 3D display

(continued)

(continued)

Author/year	Paper title	Proposal	Advantages	Tools/software
Kazuma et al. 2017	A collision-resilient hybrid P2P framework for massively multiplayer online games [3]	Most of the MMOGs work based on client/server (C/S) model whereas it lacks in scalability since the capacity of storage and performance of the server depends on the increase in the proportion of the users	A collision-resilient hybrid P2P framework for MMORPG	Games, servers, resistance, data models, scalability, cryptography, and Internet

3 Game System Design

The design of the elemental racing game involved designing of these main components [1].

3.1 The Car Movement

The movement of the car was implemented using the accelerometer values calculated by the mobile device based on its position in 3D space. The accelerometer values are then converted to force values which are applied to the game object which is used as a car as torque, while the car accelerates forward automatically until it gets to its maximum allowed velocity. A sideways friction is always applied to the car object to make it from skidding sideways.

When drifting is activated using the UI give, the sideways friction is disabled, meanwhile, a constant brake force is applied on the car object, to decrease its speed, and when the car is turned while drifting, it turns on its wheels, the way real cars do.

All the above forces are applied on the game object using Unity's physics system.

3.2 The Car Follows the Camera

A camera is placed behind and above the car and is made to follow the car. This camera is used to display the main game view to the user. The camera rotates according to the car rotation to prevent in such a way to give the user better gameplay experience.

3.3 The Wall Colliders

Invisible walls are placed all along the track to prevent the car from going off track.

3.4 The Positioning System

Waypoints are placed all along the track to calculate the positions, as well as the approximate distance between cars and the race endpoint. The waypoints are also used to respawn the car if, the car topples over, gets stuck, or goes in the wrong direction. These waypoints are also used to calculate the number of laps. The waypoint system implemented in the game is shown in Fig. 1 all the invisible mesh renderer walls place on the track and it is identified whether the car is present within the definite distance from the present waypoint.

To change the position of the car toward the present waypoint, the game engine performs a sequence of vector calculations, these calculations are based upon the initial vector created by both the position of the current waypoint as well as the current car position itself which gives the steering and breaking outputs levels of the car [1].

3.5 The Power System

In the game along with the conventional gaming mechanics, a powerup system has been implemented. In this game, the cars can collect five main types of elemental fuels, namely fire, water, wind, lightning, and earth. These fuels in combination with

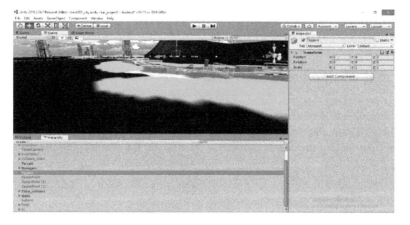

Fig. 1 Waypoints system implementation

each other can be used by the car to activate certain powerups. These powerups can be used to attack, or defend from other cars. Various amounts of fuels can give rise to various types of powerups. Whenever a powerup hits a car, the car can be affected in ways such as losing velocity, losing balance, getting its movement frozen, and losing its ability to activate or block powerups et al.

Each car object is attached with its own power control script which calculates the elemental fuel value it contains. The fuels can be increased when the car object passes through certain game objects placed throughout the racetrack and are used as elemental fuel containers. Each powerup requires certain fuel values, and when the fuel requirements are met the player can use that powerup.

3.6 LAN System

Currently, the game supports LAN multiplayer. Players can connect to a locally setup server and play the game among each other. The LAN system has been implemented using UNET, a networking module provided by Unity.

3.7 Powerup Effects

The visual look of the powerups has been implemented using Unity's particle systems (Figs. 2, 3, 4, 5, 6, 7, 8 and 9).

These were only a few powerups which we have implemented, there are a total of 14 unique powerups in the Gameplay.

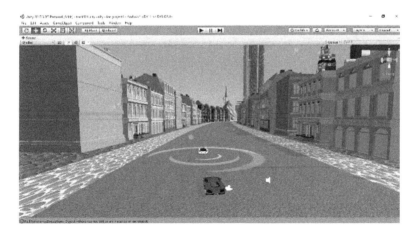

Fig. 2 Water + Earth = Water splash

Fig. 3 Fire + Water + Lighting = Lighting bolt

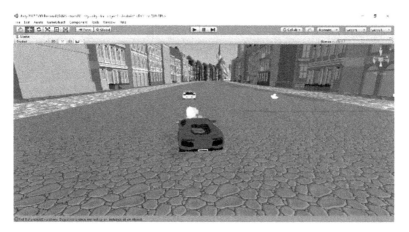

Fig. 4 Fire = Fire ball

3.8 AI

Triggers are attached to the AI cars (Fig. 10) so that they knew when to break, slow down, use powerups, and also steer. All these are done by placing large trigger colliders in front, back, and either side to the cars (Fig. 10) so that any other car enters to the perimeter of AI car (Fig. 11) all triggers gets activated and behavior according to the AI code written for the car.

Elemental Racing

Fig. 5 Lighting = Nitro

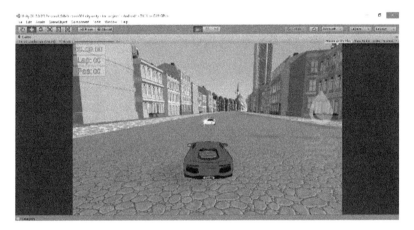

Fig. 6 Water + Wind + Lighting = Machine Gun

3.9 UML Design

For the purpose of visual representation and a better understanding of the system, we used the use case and activity diagram so that main actions and artifacts could portray crystal clear information about the project.

3.9.1 Use Case Diagram

The following use case diagram represents the graphic depiction of the interactions (Fig. 12) among the different elements of the system.

Fig. 7 Fire + Lighting = Meteor

Fig. 8 Water + Lightning = Thunder

3.9.2 Activity Diagram

To represent the series of actions (Figs. 13 and 14) and flow control of the system, we used activity diagram.

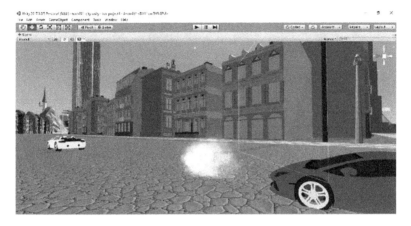

Fig. 9 Water = Ice ball

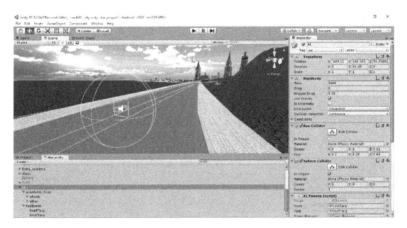

Fig. 10 Triggers attached to the car

4 Implementation of the Game System

The software used to create this game were mentioned here briefly.

4.1 Unity Game Engine

Unity is a game engine, used for the creation of 2D and 3D games (Fig. 15) and interactive content [4]. Unity makes multiplatform development smooth and effective by allowing you to create your games and interactive content in Unity and then seamlessly deploy to PC, MAC, Windows, and mobile platforms et al.

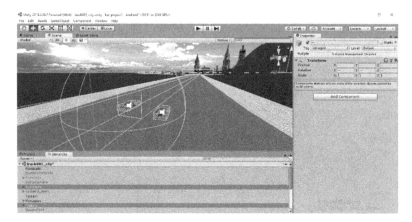

Fig. 11 Trigger and waypoint system

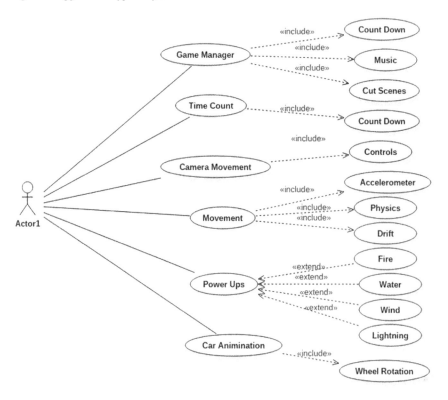

Fig. 12 The designed interaction between the elements using use case diagram

Elemental Racing

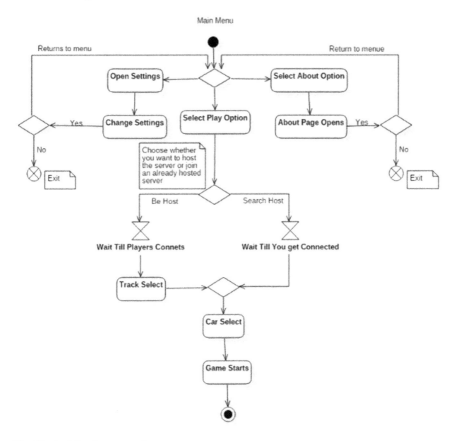

Fig. 13 Activity diagram main menu

4.2 MonoDevelop

MonoDevelop is a tool used for scripting (Fig. 16). It is an open source Integrated Development Environment (IDE) [5]. It is primarily used for the development using C#, but also supports development in JavaScript and Boo.

4.3 Unity Remote

Unity remote (Fig. 17), is an application for the Smartphones, developed by Unity. It is used to test and debug the games made for mobile devices in the Unity, rather than build the entire project and running it on the mobile you can use Unity Remote, with the help of a data cable connected with PC to Mobile, It works as a simulator

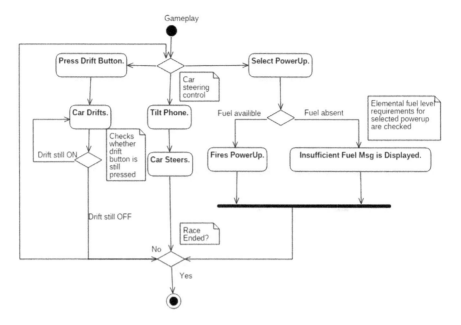

Fig. 14 Activity diagram gameplay

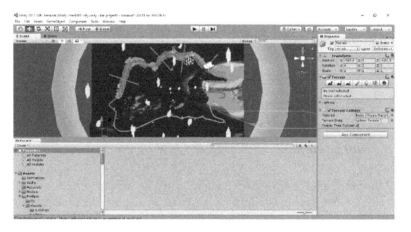

Fig. 15 Unity game engine

Elemental Racing 41

Fig. 16 MonoDevelop for coding

Fig. 17 Unity Remote for testing

for the project by running on the mobile so that we can know where to debug and all, despite building the entire project.

4.4 Autodesk Maya

Maya is a 3D modeling (Fig. 18), rendering, animation, and simulation software which provides the user a powerful toolset which can be used for character creation, motion graphics, virtual reality, environment modeling, and animation [6].

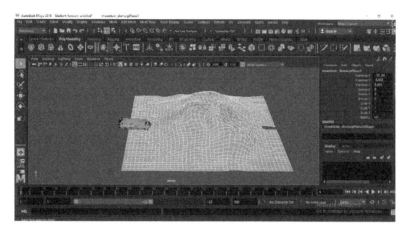

Fig. 18 Autodesk Maya

5 Sample User Run of Elemental Racing Game System

The game takes both touch and motion as input. In the racing part of the game, the main camera showing the view of the track will be placed behind and above the car. In the main HUD, details such as speed of the car in KMPH, lap count, elapsed time of the race, and position of the car will be shown. A small GPS showing only the racetrack will also be present. The amounts of the elemental fuels the car contains will also be shown. The car will accelerate on its own. The car can be maneuvered by turning the phone on the axis perpendicular to the screen. Brakes on the car can be applied by touching the half of the mobile screen on your left. The buttons to fire powerups will be displayed on the half of the screen on your right, and the players can activate them by touching them. The game can only be played in the landscape view mode of a mobile phone.

6 Conclusion and Future Work

Implementing the creature system. According to this system, the cars used in this game are not actually machines but creatures of a frictional universe. Thus, all the models of cars used in the game will have features of animals like a mouth, eyes, etc. This system has been inspired by the movie Cars. A frictional world will also be set for these creatures. The following is being done to make sense of the various fuels and abilities these cars use.

The players can customize the car, better implementation of the powerup system, Implementation of damage system, Implementation of cards system, and points system each race a player plays he will be rewarded certain amount points, multiple game modes. Game modes other than conventional racing mode will be added, such as deathmatch, racing in teams, and tournaments. More racing car models and racetracks will be added. Other than LAN, online multiplayer will also be added.

Acknowledgements We wish to acknowledge the generous support of Prof. Amiya Kumar Dash, Prof. Manjusha Pandey Faculty of Computer Science Engineering, KIIT Deemed to be University, Bhubaneswar, and Orissa, India and the reviews given by the ERCICA Chair.

References

1. Chan, M. T., Chan, C., & Gelowitz, C. (2016). Development of a cognitive vehicle system for simulation of driving behaviour. In *2016 IEEE 15th International Conference on Cognitive Informatics and Cognitive Computing (ICCI* CC)*, IEEE, 2006.
2. Sazaki, Y., Primanita, A., & Syahroyni, M. (2017). Pathfinding car racing game using dynamic pathfinding algorithm and algorithm A∗. In *2017 3rd International Conference on Wireless and Telematics (ICWT)*, IEEE, 2017.
3. Matsumoto, K., & Okabe, Y. (2017). A Collusion-resilient hybrid P2P framework for massively multiplayer online games. In *2017 IEEE 41st Annual Computer Software and Applications Conference (COMPSAC)*, (Vol. 2), IEEE, 2017.
4. Unity—Game Engine. https://unity3d.com.
5. MonoDevelop. http://www.monodevelop.com.
6. Autodesk Maya. https://www.autodesk.in.

A Survey on Existing Convolutional Neural Networks and Waste Management Techniques and an Approach to Solve Waste Classification Problem Using Neural Networks

Tejashwini Hiremath and S. Rajarajeswari

Abstract In India, waste management has become one of the major crises with population explosion, coupled with improved lifestyle of people, results in increased generation of solid wastes in urban as well as rural areas of the country. It is well known that waste management policies, as they exist now, are not sustainable in the long term. Thus, waste management is undergoing drastic change to offer more options that are more sustainable. Most of the landfills are becoming full of waste in which most part is reusable and leading to spreading of disease damaging human body and leading to unpleasant air and only 5% of whole waste is actual waste. The government of Karnataka mandated system of 2 BIN 1 BAG to be adapted at every households in Bangalore, and 2 BIN 1 BAG is a color-coded system consisting of green bin which holds garden waste, and the wastes that are compostable, reject waste can be thrown in red bin, and finally a big category called as reusable bag which holds recyclable waste. Segregation of waste at source is best solution and should be done properly. Types of waste need to be remembered by members of home in order to put them to proper bins, and this may lead to human error. So our solution can answer this in good way, what if you just click picture of waste material and application says to which category it belongs. A convolutional neural network is trained with images of waste materials, and model can be inferred by giving waste-material image as input and get the perfect category of waste material in a second. This helps society in dealing with prime problem of segregating waste materials at source.

Keywords Solid waste management (SWM) · 2 BIN 1 BAG system · Convolution neural networks (CNN) · **S**tochastic gradient descent (SGD) · Neural networks (NN)

T. Hiremath · S. Rajarajeswari (✉)
Department of Computer Science and Engineering, Ramaiah Institute of Technology, Bangalore, India
e-mail: raji@msrit.edu

1 Introduction

In India, waste management is a boastful problem, waste is generated in abundance, it is generated by industries, and economic growth is also a factor in waste production. Urbanization and the above factors increase per person solid waste generation. Managing waste is not challenge, but managing in efficient way has ample importance in cities with heavy population. The world is now experiencing population with rapid increase in number and its standards of living especially in developing country. Despite all this growth carries huge significance in socioeconomical and areas of environment, but no change in waste management and strategies followed, it is relatively same.

Anything that is of no value is termed as waste, isn't it? But most of the waste produced according to the research is actually not waste, most part of it can be reused, eventually leading to minimal waste production. Most of the landfills are becoming full of waste in which most part is reusable and leading to spreading of disease damaging human body and leading to unpleasant air.

There is a need for segregation of actual waste and reusable. The waste segregation at the landfill is vast time-consuming task, what if this is segregated at the source? Isn't this idea logically simple? But segregation should be done according to what research says. Waste can be segregated as degradable and waste that is non-degradable, organic, and the other one that is inorganic. Lot many research has been made and system has arose which effectively solves the segregation of waste, this system called as 2 BIN 1 BAG helps waste to be classified into three categories where compostable kind of waste and waste from garden falls in one category, recyclable waste in one category, and reject waste which is actually the waste falls in last category. The system of 2 BIN 1 BAG is a color-coded system consisting of green bin which holds garden waste and the wastes that are compostable, reject waste can be thrown in red bin, and finally a big category called as reusable bag which holds recyclable waste. After this segregation, it is revealed that rejected waste gathered was is only 5–10% rest can be used in some ways.

In all states of India, system is not yet implemented and is anticipated only in three languages among many languages. Not everyone at home will acclimate to system, because needs think time about waste-type. And human errors will lead in wrong categorization of waste. But taking a picture of image and application says category. Isn't this good idea? And is idea of this project. Here comes neural network ideology. Neural network is constructed to classify images to one class among three classes of 2 BIN 1 BAG. Finally, when waste-material image is given trained neural network classifies outputs the category to which it belongs. In following discussion, system design and tools used and construction of neural network a clear insight is given.

2 Literature Survey

This literature survey visualizes waste management (WM) problems faced and strategies followed solving WM problems in India and how far the results approached the waste management aim. Also survey talks about artificial neural networks (ANN), GoogleNet (InceptionV3), AlexNet, VGGNet, ZFNet, and SENet currently available and their accuracy in terms of error rate, so it will be helpful in deciding number of layer needed and operations to be carried out at each layer to design our ANN.

Currently developing nations like India are fronting environmental hitches mainly solid waste management (SWM). With population density of $382/km^2$ approximately population reached 1.32 billion in India, July 2016, with an economic growth of 7% compared to last two decades approximately. These increases lead in increasing surplus by users and instantaneously foremost to hazard for health and cleanliness. Waste production is 0.143 million tones approximately and only 23% of this waste processed daily, as reported by Central Pollution Control Board (CPCB) on municipal solid waste (MSW) [1].

According to Laura Michelle et al. [1] serves to prerequisite in effectual way, with enormous research and determination, proposed a structured waste management scheme and realized successfully in public in Alappad panchayat in Kollam, Kerala. The system scrutinized at each phase of enactment for effective assessment. System used colored containers with definite tags, this ease course of sorting for user. Following is the methodology.

Waste segregation at source by waste-receiving stations: For waste deposition, designed location of waste based on environment prerequisites are expediently located around civic. Different color bins utilize for organic matter, waste paper, or sanitary. Steadiness create practice pattern safeguarding correct split-up to bins deprived of trusting merely on marks. Like garden shrubbery waste and waste food, designated by green containers and picture marks, paper is signposted by blue bins, soft plastics indicated by orange bins, solid items fluctuating from metal waste, waste plastic, and broken goblet fit in red for reprocessing containers. Lavender color baskets for cloth bounces opportunity to utilize fabric many ways beyond burning, sanitary in pink bins, needles sharp perilous and wrecked glass to yellow containers, hair and dirt and soil in beige baskets.

Waste sorting: Steps for sorting waste are extrication organic materials and nonorganic materials, nonorganic section pre-sorting, recyclable material refined sorting.

Introduction of labels and color for bins posed quantifiable impact on growing level of sorting at source. On linking weight of waste from "source" and "after processing" perceived efficiency of source waste sorting amplified to 76% due to the color and labeling. Mistakes were only accounted to 24% of that to the model without color and label coding.

The claim for SWM in today's civilization is emerging as days passes. Essential dispute linked to environment is managing material waste, because of its global impact. Requirement of diverse practices are vital in managing and dealing fresh facilities and to adopt fresh processing means [2]. Mangesh J. Khandare et al. carried investigation on source and administration and waste materials control in construction, mechanical designing, and electronic disciplines. Expanded economic development and suburbanization created in expansion of civil industry will yield more waste. Mechanical post-customer is slight sponsor to waste incineration comprising remnant carbon produced by GHG outflow, methane landfill, water and oxides of nitrous. Electronic waste is presently prevalent developing stream in domain is delivered to developing nations like India for reusing.

1. Material Waste from Civil industry: Enormous waste after demolition work is generated, same with construction work. It is estimated by quantity surveyor that from entire waste 15–30% is of civil industry. The research objectives say, first objective, investigation will be done on reduce, reuse, recycle, and recovery technique used in the waste management system onsite to identify most used 4R techniques. Second objective, can see whether or not minimize 4R technique used provides an important impact on accumulative waste manufacture on website. Third objective, to spot variations among minimize, 4R technique used, confirm that of techniques are economical or not in manufacturing less waste.
2. Controlling Solid Waste: Solid waste management (SWM) is done by many techniques like composting, waste to energy, and biomethanation thermal process to extract energies of consumer post-waste are used like biogas plant. Output can be organic acid, soil fertility manure, gas when burnt blue flame generated and cooking need this heat.
3. Electronics Waste: It bags largest waste and hazardous throughout globe. Its management is pricey and complex. Parts of computer, institutional laboratories, PVC, etc., are e-waste examples, if disposed not correctly, harm human life.
4. Conversion Waste–Energy: Green project, separately collecting waste and recovering for reuse is prominent solution. Biogas plant for SWM organic manure by green waste, generate energy by reuse techniques and recovery techniques from e-waste.

Salman Nizarudin et al. mainly focused on reusing plastic waste as fuel or lubricant [3]. Least-expensive routine is burning plastic wastes, no petroleum required for burning since plastic itself solid fuel. The equipment consists of an electric heater, a stainless-steel reactor, a condenser, and an output collection tank. Electric heater as muffle furnace comprises heating facility for setup. Leak proof reactor fabricated with stainless steel SS316 grade. The temperature is controlled by a PID controller. Thermal degradation converted plastic waste to three components: Condensed liquid/waxy oil, uncondensed gas, and carbonaceous solid residue. Catalyst was utilized to increase reaction rate, thereby reduce temperature provided and reaction time which had a direct relation to the energy consumption.

Thermo-catalytic deprivation or pyrolysis by catalyst of household unit plastic waste exposed to heat of 550–700 °C at specific catalyst extent in sealed compartment kept up to atmospheric pressure. The ideal monetary and compelling yield was accomplished at 551 °C for specific proportion with particular catalyst. The preparatory tests propose that the slick/waxy yield could be castoff as grease or petroleum. Kerala is mainly fretted over issue of plastic. The operation is reputable in user-friendly way and hassle-free. Hence, disposal of plastic waste is linked to eruption of new resource, thereby enhancing productivity and economic value of process. Economy deciding subtleties are equipment cost, consumable items cost like washers and the fasteners, heater power, capacity of reactor volume, reaction time, catalyst used, catalyst used quantity, personnel cost of labor managing operational apparatus, maintenance and maintenance frequency cost, output storage method, shipping cost.

Suchitra Ramesh et al. reflected survey of incident readings accompanied by Reva college, Bangalore on government SWM in some capitals across India is revised to advantage perception into SWM [4]. Numerous records regarding SWM has gathered for considerate hitches being tackled by Bangaloreans. Fertilizing soil with oxygen-treating soil and vermi-treating soil and waste to energy (WTE) by incineration or by biomethanation, are two driving waste disposal systems being embraced in India. Moderately, new idea in India is WTE component for MSW disposal. Biodegradation, method for reusing squanders organically where microorganisms, parasites, bugs, worms, and different living beings complete rotting process consume passed on material and reuse as new structures, quickened with fertilizing soil. This is completed with natural waste 76% of aggregate waste. Computerizing SWM is mind boggling and includes part many variables, such as engineering, hazard examination, innovation appraisal, expenses, and client visualization. Security and medical problems are unintended components. Incineration is squander management process combusts natural surplus materials and translates the surplus into ash, vent gas, and warmth. The heat produced by incineration utilized to create electric energy. Burning is squander to vitality innovations, for example, gasification, gasification of plasma arc, pyrolysis, and processing anaerobically. Inconvenience is Plant's Expenses. Recycling is procedure of changing over materials of squander into new items to deter misuse of valuable materials, decrease use of crisp crude materials, vitality utilization, diminish air contamination from burning and the water contamination from landfilling by diminishing prerequisite for "regular" waste disposal, and lower ozone-depleting substance emissions.

Mini plants of biogas in urban/provincial muncipal territory: these biogas plants of 200 L anticipated by bio-grounded energy research laboratory. The microscopic organisms separate organic surplus in anaerobic conditions. The slurry hard matter of 5% is obligatory for observance plant in exertion and gas stored over digester. The gas delivered is exchanged over gaslight closure mounted on barrier exterior kitchen. Normal blending will build gas production. Cost is primary favorable position of digester contrasted with ordinary digester. The small bio-digester has fiber boiler dissimilar to ordinary bio-digesters two boilers and thus is economical. Its cost is around 60 $ and quite reasonable.

Discovering effectual technology for resolving encounters faced by SWM is what Government in all countries is currently focusing. Radha R C et al. in paper [5], present an appraisal on tools for isolation and SWM. The transformation in life flair of societies and innovative technologies with accumulative population with urbanization waste creation increased including post-consumer materials. If unmanaged then will be harmful to lives. To solve problem, active functioning

SWM is mandatory. In early days, waste is coped by consuming four rudimentary means. They are dumping/land fill, scorching, the recycling and waste reduction.

Technologies for segregating solid waste:

Solid waste incorporates several belongings like metal waste; plastics and so forth the significant number of reusability is open and goes about as contribution to other frameworks. By MSW fertilizing soil strategy initially squander is isolated by considering size by utilizing trammel and each sized waste independently processed. For all, paper recommends a mechanical framework. Another sort of separator by Eddy current: The detachment procedure occurs two phases, first firmly conducting elements are isolated on upper portion of drum, and afterward staying unresolved and ineffectively conducting elements get isolated at magnetic drum's lower portion. Indirect arranging technique been anticipated to the category solid waste utilizing sensor that is optical and mechanical isolating framework. The shading, shape, and waste dimension are utilized by waste separation. The mechanical gadget for sorting comprises of pressed air nozzle managed by computer; target particles perceived by radar were smothered of waste stream. Electrostatic isolation sorts gritty blend because of powers of electricity following on roughly 5.0-mm-size elements. X-ray fluorescence strategy distinguishes basic piece of materials in X-ray box fluorescent and accordance with examination materials are distinguished and isolated from material stream. Controller by programmable logic is utilized to isolate metal from squander materials. Squander is fed to belt of conveyor line with sensor clipped for detecting metal will identify metals and automated arm will separate metal from waste and store in canister.

Framework ready to oversee and isolate strong waste adequately desires GIS framework viably to gather data identified with solid waste and GPRS for successful transportation and solid waste general accumulation procedure. At last, to direct labors' simple prearrangement is camera. For isolation, utilize sensor to recognize and isolate perilous components like batteries and isolated waste accordance to dimensions and operate them autonomously. Magnetic/separator by eddy current/metallic sensor is utilized to isolate conductors. FT-IR spectroscopy and imaging with hyper-spectroscopy/NIR spectroscopy and multivariate investigation/X-ray fluorescence technique is utilized to isolate plastic, filaments and glass. This bounces advance road to innovative work of appropriate advances for SWM and isolation.

According to Predrag Milić et al. [6] optimization in vehicle routing for communal waste assembling has noteworthy role in transport cost fall. Hitches occur through optimization are primarily connected to weaknesses in contribution data. Most algorithms to route, undertake that expanse of waste at assortment spaces is known advance is deterministic value which in most cases is not. They established system of observing and lively direction-finding of automobiles for waste gathering in city. By smearing modern expertise better awareness in vehicle state on pitch with variations in waste extent on gathering places are attained. This framework apportioned into three segments. The principal speaks to apparatus introduced in RCV. It comprises of GPS-GPRS gadget, which advances evidence of vehicle position and information of weight plate position. The second part is product for information gathering and database on server. The third segment presents named routing singular host solicitations on workstations.

While taking vehicle-routing issue, various confinements occur and point is gained by fixing at least one target function. The target of vehicle-direction-finding procedure can be: minimization of aggregate transport expenses, depend on course length or principle travel time and settled expenses of vehicle and driver use, total of vehicles minimization or motorists with objective servicing all clients lastly, modifying courses relying upon way time and vehicle stack.

Dynamic routing with stochastic demands (DRSD): Routing vehicle issue dynamically (DVRP) implies that all data not pertinent to vehicle-directing procedure organizer of route starting to routing procedure and data identified with routes arrangement be altered after underlying expansion of primary route. Because of deviations in numerous algorithms are joined with conventional procedures, and hybrid procedure obtained with better execution. To accomplish DRSD in nodes of graphical forms squander sum, important to decide conveyance of alterations in measure of surplus on premise of recorded information on database server. By smearing technique for dynamic routing, vehicle path changes are accomplished amid procedure of accumulation. This guarantees ideal solutions subsequently shrinkage in absolute expenses and utilizing modem gives ideal arrangement continuously.

Business credible that narrates on waste flow management (WFM) is enormous globally but deprived of conceptualizing ecosystem to fine level, business latent might not be exposed fully [7]. Study by *Tero Peltola* et al. tailed existent perceptions in works and smeared business ecosystem thoughts into Brazilian WFM. Structure, problematic innovation players are eminent in squander stream administration and possibilities they empower, uncovered. An edge work for perceiving innovation accomplishment artists is shaped where partners in environment can be portrayed, for example, providers, central firm, complimenters, and clients. This edge effort needs in satisfactory solidness; thus, new system is raised in paper presented. Distinguishing proof of innovation performers among actors in squander stream administration environment is built on esteem blue pattern attitude toward biological community. It makes tremendous part of government and manages laws and also regulations as dynamic actors in squander biological system. In system administrative character is available through municipality and investor. Four primary innovation actors been recognized. Providers, in particular house and municipality, are not innovation actors; also, they source material through waste holders, squander pickers, and reusing distributer.

Squander is transported by various sorts of vehicles among actors and lastly it meets landfill or reusing facilitator. The fascinating finding in study is the absence of core innovation measurement in waste flow environment. The market measure, material bulks, and business openings are enormous, however, those have not produced core developments that are spread to different parts. Innovation performers be infamous reason being conceptualized squander stream business biological community yet their characters are numerous. Testing surplus business biological community show is intriguing and essential research route for future. Model be confirmed by concentrating on performer in biological system, using contextual analysis attitude, and recognizing esteem chain and plan of action of actor. The tremendous experience to catch worth in squander environment in creating nations involves helper readings with chief perspectives.

Developments in strategies and protocols for e-waste and evaluation in India, Amit Jain et al. in [8] presents scrutiny of prevailing policy and conventions and appearance of future tendencies been approved out for Indian e-waste management. Further, major carters that projected to accustom rudimentary features for expansion, institutionalization, and employment of future governing interpolation also identified.

At primary place, appraisal of reusing engineering in India been done trailed by assessments of e-squander arrangement and administrative condition and expected forthcoming patterns. Methods like Indian piece and e-surplus reusing commerce mapping utilizing "tracer procedure" took after arrangement and administrative condition survey been utilized to assess current business condition. "Tracer procedure" been utilized for outlining whole material flow since phase of accumulation, transportation, disassembling, reusing, and dumping. Further, standard vital assessment strategy of SWOT, i.e., Strength, the Weakness, the Opportunities, and the Threat investigation been connected with regard to prevailing business condition to recognize future procedure drivers.

At present, around 400 little to average waste-paper reusing units occupied with produce of paper. Around 47% of plastic waste produced in India is reused. 40,000 units and more occupied with plastic item fabricate of which 13% are in official part and rest 87% are little scale unofficial segment. Notwithstanding these units, 30,000 and more polymer processors exist in India. The formal and casual reusing areas are self-possessed at three stages of chain of expertise. (i) level 1 preparatory e-squander generators, (ii) level 2 optional e-squander producers, and (iii) level 3 tertiary e-squander producers. The contribution to prime level originates from formal market like makers, shippers, workplaces, and formal markets, where e-surplus from residential purchasers comes either return plots or disposed of things. Hence, real stakeholders are scuffle merchants/dismantlers who buy e-squander from primary level at bulk amounts. They have restricted limit of disassembling and are linked with exchanging of e-squander with higher level of dismantler/scrap merchants.

SWOT investigation of e-squander stock in India demonstrates that PCs, phones' information and communication innovations, TVs, i.e., brown products, fridges, and clothes washers, i.e., white merchandise are required to drive future development of e-squander reusing commerce in India. The consequences of scrap against e-squander reusing planning and strategy and control mapping unmistakably show without item particular regulation, casual reusing is contending formal reusing division in e-squander administration.

Alex Krizhevsky et al. in [9], proposed AlexNet, ImageNet classifier with deep convolution network of totally five convolutions with pooling layer with max operator, layer dropout, and three-layer fully connected. Mainly, network designed for grouping with 1000 categories. Training on, ImageNet input of 15 million marked images, over 23,000 categories, used nonlinearity operator RELU and augmentation technique of data, tactics that alter training information order to array representation change while protecting label same are data augmentation procedures. They are approach to artificially grow your image database with image interpretation by translation, reflecting horizontally and extraction of patch. Dropout layer, weights of system are adjusted to given preparing cases that system does not perform well when given new cases. Layer "drops out" an arbitrary arrangement of initiations to layer by fixing values zero. Exercising with gradient descent with batch stochastic used two GPU's GTX 580 for six days and fault rate 15.4% detected.

Matthew Zeiler with Fergus in [10], proposed ZFNET, ImageNet 2013 Challenge victor, and fine tuning for previous victor AlexNet with improved presentation. Used 1.3 million images for exercising network and AlexNet utilized 11*11 convolution filter at primary AlexNet layer hopped lots of evidence, so 7*7 convolution filter is utilized by ZFNet and recollects more evidence from original pixel. ReLUs for stimulation function layer raises nonlinear possessions of neural network and overall neural network without upsetting receptive pitches of layer convolution and cross-entropy as fault rectification method. Prepared for almost 12 days utilizing faster GTX-580 GPU. Methodology adopted was Deconvnet, i.e., De-convolution plotting of features to intensities with series of no pooling, rectify, and non-filter operations executed with error rate 11.2%.

Karen Simonyan et al. in [11], proposed VGGNet, not winners of ILSVRC 2014 but runners with 7.3% error rate in categorizing with 19 layers' neural network consisting of 3-by-3 convolution filter, max 2-by-2 pooling with stride2 were stride handles convolution of filter through input volume, filter may convolve everywhere on input bulk by shifting one unit. The volume of shifts by filter is stride. Advantage, volume of input does not shrink sharply at once unlike 7-by-7 convolution filter and deepness of volume also proliferates at lower neural network levels. With two convolutional levels, two ReLU layers utilized instead of one. Three layers of convolution of 3-by-3 have effectiveness more than one convolution with 7-by-7 convolution filter. One downside of VGGNet is large network and contains parameters around 160 M.

Christian Szegedy et al. in [12], proposed GoogLeNet, 22 layer CNN and was victor of ILSVRC 2014 with error rate among top 5 of 6.7%. ILSVRC 2014 challenge comprises task of categorizing image to one group of 1000 leaf-node categories in ImageNet hierarchy. About 1.2 million pictures for preparing neural network, 50,000 was for validation purpose and remaining 100,000 images testing. Each picture is connected with reference class, and execution is measured in light of the most elevated scoring classifier forecasts. Utilizations of 12 times less parameters equated with complexity to AlexNet. Amid testing, various crops of similar picture were created, served to system input, and finally softmax probabilities obtained at average and give last arrangement with couple of high-end GPUs for seven days.

Jie Hu et al. in [13], proposed squeeze-and-excitation networks (SENet). Won ILSVRC 2017 classification won first place and significantly reduced the top-five error to 2.251%, achieving a 25% relative improvement over the winning entry of 2016. Here, objective is guaranteed that system can build its useful featured neural network with goal be abused by resulting alterations, and smother not important features. It is accomplished by explicitly demonstrating channel dependencies to one another and channel reactions recalibration in two stages, squeeze and next excitation, before they are fed for next change. Squeeze includes each scholarly filters work with neighborhood receptive and unit of change yield U and cannot exploit relevant data external of this locale. This turns out an issue to sever extreme in lower layers of system whose receptive sizes are little. Excitation make utilization of assembled information in squeeze operation, moment operation which anticipates to completely catch channelwise circumstances is offered. A SENet is fabricated by stacking arrangement of SE chunks. 4 or 8 NVIDIA Titan X GPU server required (Table 1).

According to ImageNet challenge winners (based on image classification and localization), the neural network models and their accuracy on training are depicted in Table 2.

3 Existing and Proposed Method

3.1 Existing Method

In existent system, documentation of waste material is upheld and there is necessity to edify people on waste-type and class it belongs to, either green waste, reject surplus or reusable surplus. Problematic is recalling waste-material list and human blunder increases. Hence, anticipated technique is pointed to mark proficient grouping. If you just capture waste-material picture and application articulates which category it appropriates? Isn't it easier? So neural network idea arrives here and necessarily trained and deployed as handy application, just capture waste-material picture and in second neural network declares category waste-material image fits.

3.2 Proposed Method

According to study, 2 lakh plus households had executed 2 BIN 1 BAG and found reject waste finally gathered was 5.2% only. 95% lasting waste isolated as recyclable and green waste. The overall problem viewed in one sense as segregation problem. Easy solution is segregating waste at source. At every household waste should be segregated before dumping in one bin. Household people should have adequate awareness about waste material types and belonging category. Whole problem viewed as classification problem and neural network is preeminent at solving classification hitches.

A neural network is educated with number of speckled waste-material images like, plastic surplus, bottles, paper, leaf, broken glass, vegetables, blades, and capsules in exercise data set. Finally, neural network is skilled and when given image of waste

Table 1 Brief on literature survey

Author	Methodology	Results
In [1], Laura Michelle et al.	System used colored containers with definite tags	On linking weight of waste from "source" and "after processing" perceived efficiency of source waste sorting amplified to 76% due to the color and labeling. Mistakes were only accounted to 24%
In [2], Mangesh J. Khandare et al.	Carried investigation on source and administration and waste materials control in construction, mechanical designing and gadgets engineering	Green project for Civil Engineering waste, Mechanical Engineering waste management best is plants for treating solid waste, recycle and reuse methodology for e-waste
In [3], Salman Nizarudin et al.	Reusing plastic waste as fuel or lubricant, Thermo-catalytic deprivation or pyrolysis by catalyst, plastic waste exposed to heat of 550–700 °C	Slick/waxy yield could be castoff as grease or petroleum. With many economy deciding subtleties
In [4], Suchitra Ramesh et.at.	A survey of incident readings accompanied by Reva college. Various methodologies are discussed that are carried out in different cities of India	For solid waste management, a mini biogas plant is feasible for conserving energy form waste
In [5], Radha R. C. et al.	A review on waste segregation and managing solid waste. And technologies for solid waste management and segregation	GIS system and RFID/GSM/GPS/GPRS for locating solid waste and effective transportation, respectively Eddy current to separate conductors, FT-IR spectroscopy, and X-ray fluorescence method used to separate plastic, fibers, and glass

(continued)

Table 1 (continued)

Author	Methodology	Results
In [6], Predrag Milić et al.	A system that will dynamically update the routes that are loaded with waste. And updating new routes	Results proven that dynamic route updation is better than static routing
In [7], Tero Peltola et al.	A framework is proposed for identifying critical technology factors	Testing the waste business ecosystem model should be carried out. Framework would be based on synthesized ecosystems of business of several developing countries
In [8], Amit Jain et.al.	Analysis is done on policies and regulations existing for e-waste management and future trends that are emerging. Strength, Weakness, Opportunities, and Threat (SWOT) method is applied on various business contexts	The cost of recycling, recovery, and disposals are the major drivers, which is much cheaper in India in comparison with developed countries. If e-waste is included in existing policies and regulatory environments that definitely it initiates formalization of e-waste collection, e-waste transportation, and also disposal mechanisms to some extent specially influencing the e-waste produced by commercial sectors

Table 2 Neural network models and their accuracy on training

Model name	Accuracy (in terms of Error rate) (%)	Training
In [9], Alex Krizhevsky et al., proposed AlexNet	15.4	2 GTX 580 GPU's for 5–6 days
In [10], Zeiler and Rob Fergus, proposed ZFNet	11.2	Run on one GTX-580 for 12 days
In [11], Karen Simonyan and Andrew Zisserman, proposed VggNet	7.3	4 Nvidia Titan Black GPUs for two to three weeks
In [12], Christian Szegedy1 et al., proposed GoogleNet	6.7	Trained on "few high-end" GPUs within a week
In [13], Li Shen et al., SENet	2.251	4 or 8 GPU severs (8 NVIDIA Titan X per server)

material as input from analysis data set, it should classify among green waste, red waste or reusable waste.

4 System Design

System design concisely springs system outlook from high end and describes involved modules and flow of work between them. Outlining convoluted system foundations like architecture, components, and interfaces among components and data flowing in system is what complete development of system design debates about. Here, design of system designates how neural network for our problem is planned.

4.1 High-Level View of Training Model

Training any neuron network is easy, but importance should be given to construction of training data set. Training data set means collection of images for training neural network, varied collection of images for training is necessary, various data augmentation techniques be used to increase image set (Fig. 1).

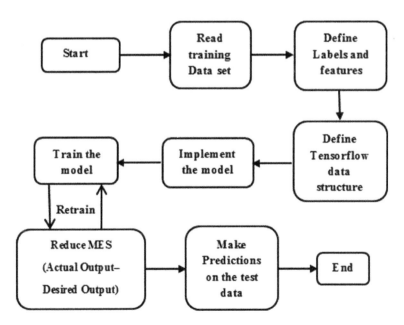

Fig. 1 Training phase

4.2 Training Phase

Training phase says training process for neural network (NN) with the images and error reduction method used and method for predicting NN performance.

1. Reading training data set:
   Data set is assembly of imageries for each categories and images in number of classes for each category. For example, green bin category means green waste may have food waste, garden waste as classes. Images can be collected with digital camera or available on the Internet and data set enhanced by data augmentation.
2. Define labels and features:
   Important part in training can be called as preprocessing stage where you insist NN about image given for processing. Otherwise, preprocessing can also include creation of feature vector and giving it as input. But in our project, input for network is labeled images, where label is category of image.
3. TensorFlow data structure:
   Utilize tensorflow for generating neural network. The neural network comprises many levels of convolution function involved with RELU, next max pooling. Defining structure is root for workflow.
4. Implement model:
   Implementation means execution of modeled tensorFlow NN and obtaining NN model and viewed graphically on tensorboard for envisioning model as graph is informal for understanding stream of tensors—the data.
5. Train model:
   Cost, or loss recognition and signify distant off system is from projected consequence. Try minimizing blunder, and slighter error margin, superior model ready. One common, very agreeable function to decide loss of model is "cross-entropy" and yields improved outcomes. Optimizers gradually alter each adjustable to diminish loss function, MES here. The gradient descent is used as simplest performance enhancer. It adjusts each variable conferring to extent of loss-derivative to variable.
6. Reduce MES:
   Difference in desired and existent output is squared and error is detected and lessening is done during training model by improved performance optimizers. The first-order optimization, the simplest is variant of gradient descent, SGD techniques performs a parameter update for each training example. These ease the computation and consume less time, fast-converging capacity on large data sets.
7. Make Prediction on test data:
   Two-phase regression by softmax: first includes proof of info belonging in definite classes, and after make modification over that confirmation into probabilities. To count up proof that given picture is in specific class, weighted total of intensities of pixel considered. Result is undesirable if pixel having more intensity, confirmation against picture belonging to class, and positive on-off chance is evidence in support.

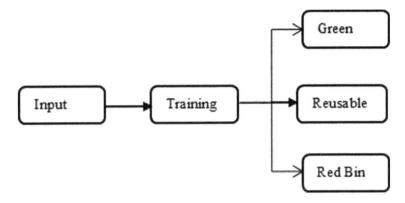

Fig. 2 Inference model

4.3 Verification Phase

Verification data set is maintained, consisting of images belonging to 3 categories of 2 BIN 1 BAG system. Finally, an image from testing data set is given as input to trained model. The model processes and recognizes pattern and try to predict pattern and outputs belonging category (Fig. 2).

5 Concepts of Neutral Network Used

Overview of CNN but we will not more specifics of conventional neural networks.

5.1 Choosing Hyper-Parameters

Hyper-parameters are very important for a neural network model, choosing right hyper-parameters give the desired output. This includes how many layers should be used for constructing our neural network, what number of convolutional layers, convolutional channel size, or qualities for stride and the padding of zeros? These are not trifling examinations and no standards are announced for these by any experts in deep learning or machine learning. This certainly is on grounds that system will and to a great extent trust upon sort of evidence that one possesses. Evidence is nothing but the data that flows within layer or from one layer to another can fluctuate by size, multifaceted nature of picture, way the picture is being processed, and sky is limit from there. When captivating a gander at data set, one approach is seeing how to pick hyper-parameters helping to mining important areas of image while neural network is training itself is to locate correct blend that makes deliberations of picture to appropriate scale.

5.2 Rectified Linear Units (ReLU) Layers

After every layer of convolution, it is a tradition to smear a task of nonlinear layer or enactment layer instantly thereafter. The motivation behind this layer after layer of convolution is to acquaint nonlinearity with framework that principally has quite freshly been registering linear actions amid the convolutional layers just componentwise duplications and the sum as whole. In past, nonlinear methods like tanh is utilized also sigmoid, however, scientists revealed ReLU layers' work far superior on grounds established that system can formulate a ton speedier in light of computational effectiveness but principally without having huge effect to exactness, sustaining has significance. It alleviates additionally fading slope issue, where subordinate layers of system prepare themselves gradually because, fact that gradient diminishes enormously exponent over layers (Clarification may off story extent of this post). The perseverance of ReLU layer smears method $f(x) = \max(0, x)$ to greater part of quantities in voluminous information. Fundamentally, ReLU layer swaps negative enactments to 0 making inactive. ReLU expands nonlinear properties of model and general system without influencing responsive fields of convolution layer.

5.3 Pooling Layers

Few ReLU layers after, software engineers may smear layer called pooling layer. It is likewise alluded as layer of down-sampling. In this class, there are likewise few more verities of layer alternatives, with max pooling being supreme prominent. This fundamentally takes channel typically with size 2×2 with stride of similar length, else we can also move with 4×4 max pooling also. At that point, system applies it on response and yields most extreme number in each subregion that channel convolves around. The instinctive thinking overdue this layer is, once we realize that particular element is in first information volume there will be high actuation esteem, its correct area is not as critical compared to relative area to alternate features.

5.4 Dropout Layers

Presently, dropout layers possess definite methodology for neural systems. Last section, we spoke about dispute of overfitting, where successive to training, weights are so adjusted to training, given illustrations then system does not succeed good when subjected to new cases. The dropout possibility is oversimplified in own behavior.

This "drops out" an asymmetrical prearrangement of initiations in layer applied by assigning zero to those. Isn't it simple? Presently, advantages of basic and seemingly superfluous and nonsensical process are in enquiry, all things measured, as it were, influencing system be repetitive, means entire system to possess ability to stretch correct categorization or yield to certain illustration nevertheless of whether portion of actuations are dropped.

5.5 Data Augmentation Techniques

At this point, we are all utmost likely numb to significance of information in ConvNets; so, we should discuss conducts that makes your current data set of images, or audio or data set videos much bigger, just due to couple simple changes. Comparably, we have said recently some time, when computer gets picture information it takes in variety of pixel intensities. Suppose that entire picture is by 1 pixel moved left. For us, this revolution is impalpable. Notwithstanding, a computer, this move be genuinely huge as classification or name of picture does not alters, while cluster does. Methodologies that modify training information as alteration exhibit portrayal while keeping mark same called techniques of augmentation. They are approaches to falsely outspread your data set. Example: the images can be flipped, rotated in several different angles to generate multiple images out of single image.

6 Neutral Network Model and Mathematical Model

A typical model for neural network for 2 BIN 1 BAG classification is shown. Input images are fed to neurons at input layer 1 and convolutional operations are done at layer 1 and layer 2, which is indeed hidden layer with multiple convolutional operation going on. Finally, the model gets trained for each category of images (Fig. 3).

Here, we discuss the mathematical model of a CNN. So, what exactly happens at convolutional layer, a simple dot product is carried out between input image I of size $(m \times n)$ and a convolution filter F of size (3×3) given by,

$$G_{\text{conv}} = \text{conv}(i, k) = (I \times F)(i, k)$$

where $\times$ is the convolutional operator between I and F, G_{conv} is the map of feature vectors obtained as output of convolution, activation is spread on each h_{ik} in G_{conv} and here, we use ReLU activation function [14], because most of the cases the output of hidden units is non-negative and in such cases, ReLU helps NN to learn faster, and given by,

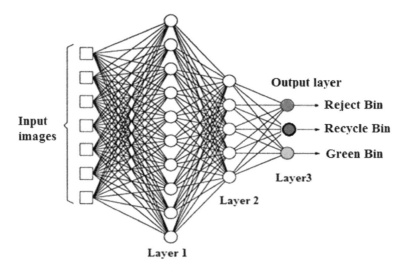

Fig. 3 Neural network model

$$G_{\text{relu}} = \max(0, h_{ik})$$

Next to activation layer, a pooling layer is added and here, we have used 2×2 max pooling, max in every 2×2 grid of G_{relu} is the output of a max pooling layer, is given by,

$$G_{\text{mp}} = \max\left[(G_{\text{relu}})_{2\times 2}\right]$$

G_{mp} is the output feature map of max pooling layer. G_{mp} is the final output obtained after a convolution and activation layer, that is in a sequence of G_{conv}, G_{relu}, and G_{mp}. The same set is repeated again and finally a "dense" layer and "softmax" activations are final output layers.

7 Current and Future Scope

7.1 Current Scope

Despite the fact that the present strong waste administration (SWM) situation is a long way from acceptable, a few results and evaluations in the examination uncovered that there are satisfactory chances to deal with and enhance the circumstance. The investigation suggests that a successful strong waste administration can be guaranteed by building up an incorporated strong waste administration rehearses and furthermore by a solid public–private organization, where partnership rose as an instrument for better administration conveyance.

- Currently according to survey done in Bangalore, two lakhs and above households adopted 2 BIN 1 BAG system.
- Garbage collectors reported that collection of garbage bags exceeded 2 lakh everyday.
- The amount of garbage landfill was drastically low, in number landfill fate avoided 180 tons and more garbage everyday.
- Along with this 40% of human error is done during classification and this can be avoided if neural network is placed as classifier.

7.2 Future Scope

SWM is experiment for authorities existent with developing countries primarily due to growing waste generation, municipal budget load posed results of expenses connected to waste controlling, absence of comprehension over assorted elements variety influencing the diverse phases of waste administration and linkages important to empower the whole taking care of framework working. So, there is scope for in SWM by neural network,

- The neural network can be deployed as application in mobiles and laptops. This becomes handy for using application wherever we move.
- The application can be used by everyone at home and just capture image picture by camera and give input for application and application says output category.
- The neural network needs to be populated with thousands of images to make it stronger classifier. The image in training date set must be increased, so neural network is trained well resulting in good accuracy.

8 Conclusion

When an image from testing data set is given as input to the trained artificial neural network model, it will be able to classify it easily to one category among green waste, reject waste, or reusable waste. No need to bother about the type of the waste and problem regarding translating this into many languages in India can be avoided. This increases the efficiency of classification and eliminates human classification errors completely. The aim of setting faster waste management technology to achieve safer environment can be accomplished.

References

1. Goris, L. M., Harish, M. T., & Bhavani, R. R. (2017). A system design for solid waste management: A case study of an implementation in Kerala. IEEE, 2017.
2. Khandare, M. J., & Khandare, S. M. (2016). Waste materials and management in civil, mechanical and electronics engineering. In *International Conference and Workshop on Electronics and Telecommunication Engineering 2016*.
3. Nizarudin, S., & Deepak, B., Thermo-catalytic degradation: Solution for plastic waste management in Kerala. In *IEEE R10.B. Smith, "An approach to graphs of linear forms (Unpublished work style)"* (unpublished).
4. Ramesh, S., Usman, A., Usman, A., & Divakar, B. P. (2013). Municipal solid waste management in Bangalore and the concept of mini biogas plant in urban localities. IEEE, 2013.
5. Siddappaji., Sujatha, K., & Radha, R. C. (2016). Technologies for segregation and management of solid waste: A review. IEEE, 2016.
6. Milić, P., & Jovanović, M. (2011). The advanced system for dynamic vehicle routing in the process of waste collection.
7. Peltola, T., & Mäkinen, S. J. (2015). Identifying critical technology actors in waste flow management.
8. Jain, A., Developments and evaluation of existing policies and regulations for E-waste in India.
9. Krizhevsky, A., Sutskever, I., & Hinton, G. E., Image net classification with deep convolutional neural networks. In*NIPS 2012: Neural Information Processing Systems*, Lake Tahoe, Nevada.
10. Zeiler, M. D., & Fergus, R. (2014). Visualizing and understanding convolutional networks. Published at ECCV 2014 and Springer International Publishing Switzerland 2014.
11. Simonyan, K., & Zisserman, A. (2015). Very deep convolutional networks for large-scale image recognition. Published as a conference paper at ICLR 2015.
12. Szegedy, C., Liu, W., Jia, Y., Sermanet, P., Reed, S., Anguelov, D., Erhan, D., Vanhoucke, V., & Rabinovich, A. (2015). Going deeper with convolutions. CVPR 2015.
13. Hu, J., Momenta, & Shen, L., Squeeze-and-excitation networks. arXiv:1709.01507v1 [cs.CV] 5 Sep 2017.
14. Dahl, G. E., Sainath, T. N., & Hinton, G. E., Improving deep neural networks for LVCSR using rectified linear units and dropout. In *2013 IEEE International Conference on Acoustics, Speech and Signal Processing*.

L1-Regulated Feature Selection in Microarray Cancer Data and Classification Using Random Forest Tree

B. H. Shekar and Guesh Dagnew

Abstract Microarray cancer data are characterized by high dimensionality, small sample size, noisy data, and an imbalanced number of samples among classes. To alleviate this challenge, several machine learning-oriented techniques are proposed by authors from several disciplines such as computer science, computational biology, statistics, and pattern recognition. In this work, we propose L1-regulated feature selection method and classification of microarray cancer data using Random Forest tree classifier. The experiment is conducted on eight standard microarray cancer datasets. We explore the learning curve of the model, which indicates the learning capability of the classifier from a different portion of the training samples. To overcome the overfitting problem, feature scaling is carried out before the actual training takes place and the learning curve is explored using fivefold cross-validation method during the actual training time. Comparative analysis is carried out with state-of-the-art work, and the proposed method outperforms many of the recently published works in the domain. Evaluation of the proposed method is carried out using several performance evaluation techniques such as classification accuracy, recall, precision, f-measure, area under the curve, and confusion matrix.

Keywords Microarray cancer · Learning curve · L1-regulated feature selection · Random Forest tree · Classification · Learning curve

1 Introduction

Cancer is an illness behaved by abnormal cell growth and division in an uncontrollable manner which destroys the healthy tissues of human beings [1]. Biological data are mostly noisy, high dimensional, and small samples size in nature. In microarray data analysis, the challenge increases as the gene sequences have large variance, which leads to overfitting problem and low efficiency during fitting data to a model. Moreover, the redundant and irrelevant data lead to distortion in model development.

B. H. Shekar · G. Dagnew (✉)
Department of Computer Science, Mangalore University, Mangalore, India
e-mail: guesh.nanit@gmail.com

© Springer Nature Singapore Pte Ltd. 2019
N. R. Shetty et al. (eds.), *Emerging Research in Computing, Information, Communication and Applications*, Advances in Intelligent Systems and Computing 906,
https://doi.org/10.1007/978-981-13-6001-5_6

Feature selection process in the domain of microarray cancer data remains to be an ill-defined and challenging problem due to the high dimensionality, noisy data, an imbalanced number of samples in each class, and small sample size [2–4].

Feature selection is a process of selecting a subset of relevant features for use in model construction. Features in microarray data are representative genes, which represent a measurement of certain traits of the typical biological cell and are characterized by high dimensionality and small sample size. There are several types of feature selection techniques which are generally categorized as ranking, wrappers, and ensemble methods. Classification of microarray medical datasets plays an important role, especially to identify those genes which contribute the most to a certain biological outcome and predict a result when a new observation arrives [5].

In this work, we propose an efficient L1-regulated feature selection method for classification of microarray cancer data using Random Forest (RF) tree. L1-regulated feature selection method is one of the model-based feature selection methods which uses classifiers such as support vector machine (SVM) and logistic regression (LR) to fit the dataset so as to remove the irrelevant features so that these selected features can be used for classification purpose on other classification models. It works by adding a penalty term to the ordinary least square methods so as to avoid those features whose values are zero in the sparse matrix.

The rest of the paper is organized as follows. Section 2 discusses literature survey related to the proposed work. Section 3 describes the dataset used to validate the proposed method. Section 4 describes the proposed work, and Sect. 5 discusses the performance metrics used to evaluate the predictive capability of the classifier. Section 6 discusses experimental results of the work, and finally, Sect. 7 provides the conclusion remarks of the work.

2 Related Work

Analysis of biological data such as microarray cancer data analysis is a hot research area across many interrelated fields. Feature selection methods have the tendency to provide useful information on the relative relevance of features for a given classification problem. Guo et al. [6] propose two-stage dimensionality reduction for classification of microarray data. In the first step, L1-regularized feature selection is followed by PLS-based feature extraction on the selected features. The informative the features that are selected, the better the results will be. Better features mean flexibility, simpler models, and better results having less complex models that are faster to run, easier to understand, and easier to maintain. Medjahed et al. [7] introduced kernel-based learning and feature selection using LS-SVM application for complete cancer diagnosis on standard microarray cancer datasets, namely Breast, DLBCL, Leukemia, Lung and Ovarian cancers. Liu et al. [8] proposed weighted extreme learning machine (ELM) method for multiclass microarray cancer data classification. Farid et al. [9] proposed combining feature selection with dissimilarity-based representation approach using decision tree (DT), Naïve Bayes (NB), and KNN classifiers. Moreover, an adaptive rule-based classifier for big biological datasets is introduced by García et al. [10] using Fisher's linear discriminant (FLD), SVM and multilayer perceptron neural network (MLP). Kumar et al. [11] introduced feature selection based on ANOVA for classification purpose using MapReduce framework. Ebrahimpour and Eftekhari [12] introduced maximum relevancy and minimum redundancy (MRMR) method of feature selection for classification using hesitant fuzzy sets.

3 Dataset Description

In this study, eight standard microarray cancer datasets are considered. Each dataset is described in terms of sample size, the original number of features before feature selection, number of classes, training, and test size as shown in Table 1. Those standard microarray cancer datasets are taken from various data repositories [13–15]. We consider both multiclass and binary class microarray cancer datasets. Colon cancer data are binary class dataset which has 2000 original number of features and 62 patients. This dataset is divided into a training set of 37 samples and test set samples of 25, whereby 18 of these test samples are class 1 and the remaining 7 are from class 2. Similarly, Leukemia_2C, Ovarian, and Prostate cancer datasets are binary classes with 7129, 15,154, and 12,600 features and 72, 253, and 102 patients. These datasets are divided into training and test set samples as 43 training samples and 29 test samples for Leukemia_2C, in the case of Ovarian data, 153 training samples, and 102 test samples. With respect to the Prostate cancer data, there are 102 sample patients each having 12,600 gene expression levels and the samples are divided into 61 training cases and 41 test cases. The rest of the datasets are multiclass datasets in which their sample size, number of features, number of classes, training and test size are presented in Table 1.

Table 1 Dataset description

No.	Dataset	Sample size	Number of features	Number of classes	Number of training samples	Number of test samples
1	Colon cancer	62	2000	2	37	25
2	Leukemia_2C	72	7129	2	43	29
3	Leukemia_3C	72	7129	3	43	29
4	MLL_3	72	7129	3	43	29
5	Ovarian	253	15,154	2	151	102
6	Prostate	102	12,600	2	61	41
7	SRBCT	83	2308	4	49	34
8	Tumor	174	12,534	11	104	70

4 L1-Regulated Feature Selection

L1-regulated feature selection is one of the model-based feature selection methods which uses classifiers such as SVM and logistic regression to fit the data so as to remove the irrelevant features so that these selected features can be used for classification purpose on other classification models. Overfitting is a typical challenge in microarray cancer data analysis. To handle this constraint, regularization methods such as L1 (LASSO) and ridge-based methods are widely used with linear models to select optimal features for classification and regression problems. The L1-regularization adds a constraint to the loss function, where α is the controller parameter and w represents the coefficients of the model.

L1-regulated feature selection method (See Eq. 3) is formulated by adding a penalty term (See Eq. 1) to the least squares (Eq. 2), where p is the penalty, α is a control variable, d is the dimension, and $|W_i|$ is the coefficient of ith sample.

$$p = \alpha * \sum_{1=1}^{d} |w_i| \tag{1}$$

$$\text{LS}(E) = \sum_{k=1}^{n} \left(y - \left(\sum_{i=1}^{d} w * X_i \right) \right)^2 \tag{2}$$

$$E(w) = \sum_{k=1}^{n} \left(y - \left(\sum_{i=1}^{d} w * X_i \right) \right)^2 + \alpha * \sum_{i=1}^{d} |w_i| \tag{3}$$

In this work, we propose an efficient L1-regulated feature selection method for classification of microarray cancer data. L1-based feature selection method is efficient in selecting optimal features by removing the irrelevant features having weights of zero. L1-regulated feature selection method uses a parameter "α" to control the number of features to be selected. As the value of the parameter "α" becomes small number such as ($\alpha = 0.0001$), few features will be selected comparing to a relatively bigger value of "α" such as ($\alpha = 0.1$).

The constraint enforces lower-valued features to have zero coefficients ultimately to select few and optimal features. It makes big sparse features with few of them relevant whose weight is nonzero, and most of the irrelevant features will get a zero weight so that these features will not be considered by the model during classification. In this work, the Random Forest tree model is used as a base classifier.

Feature scaling is essential to overcome the overfitting problem of a model. Equation (1) is used to scale down the value of features to be between zero and one [0, 1] for easy fitting of the model where f_{new} stands for the transformed values for each gene expression, X_i is the ith sample in the dataset, μ stands the mean of the feature vector, and σ stands the standard deviation of the feature vector.

$$f_{\text{new}} = (X_i - \mu)/\sigma \tag{4}$$

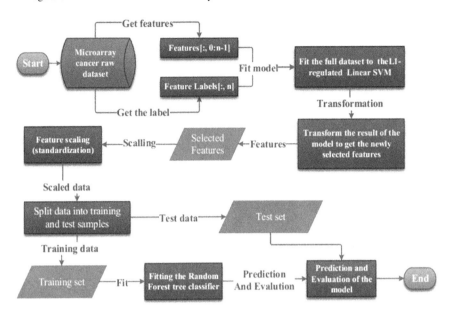

Fig. 1 Workflow diagram of the proposed model

4.1 Workflow Diagram of the Proposed Method

The workflow of the proposed method is described as follows for both the feature selection and classification process. Given the original microarray cancer dataset, the splitting of the features and labels takes place followed by L1-regulated feature selection to get the candidate relevant features to train the model. Next, splitting of the data into training and test samples is carried out. Feature scaling is conducted followed by fitting the RF classifier, and finally, the predictive capability of the model is evaluated on the test data. The performance of the method is evaluated in terms of classification accuracy, precision, recall, f-measure, ROC curve, and confusion matrix (see Fig. 1).

4.2 Random Forest Tree Classifier

Random Forest tree classifier is a bundle of decision trees working by averaging noisy and unbiased models to create a model with higher predictive accuracy. The rational to use Random Forest classifier here is that it enhances classification accuracy comparing to the use of single classifiers. First, the dataset is divided into n number of subsamples and each of the subsamples are trained in d number of decision trees (See Fig. 2). Each of these decision trees trains and predicts the class of the subsets, and the final class label prediction is carried out based on the principle of majority

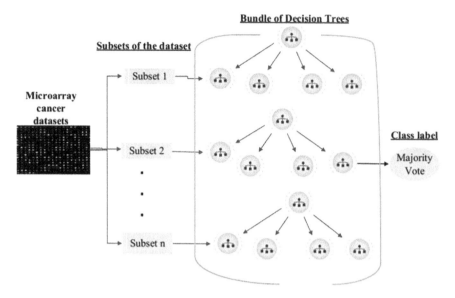

Fig. 2 Random Forest tree flow diagram

voting. The probability of a sample to be member of a certain class is expressed by the probability of each features f of a sample toward class c as shown in Eq. 5, where p is the probability and n is the number of samples.

$$p(c|f) = p_1(c|f) + p_2(c|f) + \cdots + p_n(c|f) \sum_{i=1}^{n} (p_i(c|f)) \qquad (5)$$

4.3 Algorithm of Random Forest Tree

Random Forest (RF) tree is an ensemble algorithm which is applicable for classification and regression related problems. In this work, the Random Forest tree is used as a classifier in. Random Forest classifier is an ensemble of several decision trees which takes random samples from the training data to give several predictions, and the most frequently, predicted class is considered as the final class label.

Algorithm 1 Algorithm of Random Forest

Input: Features of raw microarray cancer data
Output: Classification report of Random Forest classifier
Procedure:

1. **Randomness**: Randomly select k features from m number of total features, where k < m
2. **Apply best split**: Among the k features, compute the node d by applying best split technique
3. **Create child node**: Split the node into child nodes applying best split
4. **Repeat steps 1 to 3**: Until l number of nodes are reached
5. **Construct forest of trees**: By repeating steps 1–4 for n number times with replacement to create n number of trees
6. **Predict**: Consider the test features and apply the rules of the created decision tree to predict the target label
7. **Compute vote of each tree**: Compute the votes for each predicted target label
8. **Final voting**: The most frequently voted predicted class label is taken as the final class label

5 Evaluation Metrics

To evaluate the results of the proposed model, performance measures such as classification accuracy (CA), recall, precision, $f1$-measure, receiver operating characteristic (ROC), and area under the curve (AUV) are used. As shown in Eq. 6 classification accuracy is computed as the ratio of correctly classified test samples size in the test data. Since accuracy alone is not sufficient to measure the performance of a model, other performance metrics are also considered.

$$\text{Accuracy} = \frac{TP + TN}{TP + FP + TN + FN} \quad (6)$$

Recall also known as sensitivity is a true positive rate, which is the ratio of true positive (TP) to the sum of true positive (TP) and false negative (FN). It measures the number of true positives (TP) to the ratio of total TP and FN as shown in Eq. 7.

$$\text{Recall} = \frac{TP}{TP + FN} \quad (7)$$

Another performance measure used in this work is the precision which is also referred to as positive predictive value (PPV). Equation 8 shows the precision which computes the ratio of correctly classified samples (true positives) to the sum of true positives and false positives.

$$\text{Precision} = \frac{TP}{TP + FP} \quad (8)$$

Moreover, as shown in Eq. 9, $F1$-measure is also used as a performance metric. This metric is applied to neutralize the biases in precision and recall. $F1$-measure considers the harmonic mean of precision and recall as shown in Eq. 5.

$$F\text{measure} = \frac{2 * (\text{Precision} * \text{Recall})}{\text{Precision} + \text{Recall}} \tag{9}$$

6 Experimental Results and Discussion

In this section, detailed discussion of the experimental results of the proposed method is provided. The evaluation methods such as classification accuracy, precision, recall, f-measure, confusion matrix, and learning curve are employed to assess the performance of the Random Forest tree classifier.

As shown in Table 2, the experimental results are presented and the performance of the model was evaluated in terms of classification accuracy, precision, recall, $f1$-Measure, and AUC and confusion matrix. Accordingly, the model performs 1.00% perfect classification on two binary class datasets and one multiclass dataset, namely Leukemia_2C and Ovarian and SRBCT, respectively. The precision, recall, and f-measure on these three datasets are also 100% which shows that the proposed feature selection method is performing well in identifying the informative features as can be seen in Fig. 3. On the other hand, the model's classification accuracy on the other datasets is 0.96 on Colon cancer, 0.93 on Leukemia 3 class, and 0.95 on Prostate datasets. Furthermore, the model scores a classification accuracy of 0.97 and 0.90 on MLL 3 class dataset and 11-class Tumor dataset, respectively. For further performance measures such as precision, recall, $f1$-measure, and AUC on each dataset, see Table 2 and Fig. 3.

Table 2 Classification report using Random Forest tree classifier

Dataset	Number of features	Accuracy	Precision	Recall	$F1$ measure	AUC
Colon cancer	78	0.96	0.97	0.96	0.96	0.95
Leukemia_2C	36	1.00	1.00	1.00	1.00	1.00
Leukemia_3C	27	0.93	0.94	0.93	0.93	0.95
MLL_3	31	0.97	0.97	0.97	0.97	0.98
Ovarian	23	1.00	1.00	1.00	1.00	1.00
Prostate	45	0.95	0.96	0.96	0.95	0.95
SRBCT	61	1.00	1.00	1.00	1.00	0.97
Tumor	132	0.9	0.94	0.90	0.90	0.76

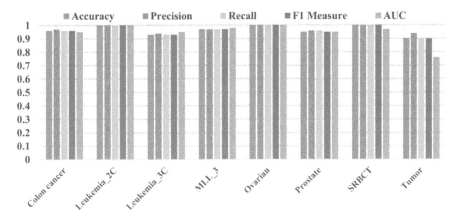

Fig. 3 Classification reports of Random Forest tree on all of the datasets using classification accuracy, precision, recall, F-measure, and AUC

6.1 The Learning Curve of Random Forest Tree Classifier on All Datasets

A learning curve is a function of the predictive error in a given training and test sets over a range of progressively increasing training set size [16]. Learning curve is used to demonstrate the improvement in accuracy with progressive increase of training data ultimately to manage the variance and bias of the model. The learning curve of a model indicates the score from a training and test samples on a various portion of the training data. Moreover, learning curve shows to what extent the model is benefitting as a result of adding more number of training data. The bias and variance errors are also identified by introducing a learning curve to classification model. Learning curve considers training examples along the x-axis, and the accuracy scored for both training and test samples is indicated on the y-axis. Microarray cancer data always suffers from lack of sufficient which leads, in learning curve case, unable to add more number of data for the model to enhance its generalization.

The model is also evaluated in terms of a learning curve which indicates the model's learning capability from a different portion of the training samples. An experimental study of the learning curve is carried out using fivefold cross-validation where the dataset is divided into five equal portions. Each training portion of the dataset has the chance of being test set as four of the folds are for training and the remaining one fold is for testing. As shown in Table 3, the learning capability of the model generally increases as the training data increases. The training accuracy of the models reaches 100% for all datasets. In the case of Colon cancer data, the average learning accuracy of the model is 0.98 at $k = 1$ where there are only four training samples and gradually reaches 0.99 when the training sample size is 46 at $k = 5$ and the maximum accuracy scored at this point is 1. The average test accuracy is initially 0.74 and reaches 0.81, with a maximum value of 0.94. Similarly, for Leukemia 2 class

Fig. 4 Learning curve for Colon cancer data

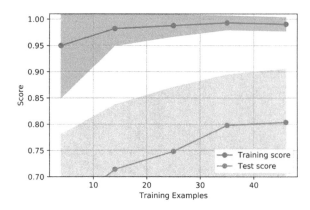

Fig. 5 Learning curve for Leukemia_2C cancer data

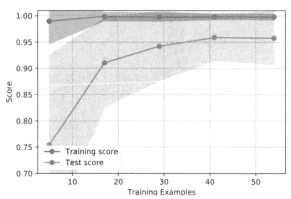

cancer data, the average learning accuracy of the model is 0.98 at $k = 1$ where the training sample size is 5 and gradually reaches 100% when the number of training sample increases to 54 at $k = 5$, the average accuracy of Leukemia 2 class data becomes 0.6 and reaches 0.96, with a maximum value of 1. Generally, the average training and test accuracy of the model depicts the bias and variance. If we look the case of Tumor 11 class dataset, the average training accuracy is 0.32 when the training sample size is only 13 at $k = 1$ and as the sample size increases to 130 at $k = 5$, the average test accuracy reaches 0.78 with a maximum value of 0.91; hence, variance error is observed here. The model is showing that the variance error can be solved by adding more number of training samples. The detail about the behavior of the model with respect to average training accuracy, average test accuracy, maximum value with increasing training size can is presented in Table 3 and Figs. 4, 5, 6, 7, 8, 9, 10, and 11 for all the experimental datasets. The hyphens in the last column of Table 3 indicate that value is not applicable.

As shown in Figs. 4, 9, and 11, for Colon, Prostate, and Tumor_11 class, there is a great gap between the training and test accuracy. However, the model's generalization is increasing as more number of training examples are added. Model's high variance (big gap between training and test score) is resolved by getting additional training

Table 3 Experimental results of a learning curve for fivefold cross-validation on all of the dataset

Dataset	Average training and test scores per portion of the dataset	$K=1$	$K=2$	$K=3$	$K=4$	$K=5$	Maximum value
Colon	Training size in each CV	4	14	25	35	46	–
	Mean of train accuracy	0.98	0.99	0.99	0.99	0.99	1
	Mean of test accuracy	0.61	0.72	0.77	0.79	0.81	0.94
Leukemia_2C	Training size in each CV	5	17	29	41	54	–
	Mean of train accuracy	0.98	1	1	1	1	1
	Mean of test accuracy	0.74	0.90	0.94	0.95	0.96	1
Leukemia_3C	Training size in each CV	5	17	29	41	54	–
	Mean of train accuracy	0.99	1	1	1	1	1
	Mean of test accuracy	0.57	0.80	0.87	0.90	0.93	1
MLL	Training size in each CV	5	17	29	41	54	–
	Mean of train accuracy	0.98	0.99	0.99	0.99	1	1
	Mean of test accuracy	0.56	0.83	0.92	0.95	0.96	1
Ovarian	Training size in each CV	18	61	103	146	189	–
	Mean of train accuracy	0.99	1	1	1	1	1
	Mean of test accuracy	0.90	0.98	0.99	0.99	1	1
Prostate	Training size in each CV	7	24	41	58	76	–

(continued)

Table 3 (continued)

Dataset	Average training and test scores per portion of the dataset	$K=1$	$K=2$	$K=3$	$K=4$	$K=5$	Maximum value
	Mean of train accuracy	0.99	0.99	0.99	0.99	0.99	1
	Mean of test accuracy	0.61	0.78	0.82	0.84	0.86	1
SRBCT	Training size in each CV	6	20	34	48	62	–
	Mean of train accuracy	0.97	0.99	1	1	1	1
	Mean of test accuracy	0.47	0.83	0.92	0.95	0.97	1
Tumor 11 class	Training size in each CV	13	42	71	100	130	–
	Mean of train accuracy	0.99	0.99	1	1	1	1
	Mean of test accuracy	0.32	0.58	0.70	0.74	0.78	0.91

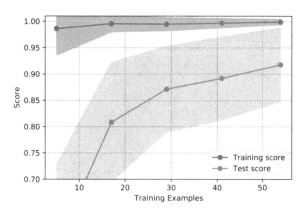

Fig. 6 Learning curve for Leukemia_3C cancer data

Fig. 7 Learning curve for MLL cancer data

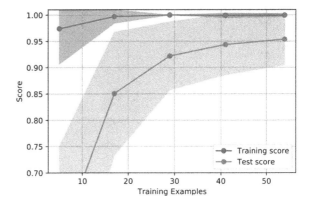

Fig. 8 Learning curve for Ovarian cancer data

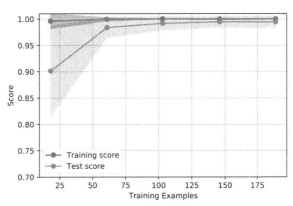

Fig. 9 Learning curve for Prostate cancer data

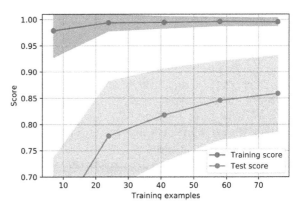

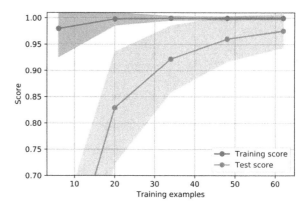

Fig. 10 Learning curve for SRBCT cancer data

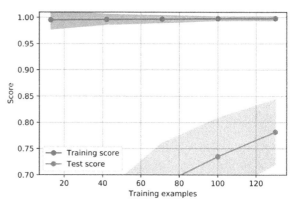

Fig. 11 Learning curve for Tumor_11 class cancer data

samples, which is not applicable in the case of microarray cancer datasets as there is always a shortage of training examples. Moreover, the learning curve indicates early convergence on the existing number of training samples and the model will not more benefit from adding more number of training examples as shown in Figs. 5, 7, 8 and 10 for Leukemia_2C, MLL, Ovarian and SRBCT datasets respectively. The learning curve for these particular datasets is converging to a similar point with existing training data and the gap between training and test accuracy to show low variance and bias; hence, the model is more generalizing even when new data points are added. In the case of Leukemia 3 class dataset, the learning curve shows a moderate convergence and the model ultimately benefitted from adding more number of training examples.

Fig. 12 ROC of Colon cancer data

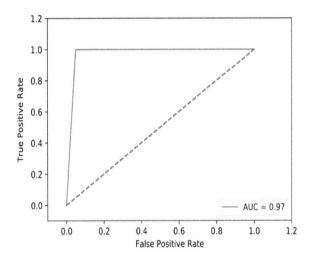

6.2 ROC Curve

ROC curve is a graph which shows the effectiveness of a classifier's predictability which is constructed considering the true positive rate on the *x*-axis against the false positive rate in the *y*-axis for different cut-off points. A perfect classifier has ROC curve which touches the upper left corner which indicates perfect classification that all true positives and false negatives are fully identified. ROC curve for the binary class is a single function which considers the two classes. In the case of multiclass data, ROC curve is computed class-wise and an average of the number of classes is taken as a final decision point. Two types of average are computed which are the micro-average and macro-average, whereby the micro-average considers a number of samples in each class, and the macro-average is assumed all classes have equal weight in the computation of ROC curve. As shown in Figs. 13 and 14, the classifier performs well by scoring an area under the curve (AUC) of 1.00 for Ovarian and Leukemia 2 class datasets. On the other hand, the model scores AUC of 0.95 and 0.97 for Colon and Prostate cancer data are indicated in Figs. 12 and 15. Furthermore, model scores an AUC of 0.98 for 3-class MLL dataset as shown in Fig. 16 and the score for Leukemia 3 class dataset is 0.95 as shown in Fig. 17. Similarly, the model scores an AUC of 0.97 on SRBCT as shown in Fig. 18. The model has shown relatively high variance (overfitting) on Tumor dataset and is scoring an AUC of 0.76 as indicated in Fig. 19.

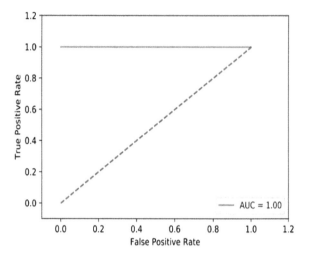

Fig. 13 ROC of Leukemia_2 cancer data

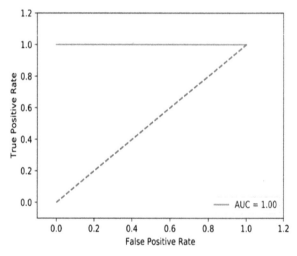

Fig. 14 ROC of Ovarian cancer data

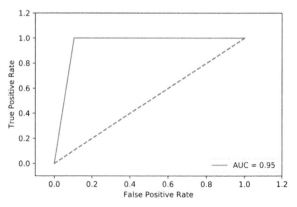

Fig. 15 ROC of Prostate cancer data

Fig. 16 ROC of MLL cancer data

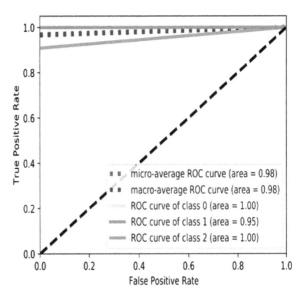

Fig. 17 ROC of Leukemia_3C cancer data

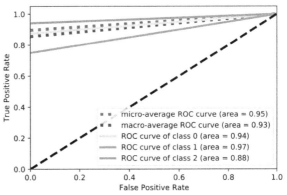

Fig. 18 ROC of SRBCT_4C cancer data

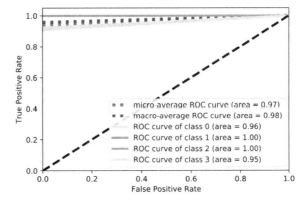

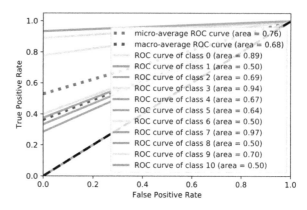

Fig. 19 ROC of Tumor_11C cancer data

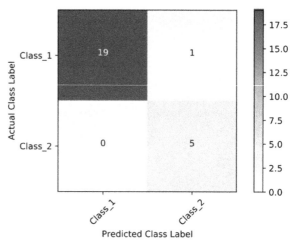

Fig. 20 Confusion matrix (Colon cancer)

6.3 Confusion Matrix

The confusion matrix is one of the performance measures which shows correctly and wrongly classified test samples. A perfect classifier displays all the true and true negative samples along the diagonal and off-diagonal elements of the confusion matrix which is filled with zeros. The model's confusion matrix of this work is presented in Figs. 16, 17, 18, 19, 20, 21, 22, 23, 24, 25, 26, and 27. The test samples along the diagonal from top left to bottom right are correctly classified test samples, and if there are nonzero values in the off-diagonal area, those test examples are misclassified.

Comparative analysis of the proposed work is carried out with state-of-the-art works in the domain. The proposed work outperforms many of the proposed methods as shown in Table 4. The hyphen-valued cells indicate that these particular works do not use the datasets in their experiment.

Fig. 21 Confusion matrix (Leukemia_2C)

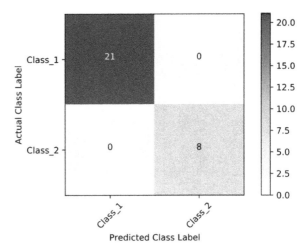

Fig. 22 Confusion matrix (Ovarian)

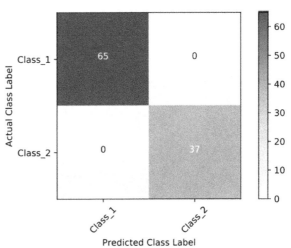

Fig. 23 Confusion matrix (Prostate)

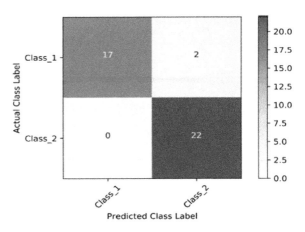

Table 4 Comparison of the proposed work with the state-of-the-art work

Times	Methods	Datasets							
		Colon	Leukemia_2C	Leukemia_3C	MLL_3	Ovarian	Prostate	SRBCT	Tumor
Dashtban et al. [17]	MOBBA-LS	–	97.1	–	–	–	94.1	85	–
Dash [18]	Hybridized harmony search and Pareto optimization + ANN	0.82	0.96	–	–	–	0.93	–	–
García et al. [19]	Relief F + multilayer perceptron neural network (MLP)	0.84	–	–	–	0.98	0.87	–	–
Bouazza et al. [20]	Filter approach + NB	–	0.44	–	–	0.97	–	–	–
Chen et al. [21]	PSO-decision tree		1.00				0.95	100	0.92
Our work	L1-Regulated + RF tree	0.92	1.00	0.93	0.97	1.00	0.95	1.00	0.9

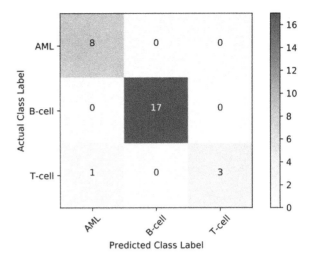

Fig. 24 Confusion matrix (Leukemia_3C)

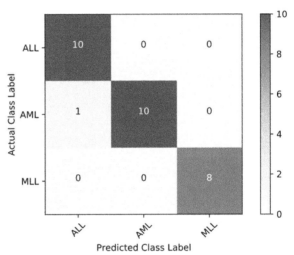

Fig. 25 Confusion matrix (MLL_3C)

7 Conclusion

Curse of dimensionality, imbalanced sample size to a number of features, and noisy data are some of the core challenges in the analysis of microarray cancer data. To alleviate these challenges, feature selection plays a vital role. In this work, we propose the L1-regulated feature selection using linear support vector machine (LSVM). The experiment is conducted on eight standard microarray cancer datasets. Feature scaling is carried out so as to enhance model's performance using the z-score normalization method. As a classifier, the Random Forest tree is used. We have explored the learning curve of the classifier by varying the number of training examples in the Forest for each dataset to get small gap (low error) between the training and test scores,

Fig. 26 Confusion matrix (SRBCT_4C)

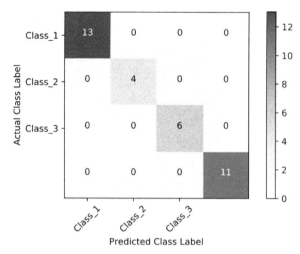

Fig. 27 Confusion matrix (Tumor_11C)

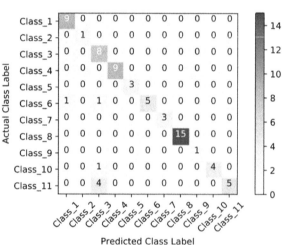

hence to control the variance-bias trade-off. An optimal learning curve is achieved for Ovarian, Leukemia_2C, MLL, and SRBCT datasets. The other datasets such as Tumor 11 class Colon cancer data are showing relatively high variance but still indicates the model can be converged by adding progressively more number of training examples, which the microarray cancer data are not enjoying due to the shortage of sample size. Comparative analysis with state-of-the-art works is conducted, and the proposed method shows better performance in almost all the datasets. Evaluation methods such classification accuracy, precision, recall, $f1$-measure, and ROC curve which has the AUC are applied. As a future work, we plan to handle the high variance in some of the datasets to converge with an existing constrained number of sample sizes.

References

1. Alberts, B., Johnson, A., Lewis, J., Raff, M., Roberts, K., & Walter, P. (2002). *Cancer as a micro evolutionary process*
2. Sharbaf, F. V., Mosafer, S., & Moattar, M. H. (2016). A hybrid gene selection approach for microarray data classification using cellular learning automata and ant colony optimization. *Genomics, 107*(6), 231–238.
3. Latkowski, T., & Osowski, S. (2015). Data mining for feature selection in gene expression autism data. *Expert Systems with Applications, 42*(2), 864–872.
4. Latkowski, T., & Osowski, S. (2017). Gene selection in the autism-comparative study. *Neurocomputing, 250,* 37–44.
5. Wang, Z., Zineddin, B., Liang, J., Zeng, N., Li, Y., Du, M., et al. (2014). cDNA microarray adaptive segmentation. *Neurocomputing, 142,* 408–418.
6. Guo, S., Guo, D., Chen, L., & Jiang, Q. (2017). A L1-regularized feature selection method for local dimension reduction on microarray data. *Computational Biology and Chemistry, 67,* 92–101.
7. Medjahed, S. A., Saadi, T. A., Benyettou, A., & Ouali, M. (2017). Kernel-based learning and feature selection analysis for cancer diagnosis. *Applied Soft Computing, 51,* 39–48.
8. Liu, Z., Tang, D., Cai, Y., Wang, R., & Chen, F. (2017). A hybrid method based on ensemble WELM for handling multi class imbalance in cancer microarray data. *Neurocomputing.*
9. Farid, D. M., Al-Mamun, M. A., Manderick, B., & Nowe, A. (2016). An adaptive rule-based classifier for mining big biological data. *Expert Systems with Applications, 64,* 305–316.
10. García, V., & Sánchez, J. S. (2015). Mapping microarray gene expression data into dissimilarity spaces for tumor classification. *Information Sciences, 294,* 362–375.
11. Kumar, M., Rath, N. K., Swain, A., Rath, S. K. (2015). Feature selection and classification of microarray data using mapreduce based anova and k-nearest neighbor. *Procedia Computer Science, 54,* 301–310.
12. Ebrahimpour, M. K., Eftekhari, M. (2017). Ensemble of feature selection methods: A hesitant fuzzy sets approach. *Applied Soft Computing, 50,* 300–312
13. Zhu, Z., Ong, Y.-S., & Dash, Manoranjan. (2007). Markov blanket-embedded genetic algorithm for gene selection. *Pattern Recognition, 40*(11), 3236–3248.
14. Tsamardinos, I.,. Statnikov, A., Aliferis, C. F.: Gene expression model selector. (Online). Available: http://www.gems-system.org/.
15. Andres Cano, S. M., & Masegosa, A. Elvira biomedical data set repository (Online). Available: http://leo.ugr.es/elvira/DBCRepository/.
16. Hess, K. R., & Wei, C. (2010). Learning curves in classification with microarray data. *Seminars in Oncology, 37*(1) (Elsevier).
17. Dashtban, M., Balafar, M., & Suravajhala, P. (2018). Gene selection for tumor classification using a novel bio-inspired multi-objective approach. *Genomics, 110*(1), 10–17.
18. Dash, R. (2018). An adaptive harmony search approach for gene selection and classification of high dimensional medical data. *Journal of King Saud University-Computer and Information Sciences.*
19. García, V., Salvador Sánchez, J. (2015). Mapping microarray gene expression data into dissimilarity spaces for tumor classification. *Information Sciences, 294,* 362–375.
20. Bouazza, S. H., et al. (2018). Selecting significant marker genes from microarray data by filter approach for cancer diagnosis. *Procedia Computer Science, 127,* 300–309.
21. Chen, K.-H., et al. (2014). Applying particle swarm optimization-based decision tree classifier for cancer classification on gene expression data. *Applied Soft Computing, 24,* 773–780.
22. Kumar, M., Singh, S., & Rath, S. K. (2015). Classification of microarray data using functional link neural network. *Procedia Computer Science, 57,* 727–737.

Automated Delineation of Costophrenic Recesses on Chest Radiographs

Prashant A. Athavale and P. S. Puttaswamy

Abstract The lung image segmentation using a model-based approach is a challenge owing to the sheer complexity and variability of the lung shape in a given data set. As a part of our effort to segment the lungs, we report a method to delineate the costophrenic (CP) recess without the human intervention. Active shape model (ASM) is used to point to the probable area of the CP recess, and a prior knowledge-based processing delineates the CP recess and hence determines the angle. The proposed method is fast and shows satisfactory results. It is intended to be used as a preprocessing step in segmenting the lungs' contour. The proposed method also can be used to initialize the model contour in any other ASM-based lung segmentation algorithms. The algorithm was tested on 45 non-nodule lung images from the JSRT database. An average accuracy of 87.02% is achieved. A comparison of the results of proposed method and gold standard which is obtained by manual delineation is given.

Keywords Costophrenic angle · Active shape model · Lung segmentation · Computer-aided diagnosis · Sensitivity · Specificity · Jaccard index

1 Introduction

1.1 Background

The contributions of digital image analysis toward faster and accurate evaluation of nodules on lungs will increase the chances of survival of the patient. The digital chest X-ray (CXR) is an important tool for early detection of abnormality in the lungs. It not only gives first-hand information about the malignancy condition, but

---

P. A. Athavale (✉)
Department of E&EE, BMS Institute of Technology & Management, Bengaluru, India
e-mail: sirprashanth@gmail.com

P. S. Puttaswamy
Department of E&EE, PES College of Engineering, Mandya, India

© Springer Nature Singapore Pte Ltd. 2019
N. R. Shetty et al. (eds.), *Emerging Research in Computing, Information, Communication and Applications*, Advances in Intelligent Systems and Computing 906,
https://doi.org/10.1007/978-981-13-6001-5_7

also the type of the disease to a diagnostician. Digital CXR requires less radiation, as compared to the analog type, and it is economical too.

The requirement of an automated diagnostic unit in a routine CXR becomes imperative in cases where mass screening is a norm, like pre-employment screening. Of the various steps in the analysis of the CXR by an automated means, the segmentation of the lung field is the first step.

Calcium in the bones, tissues of the muscle, fat, air, and other contrasting agents (if present) are the ones which create various shades of gray on the radiograph. The basis for all the interpretations on any radiograph is the differences in densities of overlapping parts of the body. The diagnostician observes the CXR with a systematic approach and looks for some key features before concluding on the findings. Some of the key features are airways, ribs, shadows of the breast, silhouette of the heart, costophrenic angle, the diaphragm, and evidence for filling up of alveolar sacs and so on. This paper presents a modified ASM-based approach to delineate the CP angle.

The paper is presented as follows: Sect. 1.2 describes the anatomy and functionality of the CP recess. A survey of some related works is reported in Sect. 1.3, and then, Sect. 2.1 explains the input images used. The proposed method is explained in Sect. 2.2. It is followed by Sect. 3 with a description of results obtained by testing the method on the data set. The conclusion is presented in Sect. 4.

1.2 CP Recess and Visibility on the CXR

The ribs enclose mainly the pair of lungs and the heart apart from some other organs of the respiratory and circulatory system. The diaphragm, a muscle which separates the thoracic cavity and the abdominal cavity, is primarily responsible for the act of breathing in which it contracts and expands. This movement cyclically creates pressure differences in the thoracic cavity. The air is inhaled and exhaled accordingly. The lungs are contained in pleural sac and are surrounded by the rib cage, with enough lubrication in the form of membranous secretions. At the bottom of the rib cage on both left and right sides, the diaphragm creates a recess with it, which is visible on a chest radiograph known as costophrenic angle. It originates from the words 'costo' meaning the ribs and phrenic meaning, the diaphragm.

Thus, the peculiarity of the CP angle is that it is a region created by the gap between the diaphragm and the rib cage at the lower part, and which is useful while inhaling, wherein the lung tissue gets extra space for expansion into this region. The space created over here could be filled with fluids, the corresponding clinical conditions known as pleural effusions. The visibility of the effusion is dependent on the position of the patient during the filming of CXR. Generally, the fluid gets collected at the lowest region of the chest cavity. If the patient is upright while capturing the CXR, then the fluids obscure the visibility of CP angle or its blunting. This becomes a very important biological marker for the diagnostician to interpret the CXR image.

1.3 Literature on CP Angle Detection

Segmenting the lungs as a preprocessing step is crucial to increase the success of image analysis methods of a computer-aided diagnostic system. Lung segmentation has been carefully studied and reported in the literature and can further be classified on various bases. The most basic is the thresholding method. Model-based methods have been used in the segmentation of the lungs owing to the repetitive form of the shape. This set of methods includes active contours, deformable templates, and level set methods. The approach that is being discussed in this paper starts with the active shape model, proposed by Cootes et al. [1]. The ASM is a set of point coordinates which define the shape of the object. It is an important algorithm because of its ability to incorporate the prior knowledge derived from the training data set about the shape under consideration and its robustness in modeling very complex shapes. There are a variety of modifications done on the ASM as central theme and thence applied to segmentation tasks.

The initial placement of the contour model on the target image is very crucial, but efforts have been made to overcome the sensitivity of it by Tsai et al. [2]. Region-based methods and edge determining are combined in the work of Paragios and Deriche [3]. They have made the active shape model more robust to noise and shown the implementation to the detection of complex curves.

Instead of an intensity gradient feature, the work of Shi et al. [4] used a modified scale-invariant feature transform (SIFT) descriptor, which modeled the feature around the pixel in an image. They have also imposed patient-specific constraints on the deformable contour in improving the accuracy of lung segmentation. Another work which uses the intensity and morphology information around the CP recess is by Maduskar et al. [5], which propose a method for localizing the CP region, wherein they have used the chest wall as a landmark structure.

A sharp and acute CP angle and symmetrical lungs are considered normal. This has been investigated in the work of Wan Ahmad et al. [6], where the abnormality features are represented as scores. The scores in connection with other parameters are used for the image analysis. The work of Campadelli and Casiraghi [7] requires that the position of the CP angle be clearly defined for detecting the bottom boundary of the lungs. The ASM-based algorithm models are more prone to become unstable due the outliers. If the outliers are clustered at a landmark point, then the search is impaired. Treating outliers in the region of CP recess is not specifically mentioned in the literature, but the works of Nahed et al. [8], De la Torre and Black [9], Behiels et al. [10] are significant. Behiels impose a penalty for the differences in neighboring possible positions to refine the cost function. Thus, it can be observed that the CP angle delineation is always a part of an algorithm of segmenting the lungs, and not reported separately. Hence, comparison of our work with other methods is difficult.

2 Material and Method

2.1 Input Images

We have used the CXR images from the Japanese Society of Radiological Technology database [11]. This database has 154 images with nodules and 93 non-nodule images, with a resolution of 2048 × 2048 matrix sizes, and the size of each pixel is 0.175 mm. Unlike a single template with rigid shape which can be used in an industrial segmentation application, the problem at hand is very much prone to the variability in the shape of the lungs due to age, sex, and/or the height of the individual.

The ASM model for segmenting the lungs was created by averaging the shapes, each of which are having 32 model or landmark points using 30 images of the JSRT database as training set. The ASM model is made to cover the outline of the entire ribcage. At this stage, the right lung and left lung are not modeled separately. Apart from the lung contour, models of the gray-level distributions at the landmark points were also created. The gray-level model at each of the landmark points has a length of 31 pixels. The ASM lungs' contour and the associated gray-level model at each point are used in searching for the best fitting curve on the target image. This coarsely finds the lungs edges and requires multiple iterations to accurately settle on to the true edge. The ground truth for some of the images of JSRT database is publicly made available by van Ginneken et al. [12].

The 32 landmark points can be described by a vector $x = (x_1 y_1, x_2 y_2, \ldots, x_{32} y_{32})$, where x_k and y_k are the coordinates of the kth point. As described by the ASM algorithm, the average of all the annotated points from the 30 training images is expressed by

$$\bar{x} = \frac{1}{30} \sum_{k=1}^{32} x_k \qquad (1)$$

and the corresponding covariance matrix S will be

$$S = \frac{1}{30} \sum_{k=1}^{32} \mathrm{d}x_k \mathrm{d}x_k^T \qquad (2)$$

where

$$\mathrm{d}x_k = x_k - \bar{x} \qquad (3)$$

is the deviation of each of the kth shape from $\bar{x}$. The model $\bar{x}$ is placed on the target image, and the model points are moved perpendicular to the edge of the lungs. Along the perpendicular direction, a best match to the gray-level model at that point in terms of a cost minimizing function is searched before updating the new position of the coordinate points.

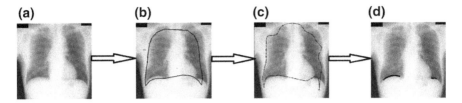

Fig. 1 Various intermediate steps for obtaining the location of the diaphragm **a** an input image from the JSRT database, **b** initial placement of the ASM model on it, **c** one iteration of the ASM algorithm to estimate the location of the diaphragm in the given image, and **d** accurate position of the diaphragm determined and marked

2.2 Proposed Method

To accurately determine the CP recess and hence its angle on a given CXR, a hybrid algorithm of the model-based and rule-based detection methods is proposed. Morphological operations are implemented locally to reduce noise and highlight the structures of interest. To highlight the subtle anatomical structures of the lungs, and suppress the foreground, all the steps mentioned in this paper are applied on the compliment of the target grayscale CXR. First, the ASM model contour as given Eq. (1) is placed on the target image and rough estimate of the CP recess point is found. The important steps starting from the placing of the model on the target image, corresponding result of one iteration of ASM and hence the detection of diaphragm are shown in Fig. 1.

At the end iterations, the ASM converges on the *best* edge lying within its search range. Figure 2 shows highlighted right CP region for two test images from the database, at the end of first iteration. In Fig. 2a, the delineation is closer to the actual boundary, and that in Fig. 2b is far from correct. The disadvantage of applying ASM alone for the detection of CP angle results in the contour model settling at points which are not true edge as shown in Fig. 2b. It needs further refinement by the rule-based approach.

To overcome the problems associated with ASM, the proposed method uses ASM for global coarse shape detection, and rule-based approach for further refinement of the detected edges at local features like the corners of lungs, notch of the aorta, etc. The local feature detection is guided by the knowledge of the anatomy and hence is more accurate. We will demonstrate the efficiency of the proposed method through experimental results.

The steps involved in the delineation of the CP recess and hence the determinations of the CP angles are as follows:

Step 1: Run the ASM algorithms' one iteration on the target CXR image.
Step 2: Determine the location of diaphragm in right and left sides of the input image, which will be used as the reference for CP recess.
Step 3: Carve out suitable region around the reference point obtained from Step 2.

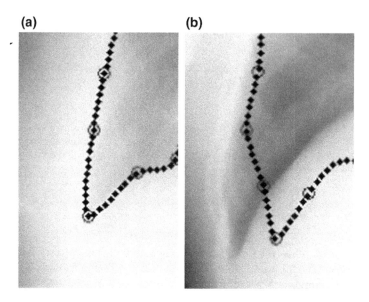

Fig. 2 **a** A portion of CXR highlighting the right-side CP recess, approximately delineated by one iteration of the ASM and **b** another example of the right-side CP recess, where ASM has settled at non-edge location in one iteration

Step 4: Divide that region vertically at the center into two parts; each part will contain the right and left portions of the diaphragm and the CP recess.
Step 5: Determine the gradient of these right and left portions by Gaussian gradient operator to highlight the lung edges at CP recess.
Step 6: Threshold each of the resulting right and left regions to obtain the prominent CP recess.
Step 7: Find the lowermost point, which is the tip of the CP angle.

The region around the CP recess has many details and possible malignancies. Also, the texture of the CP recess is different from the surrounding, and to highlight the lateral walls of the CP recess, we use Gaussian gradient operator in the horizontal direction, in which standard deviation is denoted as 'σ'.

If the parameter σ is small, then finer details are all highlighted, which would make the edge detection technique vulnerable. On the other hand, if it is large, the image is blurred. For visual comparison, the horizontal gradients computed on the right-side CP recess are presented for four values of σ viz. 0.1, 1, 5, and 10, in Fig. 3. It can be observed that as σ increases, finer details are suppressed, and the lateral edges are clear.

Three columns of images in Fig. 4 indicate the results of applying our method of delineation of the right-side CP recess. Step 1 as mentioned above is only to get an estimate of the possible location of the diaphragm. Even though ASM method is fast and simple, multiple iterations of it in finding a best fit have been avoided, thereby reducing the time required for final delineation. The other reason for omitting multiple

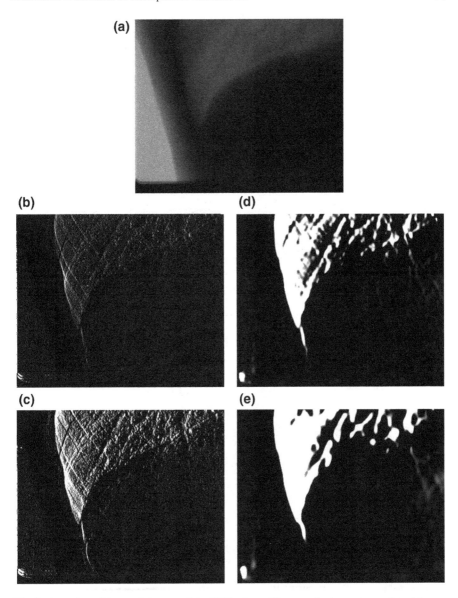

Fig. 3 A section of lower right part of the CXR where CP recess is shown in **a** original image, **b** blurring by the Gaussian kernel of $\sigma = 0.1$, **c** $\sigma = 1$, **d** $\sigma = 5$, and **e** $\sigma = 10$

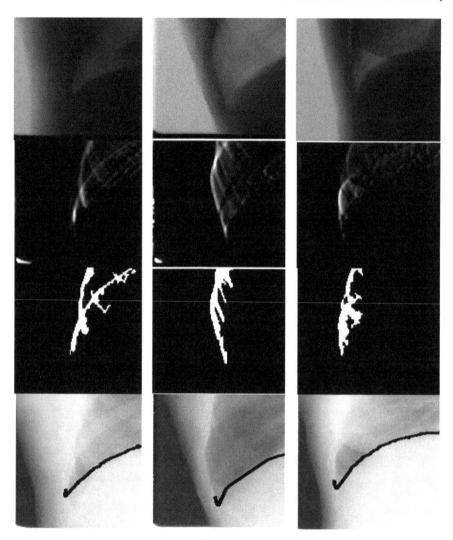

Fig. 4 The results of intermediate steps of the proposed algorithm applied on three different images taken from the JSRT database and corresponding results for the right-side CP indentation. First row is the ROI in the input image. Second row is the gradient magnitude. Only the enhanced CP recess is shown in third row. Fourth row is CP recess delineated

iterations is the tendency of the ASM model to converge at a strongest matching-profile location which could be different from the true edge. From Step 2, the possible location of the diaphragm is obtained, which is in terms of the rows of the image. Only a rectangular region is selected for further processing as shown in the first row of Fig. 4. This again reduces the handling of the large size of data. The so-obtained subimage contains the prominent looking CP recesses on either side. Apart from the CP recess, a faintly visible diaphragm and the base of the heart silhouette are part of the subimage.

A general thresholding operation will not be able to segment these multimodal images in logical groups because of the small variation in gray levels between ROI and background, and hence, adaptive thresholding is implemented on the Gaussian blurred image. The resulting binary image will have many blobs which will affect the detection of diaphragm, and hence must be omitted. Morphological operations are used to omit the background pixels and highlight the recess. Having located the diaphragm sections of the respective sides, the CP recess is obtained by further processing. As shown in Fig. 4, the section under consideration will have the CP recess at the corner. The ASM has the tendency to settle at a location which may not be the true edge. Therefore, for determining the tip of the CP recess, simple processing steps like row-wise scanning in the binary image has been applied. Thus, the tip of the CP recess is found. The CP recess is flanked by the diaphragm on one side and the lateral chest wall on the other. A dynamic program is implemented to fetch the edge points along both lateral chest wall and diaphragm.

The angle formed by the set of edge points representing the diaphragm and the edge points of the highlighted lateral chest wall are evaluated using Eq. 4. If two curves $f_1(x)$ and $f_2(x)$ intersect at a point say, (r_0, c_0), then subtended angle Ø is given as

$$\tan(\text{Ø}) = \frac{f_2'(r_0) - f_1'(r_0)}{1 + f_2'(r_0) * f_1'(r_0)} \tag{4}$$

The curves $f_1(x)$ and $f_2(x)$ can be assumed to be formed by the lateral walls flanking the CP recess. The normal range of CP angle is assumed to be thirty degrees. The salient stages of the delineation in the proposed method after obtaining the location of the diaphragm are depicted in Fig. 5a–c.

3 Results

The proposed methodology was tested on a set of 45 images from the JSRT database. In these images, the right and left CP recesses were manually segmented by an expert. Sensitivity, specificity, and accuracy defined by the standard equations numbered 5–7, in terms of True Positive (TP), True Negative (TN), False Positive (FP), and False Negative (FN), are being used to present the results of our method. The manual

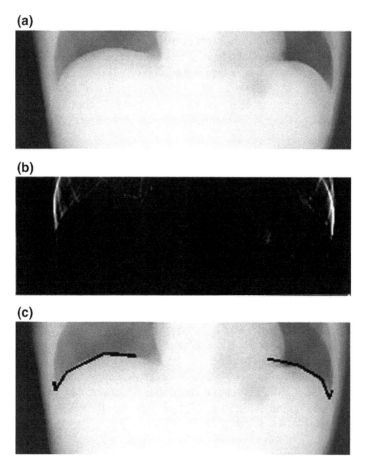

Fig. 5 Detection of chest walls on either side **a** portion of the image containing the diaphragm, **b** side edges highlighted by horizontal gradient operator with a value of $\sigma = 5$, **c** lower edges of the diaphragm detected by a combination of morphological and thresholding operations

segmentation by the expert was taken as the gold standard, and for comparison, the above validation metrics are defined as: TP is the set of pixels correctly classified as belonging to CP recess, and TN is the set of pixels correctly classified as not belonging to the CP recess. On the same lines, FP is the total number of pixels falsely classified as belonging to the CP recess and FN is the total number of pixels falsely marked as background.

$$\text{Sensitivity} = \frac{\text{TP}}{\text{TP} + \text{FN}} \qquad (5)$$

$$\text{Specificity} = \frac{\text{TN}}{\text{TN} + \text{FP}} \qquad (6)$$

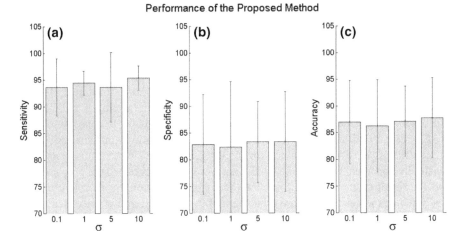

Fig. 6 A plot of the **a** sensitivity, **b** specificity, and **c** accuracy of the proposed algorithm tested on 45 images from the JSRT database

$$\text{Accuracy} = \frac{\text{TP} + \text{TN}}{\text{TP} + \text{TN} + \text{FP} + \text{FN}} \tag{7}$$

The variation of performance metrics only for the right-side CP recess is shown in Fig. 6, plotted in percentage as a function of the standard deviation of the smoothing Gaussian kernel. The proposed method was tested with standard deviation σ set to 0.1, 1, 5, and 10. The average sensitivity over the above said values of σ is 94.23, and that of specificity is 82.95. The average accuracy of the segmentation algorithm for the same values of σ is 87.02. It can be observed that the proposed method produces a consistent result over a range of σ values. The sensitivity is 93.58 $\pm$ 5.36 and specificity is 82.83 $\pm$ 9.33 for $\sigma = 0.1$. These change to 95.34 $\pm$ 2.3 and 83.35 $\pm$ 9.3, respectively, for $\sigma = 10$.

Another measure considered here for the validation of the results of our method is Jaccard index (J), which is the similarity between the segmentation results of manual and proposed methods. The average value of J over all the σ values is 82.31 $\pm$ 2.67. This measure shows that there is a very high degree of overlap between the gold standard and the segmented CP recess.

4 Discussion and Conclusion

Here, a method for delineation of the CP recess was proposed. The delineation of the selected JSRT database images was annotated manually using 'ginput' function in MATLAB by an expert. The comparison of the edges given by our method and the gold standard is very promising.

The proposed method implemented on MATLAB is able to accurately detect the CP recess and determine the angle subtended by the same in a processing time of less than 12 s for both right and left CP recess. This method is independent of the relative location of the CP recess, and issues related to initial placement of the algorithm. Physicians observing the CXR are trained to mentally subtract the occluding or overlapping anatomical structures and see through the organ of interest. The CP angle is also observed as a landmark in analyzing and interpreting the CXR. This indentation is considered as normal when angle is less than 30°. In conditions where the CP recess is filled with fluid, it appears blunt, indicating abnormal condition. Similar results have been obtained for the left CP recess. The result indicates high sensitivity and specificity, and also indicates low distance between the gold standard and the result of the proposed method. The small variations in the values of the performance indices indicate the robustness of the method against noise. The future extension of the proposed method is the segmentation of whole lung area and an analysis of the results in comparison with inputs from multiple experts. Comparison of the proposed method with other works is difficult, as no other paper has reported only the CP delineation part.

References

1. Cootes, T., Taylor, C., Cooper, D., & Graham, J. (1995). Active shape models—their training and application. *Computer Vision and Image Understanding, 61*, 38–59.
2. Tsai, A., Yezzi, A., Wells III, W. M., Tempany, C. M., Tucker, D., Fan, A., et al. (2003). A shape-based approach to the segmentation of medical imagery using level sets. *IEEE Transactions on Medical Imaging, 22*(2), 137–54. PMID: 12715991.
3. Paragios, N. & Deriche, R. (1998). *Geodesic active regions for texture segmentation*. INRIA, Sophia Antipolis, France, Res. Rep. 3440.
4. Shi, Y., Q, F., Xue, Z., Chen, L., Ito, K., Matsuo, H., & Shen, D. (2008). Segmenting lung fields in serial chest radiographs using both population-based and patient-specific shape statistics. *IEEE Transactions on Medical Imaging, 27*(4), 481–494./ https://doi.org/10.1109/tmi.2007.908130.
5. Maduskar, P., Philipsen, R. H., Melendez, J., Scholten, E., Chanda, D., Ayles, H., et al. (2016). Automatic detection of pleural effusion in chest radiographs. *Medical Image Analysis, 28*, 22–32. https://doi.org/10.1016/j.media.2015.09.004. Epub 2015 December 1.
6. Wan Ahmad, W. S. H. M., & Ahmad Fauzi, M. F., & Zaki, W. (2015). *Abnormality detection for infection and fluid cases in chest radiograph* (pp. 62–67). https://doi.org/10.1109/elecsym.2015.7380815.
7. Campadelli, P., & Casiraghi, E. (2005). Lung field segmentation in digital postero-anterior chest radiographs. In S. Singh, M. Singh, C. Apte, & P. Perner (Eds.), *Pattern recognition and image analysis* (vol. 3687, pp. 736–745). Lecture Notes in Computer Science. Springer, Heidelberg, Germany.
8. Abi-Nahed, J., Jolly, M. P., & Yang, G. Z. (2006). Robust active shape models: a robust, generic and simple automatic segmentation tool. In R. Larsen, M. Nielsen, J. Sporring (Eds.), *Medical Image Computing and Computer-Assisted Intervention—MICCAI*.
9. De la Torre, F., & Black, M. J. (2003). A framework for robust subspace learning. *International Journal of Computer Vision, 54*(1–3), 117–142.

10. Behiels, G., Maes, F., Vandermeulen, D., & Suetens, P. (2002). Evaluation of image features and search strategies for segmentation of bone structures in radiographs using active shape models. *Medical Image Analysis, 6*(1), 47–62.
11. Shiraishi, J., Katsuragawa, S., Ikezoe, J., Matsumoto, T., Kobayashi T., Komatsu, K., et al. (2000). Development of a digital image database for chest radiographs with and without a lung nodule: Receiver operating characteristic analysis of radiologists' detection of pulmonary nodules. *AJR 174*, 71–74.
12. van Ginneken, B., Stegmann, M. B. & Loog, M. (2006). Segmentation of anatomical structures in chest radiographs using supervised methods: A comparative study on a public database. *Medical Image Analysis, 10*(1), 19–40.

Using Location-Based Service for Interaction

K. M. Deepika, Piyush Chaterjee, Sourav Kishor Singh, Writtek Dey and Yatharth Kundra

Abstract Traffic is one of the biggest global factors, affecting over 100 million people. People who live in places with large content of pollutants in air have a 15–25% higher death risk from diseases like lung cancer than people who live in less polluted areas. This paper describes how Google Maps API can be used together with Google Cloud Console for location-based interaction.

Keywords Google Maps API · Google Console API · PHP

1 Introduction

The main objective of this project was to make a real-world interacting social-networking media. Often people are struck with small but devastating problems like out of petrol in the middle of a ride or, for example, someone requires physical help.

1.1 Google Maps (API)

Google Corporation presented Google Maps service on February 2005. Google Maps provides services such as directions, maps, and relevant neighborhood business listings when it was combined with Google Local. At the beginning, the USA and Canada have the access to Google Maps service. UK was the first European country to obtain satellite pictures and maps. This service is now available in majority of the European countries, Poland included. This service can distinguish between single trees or buildings and these high-level-resolution pictures are only available in big cities. It was accessed through maps.google.com Web site when the service

---

K. M. Deepika (✉) · P. Chaterjee · S. K. Singh · W. Dey · Y. Kundra
Department of ISE, Nitte Meenakshi Institute of Technology, Yelahanka,
Bangalore 560064, India
e-mail: deepika.km@nmit.ac.in

© Springer Nature Singapore Pte Ltd. 2019
N. R. Shetty et al. (eds.), *Emerging Research in Computing, Information, Communication and Applications*, Advances in Intelligent Systems and Computing 906,
https://doi.org/10.1007/978-981-13-6001-5_8

was first presented. As its popularity grown, Google Maps API was fashioned by Google Company and decisive to share it with whole world. Hence, developers are allowed to incorporate Google Maps hooked on their Web sites with their individual data points. It was allowed at free of charge to use, but needs to obtain an API key, which is bound to the Web site and directory entered when creating the key. User can start building his own application once the key is generated.

1.2 Android API

Google developed an **Android** mobile operating system, derived from the Linux kernel and designed chiefly for mobile devices for instance smartphones and tablets with touch screen. Using touch gestures that loosely characterize real-world actions, such as swiping, tapping, and pinching, to manipulate on-screen objects, in conjunction with a virtual keyboard for text input is provided for direct manipulation at the Android's user interface. Google has auxiliary urbanized Android TV for televisions, Android Auto for cars, and Android Wear for wrist watches, each with a specialized user interface in addition to touch screen devices. Notebooks, game consoles, digital cameras, and other electronic gadgets are variants of Android.

Android operating system (OS) has the largest installed base of every category. Android has been the unsurpassed business OS on smartphones, and it is dominant by any metric and on tablets since 2013.

Google bought Android in 2005 was unveiled in 2007 and it was initially developed by Android, Inc., along with the beginning of the Open Handset Alliance. Advancing open standards for mobile devices devoted to an association of hardware, software, and telecommunication companies. "Google Play store has had over one million Android applications published by July 2013, over 50 billion applications downloaded including many business-class apps that rival competing mobile platforms". As per the survey of mobile application developers, 71% of developers create applications for Android on April–May 2013 and 40% of full-time professional developers see Android as their priority target platform as per the 2015 survey, which is equivalent to Apple's iOS on 37% with both podiums far above others. Android had 1.4 billion monthly dynamic devices in September 2015.

2 The Concept

The social-networking media of this generation are Facebook, WhatsApp, Twitter, etc. These apps may have created a very successful interaction site, but the things lacking in these are real social interaction. So this becomes the most important objective of this project.

The main function of this app will be a GPS (global position service)-based application that will locate the user's location and also other users within some distance range. Suppose a person needs help, he can open the app on his mobile and post his problem. He may offer some finance in return. All the users around him will receive a notification with his statement.

This app will also help the one who seeks part-time job (very much like babysitting) as they can earn some money helping other users. This would be useful mainly to the students who study abroad and require pocket money. No commitment, no application, simply login, help people and earn. There will also be a section to socialize on the number of people one has helped and also words of thanks from the help seekers.

Socializing attracts the present generation and through this app, they will receive appreciation for the good work done and encouragements for others.

There will also be a section where some daily tasks will be provided (e.g., *let's say "Today's mission for all users is to feed 10 homeless children"*). People fulfilling the tasks may post pictures of himself/herself doing it as a proof.

Another problem people often face while visiting a new place is language. App users may also connect to local users to seek information too and solve this.

To start, the user needs to sign up and then login into their account. Once done, they need to switch on the GPS on their device. On logging in, the FCM id of the device corresponding to the username gets updated into the database.

The user can post a help or alert on the main page. When a help is posted, all the local users will get a notification containing the help/alert statement along with their location. The users can chat or call the person who posted help. The users can also see all the posts around them. They can even see the location of all the posts directly on the map.

The one who posted a help statement can see the people who had decided to help. He can call them directly or chat with them.

Our application will also integrate within itself an algorithm which can find the best user to the one who posted a help depending upon the past activities of the users in local.

3 Filling the Map

Now that we have our map, we need to fill it with overlays. We will produce a WebService in PHP, responsible for retrieving them from database and returning them to the Android application, which will place them on the map. Database should contain information about the position (latitude and longitude) and outlook of each overlay.

Listing 4

It is important to include the PHP attribute, so the WebService could be called from the application. The database is only accessed locally. It means that external programs can only send signal to PHP files which acts as a gatekeeper to the database. Data can be exchanged only through our PHP files.

Listing 5

The communication object is not made of other complex objects. Its properties are serialized to JSON format (understood by PHP and Android) and we do not have to worry about it.

Listing 6

First, we may want to declare an array for our overlays. Next, we have a function to get location from database which should be called when the map is initialized. It invokes the GetDPointsWebMethod, and if succeeded, InitDPointsArray function is called.

Listing 7

The result parameter is actually our list of ServiceOverlay objects. After filling the dpointsArray, we can finally call the InitOverlays function, which will fill the map.

4 MarkerManager

Before we continue, I should say a few words about MarkerManager class. It is used to supervise visibility of hundreds of markers on a map, founded on the map's contemporary viewport and zoom level. Our application is almost certain to have lots of overlays, so it is recommended to use the MarkerManager class, to make it more readable. Another reason for using it is that it will make our application faster, because all overlays would not be rendered together. Now, we can get back to InitOverlays function.

```
functionInitOverlays()
  {
  Map =new GMap2(document.getElementById("map")); vari = 0;
  map.setCenter(new GLatLng(51.76, 19.52), 12); map.addControl(new GLargeMapControl());

  mgr = new MarkerManager(map); var batch = [];
  for (i = 0; i<dpointsArray.length; i++)
    {
    var Icon = new GIcon("", "", "");
    Icon.image = dpointsArray[i].outlook; Icon.iconSize = new GSize(30, 30);
    Icon.iconAnchor = new Gpoint(0, 30); Icon.infoWindowAnchor = new Gpoint(5, 2); Icon.transparent = `'•
    Icon.shadow = "";

    var point=new GlatLng(
                dpointsArray[i].lat,
                dpointsArray[i].lng);
    batch.push(
      CreateMarker(point,
              dpointsArray[i].id, Icon,
              dpointsArray[i].type)
          );
    }
      mgr.addMarkers(batch, 10); mgr.refresh();
  }
functionCreateMarker(point, id, Icon ,type)
var marker = new GMarker(point,{icon: Icon,});
      marker.value = id;
      return marker; }
```

Listing 8

At the beginning, we initialize a GMap2 object, center the map, and add a standard map control. Then, we create a MarkerManager object (variable mgr) and a batch array, which holds the overlays for the MarkerManager. Next, It loops through the dpointsArray, adding markers returned by CreateMarker function to the batch. Our markers have custom icons, represented by GIcon object. Its properties are described on Google Maps API Reference Web site (see references, position 5). Finally, we fill the mgr variable with batch array, by addMarkers function and refresh it to see the effect.

5 Handling Events and Result

To continue with construction of our user interface, we must handle the events generated by the map and overlays. What we want to do is to distinguish the events based on active tab of our Tab Container control. When user clicks the map, a map dispatcher should be called:

Figure 1 gives user the log in facility to the old users and the Registration for new users or clients. Figure 2 Registration page need to be filled by a new user or any person for creating an account and thus providing them access to use the services provided in the app. Figure 3 App Home Page Provides "Map screen" Button: tap on it to fetch your current location and it will represent it in Blue dot. Figure 4 Bus stop registration in which the server has to mark up all the parking spots and nearby Bus

Fig. 1 The Login Page

Fig. 2 The Registration Page

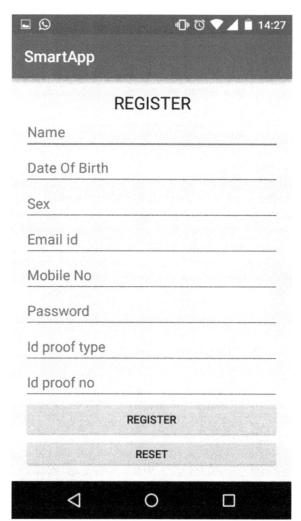

Fig. 3 App Home Page

stops for a specific Location so that users can fetch the details about the availability of parking spaces. Figure 5 Parking slot availability page allows the users get the current location, get the number of parking spaces and can allot one slot for their use.

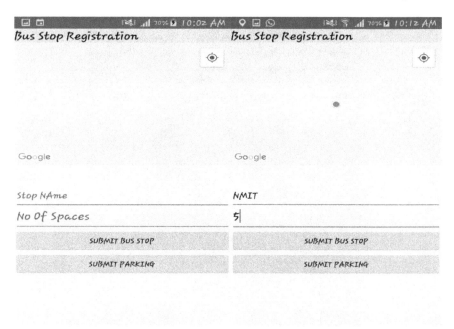

Fig. 4 Bus stop registration/marking

6 Conclusion

When you decide to use Google Maps API in your application, then you should come up with a good plan of placing the overlays on the map and handling events. It is most of the work you will have to do. As it was shown in this paper, building Web applications in Android technology based on Google Maps service is not a difficult process. It is basically about the communication between PHP and Android; it can be easily established with the usage of a WebService, where data serialization is done automatically.

Fig. 5 Parking slot availability

Modeling Implementation of Big Data Analytics in Oil and Gas Industries in India

Dilip Motwani and G. T. Thampi

Abstract Stack holders have been anxious about the quality of performance in oil and gas industries in India and recommending technology intervention to drastically improve its performance. This investigation aims to analyze existing level of information and communication technology integration in oil and gas industries in India. All such industries which generate massive revenue are preparing to leverage big data analytics (BDA) to build efficiencies and improve productivity by removing non-value adding activities. This paper also investigates to identify ways and means of applying BDA tools.

Keywords Big data analytics · Information and communication technology · Oil and gas industries

1 Introduction

Urbanization in developing markets has initiated an enormous exigency for oil and gas (O&G). The requirement for energy internationally is estimated to upsurge 1.4 times in 10 years. The rise of Asian countries has transferred monetary control to the eastern hemisphere. Trade demands are crafting commercial prospects. Energy utilization internationally is estimated to increase by 39.2% in 20 years, whereas non-O&G energy sources will increase by 50% from 2010 levels. Therefore, O&G will endure to persist as most important energy sources. By 2030, O&G's contribution globally would be 55%. Venture capital in unusual gas and augmented ordinary and shale reserve investigation have initiated a development marketplace for gas. Even though North America would be one of the dominant countries in the supply of gas, a major demand would be from Asian countries. However, there always exists a task of obtaining progressively more complex reservoirs so that profitable amounts of

D. Motwani (✉) · G. T. Thampi
Thadomal Shahani Engineering College, Mumbai 400050, India
e-mail: damotwani@gmail.com

G. T. Thampi
e-mail: gtthampi@yahoo.com

© Springer Nature Singapore Pte Ltd. 2019
N. R. Shetty et al. (eds.), *Emerging Research in Computing, Information, Communication and Applications*, Advances in Intelligent Systems and Computing 906,
https://doi.org/10.1007/978-981-13-6001-5_9

O&G can be attained. Technology has turned out to be O&G industries' essential backbone, and key modernizations are established in progressive seismic imaging expertise to improve the perception of reservoir constructions and to aim reservoirs with marketable prospects [1].

Information technology (IT) enables the radical changes in efficiencies of O&G industries. O&G industries shall use IT as force multiplier/resource multiplier o simplify core functions together with exploration and production (E&P). In general, all O&G industries are moving into the next phase of IT integration, i.e., digital oil fields. They are getting utilized by O&G industries to develop productivity gain and also improve collaborative endeavor. Furthermore, technologies like big data analytics are letting O&G industries to utilize extrapolative analytics to acquire an all-inclusive understanding of organized and spontaneous data that will help in making well-informed decisions [1].

2 Profiling O&G Industries as an Essential Industrial Sector in Contribution to GDP

O&G industries are dynamos globally engaging thousands of employees and making billions of dollars annually. They lie in the midst of top six core industries in India. Because of this industry, India has experienced a major growth in Indian economy. Natural gas and petroleum sector add approximately 15% to the country's gross domestic product (GDP). In addition, exports from petroleum are maximum with respect to overseas exchange and constitute 17% of total exports. According to statistics, Economic Affairs Committee handed over 44 O&G blocks to New Exploration Licensing Policy, thereby bringing assets worth 1.5 billion US dollars. This indicates that O&G industries are continuously growing and investing in such industries will maximize profits [2, 3].

India is a major contributor toward non-organization for economic cooperation and development petroleum consumption growth. In 2016–17, oil imports increased to 86.45 billion US dollars (i.e., 4.24% per annum). The consumption of oil in India increased to 212.7 million tons (i.e., 8.3% per annum) in 2016, thereby making India the third largest nation in the world in terms of oil consumption. Moreover, after Japan, South Korea, and China, India is the fourth largest nation that imports liquefied natural gas, which constitutes 5.8% of total global trade. The demand for domestic liquefied natural gas is estimated to rise from 64 million metric standard cubic meters per day in 2015 to 306.54 million metric standard cubic meters per day by 2021. Petroleum oil and lubricant demand grew at a compound annual growth rate of 5.6% under the twelfth five-year plan (2012–2017). Gas production in India is expected to rise from 23.09 billion cubic meters to 90 billion cubic meters by 2040. Till November 2016, the infrastructure of gas pipeline in India was 16,240.4 km. Oil and Natural Gas Corporation Limited dominates the upstream sector (i.e., E&P) by producing approximately 1847 thousand metric tons of crude oil as against 2939 million tons oil output in April 2017. Furthermore, it produces 57% of domestic crude oil as of 2016–17 [4].

3 Reasons for Delay in Integration of Production Technologies in O&G Industries

In order to upsurge functioning and lessen employment plunders because of increasing cost of assembly, growing possessions, and unavailability of suitable talents, O&G industries are under a lot of pressure. Operational excellence can be achieved via integrated planning that helps in improving security, averting mishaps and labor slowdowns, and enhancing operational and supply competence. Industry foreigners might be astonished by how often E&P workers neglect comprehensive planning of activity. Occasionally, plans are composed but not executed because of noncompliance with an industry's planned, tactical, and monetary priorities. It has been observed that activity planning is given to junior staff who are unexperienced and might have optimistic assumptions concerning time, budget, and supply necessities [5, 6].

4 Insufficient O&G Production in India

India is on the brink of an energy crisis. Millions of people in India have no access to electricity, with more than 400 million people going powerless. Currently, India depends heavily on coal with 67% of power generation coming from thermal resources. Coal is the mainstay, but its production is insufficient and depleting fast. Most coal sources are known and nearing exhaustion. India is the world's fourth biggest importer of crude oil. Domestic production of natural gas has in fact decreased in recent years. The only hope is acceleration in the discovery and tapping of oil and gas. But, Indian industries have not been able to keep pace with state-of-the-art technology.

5 Global O&G Industries Leveraging BDA

Although BDA may seem new to many, some of the few industries that have been familiar with the concept and have worked on it are O&G industries. As reserves decline rapidly, there exists tough rivalry among O&G industries to discover potential sites and find ways and means to enhance productivity. Energy industries are gathering increasingly greater amounts of data attempting to determine what remains beneath the surface of the earth and concocting means to excerpt it. In addition, O&G industries are leveraging seismic software and supplementary cardinal know-hows.

6 Availability of Faster and Cheaper Sensors Yielding Massive Data Volumes

With the availability of faster and cheaper sensors, novel potentials continue to arise that directly or indirectly facilitate O&G industries acquire increasingly additional real-time data at low prices. O&G industries have been using sensor data to supervise oil resources. Better analytics can improve the methods through which O&G industries accomplish the whole procedure of boring and linking a well. This helps not only in lessening wait time but also in reducing the total wells that are in process. For instance, broadcasting of microseismic three-dimensional imaging through optical fiber cables has the potential to enhance new well delivery performance.

7 Data Integration Reducing Technical Risks and Saving Lives

Visualization and standardization competences are significantly upgraded by integrating data into operations, thereby lowering technological risks. BDA helps in scrutinizing channels and tools and provides a precise approach toward maintenance. For example, sensors would be able to reveal when equipment is under pressure that would help workers to conduct precautionary closures, thereby avoiding accidents.

8 New Big Data Tools Improving Production, Data Collection, and Analysis

O&G industries can utilize BDA technologies through tools like the MapR converged data platform to manage, collect, and analyze drilling, seismic, and production data. Oil majors can use this data for new insights that help in increasing production and drilling performance while foiling environmental or safety issues. In addition, such technology can help petroleum industries capitalize on big data analytics to optimize business operations, reduce costs, and increase competitive edge.

9 Digital India Initiative Incomplete Without Digital Oil Fields

The government will have to ensure that the technology sitting with international energy firms is transferred to Indian oil industries. They would not be able to do it individually. Conversely, O&G industries need to invest heavily in research and development in terms of technology and software so that they can balance the growing energy needs of digital India. It is time they start leveraging new age technology such as BDA.

10 Key Areas That Can Benefit with BDA in O&G Industries

10.1 Security

With such sensitive data floating around, O&G industries find it essential to find out any type of security threat that is prevailing around and could harm the setup in any way. BDA possesses the potential to identify any type of security mishap that is possibly around and can well in time provide real-time analytics to show possible solutions to avoid them well in advance. Any type of anomaly while drilling or exploration can be identified prior to its occurrence, and hence, the unit or processing can be shut down to avoid further mishaps [7].

10.2 Continuation and Execution

A lot of sensor data gets generated through the equipment which is being used for O&G production. This data can be used through BDA to determine the health of the equipment and analyze which machines could be the reasons of failure. Its precautionary maintenance could point toward unharmed and non-hazardous processes toward an effortless procedure [7].

10.3 Assembly

To proliferate oil recovery from wells and to follow predefined assembly schedules, BDA acts as a supporter to aid industries function their best. Extricated report helps workers determine the desired modifications in programs with the intention of attaining the best of assembly [7].

10.4 Novel Predictions

Because BDA provides geographical information, it becomes simple for field managers to extract conceivable upcoming areas for oil fields, thereby increasing geographical size and stretch [7].

11 Opportunities

For O&G industries, comprehending, leveraging, and releasing the supremacy of data will aid to

- Stay viable during preparation, assessment, fabrication, and field expansion
- Increase production w.r.t continuance and predication
- Lessen time, decrease cost of operation, and upgrade asset productivity
- Guarantee unified and systematized accessibility of precise and appropriate information to workers at the right time.

Therefore, by means of big data, costs, decision making, and operational execution can be improved; higher business process efficiencies can be achieved; new insights can be gained; and new business models can be developed. This proves that big data is equivalent to return on investment [8].

12 Industry Challenges

Some of the major challenges faced by O&G industries are mining costs and unstable situation of worldwide politics. Because of such problems, many industries are approaching big data with the expectation of obtaining solutions. Big data refers to the application of forward-thinking computer exploration to the increasing amount of digital information. Businesses in each and every industry, including O&G industries, in the past few years have actively established data-led strategies for surmounting glitches and resolving confronts. Reviewing of impending sites implicates supervising low-frequency seismic waves that move across the earth caused by tectonic activities. Big data can be used to restructure transportation, sophistication, and supply of O&G. Royal Dutch Shell is precipitously unified; therefore, it is occupied in each and every attribute of a process. Processing plants have limited capacity and require a lot of fuel to reduce the cost of transportation. Convoluted algorithms consider the cost of producing both fuel and diverse data, which helps in determining requirements, assigning capitals, and establishing prices.

13 Big Data Challenges in O&G Industries

Some of the big data challenges faced by O&G industries are as follows:

(1) Exponentially rising data volumes from structured and unstructured sources.
(2) O&G industries have to heavily pay for E&P data management and handle streams of incompatible data from different phases of a well life cycle.
(3) Geologists using a combination of dissimilar software products for decision making and elucidation of data.
(4) Inconvenience in using data to efficiently respond to user needs.
(5) Outsized volume of domain-specific information embedded in each data cluster.

14 Conclusion

We arrive in a new period of unparalleled accessibility of data where digital developments are unsettling outmoded business models. These developments have permitted the occurrence of BDA, which is slowly and steadily turning out to be a huge industry. O&G industries fall back prominent businesses when it comes to broad-based adoption. Nevertheless, four main applications, i.e., digital fields, predictive plant and drilling analyses, remote operations, and reservoir modeling and seismic imaging, are emerging for big data in O&G industries. Industries can move in the right direction by espousing essential success elements and circumventing characteristic consequences. Our research work suggests that significant value can be captured from BDA if performed in the right direction.

References

1. *IT solutions for the oil and gas industry: ICT innovations to build a smarter oil and gas sector from ROLTA by frost and sullivan*. Available [Online]: http://www.rolta.com/wp-content/uploads/pdfs/resources/frostsullivan.pdf.
2. *Invest in India: Oil and gas industry*. Available [Online]: http://www.investinindia.com/industry/oil-and-gas/oil-and-gas-industry.
3. *Introduction to oil and gas industry: Uncovering the oil and gas industry*. Available [Online]: https://www.oilandgasiq.com/strategy-management-and-information/articles/oil-gas-industry-an-introduction.
4. *India brand equity foundation: Oil & gas industry in India*. Available [Online]: https://www.ibef.org/archives/detail/b3ZlcnZpZXcmMzc3MzcmNDg5.
5. *Integrated planning: The key to upstream operational excellence*. Available [Online]: http://www.bain.com/publications/articles/integrated-planning-the-key-to-upstream-operational-excellence.aspx.
6. *IT as a growth driver for Indian PES*. Available [Online]: https://www.pwc.in/assets/pdfs/publications-2011/cii_it_4.pdf.

7. *SPEC INDIA: Big data analytics provides a powerful thrust to oil & gas companies.* Available [Online]: https://www.specindia.com/blog/big-data-analytics-provides-powerful-thrust-oil-gas-companies/.
8. *Growing integration of ICT in oil and gas operations boosts efficiency.* Available [Online]: https://ww2.frost.com/news/pressreleases/growing-integration-ict-oil-and-gas-operations-boosts-efficiency-says-frost-sullivan/.

Practical Market Indicators for Algorithmic Stock Market Trading: Machine Learning Techniques and Grid Strategy

Ajithkumar Sreekumar, Prabhasa Kalkur and Mohammed Moiz

Abstract In this paper, market indicators from three different approaches for algorithmic trading are analysed (moving average convergence divergence (MACD) crossovers, machine learning (ML) label-based indicators, and grid investing strategy). Market indicators are used by traders in the stock market, to define entry and exit points of a trade. These indicators are also useful to compare different trading strategies. We take a practical stand for the approaches mentioned above, where the *same* data feed from the exchange is preprocessed to remove redundant or anomalous content. Furthermore, use of correlation data between different stocks is analysed. (i) MACD crossovers are dealt in two dimensions of variability, the dimensions being frequency of trades and length of trading intervals. (ii) The outputs of different algorithms are passed through a voting classifier to get the best possible accuracy in the ML label-based approach. Precision/Recall analysis is done to qualify the algorithms for skewed data. (iii) Finally, a grid-based trading strategy is analysed. We conclude with a trading strategy, proposed using results of indicators based on the three approaches.

Keywords MACD · Machine learning · Grid trading · Precision · Recall · Market indicators

---

A. Sreekumar · M. Moiz
Department of ECE, RVCE, Bengaluru 560059, India
e-mail: akajithkumar351@gmail.com

M. Moiz
e-mail: mohammed.moiz7@gmail.com

P. Kalkur (✉)
IISc, Bengaluru 560012, India
e-mail: prabhasa.94@gmail.com

1 Introduction

1.1 Indicators and Leverage

The momentum and direction of major securities (stocks) can be predicted by using various technical indicators, called market indicators. Many market indicators are obtained by analysing the market breadth. Market breadth indicates the number of companies that have obtained *new* high valuations, in comparison to the number that have obtained poor valuations. Usually market indicators are used to signal a *bullish* or *bearish* move. A bullish signal is got just before the stock is going to high valuations; vice versa being the bearish move. Therefore, a considerable amount of profit can be obtained on **buying during a bullish signal and selling during a bearish one**. In algorithmic trading, leverage must also be considered while acting on these binary signals.

In the context of investing, *Leverage* refers to the borrowed money. As there is no borrowed money in our proposed scheme, we make sure that the leverage in all the trading strategies is 1. This is taken care of by enabling the change in position of a *security* (buy/sell), only when there are no open orders. This also makes the strategies more practically applicable in a live trading environment.

1.2 Data Acquisition

Data is obtained using the *pandas* data-reader module of Python and Yahoo finance API. The API gives a *pandas* dataframe which can be used for storage or further processing. Free, real-time data is given by the API. A sample of the data obtained using this method is shown in Table 1.

The head of the *pandas* dataframe requested for AAPL (Apple Company) is shown in Table 1. There are seven columns in the dataframe, indexed through the column 'Date', which specifies the date at which the corresponding stock data existed. '*Open*' indicates the price at which the stock started trading (opened) on the indexed date.

Table 1 Sample data

Date (DD/MM/YY)	Open ($)	High ($)	Low ($)	Close ($)	Adj close ($)	Volume (No. of transactions)
1/3/12	58.485714	58.92857	58.42857	58.747143	52.662899	75,555200
1/4/12	58.57143	59.240002	58.468571	59.062859	52.945919	65,005500
1/5/12	59.278572	59.792858	58.952858	59.718571	53.53373	67,817400
1/6/12	59.967144	60.392857	59.888573	60.342857	54.093361	79,573200
1/9/12	60.785713	61.107143	60.192856	60.247143	54.007557	98,506100
1/10/12	60.844284	60.857143	60.214287	60.462856	54.200932	64,549100
1/11/12	60.382858	60.407143	59.901428	60.364285	54.112568	53,771200

'High' indicates the highest traded price of the stock and 'Low', the lowest price of the stock on that particular day. 'Close' indicates the price at which the stock stopped trading (closed) on the indexed day. 'Adj Close', which stands for adjusted close takes into consideration stock splits, dividends/distributions and rights offerings. This helps in historical data analysis of the stock [1]. 'Volume' indicates the total number of shares transacted in a particular time period [2]. This dataframe can now be used for further analysis.

1.3 Data Preprocessing

The dataframe obtained as discussed in Sect. 1.1 cannot be used directly for algorithmic trading as it may contain values such as Nan (Not a number) or $\pm$inf (infinity). This could be due to market holidays, or the data may be missing owing to other reasons. There is no intelligence in the dataframe to alleviate this issue. Therefore, there is a need to preprocess the data. A single faulty entry for a given index (date) renders the whole data useless. (i) In case of the stock price/volume, the Nan cells of the dataframe are first filled with 0 and $\pm$inf cells are filled with Nan. Then the remaining Nan cells are dropped with the whole row. (ii) In case of a percentage change type of data, the $\pm$inf data as well as the Nan cells are replaced with 0. This takes care of removing spurious/erroneous data before using it for different trading strategies.

1.4 Correlation

The statistic which measures the degree with which different stocks move up or down in relation to each other is called correlation. An advanced portfolio management uses correlation in addition to the market indicators. Correlation coefficient is the quantification of correlation and the value lies between 1 and -1.

Dataframes were obtained using the method detailed in Sects. 1.1 and 1.2 for top ten stocks of S&P 500 (Standard and Poor) [3]. The correlation of the stocks was calculated and a heat map of the corresponding coefficients is shown in Fig. 1. The correlation data can be interpreted as follows. A correlation of -1 indicates that the stocks are negatively correlated, and they move in opposite directions. This denoted by a shade of red in the heat map of Fig. 1. A correlation of 1 indicates that the stocks are positively correlated or that they move in the same direction. This is indicated by a shade of green in the heat map of Fig. 1. A correlation of 0 indicates that the stocks have no correlation. The swing in one stock does not affect the movement of the other stock. This is indicated by shades of yellow on the heat map. The correlation coefficients are used in conjunction with all the indicators which will be discussed in the future sections. This helps in managing a diverse portfolio.

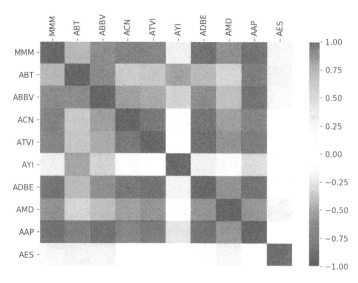

Fig. 1 Correlation between securities

2 Moving Average Convergence Divergence (MACD)

MACD helps in the price direction and trend identification as well as the momentum of price. Being a trading tool that is versatile, MACD can be used as a standalone indicator [4]. But owing to the predictive function of MACD not being absolute, MACD when used with another indicator, significantly adds to the advantage of a trader [5, 6]. Moving average lines can be overlaid on the MACD to determine the direction of a stock and its trend strength. MACD can also be viewed as a standalone histogram. A simplistic MACD calculation is required to get a market indicator which fluctuates around the zero line, giving a binary signal: either bullish or bearish. For instance, a bullish moving average crossover is detected in the following way. A 26-EMA (26-day exponential moving average) of a stock is subtracted from 12-MA (12-day moving average). This gives an oscillating indicator. If a trigger line like a 9-EMA is added, then the juxtaposition of the two indicators, gives a trading picture. When the MACD calculated is higher than the 9-EMA, a bullish moving average crossover is detected [7, 8].

Some of the best practices while using MACD are [9]:

- The centre line of the histogram must be watched for crossover or divergences. When the oscillation is above 0, then it is a buy opportunity and a sell otherwise.
- The relationship between the centre line and the moving average line crossovers also must be noted.

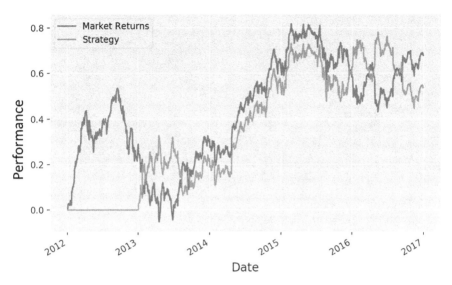

Fig. 2 Low-frequency MACD strategy

2.1 Low-Frequency MACD

Market returns specify the return obtained if the security was simply bought and held. Strategy refers to MACD crossover discussed in Sects. 2.1, and later in 2.2. The strategy is performing at 0% for the initial values of length window size, as without any prior information to preprocess, the moving average data would be zero. In Fig. 2 a low-frequency trading approach is used (weekly trading period). The MACD crossover strategy has beaten the market performance in few places (2013, and parts of 2016). For the most part, the returns are as good as the market performance.

2.2 Mid-Frequency MACD

In Fig. 3, a medium-frequency trading approach is used (daily trading period). The MACD crossover strategy has performed better than the market at some places (2013, and briefly in 2016). For the most part, it is following the trend of the market, but not as good as the market.

Fig. 3 Mid-frequency MACD strategy

Fig. 4 High-frequency MACD strategy

2.3 High-Frequency MACD

A high-frequency trading approach is used (hourly trading period), as shown in Fig. 4. The MACD crossover strategy has performed worse than the market strategy.

This is because the MACD crossovers will happen multiple times in a short window. If the data is not averaged over a long time, there can be a trend reversal within that short time and the strategy will buy/sell using the wrong signals before the trend reversals [10].

3 Machine Learning Techniques

3.1 A Four-Step Approach is Taken to Apply the Machine Learning Algorithms to Get a Trading Strategy

1. The data is processed for a required target number of days. The cumulative percentage changes of user-defined number of days are added as new columns to the dataframe and stored [11].
2. The processed data in the previous step is analysed against a user-defined percentage requirement. If the prediction is found to be performing better than the requirement in the next few user-defined days (same number of days as used in previous step), then a *buy* label is returned. If the predicted percentage return is found to be lesser than the negative of user requirement, then a *sell* label is returned. If both the above conditions of user requirement are not satisfied, then a *hold* label is returned.
3. The feature set is extracted by mapping [12] all the stocks with the labels obtained in the previous step.
4. The trading strategy is obtained based on voting classifier module of *sklearn* [13] python module. A Support vector machine (SVM) [14, 15] with a linear support vector classifier (SVC), random forest classifier and a K-nearest neighbours (K-NN) classifier are the three classifiers fed into the voting classifier. The voting classifier chooses the classifier with the best accuracy each time [16, 17].

Table 2 shows the different set of user inputs. '*Days*' column specifies how many days of cumulative returns should be considered and the '*Target*' specifies the percentage cumulative return that has to be achieved by the strategy. For example, in the case of the first user input (7 days, 2% target), the trading strategy should predict buy (1 as label) if the ML predicts that the stock will rise by target percentage (2% in this case) in the given number of days (7 in this case). Similarly, sell (−1 as label), if the ML predicts that the stock will fall by target percentage (2% in this case) in the given number of days (7 in this case). If both the conditions do not hold good in a situation, then the strategy is to hold (0 as label).

An analysis with user inputs as in Table 2 gives the results shown in Table 3. This shows that the accuracy (Whether target % was achieved or not in the given number of days) is more if the days are more (cases 3 and 4), and the accuracy decreases if the percentage requirement is high and the number of days to achieve the target percentage is less (case 5).

Table 2 User-defined inputs to the strategy

Sl. No.	Days	Target(%)
1	7	2
2	7	2.5
3	15	2
4	15	2.5
5	7	10
6	15	10

Table 3 Labels and accuracies of ML Strategy (For a total of 756 trades)

Sl. No.	Buy (1)	Sell (−1)	Hold (0)	Accuracy (%)
1	77	658	21	86.24
2	6	747	3	86.67
3	77	652	27	98.41
4	50	691	15	92.59
5	201	492	63	67.19
6	96	640	20	82.01

3.2 Precision/Recall

A definitive metric to quantify the success of an algorithm when the data is skewed towards one of the labels is precision–recall. Precision refers to the relevance of results of the algorithm while the recall represents the number of *truly* relevant results returned by the algorithm. 'Relevance' refers to the result being one of the four possibilities among true/false and positive/negative. A simple illustration can be found in [18, 19]. In our proposed approach, *positive* can be interpreted as profit, and *negative* as loss. Ideally, we would look for a measure of *true positives* as compared to *false positives* and *false negatives*.

A high precision denotes that the *false positives* are less, and a high recall denotes that *false negatives* are less. Intuitively, we would want a high precision and high recall, as both indicate the correctness of our algorithm. The equations of *precision* (ratio of number of true positives T_p to the number of true positives plus the number of false positives F_p) and *recall* (ratio of number of true positives T_p to the number of true positives plus the number of false negatives F_n) are as follows:

$$\text{Precision} = T_p/(T_p + F_p)$$

$$\text{Recall} = T_p/(T_p + F_n)$$

A precision–recall curve shows the tradeoff between the precision and recall. The curve is generated using an ISO F1 equation relating the precision and recall, as described in [20]. Figure 5 shows the precision–recall curve for the ML algorithm

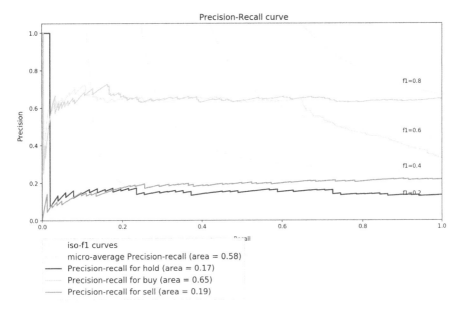

Fig. 5 Precision–recall curve

used. The data set used for the purposes of testing algorithm had a skew towards buy, and therefore it is evident that the precision–recall is high for buy. But it is evident from Fig. 5 and independently represented in Fig. 6 that the average precision–recall is high (0.58). This indicates that the algorithm performs well even with skewed data sets.

4 Grid Strategy

Grid strategy involves setting buy stop orders and sell stop orders at multiple levels with respect to the current price [21]. With each executed stop order, a take profit is set to secure the profit without setting any stop-loss. This strategy relies on the natural movement of the market to make profits.

This strategy removes the variable of knowing the direction of the price move, but this adds to the complexity and increases the margin of error as multiple trades (buy and sell) are managed at the same time.

For the strategy, one must choose several strategy levels or windows—around which buy/sell orders are placed. In this simulation, there are three levels chosen. Each level is at a distance of $3 from the current market price. Consider the share price of 'TSLA' on 7th October 2014, as shown in Fig. 7. The share closed at a price of $258.53. Now, place three buy stop orders above the current price and three sell

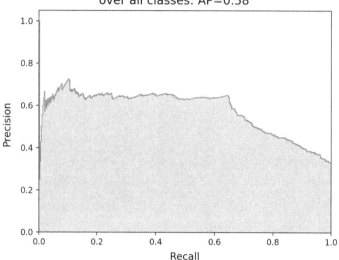

Fig. 6 Average precision score

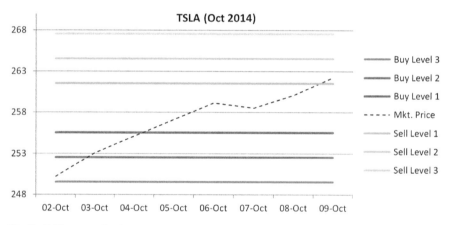

Fig. 7 Grid strategy levels

stop orders below it. Whenever either of these orders is executed, a new order is placed in the direction of profit.

By following the grid strategy on every $1000 invested on 7th October 2014, we make a profit of $207 until the 29th March 2018. However, the Market would have returned only $25 of profit. This can be observed from the graph in Fig. 8, where the losses in grid strategy are minimised compared to the market performance.

Fig. 8 Grid strategy versus market return

5 Conclusion

Three trading strategies were discussed in isolation—MACD crossover, machine learning and grid strategy—and results of these strategies were analysed. From Figs. 2, 3 and 4, it is evident that MACD crossover strategy is suitable at low frequencies. Machine learning with buy, sell and hold as labels and using voting classifier proved accurate and practical for achieving a target percentage in a given number of days. Grid strategy outperformed the market when two or more levels were used. Correlation discussed in Sect. 1.4, when used along with these three strategies will lead to a diversified portfolio. Finally, all these three strategies were discussed in a practical manner as leverage was applied (Sect. 1), precision–recall was analysed and transaction volume minimised in the algorithms leading to lower commission fees.

References

1. Koopman, S. J., Jungbacker, B., & Uspensky, E. H. (2004). Forecasting daily variability of the S&P 100 stock index using historical, realised and implied volatility measurements. *SSRN Electronic Journal*.
2. Chan, L. K., & Lakonishok, J. (1993). Institutional trades and intraday stock price behavior. *Journal of Financial Economics, 33*(2), 173–199.
3. *The S&P 500 Index, Wikipedia*, (Online). Available: https://en.wikipedia.org/wiki/S%26P_500_Index. Accessed 10 March 2018.

4. Zhang, P., & Su, W. (2012). Statistical inference on recall, precision and average precision under random selection. In *Proceedings of the 2012 9th International Conference on Fuzzy Systems and Knowledge Discovery*.
5. Almeida, R. D., Reynoso-Meza, G., & Steiner, M. T. A. (2016). Multi-objective optimization approach to stock market technical indicators. In *Proceedings of the 2016 IEEE Congress on Evolutionary Computation (CEC)*.
6. Troiano, L., Villa, E. M., & Loia, V. (2018). Replicating a trading strategy by means of LSTM for financial industry applications. *IEEE Transactions on Industrial Informatics*, 1–1.
7. Wu, M., & Diao, X. (2015). Technical analysis of three stock oscillators testing MACD, RSI and KDJ rules in SH & SZ stock markets. In *Proceedings of the 2015 4th International Conference on Computer Science and Network Technology (ICCSNT)*.
8. Kamble, R. A. (2017). Short and long-term stock trend prediction using decision tree. In *Proceedings of the 2017 International Conference on Intelligent Computing and Control Systems (ICICCS)*.
9. Staff, I. (2018). *Moving average convergence divergence—MACD, Investopedia*, 09 May 2018 (Online). Available: https://www.investopedia.com/terms/m/macd.asp. Accessed 10 May 2018.
10. Pring, M. J. (2014). *Study guide for technical analysis explained*. New York: McGraw-Hill.
11. Li, Y., Wu, J., & Bu, H. (2016). When quantitative trading meets machine learning: A pilot survey. In *Proceedings of the 2016 13th International Conference on Service Systems and Service Management (ICSSSM)*.
12. Ruta, D. (2014). Automated trading with machine learning on big data. In *Proceedings of the 2014 IEEE International Congress on Big Data*, 2014.
13. Pedregosa, et al. (2011). Scikit-learn: Machine learning in python. *JMLR, 12,* 2825–2830.
14. Hearst, M. A. (1998). Support vector machines. *IEEE Intelligent Systems*, 18–28.
15. Sadewa, C., & Harlili. (2017). Exploration and analysis of some online machine learning on GBP/USD trading simulation. In *Proceedings of the 2017 International Conference on Advanced Informatics, Concepts, Theory, and Applications (ICAICTA)*.
16. Wang, G. (2008). A survey on training algorithms for support vector machine classifiers. In *Proceedings of the 2008 Fourth International Conference on Networked Computing and Advanced Information Management-Volume 01, ser. NCM '08* (Vol. 1, pp. 123–128).
17. Deng, Y., Bao, F., Kong, Y., Ren, Z., & Dai, Q. (2017). Deep direct reinforcement learning for financial signal representation and trading. *IEEE Transactions on Neural Networks and Learning Systems, 28*(3), 653–664.
18. *True-false-positive-negative, Google Developers*, (Online). Available: https://developers.google.com/machine-learning/crash-course/classification/true-false-positive-negative.
19. Han, J., & Pei, J. (2011). *Data mining: Concepts and techniques*, Elsevier, pp. 402/740.
20. Zhang, P., & Su, W. (2012). Statistical inference on recall, precision and average precision under random selection. In *Proceedings of the 2012 9th International Conference on Fuzzy Systems and Knowledge Discovery*.
21. Forex grid trading strategy explained, Admiral markets (United Kingdom), 2018. (Online). Available: https://admiralmarkets.com/education/articles/forex-strategy/forex-grid-trading-strategy-explained. Accessed 1 March 2018.

A Review on Feature Selection Algorithms

Savina Colaco, Sujit Kumar, Amrita Tamang and Vinai George Biju

Abstract A large number of data are increasing in multiple fields such as social media, bioinformatics and health care. These data contain redundant, irrelevant or noisy data which causes high dimensionality. Feature selection is generally used in data mining to define the tools and techniques available for reducing inputs to a controllable size for processing and analysis. Feature selection is also used for dimension reduction, machine learning and other data mining applications. A survey of different feature selection methods are presented in this paper for obtaining relevant features. It also introduces feature selection algorithm called genetic algorithm for detection and diagnosis of biological problems. Genetic algorithm is mainly focused in the field of medicines which can be beneficial for physicians to solve complex problems. Finally, this paper concludes with various challenges and applications in feature selection.

Keywords Feature selection · Classification · Wrapper method · Genetic algorithm

1 Introduction

Classification is one of the essential assignments in machine learning whose purpose is to characterize each occurrence in the data set into various classes in view of its features [1]. It is often difficult to determine which features are useful for classification without prior knowledge. As a result, a large number of features are usually introduced into the data set that may be irrelevant or redundant. Feature selection is a process which eliminates the irrelevant features from original data set which contain a large set of features. In other words, we can say that feature selection algorithm removes the feature which is not required for machine learning algorithm either for classification or regression. This small subset of features may have less redundant or relevant features which will make machine learning process simple,

S. Colaco (✉) · S. Kumar · A. Tamang · V. G. Biju
Department of CSE, Christ(Deemed to Be University), Bangalore 560029, India
e-mail: savinacolaco@gmail.com

© Springer Nature Singapore Pte Ltd. 2019
N. R. Shetty et al. (eds.), *Emerging Research in Computing, Information, Communication and Applications*, Advances in Intelligent Systems and Computing 906,
https://doi.org/10.1007/978-981-13-6001-5_11

may reduce learning time complexity and increased performance. However, feature selection has another advantage that it improves prediction performance, scalability and understandability. It is also observed that the feature selection algorithm improves the generalization capability of classification algorithms.

Feature selection algorithm reduces computational complexity of classification algorithm. In addition, it offers new insights into knowledge for deciding the most applicable or relevant features. The main challenge that occurs in feature selection is large search space where for n data sets, solutions are 2^n. Feature selection comprises of complex stages that normally are exorbitant. The ideal model parameters of full feature set may be redefined for a couple of times with a specific end goal to obtain the ideal model parameters for chose feature subsets. Feature selection also contains two main objectives, which are to minimize the number of features and maximize the classification accuracy, which are both contradictory objectives. Hence, feature selection is considered as multi-objective problem with some trade-off solutions that lie in between these two objectives. Some popular methods for feature selection are information gain [2], chi-square [3], lasso [4] and Fisher score [5].

Feature selection used on gene expression data which has small sample size is called gene selection. Gene selection can be used to find key genes from biological and biochemical problems. This type of feature selection is important for disease detection and discovery such as tumour detection and cancer discovery which results in giving better diagnosis and treatment. Genomic information can be expressed as completely labelled, unlabelled or halfway marked. This prompt advancement of regulated, unsupervised and semi-administered gene selection leads to finding biological patterns in data [1]. Feature selection process can be supervised or unsupervised. However, we have another approach for feature selection which is a combination of supervised and unsupervised feature selection approach. In supervised feature selection, it uses the labelled data for feature evaluation [6]. But huge data is collected in an increasing rate. Additionally, the labelled information is expensive to acquire and might be problematic and mislabelled which may cause over-fitting in the learning procedure in regulated kind of feature determination by either expelling relevant features or utilizing irrelevant highlights. On account of supervised technique, past information is considered. Unsupervised feature selection is more difficult to work with than other two approaches because it is unaided by labelled data. But the main advantages of this type of feature selection are it is unbiased and performs well with no previous knowledge. Unsupervised feature selection method is widely used in medical discipline to determine the diseases and identify the type of diseases [7]. The disadvantage of unsupervised approach is it ignores connection between different features and it depends on a few mathematical principles with no certification that those standards are substantial for all information. Semi-supervised include blend of directed and unsupervised choice of features. Semi-supervised choice of features is likewise being utilized for quality classification by mutually utilizing both labelled and unlabelled information.

Gene expression data can be evaluated using microarray data methods is essential [8]. This approach can be grouped into supervised, unsupervised and semi-supervised methods. The microarray data has a large number of genes which are redundant. Thus, it needs to identify some important genes for better understanding of the fundamental data, also minimize the time taken for improved post-processing tasks such as classification, subset selection of genes (features) and so on [9]. Using feature selection, we get subset of features which are relevant, can be selected from the original data set consists of a large set of features. The key genes are found from enormous number of candidate genes in natural and biomedical issues using features like genes, biomarkers and so on [10]. Biomarker is a feature which gives an indication of medical condition observed from the patient externally and this can be measured as well as reproducible and different than medical symptoms which show only the signs regarding disease or health that are understood only by the patients themselves. Feature determination has a few preferences for microarray information. In the first place, reducing the dimensionality leads to computational cost efficiency. Noise reductions enhance the classification precision, more interpretable feature or attributes that can be useful to distinguish and screen the objective illnesses. Organically, just a couple of hereditary modifications relate to the harmful change of a cell. Assurance of these locales from microarray information can permit gene expression examination in these regions of high resolution, detection and classification for better diagnosis, prognosis and correct treatment for corresponding biological problems.

2 Related Work

One of the evolutionary computation techniques used on feature selection is genetic algorithms. Genetic algorithm is represented naturally as a binary string, where feature selected is represented as 1 and feature not selected as 0. There were lot of new improvements made on genetic algorithms like performance of search mechanisms, fitness function and representation. Genetic algorithms have been used in recent works to perform feature selection by evaluating the impact of size of the population, crossover administrator, change in chromosomes and reproduction administrators, yet these were led with insufficient tests.

Derrac et al. [11] had proposed a helpful co-developmental strategy for choice of feature by utilizing genetic algorithm with significant three populations, where the first population concentrated on feature selection, the second population on instance selection and third concentrated on both feature selection and pattern selection. This suggested algorithm talked about combination of two populations such as feature selection and instance selection, where the computational time was reduced. Li et al. [12] contributed a technique for GA which utilized numerous populations for choice of features, where data to build search capability is finished by each two neighbour populations that are common to two individuals. This technique was attempted with various measures of filter and wrapper and was demonstrated that it was viable for feature determination, however was tried just on data consisting of 60 features. Mao and Tsang [13] proposed an algorithm which is a two-layer cutting plane for searching feature subsets at optimal level. Min et al. Venkatraman et al. [14] proposed a measure incorporated with genetic programming of mutual information used as rank for features individually and remove irrelevant features and select subsets from remaining features.

Min et al. [15] proposed a heuristic and backtracking search algorithm using rough set theory to resolve feature selection problems. Min et al. [15] also proposed a rough set theory-based algorithm to talk about feature selection problems with limited resources constraints. Jeong et al. [16] proposed another portrayal to additionally lessen the dimensionality, where the quantity of wanted features was equivalent to length of chromosome. The chromosomes values indicates the features indexes. A limited SFFS operator was realistic when particular index appeared many times in features to select another features to escape replication. The constraint occurred in this proposed representation is the features wishes to be definite in prior, which may not be the best in size. Wang et al. [17] implemented a distance measure that evaluates the difference between the selected feature space and all feature space for finding a feature subset. Li et al. [12] provide scheme in GA with bio-encoding, where strings set was included in chromosome individually. The initial string showed features which were selected where string was paired encoded, and the feature weights were shown by another string which was encoded as real notations. The bio-encoding plan achieved better execution by consolidating with an Adaboost learning technique. Lane et al. [18] developed a method using particle swarm optimization and statistical clustering for feature selection with use of statistical feature clustering information during the search process of PSO [19]. Ke et al. [20] proposed a multi-objective ACO for feature selection of filter method, which speeds up the convergence performance.

Genetic algorithm has been used in feature selection over the years and better performance was achieved on problems which has numerous features. Genetic algorithms were usually introduced to state problems in feature selection which has lot of features which are mostly wrapper approaches. This leads to increased computational cost due to a large number of evaluations. The way operators and parameters settings in genetic algorithms are applied also matter to influence performance on feature selection.

3 Applications

Feature selection is used as tool for selecting feature subset or feature for learning algorithms. Feature selection is used for various applications in different areas such as:

(1) **Text Clustering**: we group related documents together in text clustering which is the major task. A text or document is always signified as a group of words, which causes sparse representation and high-dimensional feature space. The text clustering algorithms lower performance radically due to data sparseness and high dimensionality. Therefore, the feature space in text clustering is reduced by feature selection.

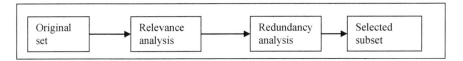

Fig. 1 Framework of feature selection

(2) *Genomic Microarray Data*: Microarray data with small sample size and high dimensionality is a main challenge for computational techniques [21]. Also, experimental complications like noise and inconsistency reduce the microarray data evaluation. Due to this issue, the dimensionality and noise are reduced and eliminated using feature selection in microarray data analysis.

(3) *Hyperspectral Image Classification*: Earth's surface reflection is recorded by hyperspectral sensors with high spectral resolution on high wavelengths, this results in high-dimensional data. However, this data contains features which are not useful and repeated. The classification of hyperspectral data reduces computational cost by selecting relevant features [22].

(4) *Sequence Analysis*: It is a method to understand a sequence's features, purposes, structure or evolution in bioinformatics. The pattern length k is varied with number of features and grows exponentially. Feature selection technique is used to select a related feature subset which is essential for sequence analysis [23].

(5) *Field of medicine*: By introducing genetic algorithm which is feature selection method in various medical disciplines such as radiology, oncology and cardiology, we can detect and diagnose various diseases faster. The use feature selection can provide new solutions in field of medicine.

4 Feature Selection

Feature selection is method of selecting features subset from original data set of large features set with elimination of features which are unrelated and redundant. The data dimensionality can be reduced using the feature selection. There are overall three approaches for feature selection: filter approach, wrapper approach and embedded approach. Figure 1 shows the feature selection framework.

4.1 Filter Methods

In filter approach, the selection of features depends on data's characteristics where learning algorithm is not used [24]. Even though the learning algorithms' heuristics and its unbiased nature are not considered, filter approach is very efficient. Hence, the intended learning algorithm may exclude relevant features. There are two steps

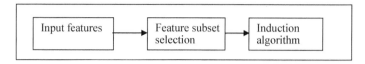

Fig. 2 Filter model for feature selection

Table 1 Filter algorithm [20]

INPUT:
D={X,L}   // a training data set with n number of features where
// X = {$f_1, f_2, f_3, \ldots f_m$} and L labels
X^1 // predefined initial feature subset (X^1 & X or X^1={Φ})
Θ
OUTPUT: X^1_{opt} // an optimal subset
Begin:
Initialize:
$X_{opt} = X^1$;
$T_{opt} = E(X^1, l_m)$; // evaluate X^1 by using an independent measure l_m
do begin
X_g = generate(X);   // Subset generation for evaluation
$T = E(X_g, l_m)$;   // X_g current subset evaluation by l_m
If ($T > T_{opt}$)
$T_{opt} = T$;
$X^1_{opt} = X_g$;
repeat (until Θ is not reached);
end
return X^1_{opt};
end;

in filter algorithm. In the first step, a certain criterion is used to rank the features. In the second step, the highest ranking features are chosen. The features' characteristics are measured with different ranking criteria. To differentiate samples within class variance from the correlation between feature-class and feature-feature, mutual information, to the various structure with various classes is feature and class label dependence. There are different filter algorithms such as Euclidean distance, correlation criteria, correlation-based feature selection, fast correlation-based feature selection and so on [25]. Figure 2 shows the feature filter model.

Table 1 shows a filter algorithm for a data set D = {X, L} (where X and L are the feature set and labels, respectively). This algorithm uses one of the following subsets of X^1 such as $X^1 = \{\Phi\}$ or $X^1 = \{NULL\}$ or X^1 & X. Each generated subset X_g is calculated by independent measure l_m and correlated with former optimal subset. The search repeats until we don't meet Θ which is the stopping criterion. As a result, the algorithm gives output X_{opt} as the current optimal feature subset.

Filter methods have variety of algorithms such as:

(1) **Correlation-based Feature Selection** (**CFS**): This is a simple filter algorithm where correlation-based heuristic evaluation function is used to rank feature

subsets. It uses function to evaluate features that are highly correlated with the class and uncorrelated with each other. The features with low correlation are ignored because they are irrelevant. Redundant features should be removed which may be highly correlated with one or more of the remaining features. If it can predict classes in areas of the instance space which have not been predicted before by other features can be accepted [26]. The CFS's feature subset evaluation function is:

$$M_S = \frac{k\overline{r_{cf}}}{\sqrt{k + k(k-1)\overline{r_{ff}}}} \quad (1)$$

where M_S is the heuristic "merit" of a feature subset S containing k features, r_{cf} is the mean feature-class correlation ($f \in S$) and r_{ff} is the average feature-feature inter-correlation as shown in Eq. (1). The numerator of the equation provides how predictive a set of features is; and the denominator tells how many redundant features are present.

(2) **Fisher score**: It is a commonly used criterion for supervised feature selection due to its general good performance. Fisher scores aims to find a subset of features by eliminating redundant features as well as maximize the data points [5].

(3) **Infinite Latent Feature Selection (ILFS)**: A training set of feature distributions X is taken such as, $X = \{\sim x1,\ldots, \sim xn\}$, where each $m \times 1$ vector $\sim xi$ is the distribution of the values of ith feature irrespective of the sample, a directed graph G is made where features are nodes and relationships among nodes are edges [27]. An adjacency matrix A is taken representing the nature of weighted edges: each element aij of A, $1 \leq i, j \leq n$, models pairwise relationships between the features. Each weight represents the likelihood that features $\sim xi$ and $\sim xj$ are good candidates. Weights are associated as binary function of the graph nodes in Eq. (2):

$$aij = \phi(\sim xi, \sim xj), \quad (2)$$

where $\phi(\cdot,\cdot)$ is a real-valued potential function. The learning framework models the probability of each co-occurrence in $\sim xi$, $\sim xj$ as a mixture of conditionally independent multinomial distributions, Given the weighted graph G, the proposed approach analyses subsets of features as paths connecting them. The cost is given by the joint probability for each path. It evaluates in the relevance of each feature with others. This approach is called Infinite Latent Feature Selection (ILFS).

(4) **Unsupervised Discriminative Feature Selection (UDFS)**: Features are selected according to labels of the training data. But discriminative information is enclosed in labels and supervised feature selection is able to select discriminative features. In unsupervised learning, there is no label information directly available, making it much more difficult to select the discriminative features. A frequently used criterion in unsupervised learning is to select the features

which best preserve the data similarity or manifold structure derived from the whole feature set. However, discriminative information is neglected though it has been demonstrated important in data analysis [28]. Unsupervised discriminative feature selection (UDFS) aims to select the discriminative features for data representation, where manifold structure is considered, making it different from the existing unsupervised feature selection algorithms.

(5) **Laplacian score**: It is a recently proposed feature selection method, which can be used in either supervised or unsupervised scenarios. It is based upon the two data points are close to each other by the observation. Laplacian score has been proved effective and efficient compared with data variance and Fisher score. It constructs a nearest neighbour graph to model the local structure and then selects features which best represents the graph structure [29].

(6) **Multi-Cluster Feature Selection** (**MCFS**): Multi-Cluster Feature Selection (MCFS) proposes to select multi-cluster structure to measure the correlation between different features without label information [30]. The MCFS uses a sparse Eigen-problem and a L1-regularized least square for solving optimization problem.

4.2 Wrapper Methods

The feature subset selected based on the performance of algorithm of clustering or classification is totally ignored in filter approach. The best feature subset depends on the learning algorithms heuristics and specific unbiased nature. Based on this statement, wrapper models estimate the quality of the selected features using specific learning algorithm [31]. Figure 3 shows general framework of wrapper model with predefined learning algorithm. A set of features is produced by feature search component based on definite search schemes. The performance is evaluated by the feature evaluation component using learning algorithms which are predefined and returned for the next iteration of feature subset selection for the feature search component. The feature set selected is based on the best performance of the final set. O (2 m) is m features of the search space. The different algorithms in wrapper approach are used in hill-climbing, branch-and-bound, best-first, genetic algorithms and so on.

Wrapper method has a few algorithms such as:

(1) **Feature Selection via concave minimization**(**FSV**): The classifiers are obtained solving FSV with feature suppression. When the distance between the 2 parallel planes defining the separating surface in the SVM problem is chosen to be the 1-norm, the resulting SVM optimization problem has the 1-norm appearing in the objective. The classifiers obtained by solving this problem did not exhibit feature selection. Similar behaviour was observed for classifiers obtained by solving the SVM 2-norm problem. In supervised machine learning method, the induction algorithm is considered as a black box and it is run on data set [32]. From the different sets of features, these data sets are partitioned into internal

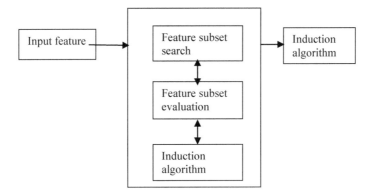

Fig. 3 Wrapper model

training and holdout sets are removed from the data. Then the feature subset will select the final set with highest evaluation. The result classified set is evaluated on an independent test set which was not used during the search. Some common wrapper methods include forward selection, backward elimination and recursive elimination.

(2) **Forward feature selection**: The model with no feature starts with this iterative method among the data set in the model. For each iteration, features are added to improve the model with new variable.

(3) **Backward feature elimination**: The model with all the features is added in the beginning and starts to remove the least significant feature during each iteration until no improvement is observed.

(4) **Recursive feature elimination**: This optimization algorithm finds the best performing feature subset. Unlike other methods, this approach constructs new model by repeatedly creating new model.

By the variation of the subset generation (X_g) and subset evaluation measure A, different wrapper algorithm can be created. An ideal subset selected for learning algorithm by the wrapper approach. Therefore, the wrapper approach performance is usually better. The wrapper algorithm is shown in Table 2.

4.3 Embedded Methods

Embedded models use the characteristics of two models where the model construction has feature selection embedded in it [33]. Thus, both filter and wrapper models advantage is taken by embedded models. Embedded models are less rigorous than wrapper methods, since features are not assessed many times by the learning models, and also include learning model interactions. The wrapper models use the candidate features to first train learning models and then the learning model uses features by

Table 2 Wrapper algorithm [20]

INPUT:
D={X,L}   // a training data set with n number of
//features where X = {$f_1, f_2, f_3, \ldots f_m$}
//and L labels
X^1 // predefined initial feature subset
//(X^1 & X or $X^1 = \{\Phi\}$)
Θ // a stopping criterion
OUTPUT: X^1_{opt} // an optimal subset
Begin:
Initialize:
$X_{opt} = X^1$;
$T_{opt} = E(X^1, A)$; // evaluate X^1 by using mining
//algorithm A
do begin
X_g = generate(X);   // Subset generation for evaluation
$T = E(X_g, A)$;   // X_g current subset evaluation by A
If ($T > T_{opt}$)
$T_{opt} = T$;
$X^1_{opt} = X_g$;
repeat (until Θ is not reached);
end
return X^1_{opt};
end;

feature selection, while features are selected during model construction process by embedded models without additional evaluation of the features to perform feature selection. The KP-SVM is an example for embedded model.

Table 3 shows a general algorithm for embedded model. An empty set X^1 is taken using sequential forward selection. All possible subsets of cardinality k + 1 are searched by adding a feature from the remaining subsets. The independent criterion l_m assesses subset which was created at cardinality k + 1 and compared with the previous ideal subset. The current optimal subset uses learning algorithm A, and performance δ is compared with the performance of the ideal subset at cardinality k. A final ideal subset is returned after stopping criterion.

Embedded methods also have a few algorithms such as:

(1) *Recursive feature elimination and support vector machines* (*RFE-SVM*): This algorithm is a wrapper feature selection method which uses backward feature elimination to generate the ranking of features. Basically, the algorithm was purposed to perform gene selection for cancer classification which aims to eliminate redundant genes and yield more compact gene subsets [34]. Elimination of the feature according to a criterion related to their support by the discrimination function, and the SVM is re-trained at each step. RFE-SVM is a weight-based method in which at each step the coefficients of the weight vector of a linear SVM are used as the feature ranking criterion.

Table 3 Embedded algorithm [20]

INPUT:	
D={X,L}	// a training data set with n number of features where // X = {$f_1, f_2, f_3, \ldots f_m$} and L labels
X^1	// predefined initial feature subset (X^1 & X or $X^1 = \{\Phi\}$)
Θ	// a stopping criterion
OUTPUT: X^1_{opt}	// an optimal subset

```
Begin:
Initialize:
X_opt = X^1;
T_opt = E(X^1, l_m);    // evaluate X^1 by using independent evaluation
                           /measure
δ_opt = E(X^1, A);      // evaluate X^1 by using mining algorithm A
C_0 = C(X^1);           //cardinality calculation of X^1
do begin
for k=C_0+1 to n
for i=0 to n-k
X_g = X_opt U {f_i}   // Subset generation for evaluation with cardinality k
T = E(X_g, l_m);        //evaluation the current subset X_g by l_m
If (T > T_opt)
T_opt = T;
X^1_opt = X_g;
end;
δ = E(X^1_opt, A);    //evaluating subset X^1_opt by A learning algorithm
If (δ > δ_opt)
X^1_opt = X_g;
δ_opt = δ;
else
break and return X^1_opt
end
return X^1_opt;
end;
```

(2) **Lasso regression**: This performs L1 regularization which adds penalty equivalent to absolute value of the magnitude of coefficients.

(3) **Ridge regression**: This performs L2 regularization which adds penalty equivalent to square of the magnitude of coefficients.

4.4 Wrapper Models Comparison with Other Models

Filter models are effective, but the learning algorithm's unbiased nature is totally ignored. The predictive accuracy estimates are better obtained by wrapper models compared to filter models, since they don't ignore the unbiased nature of the learning algorithms. However, wrapper models are very expensive. The algorithm such as classification or learning is included by the wrapper approaches in the step of feature subset evaluation. Whereas, a filter feature selection process is not dependent on any classification algorithm. There are different classification algorithms in wrapper approaches, which are used to assess the goodness of the selected features, e.g. K-Nearest Neighbors (KNN), Support Vector Machines (SVM), Artificial Neural Networks (ANNs), Decision Tree (DT), Naive Bayes (NB), multiple linear regression for classification, discriminant analysis and extreme learning machines (ELMs).

In wrapper model, there are different algorithms used for feature selection such as algorithms of sequential selection and heuristic search. In sequential feature selection algorithms [35–37], it takes set which is empty and a feature is added at the first step which provides the maximum value for the objective function. The features which are remaining are added discretely after the first step to the current subset and the different subset is evaluated. The individual features are permanently included in the subset where features give maximum classification accuracy. The process is repeated until we get required number of features. This algorithm is called a naive SFS algorithm since the dependency between the features is not taken into consideration. A sequential backward selection (SBS) algorithm is reverse of SFS algorithm [38, 39]. The algorithm starts from the entire set of variables and removes one irrelevant feature at a time whose removal gives the lowest decrease in predictor performance. The sequential floating forward selection (SFFS) algorithm introduces step called backtracking which is more adaptable than the simple SFS. The initial step is similar to SFS algorithm which adds one feature at a time based on the objective function. SFFS algorithm is applied in SBS algorithm where one feature is eliminated from the subset obtained in the first step and assesses the different subsets. The removal of feature increases the subset value and algorithm returns to first step with less subset value, else the algorithm is repeated. The whole process is repeated until the required numbers of features are obtained or required performance is reached.

The Heuristic search algorithms include genetic algorithms (GA) [40, 41], Ant Colony Optimization (ACO) [42, 43] and Particle Swarm Optimization (PSO) [44, 45]. A genetic algorithm is a search technique to find true or approximate solution to optimize and search problems used in computing. ACO is based on finding the shortest paths by real ants in their search for food sources. ACO approaches suffer from insufficient rules of pheromone update and heuristic information. PSO approach does not employ crossover and mutation operators, hence it is efficient over GA but requires several mathematical operators which may require various user-specified parameters and deciding their optimal values might be challenging for users. Though ACO and PSO algorithms perform almost identically to GA, GA has received much consideration due to its simplicity and powerful search capability upon the exponential search spaces.

5 Genetic Algorithms

Genetic algorithm (GA) represents field of study called evolutionary computation [46]. A genetic algorithm is a method of solving problems with valuable solutions.

In this method, the generation of some random solutions contains some properties for each solution. It is a probabilistic search algorithm that transforms a set of objects repeatedly, each associated with fitness value, into new offspring population objects using the Darwinian principle of natural selection and operations that imitates naturally occurring genetic operations, like crossover and mutation.

Genetic algorithms have a few basic components such as fitness function, chromosomes population, chromosomes selection, crossover operator and mutation operator.

5.1 A Fitness Function for Optimization

The algorithm is optimized using a function called fitness function [47]. The word "fitness" is taken from evolutionary theory. Each potential solution is tested and quantified how "fit" they are by fitness function. The fitness function is one of the most essential parts of the algorithm.

5.2 A Population of Chromosomes

The chromosome is referred as numerical value or values that signify a candidate solution solved by the genetic algorithm for a problem [47]. Individual candidate solution is encoded as parameter values array [48]. If a problem has N par dimensions (N parameters), then each chromosome is encoded as an N par-element array in Eq. (3)

$$\text{Chromosome} = [p_1, p_2, \ldots, p_{N\text{par}}] \quad (3)$$

where each p_i is a particular value of the ith parameter [40]. Chromosomes are translated from candidate solutions by the genetic algorithm. The parameter value is converted from the sequence of 1's and 0's and then the parameters are concatenated like genes in end-to-end manner to create chromosomes [47]. A collection of chromosomes is chosen randomly by genetic algorithm, which takes it as the first generation. Then chromosome is tested individually using the fitness function in the population to analyse how well it can resolve a problem.

5.3 Selection of Which Chromosomes Will Reproduce

Some of the chromosomes are selected by selection operator based on probability distribution for reproduction. They are selected if the chromosome seems fit. The selection operator chooses chromosomes with replacement, so the same chromosome can be chosen more than once.

5.4 Crossover to Produce Next Generation of Chromosomes

This operator is identical to natural crossing over and chromosomes recombination which occur in cell meiosis. The two chosen chromosomes are exchanged by the operator to create two offspring. A second generation of individuals resulted in crossover process in chromosomes with more diverse properties [49]. When two chromosomes and crossover point are selected, chromosomes individually have exchange values close to crossover point [49].

5.5 Random Mutation of Chromosomes in New Generation

This operator flips individual bits randomly in the new chromosomes. Mutation occurs typically with a very low probability. The mutation operator is usually applied before the selection and crossover operators by some algorithms based on the matter of preference. Finding the global peak before reaching local optimal may be difficult for the algorithm [50]. This problem can be fixed by mutation operator by keeping variety in the population, but it may make the algorithm work slowly.

These selection, crossover and mutation process are continued until the initial population is equal to number of offspring, so that the second generation replaces first generation and creates new offspring. The fitness function tests the second generation, and the cycle repeats. The maximum fitness chromosomes are noted along with its fitness value for individual generation, or the "best-so-far" chromosome [51]. Successor hypotheses created Genetic Algorithm by repeatedly mutating and recombining parts of the best currently known hypotheses. The current population is simplified at each step by offspring of the most fit current hypotheses by substituting some portion of the population. It forms a generate-and-test beam-search of hypotheses, in which the best current hypotheses variants are to be considered next Fig. 4.

A new offspring is produced by two newly generated chromosomes. The crossover process is repeated until the desired variety of individuals is made. A new configuration creates mutation using random changes in dissimilar chromosomes. The reproduction possibility is dependent upon individuals' fitness. The chromosomes with better characteristics are chosen for breeding next. The fitness values which are assigned to individuals are based on a fitness function with the fittest one. Genetic alterations occur to produce another generation in chromosomes using crossover and mutations. This process is iterated until the fittest is found or the determined number of generations is met. GAs are not identical to derivative-based, optimization algorithms. Genetic algorithms are more worked on wrappers than filters.

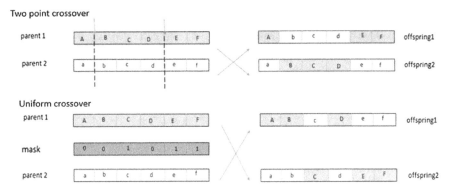

Fig. 4 Crossover process

One group of individuals in generation n is transformed by genetic algorithm into another group of individuals in generation $n + 1$. Usually, the individuals count in each generation is identical. Usually, duplication is not allowed in the 1st or 0th generation. Duplications are allowed in later generations. Some GA implementations are not generational, but adiabatic. In these implementations, a new individual is created, an old one is killed keeping the population size identical. Genetic algorithm performance depends on transformation of chromosomes from candidate solutions with certain criterion for success, or which fitness function it measures [50]. The performance of algorithm can also be increased by adjusting number crossover, number mutation, size of population and number of iterations.

Genetic algorithms are used in various applications. Some examples are automatic programming and machine learning. They are also well suited to modelling phenomena in economics, ecology, the human immune system, population genetics and social systems.

Table 4 describes a typical genetic algorithm. The inputs consist of the fitness function to rank candidate hypotheses, a threshold to define an acceptable level of fitness for terminating the algorithm, the population size to be preserved, and parameters to determine successor populations generation. A genetic algorithm can be seen as a general optimization method that examines a large space of candidate objects that performs best according to the fitness function. GAs often succeed in finding an object with high fitness.

6 Challenges

There are various challenges that occur with feature selection such as:

(1) *Scalability*: It is the main issue in feature selection algorithm, especially for online classifiers with increasing data set size. The memory cannot load large data with a single scan. But, for some feature selection, full dimensionality data

Table 4 Genetic algorithm

Genetic algorithm (Fitness, Fitness_threshold, p, r, m)

Fitness: A function that assigns an evaluation score, given a hypothesis. $F(h_i)$
Fitness threshold: A threshold specifying the termination criterion.
p: The number of hypotheses to be included in the population.
r: The fraction of the population to be replaced by Crossover at each step.
m: The mutation rate.

Initialize population: P ← Generate p hypotheses at random
Evaluate: For each h in P, compute Fitness(h)'
While [max Fitness(h)] < Fitnessdhreshold do
Create a new generation, P_s:
1. Select: Probabilistically select (1–r) p members of P to add to P_s. The probability $P_r(h_i)$ of selecting hypothesis h_i from P is given by
$$P_r(h_i) = \frac{f(h_i)}{\sum_{j=1}^{p} f(h_j)}$$
2. Crossover: Probabilistically select $\frac{r-p}{2}$ pairs of hypotheses from P, according to $P_r(h_i)$ given above. For each pair, (h_1, h_2), produce two offspring by applying the Crossover operator. Add all offspring to P_i
3. Mutate: Choose m percent of the members of P_s, with uniform probability. For each, invert one randomly selected bit in its representation.
4. Update: P ← P_s.
5. Evaluate: for each h in P, compute Fitness(h) Return the hypothesis from P that has the highest fitness

must be scanned. The feature relevance score is difficult to observe due to some density around sample.

(2) **Stability**: Using feature selection, it helps domain experts to identify similar or least related gene sets when new samples are obtained with small amount of perturbation. However, feature selection methods can select features with low stability after perturbation in training data. It is a major issue to develop accurate and stable algorithms for feature selection.

(3) **Parameter Selection**: The number of features which to be selected, are specified for feature selection. However, several features for the data set are unidentified. If too less or more features are selected, the performance may deteriorate, due to elimination of relevant features or selecting irrelevant or redundant features, respectively.

7 Experimental Setup

See Fig. 5

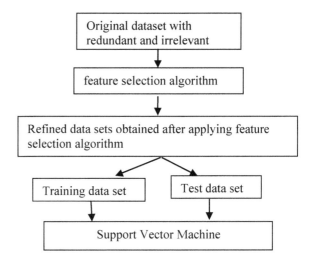

Fig. 5 Experimental setup

8 Results and Discussion

In this section, we present the results and details of experimentation parameters. We have used 5 benchmark data sets taken from the University of California, Irvin (UCI) machine learning repository [52]. The detailed description of data sets is presented in Table 5.

We have considered 11 feature selection for experimental and analysis of their results. These feature selection algorithms are ILFS [27], Fsv [32], Laplacian [29], Mcfs [30], Rfe [34], Fisher [5], UDFS [28], CFS [26], L0, mutinffs and llcfs. The details of these feature selection algorithms are presented in Sect. 3. Support vector machine is used as classifier to build a hypothesis by using data sets which include relevant features selected obtained from feature selection algorithm and samples from original data set. For the evaluation performance of all the methods, overall accuracy is considered as performance measure. A data set which includes relevant feature obtained from feature selection algorithm along with sample is given to support vector machine to build a hypothesis. The feature selection for which support vector

Table 5 Statistics of data sets

Sl No	Data set	No. of sample	No. of features	Class
1	Sonar	208	61	Binary
2	Ionosphere	351	35	Binary
3	Dermatology	366	33	Binary
4	Leukaemia	72	7130	Binary
5	Central nervous system (CNS)	60	7130	Binary

Table 6 Overall accuracy of SVM for different feature selection algorithms

	Feature selection algorithms	Data sets				
		Sonar	Ionosphere	Dermatology	Leukaemia	Central nervous system (CNS)
1	ILFS	51.22	80.00	73.97	64.29	66.67
2	Mutinffs	68.29	77.14	71.23	64.29	66.67
3	Fsv	65.85	82.86	76.71	85.71	66.67
4	Laplacian	63.41	52.86	72.60	57.14	66.67
5	Mcfs	65.85	75.56	68.49	63.23	65.32
6	Rfe	46.34	69.54	70.46	64.29	58.33
7	L0	68.29	74.81	71.91	64.29	66.67
8	Fisher	65.85	75.71	69.86	64.29	58.33
9	UDFS	56.10	75.71	73.97	62.34	60.89
10	Llcfs	58.54	81.43	73.97	64.29	75.00
11	CFS	65.85	72.86	69.86	63.23	62.36

machine gives maximum accuracy is considered to be good for that particular data set. The detailed understanding of our experimental framework can be understood from below block diagram.

Table 6 shows overall accuracy of support vector machine for feature selection algorithms, ILFS, Fsv, Laplacian, Mcfs, Rfe, Fisher, UDFS, CFS and different data sets.

Table 6 presents the overall accuracy of support vector machine for different feature selection algorithms. From our experimental result, it can be clearly observed that for Sonar data set support vector machine gives maximum accuracy when feature selection is performed by Mutinffs, L0 algorithms respectively, support vector machine gives maximum accuracy for ionosphere, dermatology and leukaemia data sets when Fsv method is applied for feature selection. Similarly for central nervous system (CNS) data set support vector machine gives maximum accuracy when llcfs method is applied for feature selection.

9 Conclusion

This paper describes feature selection and various models specifically filter, wrapper and embedded methods in detail with their pseudo code, advantages and disadvantages. In comparison to other methods, the data set with larger attributes uses the wrapper methods in general indicating considerable improvement in accuracy. Hence wrapper methods are popularly used than other methods. One such algorithm introduced in this paper is genetic algorithm which is heuristic-based algorithm under wrapper model. It also provides insights into various applications mainly in the field of medicine to which machine learning could be applied. Various feature selection algorithm under filter, wrapper and embedded techniques were executed on multiple data sets and the results were verified. The accuracy of SVM after feature selection was found to be high for mutinffs, llcfs and fsv techniques. Various applications and challenges of feature selection algorithms are also explained in this paper.

References

1. Gheyas., & Smith, L. S. (2010). Feature subset selection in large dimensionality domains. *PatternRecognition, 43*(1), 5–13.
2. Xue, B., Zhang, M. J., & Browne, W. N. (2013). Particle swarm optimization for feature selection in classification: A multi-objective approach. *IEEE Transactions Cybernetics, 43*(6), 1656–1671.
3. Yang, K., Cai, Z., Li, J., & Lin, G. (2006). A stable gene selection in microarray data analysis. *BMC Bioinformatics, 7*(1), 228.
4. Liu, H., Motoda, H., Setiono, R., & Zhao, Z. (2010). Feature selection: An ever evolving frontier in data mining. *Journal of Machine. Learning. Research Proceeding Track, 10,* 4–13.
5. PEHRO, D., & Stork, D. G. (2001). *Pattern classification.* Wiley-Interscience Publication.
6. Bo, T., & Jonassen, I. (2002). New feature subset selection procedures for classification of expression profiles. *Genome Biology, 3*(4), 1–0017.
7. Xu, R., Damelin, S., Nadler, B., & Wunsch, D. C., II. (2010). Clustering of high-dimensional gene expression data with feature filtering methods and diffusion maps. *Artificial Intelligence in Medicine, 48*(2/3), 91–98.
8. Bandyopadhyay, S., Mukhopadhyay, A., & Maulik, U. (2007). An improved algorithm for clustering gene expression data. *Bioinformatics, 23*(21), 2859–2865.
9. Maulik, U. (2011). Analysis of gene microarray data in a soft computing framework. *Applied Soft Computing, 11*(6), 4152–4160.
10. Ahmed, S., Zhang, M., & Peng, L. (2013). Enhanced feature selection for biomarker discovery in LC-MS data using GP. In *Proceedings of the 2013 IEEE Congress Evolutionary Computation (CEC)* (pp. 584–591). Cancún, Mexico.
11. Derrac, J., Garcia, S., & Herrera, F (2009). A first study on the use of coevolutionary algorithms for instance and feature selection. In *Hybrid artificial intelligence systems (LNCS 5572)* (pp. 557–564). Berlin, Germany: Springer.
12. Li, Y., Zhang, S., & Zeng, X. (2009). Research of multi-population agent genetic algorithm for feature selection. *Expert Systems with Applications, 36*(9), 11570–11581.
13. Mao, Q., & Tsang, I. W.-H. (2013). A feature selection method for multivariate performance measures. *IEEE Transactions on Pattern Analysis and Machine Intelligence, 35*(9), 2051–2063.
14. Venkatraman, V., Dalby, A. R., & Yang, Z. R. (2004). Evaluation of mutual information and genetic programming for feature selection in QSAR. *Journal of Chemical Information and Computer Sciences, 44*(5), 1686–1692.
15. Min, F., Hu, Q., & Zhu, W. (2014). Feature selection with test cost constraint. *International Journal of Appropriate Reasoning, 55*(1), 167–179.
16. Jeong, Y.-S., Shin, K. S., & Jeong, M. K. (2014). An evolutionary algorithm with the partial sequential forward floating search mutation for largescale feature selection problems. *Journal of the Operational Research Society, 66*(4), 529–538.
17. Wang, S., Pedrycz, W., Zhu, Q., & Zhu, W. (2015). Subspace learning for unsupervised feature selection via matrix factorization. *Pattern Recognition, 48*(1), 10–19.

18. Lane, M. C., Xue, B., Liu, I., & Zhang, M. (2013). Particle swarm optimisation and statistical clustering for feature selection. In *Advances in artificial intelligence (LNCS 8272)*. (pp. 214–220). Cham, Switzerland: Springer.
19. Lane, M. C., Xue, B., Liu. I., & Zhang, M. (2014). Gaussian based particle swarm optimisation and statistical clustering for feature selection. In *Evolutionary computation in combinatorial optimisation (LNCS 8600)*. (pp. 133–144). Berlin, Germany: Springer.
20. Ke, L., Feng, Z., Xu, Z., Shang, K., & Wang, Y. (2010). A multiobjective ACO algorithm for rough feature selection. In *Proceedings of the. second Pacific Asia Conference Circuits Communications and System (PACCS)* (vol. 1, pp. 207–210). Beijing, China.
21. Ghaheri, A., Shoar, S., Naderan, M., & Hoseini, S. S. (2015). November). The applications of genetic algorithms in medicine. *Oman Medical Journal, 30*(6), 406–416.
22. Hauskrecht, M., Pelikan, R., Valko, M., Lyons-Weiler, J. (2007). Feature selection and dimensionality reduction in genomics and proteomics. In *Fundamentals of data mining in genomics and proteomics*. (pp. 149–172).
23. Rui, Y., Huang, T. S., Chang, S. (1999). Image retrieval: Current techniques, promising directions and open issues. *Journal of Visual Communication and Image Representation, 10*(4), 39–62.
24. Liu, H., & Motoda, H. (2007). *Computational methods of feature selection*. Chapman and Hall/CRC Press.
25. Uysal, A. K., & Gunal, S. (2012). A novel probabilistic feature selection method for text classification. *Knowledge-Based Systems*, 36, 226–235.
26. Doshi, M., & Chaturvedi, D. S. K. (2014). Correlation based feature selection (cfs) technique to predict student perfromance. *International Journal of Computer Networks & Communications (IJCNC), 6*(3).
27. Roffo, G., Melzi, S., & Cristani,M. (2015). Infinite feature selection. In *EEE International Conference on Computer Vision*, (pp. 4202–4210).
28. Yang, Y., Shen, H. T., Ma, Z., Huang, Z., & Zhou, X. l2,1-Norm regularized discriminative feature selection for unsupervised learning. In *Proceedings of the Twenty-Second International Joint Conference on Artificial Intelligence*.
29. Xu, J., & Man, H. (2011). Dictionary learning based on laplacian score ins coding. In P. Perner (Ed.), *MLDM 2011* (Vol. 6871, pp. 253–264). LNCS (LNAI) Heidelberg: Springer.
30. Cai, D., Zhang, C., & He, X. (2010). Unsupervised feature selection for multi-cluster data. *KDD*.
31. Kohavi, R., & John, G. H. (1997). Wrappers for feature subset selection. *Artificial intelligence, 97*(1), 273–324.
32. Bradley, P. S., & Mangasarian, O. L. (2010). Feature slection via concave minimization and support vector machines.
33. Kabir, M. M., Islam, M. M., & Murase, K. (2011). A new local search based hybrid genetic algorithm for feature selection. *Neurocomputing, 74*, 2194–2928.
34. Sam, M. L., Camara, F., Ndiaye, S., Slimani, Y., & Esseghir, M. A. (2012 June). A Novel RFE-SVM-based Feature Selection Approach for Classification. *International Journal of Advanced Science and Technology*, 43.
35. Guan, S., Liu, J., & Qi, Y. (2004). An incremental approach to contribution-based feature selection. *Journal of Intelligence Systems, 13*(1).
36. Kabir, M. M., Islam, M. M., & Murase, K. (2008). A new wrapper feature selection approach using neural network. In *Proceedings of the Joint Fourth International Conference on Soft Computing and Intelligent Systems and Ninth International Symposium on Advanced Intelligent Systems (SCIS&ISIS2008)* (pp. 1953–1958). Japan.
37. Kabir, M. M., Islam, M. M., & Murase, K. (2010). A new wrapper feature selection approach using neural network. *Neurocomputing, 73*, 3273–3283.
38. Gasca, E., Sanchez, J., & Alonso, R. (2006). Eliminating redundancy and irrelevance using a new MLP-based feature selection method.*Pattern Recognition, 39*, 313–315.
39. Hsu, C., Huang, H., & Schuschel, D. (2002). The ANNIGMAwrapper approach to fast feature selection for neural nets. *IEEE Transaction son Systems, Man, and Cybernetics—Part B:Cybernetics, 32*(2), 207–212.

40. Ghareb, A., Bakar, A., & Hamdan, A. (2015). Hybrid feature selection based on enhanced genetic algorithm for text categorization. In *Expert systems with applications*. Elsevier.
41. Pedergnan, M. (2013). A novel technique for optimal feature selection in attribute profiles based on genetic algorithms. *IEEE Transactions on Geoscience and Remote Sensing, 51*(6).
42. Sivagaminathan, R. K., & Ramakrishnan, S. (2007). A hybrid approach for feature subset selection using neural networks and ant colony optimization. *Expert systems with applications, 33*, 49–60.
43. Aghdam, M. H., Aghaee, N. G., & Basiri, M. E. (2009). Text feature selection using ant colony optimization. *Expert systems with applications, 36*, 6843–6853.
44. Wang, X., Yang, J., Teng, X., Xia, W., & Jensen, R. (2006). Feature selection based on rough sets and particle swarm optimization. *Pattern Recognition Letters, 28*(4), 459–471.
45. Liu, Z., Liu, S., Liu, L., Sun, J., Peng, X., & Wang, T. (2015). Sentiment recognition of online course reviews using multi-swarm optimization-based selected features. In *Neuro-Computing*. Elsevier.
46. Kinnear, K. E. (1994). A perspective on the work in this book. In K. E. Kinnear (Ed.), *Advances in genetic programming* (pp. 3–17). Cambridge: MIT Press.
47. Mitchell, M. (1995). Genetic algorithms: An overview. *Complexity, 1*(1), 31–39.
48. Haupt, R. L., & Haupt, S. E. (1998). *Practical genetic algorithms*. New York: Wiley Interscience.
49. Hasancebi, O., & Erbatur, F. (2000). Evaluation of crossover techniques in genetic algorithm based optimum structural design. *Computer & Structures, 78*, 435–448.
50. Mitchell, M. (1996). *An Introduction to genetic algorithms*. Cambridge: MIT Press.
51. Koza, J. R. (1994). Introduction to genetic programming. In K. E. Kinnear (Ed.), *Advances in genetic programming* (pp. 21–41). Cambridge: MIT Press.
52. https://archive.ics.uci.edu/ml, 30th March, 2018.

A Review on Ensembles-Based Approach to Overcome Class Imbalance Problem

Sujit Kumar, J. N. Madhuri and Mausumi Goswami

Abstract Predictive analytics incorporate various statistical techniques from predictive modelling, machine learning and data mining to analyse large database for future prediction. Data mining is a powerful technology to help organization to concentrate on most important data by extracting useful information from large database. With the improvement in technology day by day large amount of data are collected in raw form and as a result necessity of using data mining techniques in various domains are increasing. Class imbalance is an open challenge problem in data mining and machine learning. It occurs due to imbalanced data set. A data set is considered as imbalanced when a data set contains number of instance in one class vastly outnumber the number of instances in other class. When traditional data mining algorithms trained with imbalanced data sets, it gives suboptimal classification model. Recently class imbalance problem have gain significance attention from data mining and machine learning researcher community due to its presence in many real world problem such as remote-sensing, pollution detection, risk management, fraud detection and medical diagnosis. Several methods have been proposed to overcome the problem of class imbalance problem. In this paper, our goal is to review various methods which are proposed to overcome the effect of imbalance data on classification learning algorithms.

Keywords Class imbalance · Bagging · Boosting · Classification · Ensemble · Sampling · Ensemble approach for class imbalance

1 Introduction

Creating an effective model for classification may become challenging strive if training data set available to train the model are not perfectly balanced. A data set is considered as imbalanced when the number of examples present in one class is vastly outnumber the number of examples present in other class(es). When traditional data

S. Kumar (✉) · J. N. Madhuri · M. Goswami
Department of CSE, Christ (Deemed to Be University), Bengaluru 560029, India
e-mail: kumar.sujit474@gmail.com

© Springer Nature Singapore Pte Ltd. 2019
N. R. Shetty et al. (eds.), *Emerging Research in Computing, Information, Communication and Applications*, Advances in Intelligent Systems and Computing 906,
https://doi.org/10.1007/978-981-13-6001-5_12

mining algorithms trained with imbalanced data sets it gives suboptimal classification model. Such methods would not be effective for classification of instances of minority class, which have very few instances compare to overrepresented class (majority class). Usually, it is the instances of the minority class whose misclassification cost is very high. Therefore, some special methods required in order to make sure that a model will be able to classify these important yet rarely occurring instances.

The first workshop dedicated to the class imbalance problem was held in conjunction with the American association for artificial intelligence conference 2000 (AAAI' 2000) [1]. Recently learning from imbalanced data has achieved significance concern in academia, industry, government funding agencies and many others. The class imbalance problem is quite common in many real-life applications like risk management, fault diagnosis [2–4] anomaly detection [5, 6], fraud detection, learning word pronunciations, face recognition [7], discovery of oil slicks in satellite radar images [8], content characterization, data recovery and sifting undertakings and so forth. The data sets are normally imbalanced (e.g. credit card fraud and uncommon sickness) or the data sets are not normally imbalanced but rather it is excessively costly, making it impossible to get instances of the minority class (e.g. shuttle failure) for learning. In many classification domains, misclassification can be very dangerous, for example, if cancer patient is classified into non-cancer patient by a classification algorithm may leads to death of the patients. The rest of this paper is organized as follows. In Sect. 2, we present different way of dealing with class imbalance problem. In Sect. 3, we present different methods to deal with class imbalance problem at data level. In Sect. 4, we discussed different ensemble-machine learning algorithm. In Sect. 5, we present result and discussion and finally conclusion in Sect. 6.

2 Dealing with Class Imbalance Problem

Several methods have been anticipated to overcome the consequence of imbalanced data on classifiers. Galar et al. [9] provided a brief description of several methods to handle class imbalance problem. These methods can be classified into five categories based on how they deal with class imbalance problem.

   I. Recognition-based approach
  II. Algorithm level approach
 III. Data level approach
 IV. Cost-sensitive learning
  V. Ensemble-learning methods.

2.1 Recognition-Based Approach

A recognition-based [10] or one class learning approach is an alternative solution in which the classifier is trained with the instances of the minority class in the absence of majority class instances. However, recognition-based approach is not applicable for classification algorithms such as decision tree, Naive Bayes and associative classifier. These classifiers cannot be constructed with the instances of one class.

2.2 Algorithm Level Approach

Algorithm level approaches are also called internal approaches to tackle class imbalance problem. Methods at algorithm try to bias the classifier towards minority class [11–13].

A few strategies have been proposed for the treatment of class imbalance issue at Algorithm level. Pazzani et al. [14] proposed a strategy that doles out various weights to the examples of the distinctive classes. Ezawa et al. [15] proposed a technique that tries to inclination the classifier for certain attribute relationships. Kubat et al. [16] have utilized some counter-cases to predisposition the acknowledgement procedure. Barandela et al. [17] proposed a weighted separation work for the k-closest neighbour's classifier. The primary thought behind this weighted separation is to adjust for the unevenness proportion related with the preparation informational collection without really changing the class circulation. Along these lines, weights are appointed to the particular classes as opposed to allotting weights to the individual examples. As the weighting factor is more prominent for minority class and less for greater part class, the separation for occurrences to minority class is lower than the separation to lion's share class. Along these lines, the new protests tend to discover their nearest neighbour among the instances of the minority class. The main drawback associated with this approach is that it requires information of both the relating classifier and the application area, which is effortlessly not accessible in numerous applications.

2.3 Data Level Approach

Data level is another approach to manage class imbalance issue. Data level methods are also called as external level approach because it is independent of classification algorithms. Data level (or external) approach rebalances the class distribution either by adding instances into minority class known as oversampling or removing instances from majority class known as undersampling [18, 19]. Data level approaches avoid the adjustment of classifier by attempting to diminish the impact caused by imbalanced training data set with a pre-processing step. Therefore, data level treatment for class imbalance is independent of classifier because of this it is usually more versatile.

2.4 Cost-Sensitive Learning Approach

Traditional classifiers are built using inductive learning aimed to minimize the expected number of error. Cost-Sensitive learning approach is an extension of the classical inductive learning to balance the misclassification rate. Cost-sensitive learning approach considers costs, such as misclassification costs, test cost, and waiting cost into account during training of classifier and builds a model with lowest cost. In some application domain, this approach is very important, for example, in medical science if a cancer patient is wrongly classified as a cancer-free patient then cost will be very high as it may lead to death of the patient. Several cost-sensitive learning techniques have been proposed to overcome this class imbalance problem.

In case of class imbalance treatment, cost-sensitive learning approach falls between data level and algorithm level approach. At data level, it adds cost to the training examples, and at algorithm level, it modifies the learning process to accept cost [20]. Cost-sensitive learning approach assumes high misclassification cost if instances from minority class are misclassified, hence, it biases the classification algorithms towards minority class. The main drawback associated with cost-sensitive learning is that it needs to identify misclassification costs, which are usually absent from data sets.

2.5 Ensemble-Learning Methods

Ensemble methods construct a set of classifiers which are usually called weak learners. All the classifiers are combined by using majority vote scheme to form a strong classifier. Ensemble-learning method increases the generalization ability of a classifier. The generalization capability of strong classifier will be more compared to each of the base learner which is combined to form strong classifier. In case of treatment to class imbalance problem ensemble-learning methods usually consists of a combination of learning algorithms and one of the techniques above, specially data level and cost sensitive. Several methods have been proposed by combining data pre-processing techniques and ensemble-machine learning algorithms such as RUSBoost, SMOTEBagging, SMOTEBoost, RUSMultiBoost.

3 Data Level Treatment to Class Imbalance Problem

Class imbalance is overcome by applying data pre-processing technique to original training data set to achieve desired balance ratio between majority class and minority class. Data pre-processing is achieved by applying resampling technique. Data pre-processing methods can be divided into two groups: undersampling methods, which remove instances from minority to majority class until the desired balanced ratio between majority class and minority class is achieved, oversampling methods, which add instances into minority class until the desired ratio between majority class and minority class is achieved. The main advantage of resampling technique is that they are independent of classification algorithms.

3.1 Random Oversampling

It is a non-heuristic technique which randomly adds the instances into minority class until and unless the desired ratio between majority and minority class is achieved. Since the number of instances in the sampled training data set is more compared to the original training data set, the training time of the classifier used increases. Since it makes an exact copy of existing instances; the classifier may suffer from over fitting.

3.2 Random Undersampling

Random undersampling is also a non-heuristic method which randomly removes instances from majority class until and unless the desired balance ratio between minority and majority class is achieved. Since the number of instances in sampled training data set is less compared to original training data set, the training time of the classifier used decreases. However, it may remove some of the instances which are important for classification task.

3.3 Synthetic Minority Oversampling Technique (SMOTE)

Synthetic Oversampling Technique (SMOTE) [21] is an intelligent oversampling method which is specially designed for the treatment of imbalanced data problem. The main idea behind SMOTE is to create new minority class instances by interpolating several minority class instances which are the nearest neighbour of each other.

SMOTE method takes the difference between an instance of minority class and one of its nearest neighbours, then multiplies this difference by a random number between 0 and 1 and adds the result to sample space under consideration. Since SMOTE method does not add the same instance from minority class, the overfitting problem is avoided.

Example of generation of synthetic examples (SMOTE)

Let's assume an instance (6,4) under consideration to generate synthetic example. And let (4,3) be one of the nearest neighbours of instance (6,4).

$$x1 = 6 \quad x2 = 4 \quad x2 - x1 = -2$$
$$y1 = 4 \quad y2 = 3 \quad y2 - y1 = -1$$

Then new instances new(x, y) will be generated as follows:
new(x, y) = (6,4) + rand(0−1)* (−2, −1)
rand(0−1) generates a random number between 0 and 1.

3.4 Condensed Nearest Neighbour (CNN)

In 1968, Hart [22] proposed an undersampling method to reduce the size of the training data sets for the nearest neighbour decision which is also called as "The Condensed Nearest Neighbour Rule" (CNN). The main idea behind this method is to select a subset of original training data set. Since CNN uses undersampling which requires many scan over training data set, it is slow compared to other methods. And because of the random selection, some important instances from the majority class may be discarded which may be useful for classification task. The CNN undersampling method works as follow. Initially, let's assume that instances of training data set are arranged in some order. Then we set up bins called STORE and GRABBAG.

Step1: The primary example is put in STORE.
Step2: The following occurrence is characterized by the NN administer, utilizing the present substance of STORE. (Since at first STORE has just a single case, the arrangement will be unimportant at this stage.) If the following example is ordered accurately it is set in GRABBAG else it is set in STORE. It isn't put in STORE in the event that it is grouped accurately in light of the fact that it is expected that next example is same as one of the occasions that has a place with STORE set.
Step3: Proceeding inductively, the ith example is arranged by utilizing cases in current STORE. In the event that characterized accurately, it is put in GRABBAG else it is set in STORE.
Step4: After consummation of one pass on the preparation informational collection the strategy keeps on circling to look over GRABBAG until the end, which can happen in one of two ways:

> Step4.a: The GRABBAG is depleted; with all occasions has a place with GRABBAG exchanged to STORE or on the other hand.
> Step4.b: One complete pass is made through GRABBAG with no transfers to STORE.
> Step5: Finally, STORE set is will be used as new training data set and content of GRABBAG will be removed.

3.5 Edited Nearest Neighbour (ENN)

ENN [23] is an undersampling method in which undersampling on majority class is performed by removing instances whose class label does not match with majority of its k nearest neighbours. That is undersampling is performed on majority class by removing outliers.

3.6 Repeated Edited Nearest Neighbour (RENN)

RENN is also an undersampling method in which edited nearest neighbour (ENN) method is applied repeatedly until ENN cannot remove any further instances from majority class.

3.7 Exploratory Undersampling for Class Imbalance Learning

The key idea behind "exploratory undersampling for class-imbalance learning" is to overcome drawback associated with undersampling methods, i.e. undersampling methods remove the patterns from majority class in order to achieve desire balance ratio between majority class and minority class. Hence, undersampling may cause potential loss of information from majority class.

Liu et al. [24] proposed EasyEnsemble and BalanceCascade which is an alternative to sampling methods. In case of EasyEnsemble, majority class data sets are divided into several subsets, and finally, each subset of majority class is merged with minority class and train an ensemble from each of them with AdaBoost serially. BalanceCascade is supervised strategy where undersampling process is guided by learning algorithm here the idea is to discard the patterns of majority class which are correctly classified by learning algorithm in current iteration.

4 Ensembles Technique for Solving Class Imbalanced Problem

Ensemble methods construct a set of classifiers which are usually called weak learners. All the classifiers are combined by using majority vote scheme to form a strong classifier. Ensemble-learning method increases the generalization ability of a classifier. The generalization capability of strong classifier will be more compared to each of the base learner which is combined to form strong classifier.

In case of treatment to class imbalance problem, ensemble-learning methods usually consist of a combination of learning algorithms and one of the techniques above, specially data level and cost sensitive. Several methods have been proposed by combining data pre-processing techniques and ensemble-machine learning algorithms such as RUSBoost, SMOTEBagging, SMOTEBoost, and RUSMultiBoost the details of these algorithms are presented in Sect. 5.

4.1 Ensemble-Machine Learning Algorithms

Ensemble methods construct a set of classifiers which are usually called weak learners. All the classifiers are combined by using majority voting scheme or through weighted averaging to form a strong classifier. Ensemble-learning method increases the generalization capability of classifier. Ensemble classifiers can be constructed by the following two steps: initially, all the base learners are produced, base learners can be produced either in parallel style or sequentially, and then the base learners are combined through majority voting or through weighted averaging for the prediction of output of new instances. In case of classification, the trained base learners are combined through majority voting and in case of regression, the trained base learners are combined through weighted averaging. There are several algorithms that have been proposed to construct ensemble classifier which can be classified into four categories Boosting, Bagging, Wagging and MultiBoosting.

4.1.1 Boosting

Boosting is an approach to build ensemble-machine learning algorithms; the idea of boosting is to create a highly accurate prediction rule by combining many relatively weak and inaccurate rules. Given a classification problem, the goal, of course, is to generate a rule that makes the most accurate predictions possible on new test samples.

The AdaBoost algorithm by Freund and Schapire [25] was the first practical boosting algorithm and remains one of the most widely used and studied, with applications in numerous fields. Boosting technique uses entire training data set to train each classifier.

AdaBoost

AdaBoost [25, 26] is a standout among the most normally utilized and popular boosting method. AdaBoost utilizes whole informational present in training data sets for training of the model. It gives more consideration regarding erroneously ordered example. After every iteration, it builds the weights of mistakenly ordered example, for amend characterization in the following cycle. The significance of occurrences is estimated by a weight, which is at first equivalent for all examples. The weights of misclassified sample are expanded; actually, the weights of accurately ordered examples are diminished after every iteration. Weight is additionally doled out to every individual classifier (speculation) contingent upon its general precision which is likewise called as certainty of the classifier (theory). A classifier will be doled out more weight in light of high certainty. A classifier is said to be more confident if number of misclassification is less. At long last, every one of the classifiers is consolidated through greater part voting to anticipate the yield of new occasions.

Algorithm1 AdaBoost

Input: Training set S={x_i, y_i}, i=1,....,N; and $y_i \in \{-1, +1\}$;
T:Number of iteration;
I:Weak learner
Output: Boosted classifier: $H(x) = \text{sign}(\sum_{t=1}^{T} \alpha_t h_t(x))$
where h_t, α_t are the induced classifiers (with $h_t(x) \in \{+1, -1\}$) and their assigned weights respectively.

1: $D_1(i) \leftarrow \frac{1}{N}$ for i=1,....,N
2: **for** t = 1 to T **do**
3:   $h_t \leftarrow I(S, D_t)$
4:   $\epsilon_t \leftarrow \sum_{i, y_i \neq h_t(x_i)} D_t(i)$
5:   **if** $\epsilon_t > 0.5$ **then**
6:     T ← t-1
7:     return
8:   **end if**
9:   $\alpha_t = \frac{1}{2} \log(\frac{1-\epsilon_t}{\epsilon_t})$
10:   $D_{t+1}(i) = D_t(i) * e^{(-\alpha_t h_t(x_i) y_i)}$ for i=1,....,N
11:   Normalize D_{t+1} to be a proper distribution
12: **end for**

The above Algorithm 1 shows how the ensemble classifiers can be constructed using AdaBoost algorithm. Given an integer T specifying the number of iteration, T weighted training sets $S1, S2,..., S_T$ are generated in sequence and T classifier $C1, C2,..., C_T$ are built. A final classifier C^* is formed using a majority voting scheme. The weight of each classifier depends on its performance on the training set used to build it. However, there are two extensions of AdaBoost.$M1$ and AdaBoost.$M2$. AdaBoost.$M1$ is the first extension of AdaBoost algorithm to multiclass classification with a different weight changing mechanism. AdaBoost.$M2$ is the second extension of AdaBoost to multiclass classification which makes use of base classifiers confidence rates. It is an established fact that boosting reduces bias and variance [27]. The main drawback of this method is that boosting algorithm does not work well with noisy data sets. For justification, let's assume that in first iteration an noisy instance is misclassified. In the next iteration, AdaBoost algorithm will increase the weight of noisy instance which was misclassified in the previous iteration and it will force the weak learner to learn the noisy instance by assuming the noisy instance as hard example to learn. Moreover, if the noisy instance is misclassified again in any further iteration then the algorithm will increase its weight once again and this process goes on.

4.1.2 Bagging

Breiman [28] presented the idea of bootstrap accumulation to develop groups. An arrangement of bootstrap test is created by consistently inspecting m examples from the preparation set with substitution. T bootstrap tests $B1, B2, \ldots, BT$ are produced, and a classifier is worked from each bootstrap test. A last classifier $C*$ is worked from $C1, C2, \ldots, CT$ whose yield is the class anticipated frequently by its sub classifier, with ties broken discretionarily. Bagging is well known for reducing variance, and bagging is suitable for parallel computation.

Algorithm 2 Bagging

Algorithm 2 Bagging
Input: P: it is set of samples;
  T: It represent number of classifier to be build .
  n: Bootstrap size;
  L: base learner
Output: Bagged classifier: H(x) = sign ($\sum_{t=1}^{T} S_t(x)$))
  Where $S_t \in \{+1,-1\}$ are the individual base learner
1: **for** t = 1 toT **do**
2:      $P_t \leftarrow$ Random Sample Replacement (n, P)
3:      $S_t \leftarrow$ L (P_t)
4: **end for**

4.1.3 MultiBoost

MultiBoosting is an extension of AdaBoost method which forms decision committees. MultiBoost algorithm is proposed to take the advantage of combination of any two algorithms: either AdaBoost with bagging or AdaBoost with bagging. MultiBoost is a way to make AdaBoost suitable for parallel computation. Webb [29] presented MultiBoost which unites of AdaBoost with wagging method. Zijianzaheng presented a study of multiple boost which is combination of AdaBoost and Bagging and is suitable for parallel and distributing computation.

Note that none of the ensemble algorithms discussed above itself deal with class imbalance problem directly. In order to solve class imbalance problem, these ensemble algorithms have to be changed or combined with another techniques. Since the main goal of these ensemble algorithms is to give attention to examples which is hard to learn. Recently ensemble algorithms combined with resampling techniques have gained significant attention in solving class imbalance problem. Various algorithms like SMOTEBoost, RUSBoost, RUSMultiBoost, etc. have been proposed.

4.2 Boosting-Based Ensembles to Overcome Class Imbalance Problem

Boosting-based ensemble is combination of data pre-processing and AdaBoost, depend on different pre-processing technique there different boosting-based ensemble to overcome class imbalance problem.

4.2.1 SMOTEBoost

Chawla et al. [30] proposed SMOTEBoost overcome the class imbalance problem. It is a combination of synthetic minority oversampling (SMOTE) which is specially designed to alleviate class imbalance problem and AdaBoost. Here, the data pre-processing (oversampling of minority class) is done by creating synthetic examples and the oversampled training data set is given to AdaBoost to build a model. The SMOTEBoost take the advantage of both boosting and the SMOTE method. While boosting improves the accuracy of model by giving attention to examples which is hard to learn, SMOTE improves the performance of the model by increasing number of minority class instances in training data set.

The only drawback associated with this method is that SMOTE is complex and time consuming oversampling technique compared to other resampling methods. Therefore, SMOTEBoost which uses SMOTE as oversampling technique is also complex and time consuming compared to other methods such as RUSBoost.

4.2.2 RUSBoost

Seiffert et al. [31] proposed a simple and faster method compared to SMOTEBoost to alleviate class imbalance problem. RUSBoost is a combination of random undersampling and AdaBoost. Random undersampling is applied to majority class examples to achieve desired balance ratio between minority class and majority class and then the balanced data set is given to AdaBoost to build a model. The author used two class distributions 35:65 and 50:50 as balanced ratio between majority class and minority class, where 35:65 means 35% minority class and 65% majority class. The only drawback associated with RUSBoost is that random undersampling technique randomly removes the instances from majority class which may discard instances which may be useful for classification.

4.2.3 EUSBoost

Galar et al. [32] proposed a novel method to alleviate class imbalance problem with the aim to improve the accuracy of classifiers while promoting their diversity. EUSBoost is a combination of evolutionary undersampling and AdaBoost. The working of EUSBoost is same as RUSBoost. The only difference is that it uses evolutionary undersampling instead of random undersampling.

4.2.4 RUSMultiBoost

Mustafa et al. [33] proposed a simple and efficient method to alleviate the class imbalance problem which combines random undersampling with MultiBoost algorithm. It takes the advantage of both Boosting and wagging. Boosting algorithms are known for reducing bias. MultiBoost is a combination of Boosting and wagging which a variation of Bagging. This method is specially developed to reduce the significance loss of information while undersampling. In case of RUSMultiBoost, many subsets of training set are formed from given training data set by using sampling with replacement. Some of the instances may be selected many times while some may not be selected for single time also. A weak learner is trained on each new subset of training data sets, and finally, all of them are combined together by majority voting to form a strong classifier. However, since random undersampling is involved, RUSMultiBoost suffers with loss of important information from majority class during undersampling on majority class.

4.3 Hybrid Ensembles

Hybrid ensembles method combines both boosting and bagging technique. Sometimes both boosting and bagging are also combined with a data pre-processing method.

4.3.1 EasyEnsemble

Liu et al. [24] proposed two methods: EasyEnsemble and BalanceCascade to overcome the drawback of undersampling were many majority class instances are ignored. EasyEnsemble method divides the majority class instances into several. Then it trains the learner with each subset of majority class instances and the minority class instances. Finally, it combines the output of all the learner. This can be mathematically represented as: let's assume that we have N number of instances in majority class then T number of subset will be created from majority class example namely $N1, N2, N3, \ldots, NT$. For each subset, Ni ($1 \leq i \leq T$) a classifier Hi is trained using Ni and P, where P is the Minority class instances. Finally, all generated classifiers are combined to make prediction of new instances.

Table 1 Statistics of data sets

Sl. No.	Data sets	Instances/patterns	Attributes	No. of minority	No. of majority	Imbalance ratio (IR)
1	Blood transfusion	748	5	178	570	3.2022
2	PC1	1109	22	77	1032	13.4025
3	CM1	498	22	49	449	9.1632
4	Mammographic	1035	6	445	590	1.3258
5	Wisconsin	683	11	106	590	5.5660

4.3.2 BalanceCascade

BalanceCascade is relatively similar to EasyEnsemble only difference is that BalanceCascade removes the instances of majority class from further consideration if it is correctly classified by the trained learners. The difference between EasyEnsemble and BalanceCascade is how they treat the majority class after each iteration. BalanceCascade is an supervised methods whereas EasyEnsemble is unsupervised method [24].

5 Results and Discussion

5.1 Data Sets Description

In our study, we have used five binary class data sets taken from the University of California, Irvine (UCI) machine learning repository [34]. The statistics of the data sets are presented in Table 1. The imbalance ratio (IR), i.e. ratio between number of sample in majority class and number of sample in minority class are calculated as $\frac{S_{maj}}{S_{min}}$. Tenfold cross validation is used for performance evaluation.

5.2 Performance Evaluation in Class Imbalanced Domains

Performance evaluation plays a significant role in machine learning. Performance evaluation guide and guide machine learning algorithms. Machine learning algorithms will not be able to handle class imbalance problem if choice of performance evaluation does not value minority class. The most commonly preferred choice for performance evaluation is overall accuracy. However, in case of imbalanced data set overall accuracy is not a suitable choice for performance evaluation of classification algorithm. Consider an example where classification algorithm is trained on

Table 2 Confusion matrix

		Positive	Negative
Prediction class	Positive	True positive (TP)	False positive (FP)
	Negative	False negative (FP)	True negative (TP)

imbalanced data example where and total 100 samples present in test data set and 95 samples belongs to majority, 5 samples belongs to minority class. Classifier classified all the 95 samples from majority class correctly and misclassified all 5 samples of minority class even the overall accuracy will be 95% but the fact it is not able to recognize minority class instances. Therefore, other performance evaluation metrics have been proposed which measure the performance of classifier with the help of confusion matrix (also called as contingency table). The confusion is as follow (Table 2).

There are various evaluation criteria among them, the performance criteria which are mainly used to measure performance in case of class imbalanced are precision-recall, sensitivity, F-measure, specificity and geometric mean.

5.3 Precision, Recall and F-Measure

These performance metrics are used when performance of positive class or minority class are concerned, because precision and recall are defined with respect to minority class.

- Precision of classifier defines as the percentage of positive prediction done by classification algorithm that is correct.

$$\text{Precision} = \frac{\text{TP}}{\text{TP} + \text{FP}}$$

- Recall is defined as percentage of true positive sample that is classified by classifier is correct.

$$\text{Recall} = \frac{\text{TP}}{\text{TP} + \text{FN}}$$

- F-measure is harmonic mean of precision and recall.

$$\text{F-measure} = \frac{2 * \text{Recall} * \text{Precision}}{\text{Recall} + \text{Precision}}$$

5.4 Sensitivity, Specificity and Geometric Mean

These performance measures are used when classification performance of both minority class and majority class is equally concerned and expected to be high. G-mean indicates the balance between the majority and minority class performance. G-mean takes into account both the accuracy of positive class instances (sensitivity) and the accuracy of negative class instances (specificity).

$$\text{Specificity} = 1 - \frac{\text{FP}}{\text{Total Negative}}$$

$$\text{Sensitivity} = \text{Recall}$$

$$G\text{-means} = \sqrt{\text{Sensitivity} * \text{Specificity}}$$

Table 3 present overall accuracy of EasyEnsemble, BalanceCascade, SMOTE-Boost and RUSBoost using decision tree as weak learner. EasyEnsemble outperformed other methods for three data sets and SMOTEBoost outperformed other methods for two data sets out off five data sets when decision tree is used as weak learner. Similarly from Table 4, it is observed that RUSBoost outperformed other methods for three data and SMOTEBoost outperformed other methods in two data sets when KNN is used as weak learner.

Table 5 present F-measure of EasyEnsemble, BalanceCascade, SMOTEBoost and RUSBoost using decision tree as weak learner. EasyEnsemble outperformed other methods for four data sets and RUSBoost outperformed other methods for one data set out off five data sets when decision tree is used as a weak learner. Similarly from Table 6, it is observed that RUSBoost outperformed other methods for two data, EasyEnsemble outperformed other methods for one, SMOTEBoost outperformed other methods in data sets, and for CM1 data set both EasyEnsemble and BalanceCascade gives maximum F-measure compare to other methods when KNN is used as a weak learner.

Table 3 Performance of EasyEnsemble, BalanceCascade, SMOTEBoost, RUSBoost with Decision Tree

	Data sets	EasyEnsemble (avg. accuracy)	BalanceCascade (avg. accuracy)	SMOTEBoost (avg. accuracy)	RUSBoost (avg. accuracy)
1	Blood transfusion	66.4665	62.67	62.3416	65.5452
2	PC1	76.02	77.38	90.7744	72.4222
3	CM1	72.00	60.00	86.7667	69.7083
4	Mammographic	80.73	78.13	77.7914	78.411
5	Wisconsin	77.70	74.82	77.0476	76.342

Table 4 Performance of EasyEnsemble, BalanceCascade, SMOTEBoost, RUSBoost and with KNN

	Data sets	EasyEnsemble (avg. accuracy)	BalanceCascade (avg. accuracy)	SMOTEBoost (avg. accuracy)	RUSBoost (avg. accuracy)
1	Blood transfusion	33.0886	63.8664	57.2749	64.7416
2	PC1	59.2618	58.4795	86.9157	78.7615
3	CM1	61.8667	61.8667	82.1333	75.35
4	Mammographic	59.0833	60.2853	73.7821	77.33
5	Wisconsin	71.3420	71.6277	77.6104	78.1991

Table 5 F-measures of EasyEnsemble, BalanceCascade, SMOTEBoost, RUSBoost with Decision Tree

	Data sets	EasyEnsemble	BalanceCascade	SMOTEBoost	RUSBoost
1	Blood transfusion	0.4848	0.4615	0.2118	0.3366
2	PC1	0.3117	0.3243	0.3185	0.3371
3	CM1	0.3636	0.2308	0.2098	0.2261
4	Mammographic	0.8663	0.7823	0.7312	0.7539
5	Wisconsin	0.5231	0.5205	0.2656	0.4965

Table 6 F-measures of EasyEnsemble, BalanceCascade, SMOTEBoost, RUSBoost with KNN

	Data sets	EasyEnsemble	BalanceCascade	SMOTEBoost	RUSBoost
1	Blood transfusion	0.2470	0.2514	0.2262	0.2529
2	PC1	0.2269	0.2309	0.3306	0.3059
3	CM1	0.2456	0.2456	0.2421	0.2401
4	Mammographic	0.2107	0.3564	0.6677	0.73
5	Wisconsin	0.4669	0.4698	0.3555	0.541

From our experimental result, it can be concluded that EasyEnsemble and SMOTEBoost outperformed other methods in term of both performance measure overall accuracy and F-measure. Performance of EasyEnsemble, BalanceCascade, SMOTEBoost and RUSBoost change there is change in weak learner.

6 Conclusion

In this paper present, review of data level methods and ensemble-based approach to overcome class imbalance problem. Class imbalance is an open challenge problem in data mining and machine learning researcher community. Several techniques have been proposed to overcome the effect imbalance data on classifier which can be classified into four groups as recognition-based approach, algorithm level approach, data level approach, cost-sensitive learning, ensemble-learning methods. We have to remark the good performance of EasyEnsemble and SMOTEBoost over other methods but SMOTEBoost is complex to build a model. RUSBoost is an alternative to SMOTEBoost which performance similar compare to SMOTEBoost; however, sometimes even outperform SMOTEBoost and RUSBoost have an advantage over SMOTEBoost that it is simple to build model.

References

1. Guo, X., Yin, Y., Dong, C., Yang, G., & Zhou, G. (2008). On the class imbalance problem. https://doi.org/10.1109/icnc.2008.871, IEEE.
2. Yang, Z., Tang, W., Shintemirov, A., & Wu, Q. (2009). Association rule mining based dissolved gas analysis for fault diagnosis of power transformers. *IEEE Transactions on Systems, Man, and Cybernetics, Part C, (Applications and Reviews), 39*(6), 597–610.
3. Zhu, Z.-B., & Song, Z.-H. (2010). Fault diagnosis based on imbalance modified kernel fisher discriminant analysis. *Chemical Engineering Research and Design, 88*(8), 936–951.
4. Mazurowski, M. A., Habas, P. A., Zurada, J. M., Lo, J. Y., Baker, J. A., & Tourassi, G. D. (2008). Training neural network classifiers for medical decision making: The effects of imbalanced datasets on classification performance. *Neural Networks, 21*(2–3), 427–436.
5. Khreich, W., Granger, E., Miri, A., & Sabourin, R. (2010). Iterative Boolean combination of classifiers in the roc space: An application to anomaly detection with hmms. *Pattern Recognition, 43*(8), 2732–2752.
6. Tavallaee, M., Stakhanova, N., & Ghorbani, A. (2010). Toward credible evaluation of anomaly-based intrusion-detection methods. *IEEE Transactions on Systems, Man, and Cybernetics, Part C, (Applications and Reviews), 40*(5), 516–524.
7. Liu, Y.-H., & Chen, Y.-T. (2005). Total margin-based adaptive fuzzy support vector machines for multiview face recognition. In *Proceedings of the IEEE International Conference on Systems, Man, Cybernetics*, Vol. 2, pp. 1704–1711.
8. Kubat, M., Holte, R. C., & Matwin, S. (1998). Machine learning for the detection of oil spills in satellite radar images. *Machine Learning, 30,* 195–215.
9. Galar, M., Fernández, A., Barrenechea, E., Bustince, H., & Herrera, F. (2011). A review on ensembles for the class imbalance problem: bagging-, boosting-, and hybrid-based approaches. In *IEEE Transaction on Systems, Man, and Cybernetics-Part C: Application and Review*, IEEE.
10. Nguyen, G. H., Bouzerdoum, A., & Phung, S. (2009). Learning pattern classification tasks with imbalanced data sets. In P. Yin (Ed.), *Pattern recognition* (pp. 193–208).
11. Liu, B., Ma, Y., & Wong, C. (2000). Improving an association rule based classifier. In D. Zighed, J. Komorowski, & J. Zytkow (Eds.), *Principles of data mining and knowledge discovery (Lecture Notes in Computer Science Series 1910)* (pp. 293–317).
12. Lin, Y., Lee, Y., & Wahba, G. (2002). Support vector machines for classification in non standard situations. *Machine Learning, 46,* 191–202.
13. Barandela, R., Sanchez, J. S., García, V., & Rangel, E. (2003). Strategies for learning in class imbalance problems. *Pattern Recognition, 36*(3), 849–851.
14. Pazzani, M., Merz, C., Murphy, P., Ali, K., Hume, T., & Brunk, C. (1994). Reducing misclassification costs. In *Conference on Machine Learning*, pp. 217–225.
15. Ezawa, K.J., Singh, M., & Norton, S.W. (1996). Learning goal oriented Bayesian networks for telecommunications management. In *Proceedings of the 13th International Conference on Machine Learning*, pp. 139–147.

16. Kubat, M., Holte, R., & Matwin, S. (1998). Detection of oil-spills in radar images of sea surface. *Machine Learning, 30,* 195–215.
17. Barandela, R., Sánchez, J. S., García, V., & Rangel, E. (2003). Strategies for learning in class imbalance problems. *Pattern Recognition, 36,* 849–851.
18. Batista, G. E. A. P. A., Prati, R. C., & Monard, M. C. (2004). A study of the behavior of several methods for balancingmachine learning training data. *SIGKDD Explorations Newsletters, 6,* 20–29.
19. Stefanowski, J., & Wilk, S. (2008). Selective pre-processing of imbalanced data for improving classification performance. In I.-Y. Song, J. Eder, & T. Nguyen, (Eds.), *Data Warehousing and Knowledge Discovery (Lecture Notes in Computer Science Series 5182),* pp. 283–292.
20. Zhang, S., Liu, L., Zhu, X., & Zhang, C. (2008). A strategy for attributes selection in cost sensitive decision trees induction. In *Proceedings of the IEEE 8th International Conference on Computer and Information Technology Workshops,* pp. 8–13.
21. Chawla, N. V., Bowyer, K. W., Hall, L. O., & Kegelmeyer, W. P. (2002). SMOTE: Synthetic minority over-sampling technique. *Journal of Artificial Intelligence Research, 16,* 321–357.
22. Hart, P. E. (1968). The condensed nearest neighbour rule. *IEEE Transactions on Information Theory, 14*(3), 515–516.
23. Wilson, D. L. (1972). Asymptotic properties of nearest neighbor rules using edited data. *IEEE Transactions on Systems, Man, and Cybernetics, 3,* 408–421.
24. Liu, X.-Y., Wu, J., & Zhou, Z.-H.: Exploratory undersampling for class imbalance learning. *IEEE Transactions on Systems, Man, and Cybernetics Part B, Application Review, 39*(2), 539–550.
25. Freund, Y., & Schapire, R. E. (1996). Experiments with a new boosting algorithm. In *Machine Learning: Proceedings of the Thirteenth International Conference.*
26. Cao, D. S., Xu, Q. S., Liang, Y.-Z., Zhang, L.-X., & Li, H.-D. (2010). The boosting: A new idea of building models. *Chemometrics and Intelligent Laboratory Systems, 100,* 1–11.
27. Bauer, E., & Kohavi, R. (1998). An empirical comparison of voting classification algorithms: Bagging, boosting, and variants. In Machine Learning, vv, 1, Kluwer Academic Publishers, Boston. Manufactured in The Netherlands.
28. Breiman, L. (1996). Bagging predictors. *Machine Learning, 24,* 123–140.
29. Webb, G. I. (2000). MultiBoosting: A technique for combining boosting and wagging. *Machine Learning, 40,* 159–196, Kluwer Academic Publishers, Boston.
30. Chawla, N.V., Lazarevic, A., Hall, L. O., & Bowyer, K. W. (2003). SMOTEBoost: Improving prediction of the minority class in boosting. In *Proceedings of the Knowledge Discovery Databases,* pp. 107–119.
31. Seiffert, C., Khoshgoftaar, T., Van Hulse, J., & Napolitano, A. (2010). Rusboost: A hybrid approach to alleviating class imbalance. *IEEE Transactions on Systems, Man, and Cybernetics Part A, Systems, and Humans, 40*(1), 185–197.
32. Krawczyka, B., Galar, M., Jelen, Ł., & Herrera, F. (2016). Evolutionary undersampling boosting for imbalanced classification of breast cancer malignancy. *Applied Soft Computing, Elsevier, 38,* 714–726.
33. Mustafa, G., Niu, Z., Yousif, A., & Tarus, J. (2015). Solving the class imbalance problems using RUSMultiBoost ensemble. In *10th Iberian Conference on Information Systems and Technologies (CISTI),* IEEE, Aveiro, Portugal.
34. https://archive.ics.uci.edu/ml, March 30, 2018.

"College Explorer" An Authentication Guaranteed Information Display and Management System

Sonali Majumdar, K. M. Monika Patel, Arushi Gupta and M. N. Thippeswamy

Abstract The confusion and dilemma that arises out of the unorganized plethora of information on the Internet can never help school pass outs to reach any conclusion of which college to join for higher studies. Apart from college information, internal environment and feedback from students currently studying in the college is essential part to know about an institution and its administration. This paper stresses about the need for a Web application like college explorer through the novel contributions, system model, and advantages of the Web application developed. It enables the general public to view information such as placement details, admission details, course details, etc., about the colleges. In addition, the features like class notes sharing, notice publication (separate for students as well as for faculties of respective departments of the college), application of leave facility for both faculties as well as students are developed to manage the leave application even in emergency cases for smooth internal administration. The results confirm that Web applications have the potential to address various problem statements stated using Web technology efficiently.

Keywords Web technology · Web application authentication · Information system · JDBC · MVC architecture · JSP · Servlets

1 Introduction

During this era of competition where it is tough to secure a seat in reputed institution due to the high number of students writing entrance examination for admissions, unselected students tend to find alternative institution to secure seat in the course of their choice after graduating from pre-university or high school. But the problem here is which institution to prefer.

S. Majumdar (✉) · K. M. M. Patel · A. Gupta · M. N. Thippeswamy
Department of CSE, Nitte Meenakshi Institute of Technology, Yelahanka, Bengaluru 560064, India
e-mail: sonalimajumdar80@gmail.com

© Springer Nature Singapore Pte Ltd. 2019
N. R. Shetty et al. (eds.), *Emerging Research in Computing, Information, Communication and Applications*, Advances in Intelligent Systems and Computing 906, https://doi.org/10.1007/978-981-13-6001-5_13

Following are the four problem statements for which we propose a single system of information display and management in an organized manner:

1.1. Entrance exams in various fields are written by lakhs of students every year with limited seats in best institutes/colleges. How will the remaining students decide, which college suites their requirements from pool of unreliable sources of information around them [1]. The information may include placement details, transport facilities, internal college life, and so on.
1.2. Students in college find it difficult to study from plethora of information provided through multiple books for certain subject as a result they have to visit library frequently, search and learn the topic there within limited time [2]. Not all students can study and understand content of book written by various domain experts and thus need simplification.
1.3. In cases of emergency leaves, the faculty and students need to take permission personally from higher authorities such as Head of Department (H.O.D.) which sometimes may not be possible due to circumstances.
1.4. Lecturers find it difficult to communicate and coordinate an event in college at short notice(s) for administrative purpose. Notices are published inside institutes using old methods like attaching notice sheet on notice board in college premises [1].

Therefore, there is a need for college portals which can provide smooth internal administration management for the college and also the one which can provide updated and authenticated information about the college to the general public at the same time.

In the following sections, Sect. 2 lists the novel contributions in this work. Section 3 provides a brief description of the literature survey for the development of Web application. Further Sect. 4 provides a brief explanation about the system model and Sect. 5 lists some of the advantages of the proposed system over existing one. Section 6 describes important results and Sect. 7 concludes the work.

2 Novel Contributions

The novel contributions for the problem statements by the Web application are stated as follows:

- Authenticated sign up acceptance by system admin to log into the Web application, by students, faculties, training and placement officer (T.P.O.), Head of Department (H.O.D) and Transport officer (T.O.) of that college and enjoy the functionalities provided by the system.
- Notes sharing department wise, among students and faculties of the same department.
- Department-wise notice publication.
- Leave application by students and faculties, to be approved by H.O.Ds.

- Addition of pick and drop location of college bus by T.O. of the college.
- Placement details to be added along with the year of drive and number of students placed, by T.P.O.
- Minimum and maximum ranks accepted by college in entrance exams, to admit new students is also added by T.P.O.
- Information provided through the system, by T.P.O. and T.O. will be visible to the general public for providing reliable information (hence authenticated and updated information is also provided).
- Reviews can be added by bona fide and authenticated students of the college, which is visible to the general public that too department wise.

3 Literature Survey

The current system provides a huge amount of data on the Internet in a scattered manner through Web sites of the various private and government colleges in various regions of the country, most of which are often not updated timely and are unknown to a student searching for a suitable institute. Often it is observed that students rely on external unknown sources about the particular college he/she wants to explore which has less or almost no ground for truth. Also, not all the official Web sites of all colleges are known to the student who is searching for suitable college. In the existing system, Web sites exist which provide reviews to students in an unorganized manner which allows any person to randomly post reviews without proper authentication. Another issue we identified when we observed that weak learners are unable to grasp a concept easily. The existing system contains again plethora of information, in unorganized manner. Hence, the feature of knowledge sharing was included as no system exists which can provide the concepts explained by the department faculties of college outside the class in an organized manner. The feature of leave application in colleges is manual work such that the person has to take permission from higher authorities even in emergency cases. The Web application proposes to change it and make the process less tedious.

Through some literature survey surrounding these issues, we found out that colleges need a renovation in terms of performing internal academic operations which include publish of notices and placement activities [1]. Currently, the administrative tasks are done manually in most of the colleges [3]. People face problems while locating a place to go without the use of a navigation system or location setter [4]. Educational institutions should partner up with the kind of platforms which can provide all information at the same place without overburdening students with the research work they have to pursue for their search purpose [5]. The study conducted as a survey in [2] shows that 62% of students are dissatisfied with online databases of libraries. In engineering colleges, 95% respondents are visiting for the library as the main purpose of issuing library books for their academic studies [2]. Faculties and students can share their published research papers among each other for knowledge and encouragement [6]. Small colleges can form consortium and host their college information on a common platform at lower budget [7].

In this work, the process of posting reviews has been modified and only the authenticated people related to the college are allowed to post reviews in the Web application as the process of authentication is guaranteed. Among other new features like publishing notices within the administration and the placement, activities are let known to the general public which can be updated time to time. Leave application has also been incorporated in the work for easing the administration of college. The setting of location by T.O. helps eliminate any confusion with the bus location from where the students will be picked up/dropped in case the college has transport facility. Notes sharing platform is incorporated to provide students a database of class notes to be referred to for academic study which hopes to reduce dissatisfaction among students and the students can refer library periodicals instead for carrying out research work instead of only studying.

4 System Model

The system architecture for the Web application is shown in Fig. 1 which describes the system architectural design through the interaction of different users (general public, faculty, students, H.O.D., T.P.O., and T.O.) with the system which relies on various databases maintained for information extraction. The Web application has two types of user base:

- General view for unregistered user.
- Information management application for registered users.

General view for unregistered user: This part of the project mainly deals with the display of general information [3] about the college. This includes general information such as college infrastructure, courses offered, college facilities, bus facility (if any), etc., and the student reviews for the college, which are given by bona fide students of the college.

Registered users: The list of registered users includes the following:

A. **Administrator**: This user is the first layer of security of the system in terms of permitting the authenticated members of the college such as bona fide students and currently appointed faculties to become a user of the system. The admin is supposed to be the controller of the whole system. The admin's main task is to maintain the authenticity of the system by reviewing and accepting the sign-up requests of only those users who are registered with the system database assumed to be provided by college which denotes that the user is actually the member of the college either as a student or faculty.

B. **Faculty**: The faculty is also an authenticated user of the system who can use the system if and only if he/she is registered with the system. The faculty first needs approval of admin to sign up to the system and only after the admin approves, the

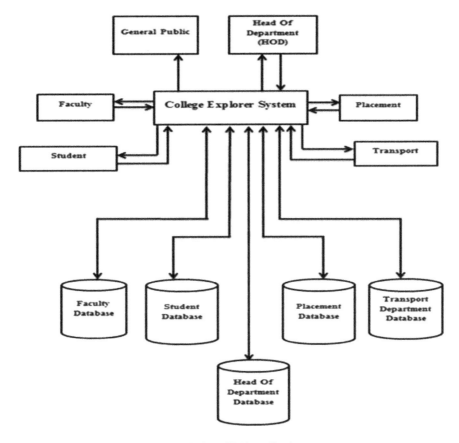

Fig. 1 Block diagram of architectural design of Web application

faculty can log into the system to perform various functionalities like uploading notes with deletion functionality, for students of his department, apply leaves which will be approved by H.O.D. and to publish emergency/urgent notice for students and faculty of the same department.

The faculty may share research papers published by them among their students to guide and encourage them through notes sharing platform [6].

C. **Student**: The student is the user who would also have a similar procedure to log into the system as that of the faculty that is through authentication. The student also enjoys various functionalities after logging into it such as upload notes for other students of the department but can't delete them, apply for leave to H.O.D., and can view the department notice also. The student is given a facility of reviewing the college which will be directly visible on the general page, to every unregistered user who uses the system.

D. **Placement officer**: He can use the system to add details about the placements in the college and also can add the details of the entrance exam ranks accepted

by the college during admission. The fee structure for these exam candidates on admission is also updated by this user only. All these information are directly displayed on the general page for helping students to get authenticated and updated information, who want to pursue higher education in that college.
E. **Transport officer**: Locating the positions in map for navigating to a nearby bus stop of the college can be helpful and done through Google Apps only when the position is known [4]. T.O. can add the bus stops through Google maps using the Web application, to add the pick/drop locations of college buses from/to where college students and faculties commute daily to/from the college. This information is also displayed directly on the general page, for keeping the general users informed about the transport facility provided by college.
F. **Head of Department**: He/she can approve the leaves which are applied by students and faculties of the department to which the H.O.D. belongs to.

5 Advantages

- Less time-consuming research for students in search of suitable institute for higher study in engineering field.
- Real-time, updated, and authenticated information is to be provided along with reviews.
- Students can make an opinion on the college easily using reviews of bona fide students of the same college.
- Additional features of the system also aim at providing faculties of the same college to utilize the system for notes uploading in order to assist students in their study.
- Faculty can also notify students in real time by publishing notice(s) of important information to be circulated among students, through this system which is helpful especially in cases of urgent information circulation.
- It also aims at providing students to view and upload the notes provided by the faculty of various subjects or classmates which will prove to be helpful for them in their studies and to keep track of notice from their side.
- Through this system, the students who are studying currently in the college can independently judge the working of the college and this will provide information about how the internal environment of the college is, to the general public (target audience include the school pass outs and parents).
- Faculty and students can apply for leave to H.O.D. directly through the system which can be approved through the system by H.O.D.

6 Implementation and Results

In this section, the results are shown for the Web application college explorer, which has been implemented using Java Server Pages (JSPs), Servlets, XML [8] on J2EE platform of java which uses JDBC connections [9, 10] to connect to MySQL database based on the MVC Architecture. Some of the important outcomes are discussed herein:

The ER diagram of the Web application, which depicts relationships between different entities of the database, is shown in Fig. 2. The list of reviews a particular student has written for the college on behalf of a department after a successful login to the system which is visible in his account is shown in Fig. 3. Figure 4 shows the summarized student reviews overall, which are visible to the general public. Figure 5 shows the list of uploaded class notes and important video links by faculties and students of a particular department being shown in the account of the student who belongs to the same department. Figure 6 shows the form which either a student or a faculty of the department on being logged into the system, has to fill up, in order to upload the contents for other faculties and students of the same department to view and refer. Figure 7 shows how the faculty currently logged into the system can maintain the notices to be published for faculty and students of the same department separately. Figure 8 shows how a T.O. can manage the information about the pick and drop location of the college buses, to be shown to the general public. Figure 9 shows the management of placement information by the T.P.O. of the college which will be shown on the general page to be showcased to the general public who are unregistered with the system. Figure 10 shows how a faculty of certain department can apply for leaves to the H.O.D. by just a click through the Web application. Based on the number of leaves available for the faculty, it will be accepted.

The H.O.D. can either accept or reject the leave application, the information about which will be notified to the same faculty through mail.

7 Conclusion

In a hope to serve students and parents in helping in their search of suitable college to pursue higher education, we have built this project, which has only the Web application part implemented. It also serves the purpose of aiding the management of college administration to function with reduced manual task to execute and to aid them in keeping record in online mode. This system is an implementation of mere a base idea to build commercial Web site which can act as a one-stop common platform for multiple colleges or schools to be searched for their academic functioning and internal environment.

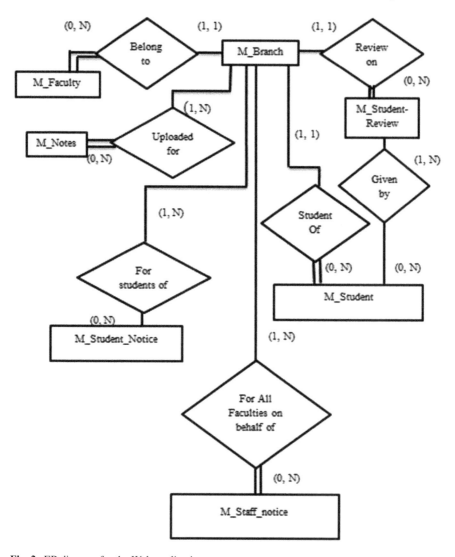

Fig. 2 ER diagram for the Web application

The future enhancements for this work are proposed as follows:

- Scale out the Web application, so as to accommodate more colleges by dynamically adding them to the common platform.
- To develop the feature of ranking the colleges if the system is able to accommodate multiple colleges based on either their overall ratings or infrastructure, etc.

Fig. 3 List of reviews by a student

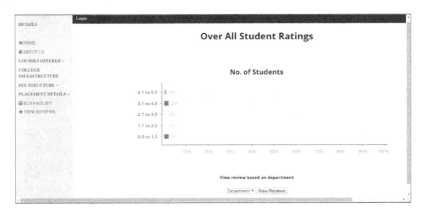

Fig. 4 Summarized view of student reviews

Fig. 5 List of uploaded notes by faculties and student of a department

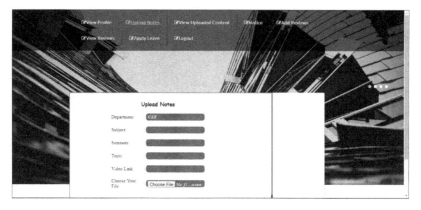

Fig. 6 Form to fill up to upload notes

Fig. 7 A faculty account can add and delete the notices published for students and other faculties of their department

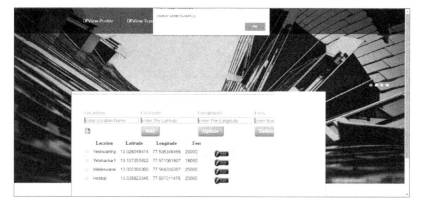

Fig. 8 Pick and drop locations can be added by transport officer through Google maps

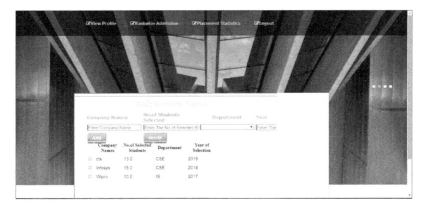

Fig. 9 Placement details can be added by placement officer

Fig. 10 Leave applied to H.O.D. by faculty account

- This system can be successful in providing real common platform for multiple colleges to showcase their college facilities and academic success.
- An android app specially designed for the system so that people can access and use it anytime and anywhere.

References

1. Rajebhosle, S., Choudhari, S., Patil, S., Vyavahare, A., & Khabiya, S. (2016, April). Smart campus—An academic web portal with android application. *International Research Journal of Engineering and Technology, 3*(4), 389–394, e-ISSN: 2395-0056.
2. Saini, P. K., & Bhakar, R. (2014, September). User satisfaction of the students of engineering college: A case study of engineering college libraries of Jaipur, Rajasthan. *International Journal of Emerging Research in Management & Technology, 3*(9), 16–26, ISSN: 2278-9359.

3. Bilawane, S., & Jambhulkar, P. (2015, March). Information system based on college campus. *International Journal of Engineering and Computer Science, 4*(3), 10852–10855, ISSN: 2319-7242.
4. Akinyemi, I. O., & Oyelami, O. M. (2013, August–September). Developing a web based location navigation system in the context of Covenant University. *International Journal of Computer Applications & Information Technology, 3*(2), 10–13, ISSN: 2278-7720.
5. *Private colleges found consortium to collaborate, negotiate with vendors* [Online]. Available https://www.insidehighered.com/news/2016/01/20/private-colleges-found-consortium-collaborate-negotiate-vendors.
6. Goel, M., Wasnik, A., Gulhane, A., Gajarlewar, S., & Rasekar, Y. (2017, March). College monitoring system. *International Journal for Research in Applied Science & Engineering Technology, 5*(II), 41–44. ISSN: 2321-9653.
7. *Private colleges found consortium to collaborate, negotiate with vendors* [online]. Available https://www.insidehighered.com/news/2016/01/20/private-colleges-found-consortium-collaborate-negotiate-vendors, Accessed April 6, 2018.
8. *XML tutorial* [online]. Available https://www.w3schools.com/xml/default.asp.
9. *How to connect to database in Java, Java database connectivity* [online]. Available https://www.javatpoint.com/steps-to-connect-to-the-database-in-java, Accessed January 20, 2018.
10. *Java JDBC Tutorial-Javatpoint* [online]. Available https://www.javatpoint.com/java-jdbc, Accessed January 20, 2018.

Activity-Based Music Classifier: A Supervised Machine Learning Approach for Curating Activity-Based Playlists

B. P. Aniruddha Achar, N. D. Aiyappa, B. Akshaj, M. N. Thippeswamy and N. Pillay

Abstract Classification of musical tracks and creation of playlists to match four primary activities such as Sleep, Party, Dinner and Workout, using concepts of machine learning (ML) and musical information retrieval (MIR), is proposed in this paper. A data set of songs using features extracted through digital signal processing (DSP) is developed for training. In this work, several prominent and distinguishing features of individual musical tracks are employed. The ML algorithms used to classify the data set are: super vector machine (SVM), kth nearest neighbour, neural networks and voting classifiers. The results show that the highest accuracy can be attained when classification is performed using the voting classifier compared to other algorithms. The increase in accuracy can be attributed to the voting classifier's ability to improve the individual classes' accuracy by utilizing multiple classifier outputs.

Keywords Discrete wavelet transform (DWT) · Kth nearest neighbour (kNN) · Logistic regression (LR) · Music information retrieval · Machine learning (ML) · Music classification · Mel-frequency cepstrum coefficient · Random forest (RF) · Spectral centroid · SVM · Spotify

---

B. P. A. Achar (✉) · N. D. Aiyappa · B. Akshaj · M. N. Thippeswamy
Department of CSE, NMIT, Yelahanka, Bangalore 560064, India
e-mail: aniruddha.achar@gmail.com

N. D. Aiyappa
e-mail: vinayaiyappa24816@gmail.com

B. Akshaj
e-mail: akshajb@gmail.com

M. N. Thippeswamy
e-mail: thippeswamy.mn@nmit.ac.in

N. Pillay
School of Engineering, UKZN, Durban 4041, South Africa
e-mail: pillayn@ukzn.ac.za

© Springer Nature Singapore Pte Ltd. 2019
N. R. Shetty et al. (eds.), *Emerging Research in Computing, Information, Communication and Applications*, Advances in Intelligent Systems and Computing 906,
https://doi.org/10.1007/978-981-13-6001-5_14

1 Introduction

Automated music classification helps in sorting, indexing, tagging and curation of petabytes of music files available on the Internet. Processing and structured storage of these songs needs a robust and effective classification mechanism. This classification of songs can be performed on various aspects like genre of the music, mood of the music, etc. In this paper, we classify music based on the activities a song is generally associated with. The target classes considered in this paper are *Dinner* activity, *Party* activity, *Sleep* activity and *Workout* activity.

In any machine learning problem, there is an emphasis on data collection and extraction and classification of data. In music classification, signal-based features [1] can be employed. These signal-based features can be classified into short-time features and long-time features. These short-time features divide the audio file into small segments and analyse these short segments. A few short-time features are MFCC [1], chroma features [2], spectral centroid [1], spectral roll-off [1] and spectral bandwidth. Long-time features are generally estimated over the entirety of the song. A few long-time features are: beat histogram [1], Daubechies wavelet coefficient histogram (DWCH) [3], etc. These signal features are just not sufficient for many music classification problems. High-level features like lyrics [4], artist details and other meta data associated with an audio file can also be used as features.

In [5], the authors focused on extracting the spectral data from a MPEG ACC audio file. The discrete wavelet transform techniques are used to extract the features. The features used are classified into two categories: Timber and Tempo [5]. The authors used SVM and RF classifiers to classify the data.

The results presented in [5] show that a highest accuracy of around 81% is achieved. The music genre classification is done using manifold learning techniques in [6]. The accuracy of these neural networks is compared with well-known learning models, such as SVM and others.

In [7], the authors have studied the automatic detection and classification of music files based on its genre with the help of various classification algorithms. Features like MFCC and FFT are used in correlation with classification algorithms like LR, kNN and SVMs. The work performed in [8] talks about the possible feature sets that can be used in retrieval of music information and the methodology used for extraction of the features and classification of music based on genre. The work carried out in [8] definitively brings out a pattern recognition problem that exists in music classification.

The authors of [9] discussed the use of different feature sets like Timberal and rhythmic features to featurize the data. The work in [9] concentrates on better classification using superior feature selection, feature reduction and the use of improved classifiers like extreme learning machines that use boosting and bagging techniques.

Classification of music can be carried out in two ways:

(i) Using unsupervised learning with no labels, i.e. by building and analysing clusters using algorithms like k-means, DB-SCAN, neural networks, etc. This method gives a group of music files with similar features put together.
(ii) Using supervised learning, where labels are used to distinguish different training examples. Once a model is trained using the training data set, the reliability of the model is tested using a testing set.

Currently, in the field of music classification, research is being conducted in areas of music genre classification, music mood classification and instrument detection. During our literature survey, we failed to find works that classified music based on human activities. Hence, in this paper, a data set is built for music files based on human activities and supervised learning algorithms are used to conduct experiments on this data set in order to classify songs based on activities.

The novel contributions are as follows: First, we build an ensemble of classifiers that accurately and quickly classify audio tracks based on human activities such as *Dinner*, *Sleep*, *Party* and *Workout*. Second, we build a data set that consists of around 45 h of music. This data set contains all the features that are discussed in this paper. This data set will help in facilitating further research in this area. Finally, we compare and analyse various machine learning algorithms, their efficiency and accuracy when trained and tested on the above-mentioned data set.

The rest of the paper is organized as follows: In Sect. 2, we discuss the feature extraction techniques. In Sects. 2.1–2.4, we discuss the various aspects of the data set such as the correlation of the features, their importance for classification and scaling the data set. In Sect. 3, we discuss the data set used for training and testing. Section 4 explains the performance of classifiers. In the last section, we discuss the conclusion.

In this work, we used raw audio files to extract features. These audio files are then labelled as one of the available target classes. This labelling is performed based on playlists extracted from Spotify. Audio features are then extracted from these songs using Librosa: a Python library for audio analysis [10]. These features are then subjected to statistical analysis and then stored onto a database. For genre classification, a combination of timbral and low-level features is used [1]. In this work, we decided to use the above combination with chroma features as they are known to improve the accuracy rate [2].

Figure 1 illustrates how an audio track when given as an input to the system is classified. The features aforementioned above are extracted using Librosa. Once these features are extracted, Numpy is used to generate the mean of the extracted features. Taking the mean of the data reduces the size of the data set reducing the problem of dimensionality. These averages are given as input to the trained models. These models then classify the data points and used to predict a class.

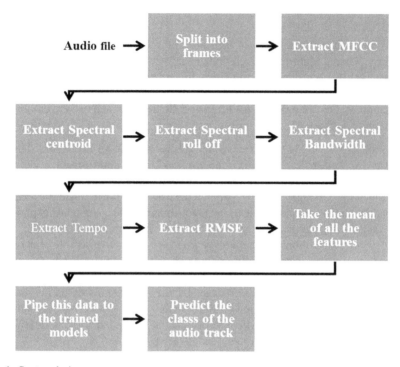

Fig. 1 System design

2 MFCC [10]

MFCC has been the dominant features used for speech recognition. The equation used for converting from frequency to mel scale is:

$$M(f) = 1125 \ln(1 + f/700) \tag{1}$$

To convert mels back to frequency:

$$M^{-1}(m) = 700\left(\exp\left(\frac{m}{1125}\right) - 1\right) \tag{2}$$

The steps to determine MFCC are as follows:

(a) Segment the audio signal into 20–40 ms frames.
(b) Estimate the logarithm of the magnitude of DFT for all signal frames.

   i. Calculate the DFT of the frame using Eq (3).

$$S_i(k) = \sum_{n=1}^{N} s_i(n)h(n)e^{-j2\pi kn/N} \quad 1 \le k \le K \tag{3}$$

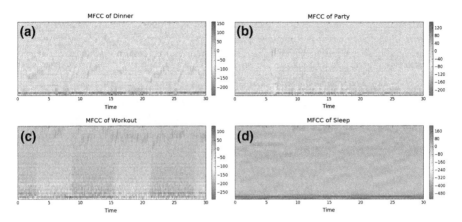

Fig. 2 MFCC graph of audio tracks labelled as Dinner (**a**), Party (**b**), Workout (**c**) and Sleep (**d**)

where $h(n)$ is an N sample long analysis window (e.g. Hamming window), and K is the length of the DFT. The periodogram-based power spectral estimate for the speech frame $s_i(n)$ is given by,

$$P_i(k) = \frac{1}{N}|S_i(k)|^2 \qquad (4)$$

(c) Filter the centre frequencies.
(d) Estimate inversely the IDFT to get all MFCC coefficients.

Figure 2 shows the MFCC extracted results from database of audios for four primary activities, i.e. *Sleep, Party, Dinner* and *Workout*. In the obtained ceptral graphs, it can be observed that activities such as *Sleep* and *Dinner* (Fig. 2a, d), which require the music to be smoother, have uniform distribution of energy throughout the graph, whereas activities such as *Workout* and *Party* (Fig. 2b, c) have regions with high and low energy unevenly spread throughout the ceptrum.

2.1 Spectral Centroid

This feature is used as a measure of brightness of a song and thus relates to music timbre. This can be used to determine the point around which most of the energy of the song is concentrated [11].

In the obtained log spectrum graphs, it can be observed that in activities such as *Sleep* and *Dinner* (Fig. 3a, b), the spectral centroid for each of the frame is in a lower frequency range. Further, it can also be observed that *Dinner* has a varied spectral centroid, whereas Sleep has a constant and smooth variation in spectral centroid and activities such as *Workout* and *Party* (Fig. 3c, d) have higher spectral centroid

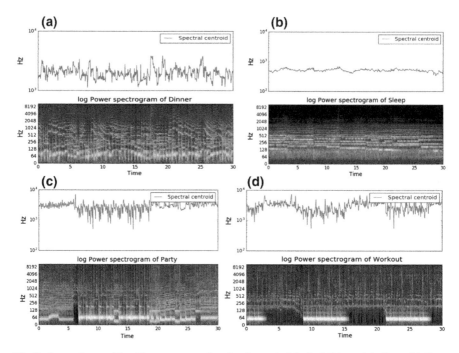

Fig. 3 Spectral centroid and log power graphs of audio tracks labelled: Dinner (**a**), Sleep (**b**), Party (**c**) and Workout (**d**)

frequency on an average. It can also be observed that the variation in the frequency is similar between the *Workout* and *Party* activities.

2.2 Spectral Roll-off

The roll-off is defined as the frequency below which 85% of the magnitude distribution of the spectrum is concentrated [1]. It is used to distinguish between voiced and unvoiced speech or music [12].

Figure 4a shows the song associated with *Dinner* activity from which it can be observed that the spectral roll-off is varied throughout the spectrum and has relatively low average roll-off frequency. Figure 4b shows the song associated with *Party* activity from which it can be observed that the spectral roll-off is smooth throughout and the spectrum and has relatively high average roll-off frequency. Figure 4c shows the song associated with *Sleep* activity from which it is can be observed that the spectral roll-off is smooth throughout and the spectrum and has relatively low average roll-off frequency. Figure 4d shows the song associated with *Workout* activity from which it is can be observed that the spectral roll-off is quite high and there are large variations in the spectral roll-off at times.

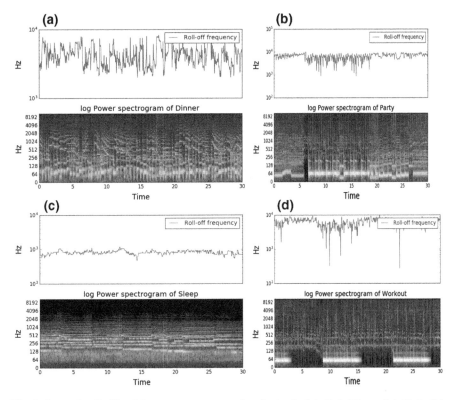

Fig. 4 Spectral roll-off and log power spectrum of audio tracks labelled: Dinner (**a**), Party (**b**), Sleep (**c**) and Workout (**d**)

2.3 Spectral Bandwidth

Bandwidth, also known as spectral spread, is derived from the spectral centroid, thus also a feature in frequency domain. Spectral bandwidth indicates the range of the interesting parts in the song [11]. The average bandwidth of a music track may serve to describe the perceived timbre. Different songs have different spread of energy among different frequency bands. Spectral bandwidth helps in identifying these features. It can be observed that songs associated with *Dinner* (Fig. 5a) have a constant rate of variation in spectral bandwidth. This periodic variation gives the songs a melodious feel. Songs associated with *Party* (Fig. 5b) have a wide spread in the spectral bandwidth. This shows that the songs have high energy at certain points and lower at other points in the song.

In songs associated with Sleep activity (Fig. 5c), the spectral bandwidth is concentrated to a very small region and is generally smooth. In songs associated with Workout activity (Fig. 5d), the spectral bandwidth is varying the spectral bandwidth in shorter periods.

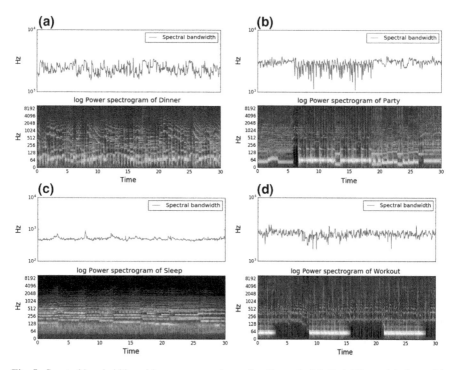

Fig. 5 Spectral bandwidth and log power spectrum of audio tracks labelled: Dinner (**a**), Party (**b**), Sleep (**c**) and Workout (**d**)

2.4 Root-Mean-Square Energy

The RMS can be defined for a continuously varying function in terms of an integral of the squares of the instantaneous values.

It can be observed that songs associated with *Dinner* (Fig. 6a) have a steady rate of variation, and the frequency is in the lower range of the spectrum. Songs associated with *Party* (Fig. 6b) have the RMS energy in marginally higher ranges, with low variation in the frequency. In songs associated with *Sleep* activity (Fig. 6c), the RMS energy is the lower range with very low variation in the frequencies. In songs associated with *Workout* activity (Fig. 6d), the RMS energy varies a lot throughout the spectrum, reaching high and low frequencies uniformly.

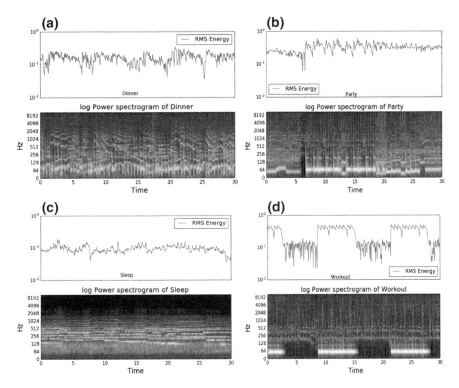

Fig. 6 Root mean square and log power spectrum of audio tracks labelled: Dinner (**a**), Party (**b**), Sleep (**c**) and Workout (**d**)

3 Data set

The data set has four target classes as described before for each activity. The data set is curated by analysing audio files of length 30 s, each of different target classes. Depending on the availability of audio files, there is a difference in the number of audio files analysed for each class. Table 1 shows the contents of the data set used for training and testing various machine learning models. The data set had the following classes: *Dinner, Party, Sleep and Workout*. The low-level, short-time and chroma information extracted was subjected to statistical analysis. Averages of these features were taken and used as the training and testing features for the algorithms.

Table 1 Data set used for training and testing

Target class	Number of audio files analysed	Number of hours of audio files
Dinner	44	3.66
Party	37	3.08
Sleep	44	3.66
Workout	49	4.08
Total	174	14.48

4 Performance Evaluation of Classifiers

Several classification methods were used for classification of the data set described above. The data set is split into a training set of size 84 samples; 21 of each target class. The rest of the data set is used as the testing set. Scikit Learn, a module in Python [13] specifically built for implementing various machine learning algorithms, is used to implement the classifiers. Multiple classifiers were used during the paper. This is done to compare between the classifiers [14–17] and to use the better performing classifiers in an ensemble of classifiers. The metrics used for measuring the accuracy of classifiers are as follows:

i. **Precision**: This is defined as the ratio of true predictions to the sum of the true predictions and false prediction.
ii. **Recall**: This is defined as the ratio of the true positive to the sum of true positive and the false negatives.
iii. **F1 score**: It is defined as the weighted score of the precision and recall.

Table 2 shows the class report of the SVM trained model. SVM's ability for kernel transformation is found to be of great importance when drawing the hyperplane during classification. It is also found that the classification between two specific classes the *Party* and *Workout* is difficult for SVM as the data points of both the classes were overlapping.

A kNN [14, 15] like SVM [15, 16] found it difficult to classify *Workout* class from a *Party* class. This behaviour further supports the notion that the data points of the *Workout* and *Party* class overlap, making it difficult for the classifiers to draw

Table 2 Class report for support vector machine

Class	Precision	Recall	F1 score	Support
Dinner	0.91	0.91	0.91	23
Party	0.50	0.81	0.62	16
Sleep	1.00	1.00	1.00	23
Workout	0.75	0.46	0.57	26
Avg/total	0.81	0.78	0.78	88

an effective decision boundary between the above-mentioned classes. Table 3 shows the class report of kNN.

A MPL's ability of varying weights and biases to find the optimal decision boundary lets it better classify between *Party* and *Workout* classes. Thus, classifying more of the Workout class testing data points correctly than both above-mentioned classifiers. There can also be observed decrease in the classification accuracy of other classes. There is always a trade-off between correctly classifying few classes and correctly classifying all the classes. Table 4 shows the class report of neural network.

A voting classifier [17] being an ensemble of classifiers combines strengths of the classifiers and tries to mitigate the weakness of the classifiers. This can be seen in the classifier report of the voting classifier that combines the above three classifiers. There is improvement in overall accuracy not only that each of the class is better classified. This meta-classifier has higher F1 score when compared to each of the individual classifiers indicating better overall performance of the classifier. Table 5 shows the class report voting classifier.

Figure 7 shows the performance of the classifiers used in this work. The accuracy of the classifiers is as follows:

i. SVM accuracy = 78.4090909090909%

Table 3 Class report of K nearest neighbour

Class	Precision	Recall	F1 score	Support
Dinner	0.88	0.91	0.89	23
Party	0.48	0.81	0.60	16
Sleep	1.00	1.00	1.00	23
Workout	0.79	0.42	0.55	26
Avg/total	0.81	0.77	0.77	88

Table 4 Class report neural network

Class	Precision	Recall	F1 score	Support
Dinner	0.79	0.83	0.81	23
Party	0.44	0.44	0.44	16
Sleep	1.00	0.87	0.93	23
Workout	0.64	0.69	0.67	26
Avg/total	0.74	0.73	0.73	88

Table 5 Class report voting classifier

Class	Precision	Recall	F1 score	Support
Dinner	0.91	0.91	0.91	23
Party	0.56	0.88	0.68	16
Sleep	1.00	1.00	1.00	23
Workout	0.82	0.54	0.65	26
Avg/total	0.85	0.82	0.82	88

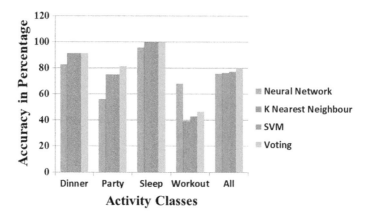

Fig. 7 Performance evaluation of classifiers; accuracy in percentage versus activity classes

ii. kNN accuracy = 77.272727272727%
iii. Neural network accuracy = 75.0%
iv. Voting classifier accuracy = 81.81818181818183%

It can also be observed that an ensemble of classifiers produces a better result when compared to a single classifier. This can be attributed to the increased accuracy. It is also observed that voting gives highest accuracy due to its ability to leverage the performances of other classifiers.

5 Conclusion

In this paper, we attempted to classify musical tracks and create playlists to match four primary activities, i.e. *Sleep*, *Party*, *Dinner* and *Workout* using concepts of machine learning and musical information retrieval (MIR). These features provided the necessary information to the machine learning algorithms to uniquely identify the activity the track could be associated with. The data set used to train the machine consisted of a total of 174 tracks, each of duration 30 s. Each of these tracks was associated with a certain activity belonging to the four primary activities. It is observed that the highest accuracy can be attained for the voting classifier. This increase in accuracy can be attributed to the voting classifier's ability to improve individual classes' accuracy by utilizing multiple classifier outputs.

References

1. Cook, P., et al. (2002). Musical genre classification of audio signals. *IEEE Transactions on Speech and Audio Processing, 10*(5), 293–302. https://doi.org/10.1109/TSA.2002.800560.
2. Ellis, D. P. (2007). Classifying music audio with timbral and chroma features. In *Proceedings of International Conference on Music, Information Retrieval*, Vol. 7, pp. 339–340. September 23–27, 2007.
3. Li, T., Ogihara, M., & Li, Q. (2003). A comparative study on content-based music genre classification. In *Proceedings of the 26th Annual International ACM SIGIR Conference on Research and Development in Information Retrieval*, pp. 282–289. New York, NY, USA: ACM Press.
4. Mayer, A. R., et al. (2008). Rhyme and style features for musical genre classification by song lyrics. In *ISMIR*, pp. 337–342.
5. Mayer, R., Neumayer, R., & Rauber, A. (2008). Rhyme and style features for musical genre categorisation by song lyrics. In *Proceedings of the International Conference on Music Information Retrieval*.
6. Kobayakawa, M., Hoshi, M., & Yuzawa, K. (2014). Music genre classification of MPEG AAC audio data. In *2014 IEEE International Symposium on Multimedia, Taichung*, pp. 347–352. https://doi.org/10.1109/ism.2014.25.
7. Rajanna, A. R., Aryafar, K., Shokoufandeh, A., & Ptucha, R. (2015). Deep neural networks: A case study for music genre classification. In *2015 IEEE 14th International Conference on Machine Learning and Applications (ICMLA)*, Miami, FL, pp. 655–660, https://doi.org/10.1109/icmla.2015.160.
8. Kumar, D. P., Sowmya, B. J., & Srinivasa, K. G. (2016). A comparative study of classifiers for music genre classification based on feature extractors. In *2016 IEEE Distributed Computing, VLSI, Electrical Circuits and Robotics (DISCOVER)*, Mangalore, pp. 190–194, https://doi.org/10.1109/discover.2016.7806258.
9. Tzanetakis, G., & Cook, P. (2002). Musical genre classification of audio signals. *IEEE Transactions on Speech and Audio Processing, 10*(5), 293–302. https://doi.org/10.1109/TSA.2002.800560.
10. Baniya, B. K., Ghimire, D., & Lee, J.: Automatic music genre classification using timbral texture and rhythmic content features. In *2015 17th International Conference on Advanced Communication Technology (ICACT)*, Seoul, pp. 434–443, https://doi.org/10.1109/icact.2015.7224907.
11. Vatin, C. *Automatic spoken language identification* (Master's thesis). University of Manchester, England.
12. Shepard, R. (1964). Circularity in judgements of relative pitch. *Journal of the Acoustical Society of America, 36*, 2346–2353.
13. *Scikit-learn: Machine learning in python*. Accessed March 12, 2017 from http://scikit-learn.org/.
14. Salari, N., Shohaimi, S., Najafi, F., Nallappan, M., & Karishnarajah, I. (2014). A novel hybrid classification model of genetic algorithms, modified k nearest neighbor and developed backpropagation neural network. *PLoS ONE, 9*(11), e112987. November 24, 2014, https://doi.org/10.1371/journal.pone.0112987.
15. Cutajar, M., Micallef, J., Casha, O., Grech, I., & Gatt, E. (2013). Comparative study of automatic speech recognition techniques. *IET Signal Processing, 7*(1), 25–46.
16. Zahid, S., Hussain, F., Rashid, M., Yousaf, M. H., & Habib, H. A. (2015). Optimized audio classification and segmentation algorithm by using ensemble methods. *Mathematical Problems in Engineering, 2015*(209814), 11. https://doi.org/10.1155/2015/209814.
17. Geiger, J. T., Schuller, B., & Rigoll, G. (2013). Large-scale audio feature extraction and SVM for acoustic scene classification. In *2013 IEEE Workshop on Applications of Signal Processing to Audio and Acoustics*, New Paltz, NY, pp. 1–4, https://doi.org/10.1109/waspaa.2013.6701857.
18. Müller, M. (2007). Information retrieval for music and motion. Heidelberg: Springer. ISBN: 978-3-540-74048-3, https://doi.org/10.1007/978-3-540-74048-3.

19. Schindler, A. Music information retrieval [Online]. Available http://www.ifs.tuwien.ac.at/mir.
20. Goto, M., & Muroaka, Y. (1997). Issues in evaluating beat tracking systems. In *Workshop on Issues in AI and Music*.
21. Casey, M. A., Veltkamp, R., Goto, M., Leman, M., Rhodes, C., & Slaney, M. (2008). Content-based music information retrieval: Current directions and future challenges. *Proceedings of the IEEE, 96*(4), 668–696. https://doi.org/10.1109/JPROC.2008.916370.
22. Kobayakawa, M., Hoshi, M., & Yuzawa, K. (2014). Music genre classification of MPEG AAC audio data. In *2014 IEEE International Symposium on Multimedia*, Taichung, pp. 347–352. https://doi.org/10.1109/ism.2014.25.

New Password Embedding Technique Using Elliptic Curve Over Finite Field

D. Sravana Kumar, C. H. Suneetha and P. Sirisha

Abstract In the present sophisticated digital era, safe communication of user password from one source to the other is quite difficult in client/server system. Also storing the password as it appears increases the potential risk of the security. Protection of the password is at most important in group communications to avoid the access of the illegal person to group resources. In addition, a roaming user who uses the network from different client terminals requires access to the private key. The present paper explains secure communication of password from one entity to the other. Here the password is encrypted using elliptic curve over finite field, embedded in a large random text at different selected positions, and communicated to the receiver via public channel.

Keywords Encryption · Decryption · Elliptic curve over finite field

1 Introduction

Since the digital network is growing explosively, the main difficulty in password communication mechanisms arises due to the fact that password is generally very small in length and simple. Exhaustive search attack will break the security of the password easily. Though the password is very small in size, it remains safe if it is communicated as cryptographic key. Diffie–Hellman key exchange protocol is an asymmetric cryptographic technique to negotiate session key over a secure channel which is revolutionary in the history of public key cryptography. But Diffie–Hellman

---

D. S. Kumar
Dr. V. S. Krishna Government Degree College, Visakhapatnam, India
e-mail: skdharanikota@gmail.com

C. H. Suneetha (✉)
GITAM University, Visakhapatnam, India
e-mail: gurukripachs@gmail.com

P. Sirisha
Faculty in Mathematics, Indian Maritime University, Visakhapatnam, India
e-mail: sirinivas06@gmail.com

© Springer Nature Singapore Pte Ltd. 2019
N. R. Shetty et al. (eds.), *Emerging Research in Computing, Information, Communication and Applications*, Advances in Intelligent Systems and Computing 906,
https://doi.org/10.1007/978-981-13-6001-5_15

key exchange protocol suffers from the defect of man-in-the-middle attack and does not establish the identity of the entity. Most of the modern research in cryptography is done in the direction of removing the weakness of Diffie–Hellman key exchange algorithm using public key certificates and digital signature.

2 Literature Survey

Several researches have been done pertaining to the attacks on key exchange protocols. Ku and Wang [1] established attacks on backward reply without modification and backward reply with modification. Aziz and Diffie [2] proposed an authentication protocol with requirement of pre-communication. This protocol uses cryptographic techniques of off-line certification procedures. But conventionally the cryptographic techniques need online certification of users with encrypted secret key. Later on, several modifications have been done by researchers to overcome the attacks on key exchange protocol. In this direction Authenticated Key Agreement (AK) protocols [3–7], Authenticated Key Agreement protocols with key confirmation [8–10] (AKC) came into picture. Perrig [11] proposed three different protocols suitable for applications with different types of security levels. Moreover, he intensified the strength of the efficiency of existing key agreement protocols using binary tree. In that chapter, he emphasized that these protocols reduce the number of encryption rounds from n to $\log(n)$ and reduce the bandwidth requirements from quadratic number to linear number. Key agreement protocols are contributory by all the individuals in the group. Burmster and Desmedt [12] suggested collaborative group key agreement protocols. They describe star-based, tree-based broadcast and cyclic protocols. But, all these protocols are only variations on Diffie–Hellman key exchange protocols. They have not addressed the problems of dynamic groups. Aydos et al. [3] proposed key agreement protocols for wireless communications using Elliptic Curve Cryptographic (ECC) techniques. Key agreement protocols basing on ECC produce considerable development to protect the integrity and confidentiality of the data like RSA and DSA.

3 Elliptic Curve Cryptography (ECC)

An affine equation $E: y^2 + b_1 xy + b_3 y = x^3 + b_2 x^2 + b_4 x + b_6$ over the set of real numbers is said to be Weierstrass equation, where b_1, b_2, b_3, b_4, b_6 and x, y are real numbers. An elliptic curve for cryptographic purpose is defined by the equation $y^2 = x^3 + ax + b$, $4a^3 + 27b^2 \neq 0$ over the finite field F_q.

Group laws of elliptic curve: Let E be an elliptic curve defined over the finite field of integers K. Addition of two points uses chord-and-tangent rule to get the third point [13–15]. The set of all points on the elliptic curve over the finite field with

Fig. 1 Geometric addition

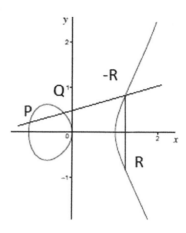

addition as binary operation forms an abelian group with ∞, the point at infinity as identity element.

Geometric rules of Addition: Let $P(x_1, y_1)$ and $Q(x_2, y_2)$ be two points on the elliptic curve E. The sum of the two points P and Q is $R(x_3, y_3)$ which is the reflection of the point of intersection of the line through the points P, Q and the elliptic curve about x-axis. The same geometric interpretation also applies to two points P and $-P$, with the same x-coordinate. Here the points are joined by a vertical line, which is considered as the intersecting point on the curve at the point infinity. $P + (-P) = \infty$, the identity element which is the point at infinity [13–15].

3.1 Doubling the Point on the Elliptic Curve

If $P(x_1, y_1)$ is point on the elliptic curve, then $2P$ is the reflection of the point of intersection of the tangent line at P and the elliptic curve about x-axis [13–15]. Example of addition of two points and doubling of a point are shown in the following Figs. 1 and 2 for the elliptic curve $y^2 = x^3 - x$.

Identity: $P + \infty = \infty + P = P$ for all E, where ∞ is the point at infinity [13–15].

Negatives: If $P(x, y)$ is a point on the elliptic curve, then $(x, y) + (x, -y) = \infty$, where $(x, -y)$ is the negative of P denoted by $-P$ [13–15].

Point Addition: If $P(x_1, y_1)$, $Q(x_2, y_2)$ are two points where $P \neq Q$, then $P + Q = (x_3, y_3)$ [13, 14], where $x_3 = \left(\frac{y_2 - y_1}{x_2 - x_1}\right)^2 - x_1 - x_2$ and $y_3 = \left(\frac{y_2 - y_1}{x_2 - x_1}\right)(x_1 - x_3) - y_1$.

Point Doubling: Let $P(x_1, y_1) \in E(K)$, where $P \neq -P$, then, $2P = (x_3, y_3)$ [13, 14] where $x_3 = \left(\frac{3x_1^2 + a}{2y_1}\right)^2 - 2x_1$ and $y_3 = \left(\frac{3x_1^2 + a}{2y_1}\right)(x_1 - x_3) - y_1$.

Point Multiplication: Let P be any point on the elliptic curve over finite field of integers. Then the operation multiplication of P is defined as repeated addition [16].

Fig. 2 Geometric doubling

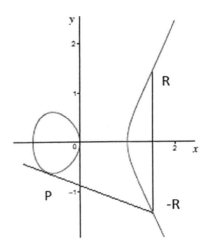

$$kP = P + P + \cdots \text{ktimes}$$

3.2 Elliptic Curve Discrete Logarithmic Problem (ECDLP)

The robustness of ECC relies on the solution of Elliptic Curve Discrete Logarithmic Problem [16–18]. For two points $P, Q \in E_q(a, b)$, the elliptic curve over the finite field, it is hard to solve $x = \log_p Q$ for given $Q = xP$, where $x \in \{1, 2, \ldots, q-1\}$.

4 Proposed Method

In view of the above survey, here authors designed secure password transmission protocol to overcome all the difficulties of key communication protocols. Here the password characters are encrypted embedded at different selected positions of a large random text using elliptic curve over finite field and transmitted to the intended recipient through public channel. This method provides the confidentiality of the password as well as the facility to store the password without any threat of stealing and can be retrieved whenever it is forgotten.

If Alice and Bob want to communicate with each other before communicating e messages, Alice selects an elliptic over finite field $E_q(a, b)$ having large number points on it. In selecting the elliptic curve Alice takes precautions that the elliptic /e is secure for cryptographic purpose to escape from main physical attacks like channel analysis and fault analysis. The desirable features of elliptic curves for otion purpose to avoid the risk of attacks are 1. The order of the curve # $E_q(a, b)$ not be factorized into small primes to defend against Pohlig–Hellman attack

2. The curve should be non-super singular curve 3. It should be non-anomalous curve, i.e., the order # $E_q(a, b) \neq q$. In addition, Alice selects the elliptic curve $E_q(a, b)$ such that the order of the elliptic curve # $E_q(a, b)$ is a prime number because if the order of the group is prime, it is cyclic group. Alice chooses a generator $G(x, y)$ on the elliptic curve and also selects a random function $f(x, y, n)$ of the co-ordinates of generator $G(x, y)$ and a large integer n less than the order of the generator. She publishes the elliptic curve $E_q(a, b)$ and the function $f(x, y, n)$ on public channel. She shares the generator $G(x, y)$ with Bob as the secret key specific to Bob only for their communication. Again she selects a large random text having all types of characters (text/numerical/special) as medium for inserting the encrypted password. Since all the points on the elliptic curve are generators of the cyclic group of finite order, she shares another generator $G_1(x_1, y_1)$ with other communicating party Charles, $G_2(x_2, y_2)$ with Dickens, and so on.

Encryption:

If Alice wants to communicate the password having the characters or numbers $M_1 M_2 M_3 M_4$.

1. She selects large random numbers $n, n_1, n_2, n_3, n_4, n_5$ less than the order of the generator and calculates the points $P = nG, P_1 = n_1 G, P_2 = n_2 G, P_3 = n_3 G, P_4 = n_4 G, P_5 = n_5 G$.
2. Again she calculates $m_1 = [f(x, y, n)]_{\mod n_1}, m_2 = [f(x, y, n)]_{\mod n_2}, m_3 = [f(x, y, n)]_{\mod n_3}$, and $m_4 = [f(x, y, n)]_{\mod n_4}$. In computing these values, Alice takes care that no two m_j are equal for $j = 1, \ldots, 4$.
3. Alice encrypts the password $M_1 M_2 M_3 M_4$ using logical XOR operation. She performs logical XOR operation between ASCII binary equivalents K_1, K_2, K_3, K_4 of M_1, M_2, M_3, M_4 and 8421 BCD codes of $[(n_5 + n_1)]_{\mod 100}, [(n_5 + n_2)]_{\mod 100}, [(n_5 + n_3)]_{\mod 100}, [(n_5 + n_4)]_{\mod 100}$ to obtain the 8-bit binary numbers $K_{1E}, K_{2E}, K_{3E},$ and K_{4E}.

$$K_{1E} = K_1 \text{ XOR}[(n_5 + n_1)]_{\mod 100}, K_{2E} = K_2 \text{ XOR}[(n_5 + n_2)]_{\mod 100}$$

$$K_{3E} = K_3 \text{ XOR}[(n_5 + n_3)]_{\mod 100}, K_{4E} = K_4 \text{ XOR}[(n_5 + n_4)]_{\mod 100}$$

Here mod100 is considered for performing logical XOR operation between 8-bit binary numbers K_1, K_2, K_3, K_4 and 8421 BCD codes of $[(n_5 + n_1)]_{\mod 100}, [(n_5 + n_2)]_{\mod 100}, [(n_5 + n_3)]_{\mod 100}, [(n_5 + n_4)]_{\mod 100}$.

4. The binary numbers $K_{1E}, K_{2E}, K_{3E},$ and K_{4E} are coded to characters using ASCII code table to obtain the cipher characters $C_1, C_2, C_3,$ and C_4 of the password to be transmitted. These characters $C_1, C_2, C_3,$ and C_4 are inserted at the positions $m_1, m_2, m_3,$ and m_4 respectively, in the selected random text. For example, if $m_1 = 1256$, then the 1256th character of the random text is replaced by the first character C_1. Similarly, the character at m_jth position is replaced by C_j $j = 1, 2, 3, 4$.

5. Alice communicates the points $P, P_1, P_2, P_3, P_4, P_5$ and the random text in which the encrypted password characters are embedded to Bob via universal channel.

Decryption: Before decrypting the password, Bob constructs table of all points on the elliptic curve as multiples of generator G. After receiving the random text and points $P, P_1, P_2, P_3, P_4, P_5$ first picks up the values $n, n_1, n_2, n_3, n_4, n_5$ from the constructed table. $P = nG, P_1 = n_1G, P_2 = n_2G, P_3 = n_3G, P_4 = n_4G, P_5 = n_5G$.

1. He calculates the function $f(x, y, n)$ and $m_1 = [f(x, y, n)]_{\text{mod} n1}$, $m_2 = [f(x, y, n)]_{\text{mod} n2}$, $m_3 = [f(x, y, n)]_{\text{mod} n3}$, and $m_4 = [f(x, y, n)]_{\text{mod} n4}$.
2. Then Bob locates the encrypted password characters $C_1, C_2, C_3,$ and C_4 which are at the positions $m_1, m_2, m_3, m_4,$ respectively, in the received random text.
3. Bob writes equivalent 8-bit binary numbers for the characters of $C_1, C_2, C_3,$ and C_4 using ASCII code table to get $K_{1E}, K_{2E}, K_{3E}, K_{4E}$. He decrypts $K_{1E}, K_{2E}, K_{3E}, K_{4E}$ by executing logical XOR operation between $K_{1E}, K_{2E}, K_{3E}, K_{4E}$ and 8421 BCD codes of $[(n_5 + n_1)]_{\text{mod}100}, [(n_5 + n_2)]_{\text{mod}100}, [(n_5 + n_3)]_{\text{mod}100}, [(n_5 + n_4)]_{\text{mod}100}$ to obtain the binary numbers $K_1 K_2 K_3 K_4$.

$$K_1 = K_{1E} \text{ XOR}[(n_5 + n_1)]_{\text{mod }100}, K_2 = K_{2E} \text{ XOR}[(n_5 + n_2)]_{\text{mod }100}$$

$$K_3 = K_{3E} \text{ XOR}[(n_5 + n_3)]_{\text{mod }100}, K_4 = K_{4E} \text{ XOR}[(n_5 + n_4)]_{\text{mod }100}$$

4. Corresponding ASCII characters of K_1, K_2, K_3, K_4 is the password sent by Alice.

Example: Consider an elliptic curve $y^2 = x^3 + 3x + 15$ and let $q = 37$. This curve satisfies all the above-listed required properties. Since the order of the curve $E_{37}(3, 15)$ is a prime number 47, the group is cyclic. The graph of the curve is,

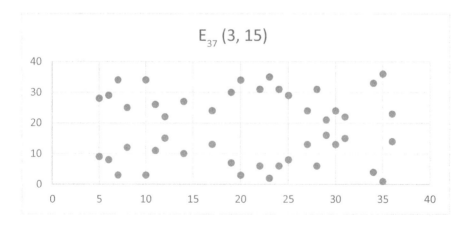

Let the generator be $G(19, 30)$. The table expressing all the points as the multiples of generator is

$1G = (19, 30); 2G = (11, 26); 3G = (35, 36); 4G = (8, 25); 5G = (31, 15); 6G = (14, 10);$

$7G = (20, 3)$; $8G = (24, 31)$; $9G = (34, 4)$; $10G = (12, 22)$; $11G = (36, 14)$; $12G = (30, 13)$;

$13G = (28, 31)$; $14G = (6, 29)$; $15G = (5, 28)$; $16G = (10, 3)$; $17G = (17, 13)$; $18G = (27, 13)$;

$19G = (25, 29)$; $20G = (29, 21)$; $21G = (22, 6)$; $22G = (23, 2)$; $23G = (7, 34)$; $24G = (7, 3)$;

$25G = (23, 35)$; $26G = (22, 31)$; $27G = (29, 16)$; $28G = (25, 8)$; $29G = (27, 24)$; $30G = (17, 24)$;

$31G = (10, 34)$; $32G = (5, 9)$; $33G = (6, 8)$; $34G = (28, 6)$; $35G = (30, 24)$; $36G = (36, 23)$;

$37G = (12, 15)$; $38G = (34, 33)$; $39G = (24, 6)$; $40G = (20, 34)$; $41G = (14, 27)$; $42G = (31, 22)$;

$43G = (8, 12)$; $44G = (35, 1)$; $45G = (11, 11)$ $46G = (19, 7)$; $47G = \infty$.

Encryption:

1. Let the function $f(x, y, n)$ selected by Alice be $f(x, y, n) = (8x + y)^n$. Alice publishes the elliptic curve $E_{37}(3, 15)$ and the function $f(x, y, n) = (8x + y)^n$ on public channel and shares the generator $G(19, 30)$ as the secret key with Bob. Let $n = 3, n_1 = 9, n_2 = 11, n_3 = 17, n_4 = 19, n_5 = 16$.
2. Alice calculates the points $P = nG = (35, 36)$, $P_1 = n_1 G = (34, 4)$, $P_2 = n_2 G = (36, 14)$, $P_3 = n_3 G = (17, 13)$, $P_4 = n_4 G = (25, 29)$, $P_5 = n_5 G = (10, 3)$ and the function $f(x, y, n) = (8x + y)^n = f(19, 30, 3) = 6, 028, 568$.
3. Again she calculates $m_1 = (6,028,568)_{\mod 9} = 8$, $m_2 = (6,028,568)_{\mod 11} = 7$, $m_3 = (6,028,568)_{\mod 17} = 11$, $m_4 = (6,028,568)_{\mod 19} = 1$.
4. Let the password to be communicated is 'SSSS'. These characters are encrypted by performing logical XOR operation between ASCII binary equivalent of S and 8421 BCD codes of $[(n_5 + n_1)]_{\mod 100}$, $[(n_5 + n_2)]_{\mod 100}$, $[(n_5 + n_3)]_{\mod 100}$, $[(n_5 + n_4)]_{\mod 100}$ to get 8 bit binary numbers $K_{1E}, K_{2E}, K_{3E}, K_{4E}$. These numbers are coded to corresponding ASCII characters to get encrypted password characters $C1, C2, C3,$ and C_4.

$$K_{1E} = K_1 \text{ XOR}[(n_5 + n_1)]_{\mod 100}, K_{2E} = K_2 \text{ XOR}[(n_5 + n_2)]_{\mod 100}$$

$$K_{3E} = K_3 \text{ XOR}[(n_5 + n_3)]_{\mod 100}, K_{4E} = K_4 \text{ XOR}[(n_5 + n_4)]_{\mod 100}$$

The encrypted password for 'SSSS' is 'JHrp'. Alice selects a random plain text vachk#momn8pq125jabc#46)$WRU. The encrypted characters 'JHrp' are inserted at the positions 8, 7, 11, 1 of the random text. Then Alice communicates the random text 'pachk#HJmnrpq125jabc#46)$WRU' in which the encrypted password is embedded at the positions 8, 7, 11, 1 and the points (35, 36), (34, 4), (36, 14), (17, 13), (25, 29), (10, 3) to Bob in public channel.

Decryption:

1. Bob after receiving plain text 'pachk#HJmnrpq125jabc#46)$WRU' in which the encrypted password is embedded and the points (35, 36), (34, 4), (36, 14), (17,

13), (25, 29), (10, 3) first picks up the values $n = 3$, $n_1 = 9$, $n_2 = 11$, $n_3 = 17$, $n_4 = 19$, $n_5 = 16$ from the table of points he constructed as multiples of generator.
2. He calculates $f(x, y, n) = (8x + y)^n = f(19, 30, 3) = 6,028,568$, the values $m_1 = (6,028,568)_{\text{mod}9} = 8$, $m_2 = (6,028,568)_{\text{mod}11} = 7$, $m_3 = (6,028,568)_{\text{mod}17} = 11$, $m_4 = (6,028,568)_{\text{mod}19} = 1$ where the encrypted password characters are inserted.
3. Bob locates the encrypted password characters 'JHrp' in the received random text.
4. He performs logical XOR operation between ASCII binary equivalents K_{1E}, K_{2E}, K_{3E}, K_{4E} of the characters J,H,r,p and 8421 BCD codes of $[(n_5 + n_1)]_{\text{mod}100}$, $[(n_5 + n_2)]_{\text{mod}100}$, $[(n_5 + n_3)]_{\text{mod}100}$, $[(n_5 + n_4)]_{\text{mod}100}$ to get 8-bit binary numbers K_1, K_2, K_3, and K_4.

$$K_1 = K_{1E} \text{ XOR}[(n_5 + n_1)]_{\text{mod } 100}, K_2 = K_{2E} \text{ XOR}[(n_5 + n_2)]_{\text{mod } 100}$$

$$K_3 = K_{3E} \text{ XOR}[(n_5 + n_3)]_{\text{mod } 100}, K_4 = K_{4E} \text{ XOR}[(n_5 + n_4)]_{\text{mod } 100}$$

The equivalent ASCII characters of K_1, K_2, K_3, and K_4 reveal the password 'SSSS'.

5 Usage of the Protocol in Client/Server System

In the present paper, encrypted password embedding technique using elliptic curve over finite field is narrated. The password characters are encrypted and embedded in a large random text. The elliptic curve $E_q(a, b)$ and the function $f(x, y, n)$ are public. The encryption algorithm proposed here is more appropriate in client/server systems to send one-time password (OTP) to the clients for their transactions by central servers of banks or corporate. Initially, the communicating parties agree upon to use a point $G(x, y)$, the generator of $E_q(a, b)$ for transmission of different passwords for different transactions at different times. Here $G(x, y)$ is vital in this technique that acts as the secret key for their communication. If the elliptic curve is selected so that it has huge number of points on it, then every point is the generator of the cyclic group and acts as secret key. So, multiple numbers of persons can communicate with one another simultaneously. This provides ease and secure communication among multiple peers over universal channel. This is the reason that the password-embedded technique designed here is well suited for transactions in client/server system. As the central or bank server is powerful, it provides the information to workstations or clients. Enormous number of elliptic curves over finite fields in the form (a_1, b_1, q_1), (a_2, b_2, q_2), ..., (a_i, b_i, q_i) having large number of points on each curve in the form (x_{11}, y_{11}), (x_{12}, y_{12}), ..., (x_{1n}, y_{1n}); (x_{21}, y_{21}), (x_{22}, y_{22}), ..., (x_{2n}, y_{2n}), ..., (x_{i1}, y_{i1}), (x_{i2}, y_{i2}), ..., (x_{in}, y_{in}) [$i = 1, 2, 3, ...$] are stored in the central server. The client first approaches the bank for registration, and then the client requests for the activation of the client's application server. At that stage, the central server allots an elliptic curve $E_q(a, b)$ at random and a few points on that curve to the client for choosing a point $G(x, y)$ as the secret key for future communication with the bank. Then central server stores the client's details and the secret key $G(x, y)$ chosen by the client in it. This process is called activation of the client's application. When the client needs transaction OTP, he/she request the central server to generate password. Then central server confirms the client's details, encrypts the OTP inserts in a large random text, and communicates to the client through public channel. Client server does the decryption procedure and shows the OTP to the client which is valid for a few seconds. As the central server is well built, many numbers of elliptic curves with large number of points on each curve are fed to it. Different curves with different points as secret keys can be allotted to several clients. So, various transactions from different clients can be performed by the central server simultaneously without overlapping.

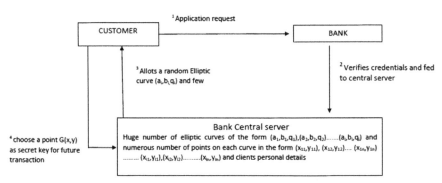

6 Conclusions and Cryptanalysis

The main difficulty in password communication mechanisms arises due to the fact that length of the password is small and much easier to crack. Here in the present algorithm, the password characters are encrypted using elliptic curve over finite field and inserted at different selected positions of a large random text. Elliptic curve cryptography is more secure than classical methods by the strength of hard problem Elliptic Curve Discrete Logarithm Problem (ECDLP). The well-known attack to solve ECDLP is Pollard's Rho attack, i.e., finding $x = \log_P Q$ for $Q = xP$ where P, Q are points on elliptic curve using random walk technique and collision detection algorithm. Here the elliptic curve used and the function $f(x, y, n)$ are public, and the generator point $G(x, y)$ is secret or private key between two individuals. Sender transmits the large random text in which the encrypted key characters are inserted and the points $P, P_1, P_2, P_3, P_4, P_5$ through public channel. Since the generator G is secret and $n, n_1, n_2, n_3, n_4, n_5$ values are calculated by the receiver, the context of Pollard's Rho attack does not arise here. The password embedding technique presented here achieves all the required qualities of standard cryptographic algorithms.

References

1. Ku, W. C., & Wang, S. D. (2000). Cryptanalysis of modified authenticated key agreement protocol. *Electronic Letters, 36*(21), 1770–1771.
2. Aziz, A., & Diffie, W. (1994). A secure communications protocol to prevent unauthorized access: Privacy and authentication for wireless local area networks. In *IEEE Personal Communications*, pp. 25–31, first quarter.
3. Aydos, M., Sunar, B., & Koc, C. K. (1998). An elliptic curve cryptography based on authentication and key agreement protocol for wireless communications. http://www.researchgate.net.
4. Lee, C.-Y., Wang, Z.-H., Harn, L., & Chang, C.-C. (2011). Secure key transfer protocol based on secret sharing for group communications. *IEICE Transactions, 94-D*(11), 2069–2076.
5. Diffie, W., & Hellman, M. E. (1976). New directions in cryptography. *IEEE Transactions on Information Theory, IT-22*(6), 644–654.
6. Washington, L. C. (2008). *Elliptic curves: Number theory and cryptography* (2nd ed.). Boca Raton, FL: Chapman and Hall.
7. Baalghusun, A. O., Abusalem, O. F., Al Abbas, Z. A., & Kar, J. (2015). Authenticated key agreement protocols: A comparative study. *Journal of Information Security, 6*, 51–58.
8. Juels, A., Molnar, D., & Wagner, D. (2005). Security and privacy issues in E-passports. In *IEEE SecureComm'05*, pp. 74–88.
9. Black, U. (2009). "Other key security protocols" book. Teach yourself networking in 24 hours, 332p.
10. Bellare, M., Kilian, J., & Rogaway, P. (2000). The Security of the cipher block chaining message authentication code. *Journal of Computer and System Sciences, 61*, 362–399.
11. Perrig, A. (1999). Efficient collaborative key management protocols for secure autonomous group communications. http://semanticscholar.org/18a4/25717de52e3981d67dd710a05ba2c926d2.pdf.
12. Burmster, M., & Desmedt, V.O. (1994). A secure and efficient conference key distribution system. In A. De Santis (Ed.), *EUROCRYPT 94*, LNCS 950, pp. 275–286.

13. Koblitz, N. (1987). Elliptic curve cryptosystems. *Mathematics Computation, 48*(177), 203–209.
14. Miller, V. (1985). Uses of elliptic curves in cryptography. In *Advances in Cryptography (CRYPTO 1985)*, Springer LNCS, Vol. 218, pp. 417–426.
15. Maurer, U., Menzes, A., & Teske, E. (2002). Analysis of GHS weil decent attack on the ECDLP over characteristic two fields of composite degree. *LMS Journal of Computation and Mathematics, 5,* 127–174.
16. Menzes, A., & Vanstone, S. (1997). Hand book of applied cryptography. In *The CRC-Press Series of Discrete Mathematics and its Applications*. CRC-Press.
17. Blumenfeld, A. (2011). *Discrete logarithms on elliptic curves*.
18. Miyaji, N., & Takano, S. (2006). Elliptic curves with low embedding degree. *Journal of Cryptology, 19*(4), 553–562.

Performance of Wind Energy Conversion System During Fault Condition and Power Quality Improvement of Grid-Connected WECS by FACTS (UPFC)

Sudeep Shetty, H. L. Suresh, M. Sharanappa and C. H. Venkat Ramesh

Abstract The demand for the power generation from wind is constantly growing. This situation forces the revision of the grid codes requirements, to remain connected during grid faults. Immediately, the voltage level will drop below 80% when fault occurs at PCC (point of common coupling) and the rotor speed of IG (induction generators) becomes unstable. In this work, UPFC are used under fault condition to improve the low voltage ride-through (LVRT) of wind energy conversion system (WECS) and damping of rotor speed oscillations of IG. Furthermore, after the fault UPFC acts as virtual inductor and leads to increase in terminal voltage of WECS. WECS with DFIG-based system is considered for analysis here. *By simulating DFIG-based WECS with UPFC indicates the improvement in LVRT* and remains and WTGs continues to operate with grid at certain voltage fluctuations, near grid. Also, indicates voltage improvement at PCC under fault conduction, and voltage is recovered easily to 1 pu at PCC.

Keywords DFIG-WECS · UPFC · Indian electricity grid code · LVRT · HVRT

S. Shetty · C. H. Venkat Ramesh
Department of EEE, NMIT, Bangalore, India

M. Sharanappa
Department of EEE-Renewable Energy, NMIT, Bangalore, India

H. L. Suresh (✉)
Department of EEE, SIRMVIT, Bangalore, India
e-mail: hlsuresh69@gmail.com

1 Introduction

The energy demand in world increases day by day. To meet this energy demand, we should consider conventional and nonconventional sources of energy. The nonconventional sources of energy have become more important to meet demand in recent years. The nonconventional energies are nothing but renewable energy sources like wind, solar, and biomass, etc. Compared to all energy sources Wind & Solar abundantly available in nature, but availability wind energy more [1, 2].

Availability of wind energy source in nature is abundant. Due to this, wind energy conversion system (WECS) became more popular in renewable energy sources [3]. The most popular wind turbine used in WECS is DFIG (Doubly-Fed Induction Generator), it as variable-speed operation. It can operate in sub- and super-synchronous speed and can control active and reactive power separately. DFIG uses the induction generator, so it requires reactive power from the grid during starting, which leads to the disturbance in grid voltage. It becomes sensitive to voltage dips, symmetrical and unsymmetrical faults, under disturbance it requires voltage compensation to keep voltage within the limits of LVRT and HVRT of the Indian Electricity Grid Codes (IEGC). Under fault condition, wind power generator is made compulsory with Fault ride-through (FRT) condition, to keep it steady and connected to grid. If grid is disconnected from WECS during penetration of high wind, which leads to the worst situation & make power system insecurity. During and after occurrence of disturbance in distribution/transmission network, WECS should operate with grid satisfactorily without tripping, for certain duration under voltage swell (HVRT) or voltage drop (LVRT) at PCC [4]. FACTS devices (flexible AC transmission system) are used for continuous operation of WTGs access to electric grid under fault and varying wind speed condition. FRT of wind turbine capability can be improved by using unified power flow controller (UPFC) with the agreement of IEGC. FACTS devices can be used to compensate or increase voltage, phase shift, real and reactive power improvement.

In different FACTS devices, UPFC is used to analyze the DFIG-WECS system which is connected to grid, which has capability of series as well as shunts compensation [5].

2 Problem Statement

The integration of WECS to grid can negatively affect the power system due to variation in wind power. In wind turbine controlling of active and reactive power in steady state and transient to improve capability of FRT to manage the grid codes. In IEC 61400-21, they mention power quality assessment for grid-connected wind farm.

By integration of large-scale wind farms, leads to power quality issues are:

1. Performance of wind turbine during normal and faulty condition.

2. How to overcome grid-side power quality problems under fault condition like voltage sag or swell by using unified power flow controller (UPFC).

3 MATLAB/Simulink Model of WECS Without UPFC System

Figure 1, shows the WECS with Grid-Connected, it consists of 6 nos of 1.5 MW DFIG with Grid-Connected forms 9 MW. Simulation is carried out in ideal mode three-phase voltage sources and frequency through two transformers and 20 km transmission line.

A 9 MW WECS formed by 6 nos of 1.5 MW wind turbines coupled to 25 kV distribution system export power to 120 kV grid through 20 km transmission line, by using 25 kV feeder. A "R" load of 500 KW with 0.9 MVAr filter, $Q = 50$ connected to generation bus of 575 V.

The type of model used is phasor and is to facilitate transient stability nature study with long-time simulation. In this study, the simulation is done for the duration of 30 s. The 9 MW wind farm is obtained by using 1.5 MW wind-turbine model. In 1.5 MW wind-turbine block, the necessary parameters are multiplied by 6. Multiplying is done by 6, and 9 MW wind farm is obtained as below:

- Nominal wind turbine and mechanical output power (NWTMOP): 6 × 1.5 MW, in wind turbine data menu.
- Rated power of generator: 6 × 1.5 × 0.9 MVA (6 × 1.5 MW at 0.9 Pf).
- Nominal DC-link bus capacitor: 6 × 10,000 μF.

In above model, select voltage regulation in control parameter dialog box. Then, voltage at terminal is restricted to $V_{ref} = 1$ pu (reference voltage) and voltage drops to $X_s = 0.02$ pu. A constant wind speed is maintained at 14 m/s. And torque controller is formed by using control system, which limits speed to 1.09 pu. The reactive power created in wind turbine is regulated to 0 MVAr. At wind speed 14 m/s, output power be 0.55 pu, which gives the rated power 0.55 × 9 MW = 4.95 MW and generator synchronous at speed 1.09 pu. Under normal operation, the wind turbine reactive power is regulated to 0 MVAr and the system operation is achieved to unity power factor (UPF). When wind speed 14 m/s, turbine output is active power which is 1.0 pu, when generator speed is 1.0 pu. FRT of WECS can be improved by connecting UPFC with WTGs; by connection of UPFC we can control active and reactive power at various buses. Mathematical simulation is done to identify the voltage regulation problems; voltage compensates by the sudden load connection at PCC.

The complete power system MATLAB/Simulink model used for study is shown in Fig. 1, which contains the controllers with the control strategy; MATLAB/Simulink software is used to implement control strategies.

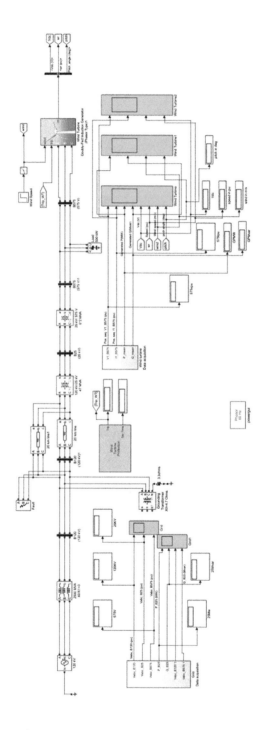

Fig. 1 Complete MATLAB/Simulink model of WESC

4 Performance of Wind Energy Conversion System During Different Conditions

4.1 Performance of WECS During Normal Condition for Change in Wind Speed

At the starting, speed of wind is set to 8 m/s, at $t = 5$ s, and wind speed is slowly increased from 8 to 14 m/s. Start the simulation observe various signal wave forms in scope monitoring and this signals are given by the "Wind Turbine and Grid Data Acquisition System" of the model. The different signals are wind turbine current, voltage, active and reactive power generated, DC-Link voltage, pitch angle, and turbine speed as shown in Figs. 2, 3, 4 and 5. At $t = 5$ s, the generated active power starts increasing uniformly (together with the turbine speed) to reach rated value of active power 9 MW, and it takes roughly 15 s. At this time, turbine speed is increased from 0.8 to 1.4 pu. At the beginning, the turbine blades pitch angle is 0° and the operating point of turbine follows the curve of turbine power characteristics. And pitch angle of turbine blades increased from 0° to 0.76° limits the mechanical power wind turbine.

Voltage and reactive power are generated. The reactive power is maintained at 1 pu voltage by using the control of the controller. When speed attains 14 m/s, the reactive power delivered to grid from wind generator becomes greater than 0.8 MVAr, and PCC voltage becomes 105% which is legally responsible for pay penalty, according to Indian Electricity Grid Code (IEGC). In the absence of UPFC, voltage at PCC will swell and also violate the safety margin of high voltage ride-through (HVRT) of the IEGC, due to these WTGs which should be disconnected from the grid. The output data of grid and wind turbine are shown in Table 1.

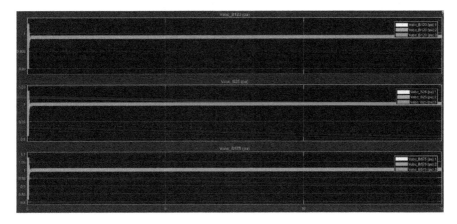

Fig. 2 Graph of output voltages at various buses to change in wind speed

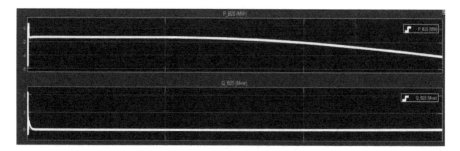

Fig. 3 Graph of real and reactive power at 25 kV bus to change in wind speed

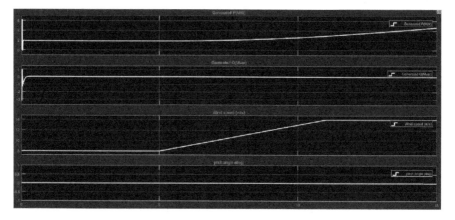

Fig. 4 Graph of real and reactive power, wind speed, and pitch angle for change in wind speed

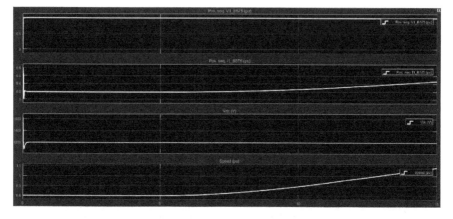

Fig. 5 Graph of positive sequence voltage and current at 575 V bus, V_{dc}, and speed (pu) at PCC for change in wind speed

Performance of Wind Energy Conversion System During Fault ...

Table 1 Grid data and turbine data under normal operation

Grid data under normal operation	Wind turbine data under normal operation
Voltage at different buses in per unit (pu): At 575 V bus, $V_a = V_b = V_c = 1$ pu At 25 kV bus, $V_a = V_b = V_c = 0.9995$ pu At 120 kV bus, $V_a = V_b = V_c = 0.999$ pu Active and reactive power at 25 kV bus, $P = -3.668$ MW, $Q = 0.0488$ MVAr	Positive sequence voltage and current at 575 V bus in pu, $V_{abc} = 1$ pu, $I_{abc} = 0.4194$ pu DC-link Voltage: $V_{dc} = 1200$ V Generated active and reactive power, $P = 4.194$ MW, $Q = 0.007684$ MVAr Speed in pu: 1.044 pu Wind speed in m/s: 14 m/s Pitch angle in degree: $0°$

Note In grid data under normal operation, the active power at 25 kV bus is negative, i.e., −3.668 MW. This is because of power flow from grid to generator to keep it continuous in operation. This may happen some time when the engine/turbine is not able to produce enough power to overcome the friction/windage losses; so in this condition or situation, grid-side electrical bus should supply power to generator to keep it running. This type of situation, we face generally in WECS with DFIG (Doubly-Fed Induction Generator)

4.2 Performance of WECS During Fault Condition at 120 kV Bus at Grid Side to Change in Wind Speed

Now, simulation is carried out by applying fault at 120 kV bus. The Fault in model is applied by double clicking on the fault block and select the "Phase A to Ground Fault," i.e., LG fault at 120 kV line at B120 bus. And ensure the fault is programmed to 9-cycle LG fault at $t = 5$ s. From graphs of Figs. 5 and 6, we monitor the wind turbine in "Voltage regulation" condition, +Ve Sequence voltage at V1_B575 bus, and terminal voltage of wind turbine drops to 0.8 pu under fault operation, at this condition voltage drop is beyond the under voltage protection threshold value of 0.75 pu for a $t > 0.1$ s. As a result, wind farm stays in operation. In the next situation compared with HVRT of IEGC as shown in Fig. 7. The graph shows the voltage at PCC break IEGC of HVRT level; in this situation, WTGs are disconnected from grid in order to avoid damages to wind turbine (Figs. 8 and 9).

Further, WECS model set to VAr regulation mode with $Q_{ref} = 0$, when voltage drops below 0.7 pu and under voltage protection is used to trip wind farm. Due to this, turbine speed increases. When $t = 10$ s, in order to limit the speed, pitch angle starts to increase. Table 2 shows the grid and wind data at- and after-fault operation.

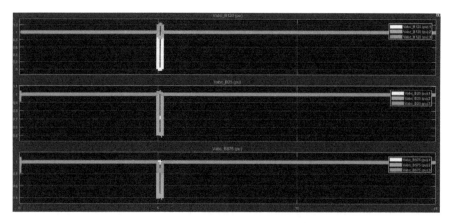

Fig. 6 Voltage at different buses on grid-side before-fault, at-fault, and after-fault

Fig. 7 Graph showing real and reactive power at 25 kV bus before-fault, at-fault, and after-fault

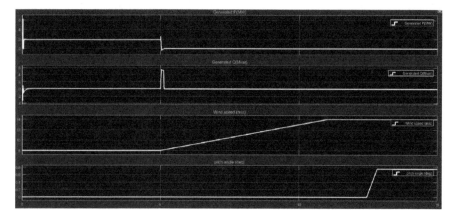

Fig. 8 Graph showing reactive and active power, wind speed, and pitch angle at PCC, before-fault, at-fault, and after-fault

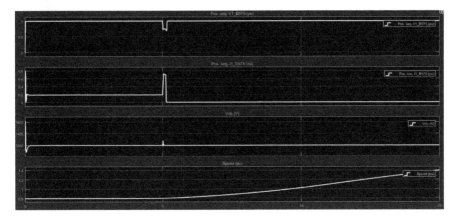

Fig. 9 Graph showing positive sequence voltage and current at 575 V bus, V_{dc}, and speed (pu) at PCC, before-fault, at-fault, and after-fault

Table 2 a Grid data and grid data exactly at fault operation at $t = 5.15$ s, **b** Grid data and wind turbine after fault operation

(a) Grid data exactly at-fault operation at $t = 5.15$ s	Wind turbine data exactly at-fault operation $t = 5.15$ s	(b) Grid data after-fault operation	Wind turbine data after-fault operation
Voltage at different buses in per unit (pu): At 575 V bus, $V_a = 0.9515$, $V_b = 0.3921$, $V_c = 0.8172$ At 25 kV bus, $V_a = 0.6954$, $V_b = 0.4959$, $V_c = 0.9989$ At 120 kV bus, $V_a = 0.0003353$, $V_b = 0.8592$, $V_c = 1.204$ Active and reactive power at 25 kV bus, $P = 0.3017$ MW, $Q = 0.0006286$ MVAr	Positive sequence voltage and current at 575 V bus in pu, $V_{abc} = 0.6865$ pu, $I_{abc} = 0$ pu DC-link Voltage: $V_{dc} = 1199$ V Generated active and reactive power, $P = 0$ MW, $Q = 0$ MVAr Speed in pu: 0.8032 pu Wind speed in m/s: 8.15 m/s Pitch angle in degree: $0°$	Voltage at different buses in per unit (pu): At 575 V bus, $V_a = V_b = V_c = 0.9988$ pu At 25 kV bus, $V_a = V_b = V_c = 0.9989$ pu At 120 kV bus, $V_a = V_b = V_c = 0.9988$ pu Active and reactive power at 25 kV bus, $P = 0.5228$ MW, $Q = 0.001089$ MVAr	Positive sequence voltage and current at 575 V bus in pu, $V_{abc} = 0.9988$ pu, $I_{abc} = 7.105 \times 10^{-15}$ pu DC-link Voltage: $V_{dc} = 1199$ V Generated active and reactive power, $P = 0$ MW, $Q = 0$ MVAr Speed in pu: 1.395 pu Wind speed in m/s: 14 m/s Pitch angle in degree: $0.76°$

5 Performance of Wind Energy Conversion System During Fault Condition with UPFC

5.1 UPFC

The wide global growth in electrical power demand, there is challenge to deliver the required electrical power considering the quality sustainability and reliability of the delivered power. To achieve this goal, it is essential to control the existing transmission systems for resourceful utilization and to avoid new plant installations cost [6]. In recent years, FACTS devices became more popular as they improve the utilization of power system that already exists and also give technical solution for improving the power system production [7]. There are many FACTS device in that, use of UPFC in more flexible in power system, which gives fast response and compensate active and reactive power in system. Installation of UPFC in the transmission system at exact locations increases the power supply from generator to grid, and this leads to the increase in generator power rating, thermal limits, transformer, line conductor, and stability level of the system. Active and reactive power control in four quadrants is done by using UPFC evenly, quickly, and independently, which consist of series and shunt converter [8].

A diagram of UPFC shown in Fig. 10 mainly contains two voltage source inverters (VSIs). In that, one acts as shunt and other acts as series VSI. And two VSIs are connected through DC link. A DC link includes DC capacitor (C). The bus and DC capacitor voltage are controlled by shunt converter of UPFC, which is linked to UPFC [9–13]. And series converter of UPFC controller is used to control line reactive and active power flow, by injecting series voltage of variable phase angle and magnitude.

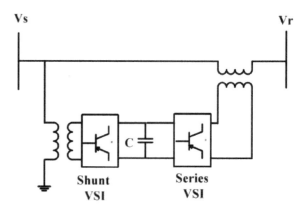

Fig. 10 Diagram of UPFC

5.2 MATLAB/Simulink Model of WECS with UPFC System

See Fig. 11.

5.3 Performance of WECS During Fault with UPFC Condition at 120 kV Bus at Grid Side for Change in Wind Speed

Simulation is carried out by applying a fault in fault block at grid which causes voltage sag at PCC bus, $t = 3.1$ s for a period of 0.6 s. Then at PCC, the performance of voltage is carried out under fault, by connecting UPFC at PCC bus (Fig. 12).

Figure 13 shows when fault occurs at grid side and voltage decreases lower than 0.5 pu at PCC. By referring LVRT of IEGC, WTGs have to be removed from grid, due to the violation of lower permissible voltage limit.

The connection of UPFC at PCC bus in power system, voltage sag reaches to its safety level of the IEGC as shown in Fig. 13, hence WTGs are not disconnected from grid.

By considering the US grid code, Fig. 6, shows the violation of voltage sag form the security level in the absence of UPFC at PCC. In the presence of UPFC, voltage sag is maintained to a safe level, WTGs are connected to grid under fault condition as shown in Figs. 12 and 14 which show the voltage across the DC-link capacitor of the WTG V_{dc} with and without the connection of the UPFC. The connection of UPFC reduces settling time and overshoots substantially. When fault occur at PCC, immediately, the UPFC will act to exchange reactive power with AC system and regulates the voltage. Under normal operating conditions, there is no reactive power in the system; this means the generation of reactive power is zero. When reactive power is zero under normal operation, no reactive power exchanges between the UPFU and power system. This means by using UPFC, we can achieve zero reactive power generation for WTG.

At the time of fault I_d (direct axis) and I_q (quadrature axis) are current of UPFC, there waveforms shown in Fig. 15. I_d and I_q currents set to zero under normal operation; therefore, no power exchange takes place between power system and UPFC At the time of fault occurrence, consequently change in the level of I_d and I_q, to support the reactive power of the system under the fault. I_d and I_q currents of UPFC reach to initial level i.e. zero again when fault is cleared. Table 3 shows the grid and wind turbine data at-fault operation with UPFC.

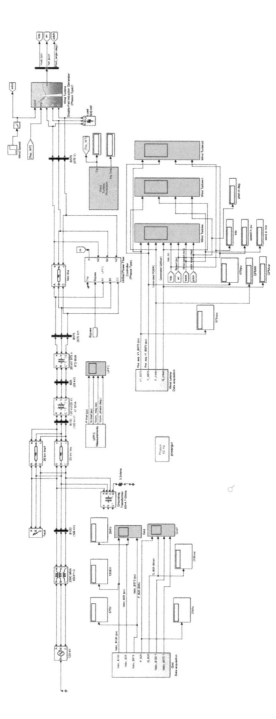

Fig. 11 WECS with UPFC system MATLAB/Simulink model

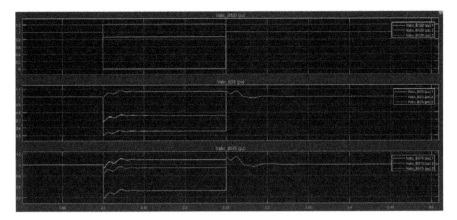

Fig. 12 Voltage at various buses at grid-side before-fault, at-fault, and after-fault with UPFC system

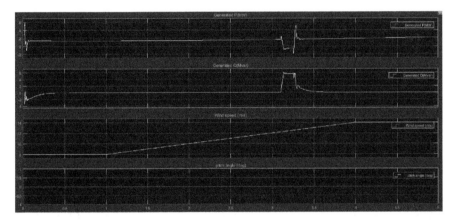

Fig. 13 Real and reactive power at PCC, before-fault, at-fault, and after-fault with UPFC system

Table 3 Grid data and wind turbine exactly at-fault operation at $t = 3.1$ s

Grid data exactly at-fault operation at $t = 3.1$ s	Wind turbine data exactly at-fault operation $t = 3.1$ s
Voltage at different buses in per unit (pu): At 575 V bus, $V_a = 1.088$, $V_b = 0.5249$, $V_c = 0.942$ At 25 kV bus, $V_a = 0.7576$, $V_b = 0.5533$, $V_c = 1.066$ At 120 kV bus, $V_a = 0.000337$, $V_b = 0.8648$, $V_c = 1.212$ Active and reactive power at 25 kV bus $P = 0.6592$ MW, $Q = -12.65$ MVAr	Positive sequence voltage and current at 575 V bus in pu $V_{abc} = 0.8227$ pu, $I_{abc} = 0.6825$ pu DC-link voltage: $V_{dc} = 1199$ V Generated active and reactive power, $P = -0.5431$ MW, $Q = 5.594$ MVAr Speed in pu: 0.8745 pu Wind speed in m/s: 14 m/s Pitch angle in degree: 0°

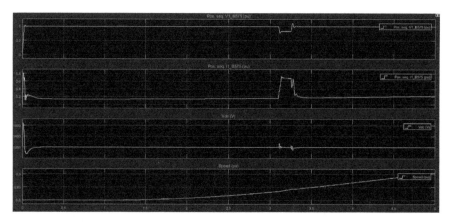

Fig. 14 Voltage across DC-link capacitor of the WTG, before-fault, at-fault, and after-fault with UPFC system

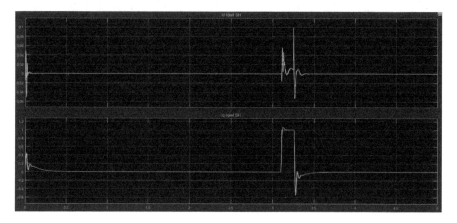

Fig. 15 I_d and I_q UPFC current during fault

6 Conclusion

This paper work is carried out for understanding the complete behavior of WECS at normal and fault at 120 kV bus. And waveforms for both normal and fault conditions are observed. Whole paper is simulated in MATLAB/Simulink software. When fault occurs, there will be voltage sag or swell that takes place due to these WTGs that require reactive power from the gird. And this voltage sag or swell should be within the safety margins of IEGC. If the voltage sag or swell is not in safety margins, we should compensate by using FACTS devices.

In this paper, the FRT can be improved by using UPFC with WECS and compared with IEGC and US grid code (USGC). Results show that, when fault occurs at PCC in the absence of UPFC, the voltage swell or sag takes place, when voltage swell or sag happens the WTGs should be disconnected from the grid to prevent damages to turbine. The voltage swell or sag at PCC violates safety margins necessary for grid codes.

When WESC is connected with UPFC, it improves the FRT ability of the WTGs; so there is no need to disconnect WTGs and operation of WECS to support the grid at the time of fault occurrence and continued to deliver power to grid. In this work, we can suppress unwanted electromechanical oscillation with the help of UPFC control strategy in power. Usage of UPFC improves the stability of voltage and also leads to the improvement in transfer capability of power into power system by WECS.

References

1. Salem, Q., & Altawil, I. (2014). Stability study of grid connected to multiple speed wind farms with and without FACTS integration. *International Journal of Electronics and Electrical Engineering, 2*(3), 168–174.
2. Amini, S., Bina, M. T., & Hajizadeh, A. (2014, March). Reactive power compensation in wind power plant using SVC and STATCOM. *International Journal of Emerging Science and Engineering (IJESE), 2*(5), 18–21, ISSN: 2319–6378.
3. Angalaeswari, S., Thejeswar, M. G., Poongodi, R. S., Basha, W. V., & Sasikumar, P. (2014). Power quality improvement of standalone hybrid solar-wind power generation system using FACTS devices. *International Journal of Advanced Research in Electrical, Electronics and Instrumentation Engineering, 3*(3), 7987–7992.
4. Deepa, S. N., & Rizwana, J. (2013). Multi-machine stability of a wind farm embedded power system using FACTS controllers. *International Journal of Engineering and Technology (IJET), 5*(5), 3914–3921.
5. Rajan, S. R. (2013). Power quality improvement in grid connected wind energy system using UPQC. *International Journal of Research in Engineering & Technology (IJRET), 1*(1), 13–20.
6. Kaldellis, J. K., & Zafirakis, D. (2011). The wind energy (r)evolution: A short review of a long history. *Renewable Energy, 36*, 1887–1901.
7. Baskar, S., Kumarappan, N., & Gnanadass, R. (2010). Switching level modeling and operation of unified power flow controller. *Asian Power Electronics Journal, 4*(3), 109–115.
8. Tumay, M., & Vural, A. M. (2004). Analysis and modeling of unified power flow controller: Modification of Newton-Raphson algorithm and user-defined modeling approach for power flow studies. *The Arabian Journal for Science and Engineering, 29*(2B), 135–153.
9. Hingorani, N. G. (1998, July). High power electronics and flexible AC transmission system. In *IEEE Power Eng.* (pp. 3–4).
10. Kumar, V., Pandey, A. S., & Sinha, S. K. (2016). International conference on emerging trends in electrical, electronics and sustainable energy systems (ICETEESES–16). In *IEEE explorer "Grid Integration and Power Quality Issues of Wind and Solar Energy System"*.
11. Tsili, M., & Papathanassious, S. (2009). A review of grid code technical equipments for wind farm. *IET Renewable Power Generation, 3*(3), 308–332.
12. Miller, N. W., Sanchez Gasca, J., & Delmerico, R. W. (2003). Dynamic modeling of GE 1.5 and 3.6 MW wind turbine-generators for stability simulations. *IEEE Power Engineering Society General Meeting*.
13. Abbas, F. A. R., & Abdulsada, M. A. (2010). Simulation of wind-turbine speed control by MATLAB. *International Journal of Computer and Electrical Engineering, 2*(5), 1793–8163.

Comprehensive Survey on Hadoop Security

Maria Martis, Namratha V. Pai, R. S. Pragathi, S. Rakshatha and Sunanda Dixit

Abstract The new emerging technologies have provided a way for a large amount of data generation. Secure storage of such a huge data is of prime importance. Hadoop is a tool used to store big data, where security of it is not assured. In this paper, we have considered a survey on various approaches which helps in providing secure storage of files in Hadoop. Hadoop framework is developed for the support of processing and storage of Bigdata in a distributed computing environment. Usage of Bigdata has become a key factor for the companies as they can increase their operating margin. Bigdata contains user-sensitive information and bring forth many privacy issues. Bigdata is a larger and a more complex datasets obtained from a variety of network resources. These datasets are beyond the ability of traditionally used data processing software to capture, manage, and process the data within the given time frame. These massive volumes of data are used by many of the organizations to tackle the problem that could not be done before. Since the data holds a lot of valuable information, these data need to be processed in short span of time by which companies can boost their scale and generate more revenue, traditional system resources are not sufficient for processing and storing, and this is where Hadoop comes into picture. The main objective of Hadoop is running of application of bigdata. Hadoop being a great tool for data processing, it was initially designed for internal use (i.e., within local cluster) without any security perimeter of organization, so they were easily hackable and exposed to threats.

Keywords Hadoop · Data security · Big data · Authentication · Authorization

1 Introduction

In the digital universe, due to the rapid growth, expansion of new services and number of online users on the Internet has lead to the significant increase in amount of data produced every millisecond. According to the Internet World Stats in the month

---

M. Martis (✉) · N. V. Pai · R. S. Pragathi · S. Rakshatha · S. Dixit
Department of ISE, Dayananda Sagar College of Engineering, Bangalore 560078, India
e-mail: mariamartis4@gmail.com

of December 2017, there were 4.157 million data users, i.e., 54.4% of total world population [1]. An open-source framework named Hadoop is provided by Apache in order to handle enormous amount of data efficiently. Hadoop is a setup, and HDFS is a framework. Hadoop consists of name node and data node which will be master–slave configuration. The job trackers that run on name node accept the jobs and allocate it to the task tracker. Task tracker that runs on the data node is responsible for performing the task and reporting the completion of task to the job tracker. The input file is broken down into blocks by name node, and each of the blocks is allocated to the data node. Replicas are used to provide the fault tolerance. More than 50% of Big Data experiments are executed on Hadoop, so securing these data has become a necessity. Hadoop does not contain any security framework; a foreign user can easily attack the data and access the contents.

The Present Hadoop security level is as follows:

(1) There is no encryption technique between host and Hadoop.
(2) The files which are stored in text format are controlled by name node.
(3) There is no strong security for communication between data nodes and clients.

Some of the known general procedures to provide security are Apache Knox gateway, it provides single point authentication, and main protocols used in this are HTTP or HTTPs to Hadoop cluster that provide single access point, central authentication, authorization, and hides topologies. Next level is authentication which allows identifying who the user is. It is provided by Kerberos [2] which allows nodes communicating with each other to prove their identity among themselves in much secured manner. Authorization ensures that user can access only those data for which they are entitled. Knox gateway provides authorization by evaluating user, group, and IP address. Hadoop (Highly Archived Distributed Object Oriented Programming) Hadoop actually came into, the picture in 2005 is a package that offers documentation, source code, location awareness, work scheduling.

Hadoop addresses bigdata challenge [3] which is explained below.

1. Hadoop framework allows user to store bigdata in distributed environment where the data is stored in blocks across the data nodes and block size can also be specified.
2. Extra data nodes can be added to HDFS cluster; thereby, scaling problem is resolved.
3. The data written once can be read multiple times since there is no predumping schema validation; hence, a variety of data can be handled in Hadoop.

Hadoop being a great tool for data processing, it was initially designed for internal use (i.e., within local cluster) without any security perimeter of organization, so they were easily hackable and exposed to threats. Hence, encryption is used to protect the data stored in Hadoop, which is must in government and finance sectors worldwide to meet privacy and other security.

Hadoop is a efficient, reliable distributed platform that provides scalable storage and computing on large datasets. Hadoop offers framework to process a large volume of data by running a large number of jobs in parallel on a cluster of machines.

Organizations use Hadoop because of its ability to store and process huge amounts of data quickly. But when they started storing confidential sensitive data on Hadoop clusters, a need for strong security mechanisms to protect these data is observed.

The user data stored in the cloud is not a controllable domain, and in order to protect the important data of user, confidentiality is an issue of most concern. In traditional public encrypt mechanism, the encryption resource provider needs to obtain all relevant information of user, and it will damage the user's privacy and requires more bandwidth and large processing overhead. The CP-ABE method has been proposed to solve the above issue.

The 4 Modules of Hadoop:

- Distributed File System

The most important two are the distributed file system, which allows data to be stored in an easily accessible format, across a large number of linked storage devices, and the MapReduce, which provides the basic tools for poking around in the data. (A "file system" is the method used by a computer to store data, so it can be found and used. Normally, this is determined by the computer's operating system; however, a Hadoop system uses its own file system which sits "above" the file system of the host computer—meaning it can be accessed using any computer running any supported OS).

- MapReduce

MapReduce is named after the two basic operations this module carries out—reading data from the database, putting it into a format suitable for analysis (map), and performing mathematical operations, i.e., counting the number of males aged 30+ in a customer database (reduce).

- Hadoop Common

The other module is Hadoop Common, which provides the tools (in Java) needed for the user's computer systems (Windows, Unix, or whatever) to read data stored under the Hadoop file system.

- YARN

The final module is YARN, which manages resources of the systems storing the data and running the analysis. Various other procedures, libraries, or features have come to be considered part of the Hadoop "framework" over recent years.

2 Hadoop Security Methodologies

This section discusses various Hadoop security methodologies.

2.1 Data Leakage Detection Using Haddle Framework

Gao [4, 5] proposed Haddle, a forensic framework that helps us to find the illegally copied and offender who stole the data. Haddle can improve the inspection of financial records of Hadoop. Haddle uses data collector and data analyzer to collect Hadoop logs, Fsimage files are sent to server, and automatic analytic method is used to find stolen data and the offender to recreate the crime scene, respectively. Investigating a big Hadoop cluster to identify the attacked node is very difficult, but this can be achieved by using automatic detection algorithm and abnormal condition can be noticed. Hadoop progger provides a lot of evidences even if Hadoop logs or Fsimage is compromised. Better techniques for collecting files are to make data contingent and reduce the performance issues. More improvement must be done on attack detection algorithm.

2.2 CP-ABE Security Access Mechanism

Zhou [6] proposed CP-ABE; it is a type of encryption where both private key and encrypted text depend on attribute. The system includes users and the intersection occurs when user and the encrypted text which is to decrypted consists pack of attributes which is greater than or equal to threshold and its user-specific key helps in regaining the original text. The proposed method prevents obtaining user information and reduces the chances of violating user rights. The approach provides data security accessing in the Hadoop throughout an area. The efficiency of implementation is yet to be achieved.

2.3 Data-at-Rest Security Using SDFS

Petros Zerfos [7, 8] proposed a method in which data at rest is provided security by using Hadoop in the cloud service by developing a new Hadoop file system called secure distributed file system (SDFS). The performance bottleneck that arises from the key distribution lowers the storage requirement for the secure data by using secret sharing and information dispersal technique. End-to-end security and controlled access of the data stored in enterprise are provided by SDFS which is used to minimize computational overhead and cost.

2.4 Modified RC4 Technique Using MapReduce

Jayan [9] proposed a method which uses parallel sections and modified RC4 algorithm to encrypt the data. The input is split into blocks and applied the encryption algorithm. The output is combined to get text in non-readable form. The algorithm is made parallel by combining the keys with the number of threads. This enhances the security. The key scheduling part and the random number generation part are executed parallel. Because of the parallel generation of keys, modified RC4 algorithm shows better performance and is more cost-effective than the MapReduce algorithm. Although modified RC4 algorithm is parallel, it is not capable of performing according to the expected level.

2.5 Cloud Disk Security Based on the Hadoop

Jing [10] proposed cloud disk storage which upholds the confidentiality. The encryption is done by the symmetric encryption algorithm and RSA algorithm which are operated on Hadoop cluster, and cloud is used for storage of data to give security. Performance was evaluated by comparing the expenditure and the files loaded into storage. The security provided by the cloud is not efficient enough.

2.6 Data Security Based on Hash Chain Technique

Jung [11–13] proposed a system which introduces an additional hash chain with one-way hash function. The hash function h of the PK scheme is utilized, and the output of h is again hashed; it takes only one hash value; thus, two values are required for the operation. It provides improved performance compared to PK scheme. PK scheme is based on one-time token method and prevents vulnerability of the block access token, which acts as a proof of user access rights on the data blocks. If a block access token is exposed to an attacker during its usage, the token cannot be used in a replay attack. It also offers high fault tolerance and high availability.

2.7 Security in G-Hadoop

Jam [14, 15] proposed the technique G-Hadoop that runs on many collections in a grid working area. G-Hadoop uses Secure Shell protocol between the client and the Hadoop cluster, it uses Globus Security Infrastructure (GSI) for providing security. GSI is one of the standards which is used for the grid security. Setting up GSI architecture provides security to each Hadoop cluster. It provides single sign on process, many other authentication mechanisms and upholds the integrity of messages sent over the grid. With the help of SSL hand shaking mechanism, communication between job node and the data node is securely established. The above approach provides better scalability.

2.8 Fully Hamomorphic Encryption

Jin [15] proposed work uses two technologies, namely fully homomorphic encryption technology and authentication agent technology. Former allows many users to experiment on data in form of unreadable format with different operations and yields same result as the data had been unlocked. Homomorphic encryption technology encrypts data first and then stores in data storage and latter combines many access control mechanism. The main drawback is the large increase in data and high complex computation.

2.9 Hadoop Security in Public Cloud

Yu [12] proposed SEHadoop [16, 17] model; this is developed to make Hadoop recover quickly from any malicious attack. This can be achieved by isolating Hadoop components, limit the range of data that can be accessed and gives entry license for each process. Setup for security was made on name node resource manager Kerberos using SE block token. In SEHadoop, there is no over usage of authenticated key, and it provides access control with minimum expenditure. SEHadoop includes SEHadoop block and delegation token, where SEHadoop block token does not possess any heavy burden or depreciation, whereas SEHadoop delegation token shows very limited accomplishment impact on existing Hadoop.

2.10 OTP Authentication Technique

Somu [18] proposed one-time pad technique that prevents the transfer of password in between servers by using one-time pad algorithm. It involves two steps:
1. Registration process,
2. Authentication process.

In the registration process, user enters the username and password. The password is encrypted two times using the OTP algorithm and mod 26 operation and stored in the registration server. Once again the password is encrypted using the symmetric cipher technique and sent to the backend server along with the username. In authentication process, the user needs to enter the username, and the backend server sends the cipher to the user via the registration server. Registration server decrypts with the key returned by the user. The registration server compares username stored in the backend with the username entered by the user. If it matches, authentication of the user is successful. This technique helps to store the process data related to the credit cards and healthcare. User must register and obtain OTP as the security code to access the resource on Hadoop. The OTP is not much secure which makes the outsider to get it easily and get access to the individual system.

2.11 Hadoop-Based Cloud Data Security Using Triple Encryption Scheme

Chau Yang proposed a method which combines HDFS file encryption using DEA and data key encryption with RSA and encrypts the user RSA private key using IDEA. The hybrid encryption consists of choice against two forms of encryption. The proposed system uses DEA and RSA algorithm to encrypt the data and get the data key and then encrypts the data key. The private key is kept with the user in order to decrypt the data key. It increases the performance through parallel processing of encryption and decryption using Hadoop. The performance factor is calculated by comparing the speed of writing and file size which is given as input to the HDFS system. The future work aims to achieve the parallel processing using MapReduce framework.

Table 1 gives a detailed survey of various security methods for Hadoop.

3 Conclusions

In any storage environment, protection of data is of almost importance. The above-listed methodologies provide various levels of security on Hadoop. The above-listed methodologies do not provide security measure at a very large scale. The techniques specified in this paper yield the performance which is not feasible to the expected level. In future, there is a need of producing and implementing a variety of powerful security measure to tackle the problem of danger or threat for data loaded in Hadoop and let the enormous data to be in a state of feeling safe.

Table 1 Survey of different Hadoop security methodologies

Year	Title	Encryption/scheme	Methodology	Pros	Cons
2017	Modified RC4 technique using MapReduce	Modified RC4 algorithm	Parallel section and modified RC4 algorithm	Better performance and cost-effective than MapReduce	Not flexible as per the expected level
2015	Data leakage detection using Haddle	Automatic detection algorithm	Security is provided using data collector and data analyzer	Identifying abnormal condition in node can be determined	Enhancement of detection algorithm to analyze the real scene and data in efficient manner
2015	Data security based on Hash chain technique	Hash chain technique	Two hashing functions required	Improved performance compared to PK scheme	NA
2015	Hadoop security in public cloud	SE Hadoop model	Isolating Hadoop components and limiting range of data	Access control over minimum overhead	SEHadoop delegation token shows limited performance
2015	Data-at-rest security using SDFS	SDFS to secure data at rest by using Hadoop in cloud service	Secret sharing and informal dispersal	Minimal computational overhead and cost	NA
2014	CP-ABE security access mechanism	Encryption done based on attributes	Encryption done where both user-specific key and encrypted text depend on attribute	Prevents obtaining complete user information	Efficiency of implementation is yet to be achieved
2014	OTP authentication technique	One-time pad algorithm	Multiple password encryptions is done and sent to backend server	Efficiently used in credit card, health care, etc.	OTP is not much secure and cracked easily

(continued)

Table 1 (continued)

Year	Title	Encryption/scheme	Methodology	Pros	Cons
2014	Security in G-Hadoop	G-Hadoop security architecture	Setting up GSI architecture provided security to cluster	Provides authentication communication using asymmetric cryptography	NA
2014	Fully homomorphic encryption	Authentication agent technology	Encryption is done by fully homomorphic encryption and then stored in HDFS	Combines any access control mechanisms	Complex computation and high cost
2013	Hadoop-based cloud data security using triple encryption scheme	Triple encryption scheme	HDFS file encryption using IDEA, data encryption with RSA, RSA private key using IDEA	Increases performance through parallel processing	NA
2013	Cloud data security based on Hadoop	BANLOGIC	Symmetric encryption algorithm and RSA algorithm used to secure Hadoop cluster	Ability to provide flexible and low-cost services	Security was not sufficient enough

References

1. www.internetworldstats.com.
2. Park, S. H., & Jeong, I. R. (2013). A study on security improvement in Hadoop distributed file system based on Kerberos. *Journal of the Korea Institute of Information Security and Cryptology, 23*(5), 803–813.
3. Abouelmehdi, K., Beni-Hssane, A., Khaloufi, H., & Saadi, M. (2016). Big Data emerging issues: Hadoop security and privacy. In *2016 5th International Conference on Multimedia Computing and Systems (ICMCS)*.
4. Gao, Y., Fu, X., Luo, B., Du, X., & Guizani, M. (2015). Haddle: A framework for investigating data leakage attacks in Hadoop. In *IEEE 2015*.
5. Chen, C. L. P., & Zhang, C. Y. (2014). Data intensive applications, challenges, techniques and technologies: A survey on Big Data. *Information Sciences, 275,* 314–347.
6. Zhou, H., & Wen, Q. (2014). A new solution of data security accessing for Hadoop based on CP-ABE. In *IEEE 2014*.
7. https://elastic-security.com.
8. Clouder Inc. (2015). *HDFS data at rest encryption*. Retrieved July 12, 2015 from http://www.cloudera.com/content/cloudera/en/documentation/core/latest/topics/cdhsghdfsencryption.html#xd583c10bfdbd326ba--5a52cca-1476e7473cd--7f85.
9. Jayan, A., & Upadhyay, B. R. (2017). RC4 in Hadoop security using mapreduce. In *2017 International Conference on Computational Intelligence in Data Science (ICCIDS)*.
10. Jing, F. A. H., Renfa, S. B. L., & Zhuo, T. C. T. (2013). The research of the data security for cloud disk based on the Hadoop framework. In *2013 Fourth International Conference on Intelligent Control and Information Processing (ICICIP)*.
11. Jung, Y.-A., & Woo, S.-J. (2015). A study on Hash Chain-based Hadoop security scheme. In *IEEE 2015*.
12. Yu, X., Ning, P., & Vouk, M. A. Enhancing security of Hadoop in a public cloud. In *2015 6th International Conference on Information and Communication Systems (ICICS)*.
13. Dean, J., & Ghemawat, S. (2004, December). MapReduce: Simplified data processing on large clusters. In *Proceedings of the 6th Conference on Symposium on Operating Systems Design and Implementaton*, pp. 137–150.
14. HadoopGIS on the FutureGrid.
15. Jam, M. R., Khanli, L. M., & Akbari, M. K. (2014). A survey on security of Hadoop. In *2014 ICCKE*.
16. O'Malley, O., Zhang, K., Radia, S., Marti, R., & Harrell, C. (2009). Hadoop security design. In *Yahoo, Inc., Tech. Rep*.
17. Yuan, M. (2012). Study of security mechanism based on Hadoop. *Information Security and Communications Privacy, 6,* 042.
18. Somu, N., Gangaa, A., & Sriram, V. S. S. (2014, April). Authentication service in Hadoop using one time pad. *Indian Journal of Science and Technology, 7*(4), 56–62.

Descriptive Data Analysis of Real Estate Using Cube Technology

Gursimran Kaur and Harkiran Kaur

Abstract With the progress and application of data analysis and processing expertise, the technologies such as data warehouse and cube technology had become the research spot of each business area. Online analytical processing technology has applications in the areas of merchandising system, college decision support system, enterprise marketing management system, knowledge data warehouse, and to analyze the curriculum chosen by students and many more. If we consider one business domain that is real estate, it is observed that online transactional processing technology has been applied. This paper focuses on application of online analytical processing and specifically descriptive data analytics, to extract more information from the traditional real estate datasets.

Keywords Cube · Descriptive analysis · Real estate analysis · OLAP · Visualization

1 Introduction

Traditional online transactional processing (OLTP) systems are outdated and replaced with online analytical processing (OLAP) systems because of their purposeful and routine necessities. OLAP technology is the need of an hour for every business domain to apply aggregations on large datasets wherein these datasets are residing in large repositories called data warehouses (DW). OLAP systems are used to organize business domains and decision-making process using aggregations that describe descriptive analytics. OLAP databases can be broken down into one or more cubes which are organized by cube administrator according to the way it may retrieve the data. These cube structures are all about aggregating measures based on dimen-

---

G. Kaur (✉) · H. Kaur
Department of Computer Science and Engineering, Thapar Institute of Engineering and Technology, Deemed to be University, Patiala 147002, India
e-mail: gkaur.me16@thapar.edu

H. Kaur
e-mail: harkiran.kaur@thapar.edu

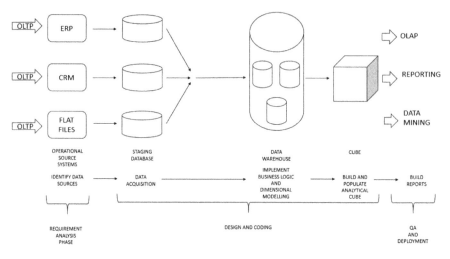

Fig. 1 Extract, transform, and load (ETL) process [1]

sions and hierarchies. Finally, OLAP makes decisions based on these pre-aggregated values of cube structures. Following sections and Fig. 1 explain the OLAP process.

1.1 Operational Source System

Basic but an important role is played by data in computing. In computing, data is a collection that has been apprehended in a way that is effective for demonstration or distribution. In comparison with today's transmission media, data is info converted into dualistic mathematical form. Raw data is a term used to identify data in its most primary numerical form. There are two types of data sources: internal and external. Internal data refers to the information that is gathered and collaborated from different branches inside an organization. While external data is the data which is not collected from internal sources of the organization, but acquired from external sources. This primary data is collected from operational source systems. An operational source system is a chunk used in data warehousing to represent an organism that is used to practice everyday transactions of an institution. The design of these systems is done in a way that the distribution of day-to-day transactions is executed adequately and the certainty of the transactional data is conserved. These systems maintain transactional or operational data and are outside the data warehouse. There could be any number of such systems feeding data to the data warehouse but they may maintain little historical data. Operational systems include DB's, ERP, CRM, and flat files. The benefit of maintaining less historical data is being simple (keeping in mind that these are typically used by clerks, cashiers, clients, and so on) and being efficient so that it allows its users to read, write, and delete data quickly, due to which these systems were designed to support online transactions and query processing. While on considering the security issues or the decision-making process, OLTP system lacks behind. So, online analytical processing systems came into novelty.

1.2 Staging Database

To convert OLTP systems to OLAP systems, extract, transform, and load (ETL) process is used which includes requirement analysis followed by design, coding, quality assurance, and deployment phases of software development life cycle (SDLC) [1]. The process begins with getting data into the data warehouse environment. This is called extraction. The next step after extraction is transforming the heterogeneous form of datasets to homogeneous form. The transformation process enables the dataset to be loaded into the multidimensional model.

1.3 Data Warehouse

DW is a subject-edified, assimilated, and permanent assortment of data to act as decision support of management [1]. Data warehouse consists of operational source systems, data staging area, and presentation area. Data from operational systems flow into the staging area where it undergoes transformation and is then put into the presentation area. The data can be accessed from the presentation areas using accessing tools.

1.4 Cube

Cube mapping and validating the same is done in analytical workspace manager (AWM). AWM is the basic application software for forming, evolving, and conducting dimensions in the Oracle DB. The first step in AWM is creation of dimensional data store. After this step, cubes, measures, dimensions, levels, hierarchies, and attributes are defined followed by mapping these objects to their respective relational data objects. The next step involves loading the data from relational DB to dimensional DB.

Creating Dimension: Dimensions are the distinctive values which recognize and classify the dataset [1]. These are located at the edges of multidimensional cube. Each dimension is defined by either levels or values [1].

Creating Levels: There is a level-based hierarchy where levels have base-derived relationships [1]. This hierarchical structure discloses the trends at the abstract levels and then rolls down to the detailed levels for recognizing the indicators contributing toward the trend.

Creating Hierarchies: Hierarchies ensure that if the relational table columns and levels of the respective dimensions in the multidimensional model are in the same sequence or not.

Mapping Dimension: The relational data tables are mapped with cube model consisting of dimensions. Mapping supports the process of loading data into the respective dimensions.

Creating Cube: Cubes are defined by the orderly arrangement of dimensions. Measures incorporate the facts and figures compiled for the selected dataset. Measures act as the outcomes of cube formation process, as these include all the calculations related to the said dataset.

2 Related Work

He et al. in [2] proposed the application of OLAP and data warehouse in merchandising system. In this paper, merchandise sales data warehouse is built on the basis of merchandise sales system data resources, which further establishes merchandise sales multidimensional datasets on which data analysis is conducted, which helps in significant decision making. On the basis of this system and using the analysis services, this paper builds DW and proposes on-line analytical processing system for judgments, giving a suitable method for enterprise to route gigantic volumes of information and to launch an operative verdict sustenance structure.

Haiyan et al. in [3] discussed the idea of multidimensional methodical approach based on online analytical processing. Also, this paper blends efficiently the development platform of MS-BI and trains the application of online analytical processing technology in AWC Company. Online analytical processing technology does not just merely examine queries but also counts omnidirectional research manufacture and gather these data to guide everyday difficulties and even to estimate future undertakings based on predictable data. In this paper, unnaturally presentation of MS's BI development platform determines knowledge from data warehouse of AW's DW. With the online analytical processing technology becoming progressively corporate, the range of its application has been discovered and prolonged progressively. Using precise procedures, it can deal with the unambiguous data and realize tangible purpose and its dependability of the acquaintance discovered is extraordinary.

Ying-Ping et al. in [4] described the OLAP bid in enterprise marketing management (EMM) system. According to the author in [4], with the growth of budgetary globalization and extraordinary mechanism, marketing is experiencing insightful variations. The marketing process is an administration verdict creation process. The crucial aim of EMM is to figure out the data storeroom to deliver confirmation provision for judge to make a verdict. Use of OLAP and MDX technology in the marketing management systems apprehends the erection of multidimensional data and fact analysis.

Wang et al. in [5] defined that the assemblage of DW technology and OLAP opens DW and this is the innovative way for DSS. This paper examines the substantial knowledge of DW and OLAP and discerns the application of online analytical processing in apprentice enactment investigation using MS SQL Server and Analysis Services. The authors determined the appropriate evidence by examining exhibited teaching value and enhancing teaching possessions for the decision maker. OLAP emphasizes on the verdict building provision for verdict building executive. The MD analysis model is suitable for analysis demand for data warehouse and is the most essential methodical factor to make the DW successful. This article applies MD analysis model of DW to apprentice enactment investigation of colleges and universities, and this model which has appropriate manifestation ability and quicker analysis speed can fulfill numerous analysis application demand of the apprentice enactment information.

Zhao in [6] explains that with the expansion of the scale of advanced teaching, surplus data about set of courses chosen by learners has been produced. This paper analyzes the application of DW and OLAP for the analysis of set of courses selected by learners, undertakes the design of DW about universities set of courses chosen by learners, and applies ETL process upon set of courses chosen by learner's data. This paper builds multidimensional cube analysis data model by taking use of OLAP technology on set of courses chosen by learner's data and captures the query and show analysis of set of courses chosen by learner's multidimensional data, so it can analyze the set of course's formation situation from various angles and assist the university teaching DSS. This paper applies the STAR model of DW to the analysis of set of courses chosen by learners and determines that this model can mollify to various analytical application demands for set of courses chosen by learner's information.

Quafafou et al. in [7] find OLAP application in knowledge data warehouses. This paper is a profound analysis of both data warehouse and knowledge discovery and the necessities basic for profound assimilation. The authors suggested a data warehouse model. Also, they proposed a STAR schema and established a consideration using operatives. OLAP analysis depends upon the observed data and a set of OLAP operators for restructuration and granularity. Main aim is to discover concealed outlines in data. The combination of knowledge into DW conduces to supplemented analysis circumstance where objects and their relations are unambiguously handled and visualized. The authors in [7] investigate profound analysis where the simple DW operators cogitate both data and knowledge. This paper applies knowledge DW concept to network usage analysis.

Mitsujoshi et al. in [8] proposed an emotion concept model (ECM) which combines the desire and annoyance dimension and numerous feelings, to obtain a descriptive analysis on sentiment and mood in voice. For descriptive analysis, the main objective of this paper is to attain ample quantity of vocal sound data and to analyze the correlation between sentiments and moods and improving the accuracy of voice emotion recognition. By using prior filtering of desire and annoyance sentiments, moods will be categorized more accurately. It is not always the case that desire and annoyance tally to each mood. There can be a case where basic negative moods are generated from desire sentiments and basic positive moods are generated from annoyance sentiments. Mitsujoshi et al. in [8] explain that emotion concept model recommends a conceivable link between desire and annoyance sentiments and numerous moods by defining the effect of sentiments on moods. To be precise, data that evidences the directions of sentiment and moods is dissimilar. Further, the authors congregated samples from a footage of a free discussion between applicants. Their tactic to acquire voice data that the desire and annoyance direction did not correspond to the mood was confirmed, by associating the results with those reported earlier. Their voice recording experiments and descriptive analysis of the relationship between the desire and annoyance sentiments and numerous moods revealed that the desire and annoyance direction did not essentially correspond to the feeling.

Wang et al. in [9] proposed a descriptive model that explains with respect to power-driven vitality costs, reinforcement costs show assorted attributes. While the show of power-driven energy costs has been generally considered in the writing, such investigation on supplementary facilities costs is incomplete. This paper investigates the utilization of trustworthy stochastic methodologies for representing the conduct of operating reinforcement costs in power market. Wit h the de-control of in power frameworks, helper facilities arcades have emerged to get these facilities through sensible sell-offs. In general, the cost of providing assistant facilities is lower than fabricating vitality. In this way, however the arcade volume for operating stores is usually lower than energy market, and the incomes from retailing these facilities can be equivalent to energy. Hourly save costs are more eccentric than the day by day reinforcement costs. In view of the attributes of the hourly reinforcement cost, BSD display is connected and coordinated with a BS show. The replication outcomes demonstrate that the BSD show has taken key highlights of the concentrated hourly backup cost and beaten the BS display. Despite the fact that just a single backup cost is arranged, the results can be drawn out to other backup costs that have indicated alike highlights, extraordinary erraticism, and rehashed substantial spikes.

3 Need of Present Research Study

Presently, OLTP projects for the real estate dataset are available that does not support descriptive analytics upon business-related datasets because nothing has been done to analyze and query such data.

One way out to this problem is working upon OLAP database technology that is optimized for querying and reporting, aiming to create applications according to today's need.

4 Proposed Implementation

Also these projects work in a formal two-dimensional view using data definition language (DDL) and data manipulation language (DML) commands. So, another technique called online analytical processing (OLAP) resolves this issue of analyzing large datasets related to business in multidimensional structure. These projects work in three-dimensional or multidimensional view. Here, multidimensional view or cube technology is comprised of dimensions, levels, hierarchies, mappings, measures, and lastly the cube creation. To create a multidimensional structure or a cube, some sequential steps have been followed.

4.1 Extract, Transform, and Load (ETL)

Extract, transform, and load (ETL) process is applied on heterogeneous datasets to convert it into homogenous form.

Extract: Extract real estate dataset in the form of .csv files, .xls files, and .xlsx files from [10] and [11].

Transform: Transformations are applied for converting heterogeneous datasets to homogenous format files (.xlsx in this work) where transformations could be data type conversions, formatting, merging and splitting applied on columns, etc.

Load: Loading the homogeneous format (.xlsx) files into SQL Developer tool to map these datasets with multidimensional analysis model.

4.2 Creating Data Sources in Oracle iSQLplus

Creating data sources involves data entries to some particular columns selected for various tables. For creating a particular database related to real estate, first step is to select various dimensions/tables to be created. In this research work, the tables nominated include (A) customer profile dimension, (B) property dimension, and (C) document dimension, as shown in Fig. 2. The tables are created in Oracle iSQLplus using create table query. The data entries to these tables is made using insert query. And lastly, the table is viewed using select query.

4.3 Creating Cube

Cube creation, mapping, and the validation of ETL process are done in analytical workspace manager (AWM). Main aim in using AWM is the formation of multidi-

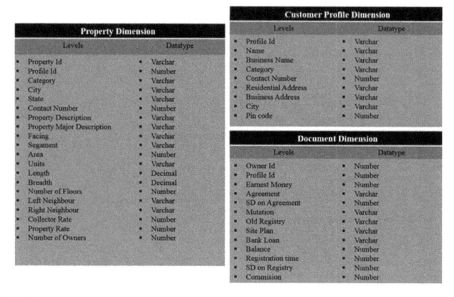

Fig. 2 Dimensions, levels, and datatypes

mensional data mart which provisions the task of real estate data analysis. So, the whole process is carried out in the following steps:

Creating Dimensions: Dimensions selected for a particular business domain are unique and are level-based dimensions. The levels selected for our domain are: (a) customer profile, (b) property dimensions, and (c) document.

Creating Levels: In our domain, customer profile dimension consists of nine levels named as profile id, name, business name, category, contact number, residential address, business address, city, and pin code. Property dimension consists of twenty levels, namely property id, profile id, category, city, state, contact number, property description, property major description, facing, segment, area, units, length, breadth, number of floors, left neighbor, right neighbor, collector rate, property rate, and number of owners. Document dimension consists of thirteen levels named as owner id, profile id, earnest money, agreement, stamp duty on agreement, mutation, old registry, site plan, bank loan, balance, registration time, stamp duty on registry, and commission.

Creating hierarchies: Dimensions selected for this particular domain have one hierarchy and are a level-based hierarchy. Likewise, the hierarchy for customer profile follows the sequence: profile id, name, business name, category, contact number, residential address, business address, city, and pin code; the hierarchy for property dimension be property id, profile id, category, city, state, contact number, property description, property major description, facing, segment, area, units, length, breadth, number of floors, left neighbor, right neighbor, collector rate, property rate, and number of owners; and the hierarchy for document dimension be owner id, profile

Descriptive Data Analysis of Real Estate Using Cube Technology

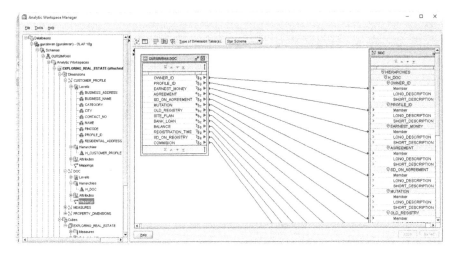

Fig. 3 Mapping relational model with analysis model

id, earnest money, agreement, stamp duty on agreement, mutation, old registry, site plan, bank loan, balance, registration time, stamp duty on registry, and commission.

Mappings: In mapping part, each column of relational database table is mapped to the member part of each level in the analysis model as shown in Fig. 3.

Loading data into dimensions from SQL developer: A data connection is built between AWM and SQL Developer tools. The dataset is loaded in SQL Developer and further accessed in AWM for the cube creation.

Cube Creation: Cubes is created by firstly creating the measures and then mapping the measures to the relational database table to create a final design. In this research work, STAR schema design has been implemented for the selected dataset.

5 Results and Findings

A desired STAR schema is successfully designed which is required to create the cube, the main aim of this research. Further operations are applied on the cube and finally creating dashboards.

5.1 Measures Defined

The selected real estate domain has the nine measures named as balance, collector rate, commission, earnest money, property rate, registration time, stamp duty on agreement, stamp duty on registry, and total area.

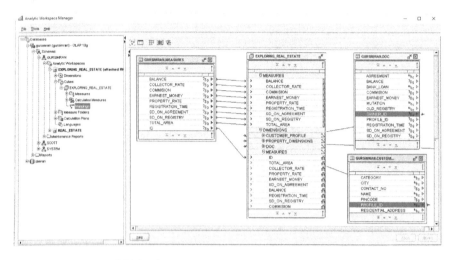

Fig. 4 Cube mappings/STAR schema design of cube

5.2 Cube Mapping

Also, for this problem domain, the main aspect is to choose the appropriate schema design that is either STAR schema or SNOWFLAKE schema. Basic reason for choosing STAR schema is that firstly, it is simplest of data warehousing schema. Secondly, it consists of large central table called fact table, where central table is referred by number of dimension tables. Lastly, STAR schema is very effective in handling queries because of very less normalization (Fig. 4).

5.3 Cube

Cube is created when mapped measures are loaded using maintain cube routine and later on viewed using view cube utility. So this final cube is created in Power Pivot that is further an add-in in Excel 2016. Cube creation is shown in Fig. 5.

5.4 Cube Operations

Major challenge is turning data into valuable insights in order to increase performance and efficiency. OLAP cubes cover wide variety of data, display them, and apply aggregations on them. Also, users can simplify their search by using slicing and dicing operations on cube that provides them searchable access to data points.

Descriptive Data Analysis of Real Estate Using Cube Technology

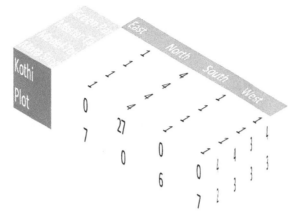

Fig. 5 CUBE creation

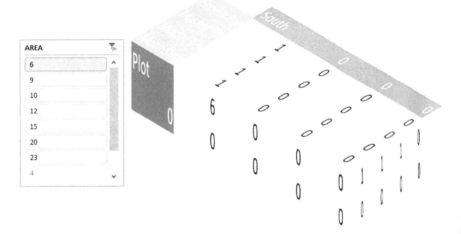

Fig. 6 Slicing

Slicing: For demonstrating the slicing process applied, Fig. 6 describes a single slice which is cut based on one parameter that is area = 6 marlas. So, the cube shows a plot of 6 marlas in south facing where it further shows 1 plot in Green Park, 1 plot in Master Tara Singh Nagar, 1 plot in Model town, and lastly, 1 plot in Mota Singh Nagar.

Dicing: For dicing process, multiple slices can be cut at same moment of time. So, the facing, segament, and property dimensions vary with parameter area = 6,9,10,15,20,23. This process is described in Fig. 7.

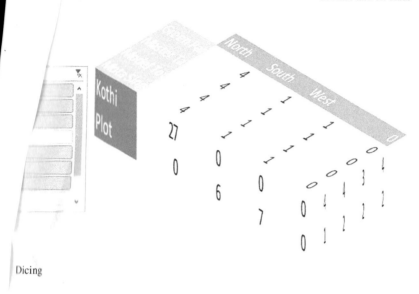

Dicing

Visualizations of Cube

...lizations display insights that have been discovered in the data in the form of ...boards. Tool used to represent visualizations is Power BI, which might have ... page with one visual or might have pages full of visuals. Different types of ...lizations can be created using Power BI tool, such as bar chart, combo chart, ...hart, gauge chart, area chart, funnel chart, and waterfall chart.

...r chart in visual 1 shows the specific values across different parameters such ...pin code by category and city.
...mbo chart in visual 2 associates the column chart and the line chart. So, here ...escribes the count of property description by city.
...e chart in visual 3 gives the inclusive shape of the complete sequence of ...nciples based on area by property description.
...ge chart in visual 4 gives the current status.
...a chart in visual 5 covers the area between the axis and the line such as city ...roperty description and category.
...el chart in visual 7 helps to picture a process that has phases and substances ...uccessively from one phase to the next. So, here it shows property major ...iption and property description funnel.
...fall chart in visual 8 shows the running total as soon as the values are ...or subtracted. So, this visual gives values of registration time by property ...tion (Fig. 8).

Descriptive Data Analysis of Real Estate Using Cube Technology 249

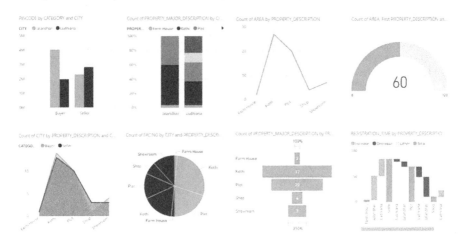

Fig. 8 Visualizations on various parameters

6 Conclusion

In this paper, the OLAP technology is applied in the selected field of business domain: the real estate domain. It has been identified that the OLAP operations lead to summarization of data in data warehouse. Also, cube technology leads to multidimensional view of data. Here, STAR schema design is used because it finds its usage in a variety of applications and it suits our domain area.

The main agenda of the paper is to create a cube, and further its base lies in defining and deciding dimensions, levels, hierarchies, attributes, mappings, measures, cube mapping, etc. The cube is formed after validating the ETL process in AWM. Using any of the schemas available, various cubes can be built considering different parameters. Snowflake schema finds its application in the area that comprises complex modules, but since our business domain comprises single fact table, so STAR schema is of greater use in this selected domain. Also, it has been concluded that slicing and dicing operations applied on cube are of greater use since they explain the variations occurring in dataset.

Lastly, visualizations are created to explain the concept in an easy manner by creating dashboards. In short, this paper completes the descriptive analysis of real estate dataset.

Future work lies in defining the predictive analysis of the same domain where predictions would be done based on various KPI's to identify the rise or fall of property in certain area or market. This would later on become helpful in identifying whether to invest or not in that certain period of time.

References

1. Kimball, R. (1996). *Data warehouse toolkit* (3rd ed.).
2. He, Z., & Lin, B. (2011). Application of data warehouse and OLAP in merchandising system. In *Proceedings-2011 International Conference on computational and Information Sciences*, ICCIS 2011, pp. 449–451.
3. Zhenyuan, W., & Haiyan, H. (2010). OLAP technology and its business application. In *2010 Second WRI Global Congress on Intelligent Systems*, pp. 92–95.
4. Ying-Ping, A., & Hu, J. (2017). OLAP application in enterprise marketing management system. In *2016 3rd International Conference on Systems and Informatics*, ICSAI 2016, pp. 588–592.
5. Li, C., Wang, Y., & Guo, X. (2010). The application research of OLAP in college decision support system. In *2010 International Conference on Multimedia and Information Technology*, MMIT 2010, pp. 74–77.
6. Zhao, H. (2008). Application of OLAP to the analysis of the curriculum chosen by students. In *2nd International Conference on Anti-counterfeiting, Security and Identification*, ASID 2008, pp. 97–100.
7. Quafafou, M., Naouali, S., & Nachouki, G. (2005). Knowledge datawarehouses: Web usage OLAP application. In *Proceedings—2005 IEEE/WIC/ACM International Conference on Web Intelligence*, WI 2005, pp. 334–337.
8. Shimura, M., Monma, F., & Mitsuyoshi, S. (2010). Descriptive analysis of emotion and feeling in voice. In *Proceedings of the 6th International Conference on Natural Language Processing and Knowledge Engineering*, NLP-KE 2010, pp. 4–7.
9. Wang, P., Zareipour, H., & Rosehartm, W. (2014). Descriptive models for reserve and regulation prices in competitive electricity markets. *IEEE Transactions on Smart Grid, 5*(1), 471–479.
10. Retrieved from Collector Rate in Punjab, http://punjabrevenue.nic.in/collectorrateframe.htm, (n.d.).
11. Retrieved from Justdial—Local Search, Order Food, Travel, Movies, Online Shopping, https://www.justdial.com/, (n.d.).

Temporal Information Retrieval and Its Application: A Survey

Rakshita Bansal, Monika Rani, Harish Kumar and Sakshi Kaushal

Abstract With an advent of the Web, a tremendous amount of information is available online. Information can be organized and explored in the time dimension. This temporal information has to be distilled out, so as to extract the temporal entities such as temporal expressions and temporal relations out of it. Temporal information processing is an ongoing field of research that deals with natural language text, temporal relations, events or temporal queries. This paper presents a detailed analysis of the work carried out under temporal information retrieval (TIR) highlighting its subtasks like information extraction, indexing, ranking, query processing, clustering and classification. Also, it presents various challenges while dealing with temporal information. To the end, various application areas are elaborated such as temporal summarization, exploration and future event retrieval.

Keywords Temporal information · Temporal events · Temporal expressions · Temporal queries

1 Introduction

Information retrieval (IR) is the process of obtaining the relevant information to satisfy the need from a collection of information resources. An information need is a topic about which a user desires to know and is expressed in the form of a short textual query. IR facilitates unstructured or semi-structured searching which can be based on either a full-text or content-based indexing.

With the rapid growth of information on the Web, the information can be organized and explored along the time dimension. Metzger [1] considered the timeliness as one significant measure to determine the quality of a document. The others are accuracy, coverage, objectivity and relevance. The quality and value of information retrieval are primarily dependent on time. Information on Web persists to have different versions with time which leads to a new search area known as temporal

R. Bansal (✉) · M. Rani · H. Kumar · S. Kaushal
Department of CSE, UIET, Panjab University, Chandigarh 160014, India
e-mail: bansal.rakshita@gmail.com

© Springer Nature Singapore Pte Ltd. 2019
N. R. Shetty et al. (eds.), *Emerging Research in Computing, Information, Communication and Applications*, Advances in Intelligent Systems and Computing 906,
https://doi.org/10.1007/978-981-13-6001-5_19

Fig. 1 Google search result for the queries "*who is Miss India*" (left) and "*who is Miss India 2017*" (right)

information retrieval (TIR). Figure 1 shows an example of Google search for two queries different in time context. The first query "*who is Miss India*" provides the general information about the Femina Miss India contest, whereas the second query "*who is Miss India 2017*" provides the more precise and time-specific information and delivers the name of Miss India 2017 "Manushi Chillar". A tremendous amount of time-related information which possesses the time-varying attributes exists in Web documents. Temporal taggers are used to extract temporal expression from the text and to normalize them in some standard format. It can be exploited by the content analysis, query analysis, retrieval and ranking models. Temporal information accessible through metadata, document creation time and date of last modification is used in several tasks such as time-aware search or temporal clustering. Time dimension has a strong influence in many domains like information extraction, topic detection and tracking, question-answering, query log analysis and summarization.

The remainder of the paper is organized as follows. In Sect. 2, time information available in documents and how time and events are related are discussed. In Sect. 3, the framework of temporal information process and its subtasks are discussed in detail. In Sect. 4, the review of diverse real-world applications of TIR is done and in Sect. 5, various challenges associated with TIR are highlighted. Finally, Sect. 6 concludes the paper.

2 Time in Document

There is plenty of temporal information in text documents. The extraction of temporal information is important for numerous IR tasks. It is useful for many real-world applications, viz. the ordering of clinical records, finding burst events in the news,

Web searching and reservation systems. In this section, the basic concepts related to temporal entities and temporal relations have been introduced. A temporal entity portrays a particular point in time, or event, or a time period. Different approaches based on name entity extraction are performed on the document to identify temporal entities and to determine the temporal expressions.

Temporal expressions can be defined as phrases and words with temporal importance. To better understand the natural language processing tasks, it is important to recognize and normalize these expressions to some standard format. Schilder and Habel [2] classified temporal expressions into three categories: explicit, implicit and relative temporal expressions. The explicit temporal expression describes a specific point in time which can be mapped to the timelines without any further knowledge, e.g. the token sequence "November 2017" or "3.12.2017". The implicit temporal expression refers to temporal information which is not specified explicitly, and therefore, it is difficult to map these expressions on the timeline but names of holidays or events can be mapped easily with the help of prior knowledge, e.g. The token sequence "Christmas day" can be mapped to the expression "25th December". The relative temporal expression represents those temporal entities which are mapped to the timeline with reference to already mapped temporal expression, e.g. "Next month" or "last Monday" is mapped with reference to the document creation time or the absolute dates close-by in the content. These three categories of expressions are further classified into four types. The initial two types, viz. DATE and TIME, are single entities and refer to the particular point of time and date, unlike the rest two DURATION and SET, which convey information about the length of interval and repetitiveness. These temporal expressions refer to a date or duration or periodical aspect of an event.

Time and events are closely associated with each other. The event is a term which refers to a situation that happened or is happening in time like resigning, meetings, capturing, holding, etc. They can be scheduled to a specific time or a duration, e.g. in the news "A 22-year-old woman miraculously escaped serious injury when she passed out and **fell** onto subway tracks in Brooklyn **on Saturday**". The temporal expression used to define the event "*fall*" is "*on Saturday*". Events can also be expressed in the form of a situation, occurrence, circumstance, state, action, *etc*.

Pustejovsky et al. [3] expressed events by using tenses, verbs, adjectives, prepositional phrases or normalizations present in a document and classified events as OCCURRENCE, REPORTING, ASPECTUAL, PERCEPTION, STATE, I-STATE or I-ACTION. In natural language, various relationships occur between temporal expressions and events. Allen [4] proposed seven types of temporal relations: *before, after, including, at, start, finish* and *excluding* which are used to classify all temporal relations.

3 Framework

TIR undergoes a sequence of phases to retrieve important information from a set of documents. This section presents the temporal framework for basic supporting structure beneath TIR. As shown in Fig. 2, there are six subsections related to, i.e., the temporal information extraction, indexing, temporal query, temporal ranking, temporal clustering and temporal classification processes. The details are presented in the next subsection.

3.1 Temporal Information Extraction (TIE)

Temporal information extraction is an emerging branch of information extraction in which the desired information is extracted from the documents for many artificial intelligence systems. Temporal tagging is a prerequisite for the whole temporal information extraction, which covers the detection and interpretation of temporal expressions, events and the temporal relations, but the temporal taggers only address the extraction and normalization of temporal expressions present in the content to some standard format. The extraction task decides whether a token is part of a temporal expression or not. Therefore, it can be regarded as a typical classification problem, whereas the normalization task requires semantic knowledge to assign a rule-based value to a temporal expression. In the last two decades, the temporal tagging has become the major focus in the field of research. Different research competitions like ACE, TEMPEVAL, EVENTI, I2B2, CLINICAL TEMPEVAL and QA REMPEVAL have been organized to address the temporal tagging challenge based on the annotation standards. The annotation standards with precise specifications are required to perform temporal tagging. TimeML [5], an XML-based annotation lan-

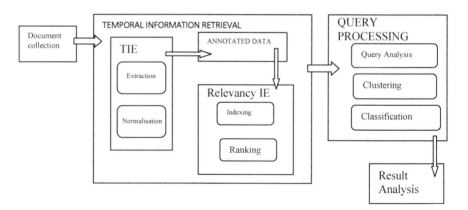

Fig. 2 Temporal information retrieval framework

Table 1 Description of TimeML tags

Tag	Description
TIMEX	Timex tag represents time, date and duration in the text
EVENT	Event tag is used to highlight different classes of events
TLINK	Temporal Link represents a relationship that occurs between events and time
SLINK	Subordinate Link is used to annotate the directional relation between predominant and inferior event
ALINK	Aspectual Link represents a relationship that occurs between a conditional event and its conflicting event

guage, presents precise guidelines for the annotation of temporal entities. Table 1 describes the TimeML tags used to annotate different types of information and their relationships.

Two types of approaches, namely rule-based approaches and machine learning approaches, are mostly followed for temporal information extraction. The approach requires great human efforts to formulate domain specific extraction rules. Mani and Wilson [6] presented a rule-based system TempEx for the temporal annotation of multiple languages. It annotates in the form of TIMEX2 tags. Another system Chronus [7] was developed as a part of ACE TERN 2004 which also applied the TIMEX2 tag on the Italian and English texts. Later on, Perl temporal tagger GuTime was developed at Georgetown University which was evaluated on ACE TERN 2004 training data. It used TIMEX3 tags to annotate and achieve competitive results. It is a part of TARSQI TOOLKIT [8]. Another tagger SUTIME [9], which is a part of Stanford CoreNLP pipeline, is also available as Java library which also annotates in the form of TIMEX3 tags. HeidelTime [10, 11] was developed at the University of Heidelberg as UIMA component. It is a multilingual temporal tagger which also annotates in TIMEX3 tags format. It has achieved the best result in the TempEval-2 challenge for English. Ramrakhiyani and Majumder [12] proposed a temporal rule-based annotation system for text in Hindi. The limitation of the rule-based system is its domain dependency which leads to lack of portability. So, to minimize the dependency on rules, the machine learning approach came into the picture. Machine learning approach is feature based and is implemented using training data for temporal tagging. Machine-based system ClearTk was developed by Bethar [13] to perform temporal annotation using SVM classifier, and TIPSem [14] was developed at the University of Alicante to perform temporal annotation using conditional random fields according to the TimeML [3] specifications. Although machine learning approach saves the human energy, a significant limitation in the system is that there is difficulty in modifying the training models.

3.2 Temporal Indexing

Temporal indexing is used to manage the documents and to allow fast searching of enormous data in time. Indexing is done after the temporal information extraction on the pre-processed documents. In pre-processing, the documents are converted into some standard format. Temporal searching over past documents has been identified as a challenging task by the researchers. Tang et al. [15] categorized temporal indexes into two types, namely single temporal index and bitemporal index. The former is based on the valid time or the transaction time. Indexes based on valid time are: ST-tree, interval tree, time index. Indexes based on transaction time are: the Snapshot Index and HR-Tree. On the other hand for bitemporal index, the valid time and the transaction time are considered simultaneously, e.g. GR-Tree, 2LBIT, IVTT and M-IVTT. In the year 2006, Zobel and Moffat [16] conducted a survey which explained that indexing usually adopts inverted index structure. In 2007, Berberich et al. [17, 18] suggested an extended inverted file index-based solution for time-based text search. In 2008, authors [19] proposed TISE, a hybrid two-layer temporal structure where the first layer is associated with the inverted indexing of keywords and the second layer is time-based index using MAP21 [20]. A few years later, inverted index technique was modified by Anand et al. [21] in 2012, which allowed the incremental inclusion of latest document versions without reconstructing the index structure. In 2017, a temporal search engine GIGGLE [22] depends on temporal indexing that distinguishes the novel and unexpected relationship between a query and interval set.

3.3 Temporal Ranking

The rank of the document is according to its relevance to the query. To address the temporal queries and to increase the relevancy of retrieved documents, the time dimension has been added to the ranking model. Temporal ranking amends position of a document with respect to time objective stated in queries. Therefore, both the textual and temporal similarities are considered while ranking. Based on the temporal query categorization, the ranking process is divided into two types: recency-based and time-dependent. In the former type, the fresh documents are preferred. The freshness is directly proportional to the time passed since documents' creation or modification. In the latter type, the results are adjusted with respect to the time of the query. In time-dependent ranking, the creation time of resulted document is close to the time period underlying in the temporal query.

Li and croft [23] suggested a time-based language which assigns high priority to the recently created document. Jatowt et al. [24] suggested an approach that used document freshness and its relevance. However, Zang et al. [25] ranked the time-sensitive query which adopted the more recently dated document. Further, Stronger and Gertz [26] considered temporal terms and geographical terms in the text and their proximity to answer the spatiotemporal query. To enhance the search effectiveness of Web archive, author [27] used hybrid ranking model for successive versions of the archive.

3.4 Temporal Query Processing

Temporal query processing is related to the assessment of temporal aim hidden in search queries of users. In 2013, Cheng et al. [28] focused on a class of queries called timely queries. Based on the correlation of users' perception of time and distortion in term distribution of results of timely queries, they estimated the query specifications and how to consolidate document freshness into the ranking model. Following their work, Campos et al. [29] in 2014 categorized queries into two classes, namely recency-sensitive queries and time-sensitive queries. Recency-sensitive queries result in fresh and topically relevant documents. Recency of the document is inversely proportional to the time span from its creation time. Dong et al. [30] proposed the idea of "recency ranking", by using a high-precision classifier to rank document by relevance and distinguished queries containing breaking news.

The time-sensitive queries focus on the change over time. Dakka et al. [31] identified topically relevant time intervals based on the query from a news archive automatically.

Rosie [32] classified the temporal queries into three types on the basis of common patterns of the retrieval documents which are atemporal, temporally unambiguous and temporally ambiguous. The first query type corresponds to a topic which is ongoing. The second query type corresponds to a specific period of time. The third query type refers to a combination of several possible events.

3.5 Temporal Clustering

Temporal clustering is an unsupervised classification process of grouping large data sets according to their temporal similarity into meaningful clusters. It is a challenging task due to rapidly increasing data collected from Internet and remote sensing devices. It is performed to dissect the collection of the documents by time. Few studies in this field have been proposed so far. TCluster, introduced by Alonso and Gertz [33], is an overlapping clustering algorithm, where a list consisting of 3-tuples has been maintained containing the information about the types of expression, its normalized value and set of its positions. Clusters have been formed on the basis of the more relevant overlapping document. However, the documents in a cluster may not be topically related.

Another technique is based on hit list clustering [34] where query outcomes are classified based on real-time search. The user does not need to look into analogous items as similar documents are clustered along the timeline. The engines based on this technique, e.g. Vivisimo [35], construct clusters based on web page title,

URL, snippets and temporal expression. To detect the future events, a model-based clustering algorithm [36] was proposed by Jatowt and Yeung.

3.6 Temporal Classification

It is the process of classifying the temporal relations between temporal entities and events and building the classification model by adopting the proper feature vectors. This process is essential to frame event timelines and to recreate the episode of events which can be utilized by decision support systems and document archiving applications. Mani et al. [37] built a classifier based on supervised classification technique, MaxEnt. It annotated temporal links applying temporal closure over training data. Later on, authors [38] used previously learned attributes for classifying the relationship type of event–event pair in time. In 2011, UzZaman and Allen [39] proposed classification as a part of the TempEval-3 shared. In 2013, D'Souza and Ng [40] and Laokulrat et al. [41] addressed this task by using features based on deep parsing, semantic role labelling and discourse parsing. Wu and Zhu [42] proposed a recurrent fuzzy neural network for the temporal classification. To enhance the ability to memorize information, fully connected feedback topology is used. Furthermore, it considers minimum-classification-error and minimum-training-error which enhance the discriminability factor and results in excellent classification performance.

4 Applications

With the progression in the research in temporal information retrieval, various streams of applications have focused on exploiting the use of time, e.g. temporal search engines to improve search processes. In 2012, Campos et al. [43] presented a temporal search interface, GTE-Cluster, which performs topic-based searching in a temporal perspective. On the basis of document relevance and temporal relevance, it returns the output of the user query. Google is the most popular and worldwide used search engine. It supports various time-based tools, but still it has not addressed the challenge of incorporating temporal relations in the retrieval process. Later, Uma et al. [44] proposed a search plug-in, RaTeR, to support searching of the Web using temporal operators. Search engines use summarization technology to help users in deciding whether a document is relevant as an output for a query or not. Temporal summarization is another emerging field where temporal information is used to enhance the storytelling. It can be termed as sentence scoring problem where each sentence is given a score and on that basis, the decision is built whether to include or emit that sentence. It highlights the prevalence of event occurrence and assists users in observing a current document. The goal is to develop a system to keep users up to date with the topic of their interest. One of the main challenges in temporal summarization is to maintain quality when the new aspects are added dynamically at any time and to impose word limit for a summary. In 2013, TS was introduced as a part of the Text REtrieval Conference (TREC) to return the updates for large events in news stream and social content. Xu et al. [45] proposed a system HLTCOE, which used a simple linear model and features such as document relevance, sentence relevance and topical salience information to select the sentence for summarization. In 2014, McCreadie et al. [46] proposed a novel incremental update summarization approach to rank sentences using features such as relevance, quality and novelty and to create timelines of top-ranked results using a supervised regression model. In 2017, a monitoring system TOTEMSS was proposed by Boluang et al. [47] for summarization of sentiments based on temporal properties by performing clustering technique on the tweets containing same sentiments towards a topic. Later in 2018, Richard et al. [48] proposed a novel approach in which they used explicit event aspects and different types of user interesting information to select the sentences for summarization. Another popular application that makes use of time-based IR techniques is future-related information retrieval and prediction. It helps in decision-making process, e.g. a person wishes to buy a laptop and would like to know the upcoming models before making the decision of purchase. In 2013, authors [49] proposed a method to anticipate the upcoming events from large news corpus. Later on, Zang et al. [50] used a recurrent neural network technique on multivariant time series data containing heterogeneous variables to predict the future events. After the analysis of temporal information retrieval and its application, the following real-world challenges are identified as presented in next section.

5 Challenges to Temporal Information Retrieval

Temporal information retrieval faces many challenges starting from the collection and pre-processing of the data to the implementation of solutions in various phases of the framework. There are many unsolved problems arising due to the linguistics. In some languages, the extraction of temporal information is harder than others, e.g. in Chinese and Japanese, because these languages have no express limit between words. Therefore, the word segmentation has been a challenging issue for them. While in some languages the time words are utilized in contrast to their expected utilization, *e.g.* in the Hindi language "Sham" is a very common name; however, it additionally signifies a time (evening). Even after extracting the temporal expressions, many new challenges arise while performing the normalization. Identifying the reference time and its temporal relation has been the main issue in normalization of temporal expression, e.g. the phrase "on Monday" in the document can either refer to previous Monday or to the next Monday.

6 Conclusion

One of the key measurements of our lives is time. With the help of timeliness, the quality of information is highlighted and therefore the researchers are using the temporal information in various applications, e.g. generation of temporal summaries of events or predicting future events. The purpose of this paper is to provide a better understanding of temporal information retrieval. In this paper, a framework of TIR and its subtasks, namely information extraction, indexing, ranking, query processing, clustering and classification, are presented. In the end, this paper covers the challenges and various applications of temporal information retrieval.

References

1. Metzger, M. (2007). Making sense of credibility on the web: Models for evaluating online information and recommendations for future research. *Journal of the American Society for Information Science and Technology, 58,* 2078–2091. https://doi.org/10.1002/asi.20672.
2. Schilder, F., & Habel, C. (2001). From temporal expressions to temporal information: Semantic tagging of news messages. In *ACL01 Work Temporal Spat Information Processing* (pp. 65–72). https://doi.org/10.3115/1118238.1118247.
3. Ingria, R., Saur, R., Pustejovsky, J., et al. (2002). TimeML: Robust specification of event and temporal expressions in text. In *Proceedings of the AAAI Spring Symposium on New Directions in Question Answering* (pp. 28–34).
4. Allen, J. (1983). Maintaining knowledge about temporal intervals. *Communication of the ACM, 26,* 823–843. https://doi.org/10.1145/182.358434.
5. Pustejovsky, J., Lee, K., Bunt, H., & Romary, L. (2010). ISO-TimeML: An international standard for semantic annotation. In *LREC*, May 2010, La Valette, Malta.
6. Mani, I., & Wilson, G. (2000). Robust temporal processing of news. In *Association for Computational Linguistics* (pp. 69–76).
7. Negri, M., & Marseglia, L. (2005). *Recognition and normalization of time expressions: ITC-irst at TERN 2004* (Tech. Report WP3.7, Information Society Technologies).
8. Uzzaman, N., & Allen, J. F. (2010). TRIPS and TRIOS system for TempEval-2: Extracting temporal information from text. In *Proceedings of the 5th International Workshop on Semantic Evaluation ACL* (pp. 276–283).
9. Chang, A. X., & Manning, C. D. (2012). SUTime: A library for recognizing and normalizing time expressions. In *LREC* (pp. 3735–3740).
10. Strötgen, J., & Gertz, M. (2010). HeidelTime high-quality rule-based extraction and normalization of temporal expressions. In *Proceedings of the 5th International Workshop on Semantic Evaluation, ACL* (pp. 321–324).
11. Strötgen, J., & Gertz, M. (2013). Multilingual and cross-domain temporal tagging. *Language Resources and Evaluation, 47,* 269–298. https://doi.org/10.1007/s10579-012-9179-y.
12. Ramrakhiyani, N., & Majumder, P. (2015). Approaches to temporal expression recognition in Hindi. *ACM Transactions on Asian and Low-Resource Language Information Processing, 14,* 1–22. https://doi.org/10.1145/2629574.
13. Bethard, S. (2013). ClearTK-TimeML: A minimalist approach to TempEval 2013. *Seventh International Workshop on Semantic Evaluation, 2,* 10–14.
14. Llorens, H., Saquete, E., & Navarro, B. (2010). Tipsem (English and Spanish): Evaluating crfs and semantic roles in tempeval-2. In *Proceeding SemEval'10 Proceedings of the 5th International Workshop on Semantic Evaluation* (pp. 284–291).

15. Tang, Y., Ye, X., & Tang, N. (2011). *Temporal information processing technology and its application*, Chap. 8, (pp. 51–158).
16. Zobel, J., & Moffat, A. (2006). Inverted files for text search engines. *ACM computing surveys (CSUR), 38*(2), 1–56. https://doi.org/10.1145/1132956/1132959.
17. Berberich, K., Bedathur, S., Neumann, T., & Weikum, G. (2007). A time machine for text search. In *Proceedings of the 30th Annual International ACM SIGIR Conference on Research and Development in Information Retrieval—SIGIR'07*, (pp. 519–526). https://doi.org/10.1145/1277741.1277831.
18. Berberich, K., Bedathur, S., Neumann, T., & Weikum, G. (2007). FluxCapacitor: Efficient time-travel text search. In *Proceedings of the VLDB '07* (pp. 1414–1417).
19. Jin, P., Lian, J., Zhao, X., & Wan, S. (2008). TISE: A temporal search engine for web contents. In *Proceedings—2008 2nd International Symposium on Intelligent Information Technology Application*, IITA 2008 (pp. 220–224). https://doi.org/10.1109/iita.2008.132.
20. Nascimento, M. A., & Dunham, M. H. (1999). Indexing valid time databases via B+-trees. *IEEE Transactions on Knowledge and Data Engineering, 11*, 929–947. https://doi.org/10.1109/69.824609.
21. Anand, A., Bedathur, S., Berberich, K., & Schenkel, R. (2012). Index maintenance for time-travel text search. In *Proceedings of the 35th International ACM SIGIR Conference on Research and Development in Information Retrieval—SIGIR'12* (pp. 235–243). https://doi.org/10.1145/2348283.2348318.
22. Layer, R. M., Pedersen, B. S., Disera, T., Marth, G. T., Gertz, J., & Quinlan, A. R. (2018). GIGGLE: A search engine for large-scale integrated genome analysis. *Nature Methods, 15*, 123–126. https://doi.org/10.1038/nmeth.4556.
23. Li, X., & Croft, W. B. (2003). Time-based language models. In *Proceedings of the 12th International Conference on Information and Knowledge Management* (pp. 469–475).
24. Jatowt, A., Kawai, Y., & Tanaka, K. (2005). Temporal ranking of search engine results. In *Proceedings of the 6th International Conference on Web Information Systems Engineering* (pp. 43–52).
25. Zhang, R., Chang, Y., & Zheng, Z. (2009). Search result re-ranking by feedback control adjustment for time-sensitive query. In *Proceedings of NAACL HLT* (pp. 165–168).
26. Strötgen, J., & Gertz, M. (2013). Proximity 2-aware ranking for textual, temporal, and geographic queries. In *Proceedings of the 22nd ACM International Conference on Conference on Information & Knowledge Management—CIKM'13* (pp. 739–744).
27. Costa, M., Couto, F., & Silva, M. (2014). Learning temporal-dependent ranking models. In *Proceedings of the 37th International ACM SIGIR Conference on Research & Development in Information Retrieval—SIGIR'14* (pp. 757–766).
28. Cheng, S., Arvanitis, A., & Hristidis, V. (2013). How fresh do you want your search results? In *Proceedings of the 22nd ACM International Conference on Conference on Information & Knowledge Management—CIKM'13* (pp. 1271–1280). https://doi.org/10.1145/2505515.2505696.
29. Campos, R., Dias, G., Jorge, A., & Jatowt, A. (2014). Survey of temporal information retrieval and related applications. *ACM Computing Surveys, 47*, 1–41. https://doi.org/10.1145/2619088.
30. Dong, A., Chang, Y., Zheng, Z., et al. (2010). Towards recency ranking in web search. In *Proceedings of the Third ACM International Conference on Web Search and Data Mining—WSDM'10* (pp. 11–20). https://doi.org/10.1145/1718487.1718490.
31. Dakka, W., Gravano, L., & Ipeirotis, P. (2012). Answering general time-sensitive queries. *IEEE Transactions on Knowledge and Data Engineering, 24*(2), 220–235.
32. Jones, R., & Diaz, F. (2007). Temporal profiles of queries. *ACM Transactions on Information Systems, 25*(3), 14. https://doi.org/10.1145/1247715.1247720.
33. Alonso, O., & Gertz, M. (2006). Clustering of search results using temporal attributes. In *Proceedings of the 29th Annual International ACM SIGIR Conference on Research and Development in Information Retrieval—SIGIR'06* (pp. 597–598). https://doi.org/10.1145/1148170.1148273.

34. Zamir, O., & Etzioni, O. (1998). Web document clustering. In *Proceedings of the 21st Annual International ACM SIGIR Conference on Research and Development in Information Retrieval—SIGIR'98* (pp. 46–54). https://doi.org/10.1145/290941.290956.
35. Toda, H., & Kataoka, R. (2005). A search result clustering method using informatively named entities. In *Proceedings of the Seventh ACM International Workshop on Web Information and Data Management—WIDM'05* (pp. 81–86). https://doi.org/10.1145/1097047.1097063.
36. Jatowt, A., & Au Yeung, C. (2011). Extracting collective expectations about the future from large text collections. In *Proceedings of the 20th ACM International Conference on Information and Knowledge Management—CIKM'11* (pp. 1259–1264). https://doi.org/10.1145/2063576.2063759.
37. Mani, I., Verhagen, M., Wellner, B., et al. (2006). Machine learning of temporal relations. In *Proceedings of the 21st International Conference on Computational Linguistics and the 44th Annual Meeting of the ACL—ACL'06* (pp. 753–760). https://doi.org/10.3115/1220175.1220270.
38. Chambers, N., Wang, S., & Jurafsky, D. (2007). Classifying temporal relations between events. In *Proceedings of the 45th Annual Meeting of the ACL on Interactive Poster and Demonstration Sessions (ACL'07)*, Association for Computational Linguistics, Stroudsburg, PA, USA (pp. 173–176).
39. UzZaman, N., & Allen, J. (2011). Temporal evaluation. In *Proceedings of the 49th Annual Meeting of the Association for Computational Linguistics: Human Language Technologies* (pp. 351–356).
40. Souza, J. D., & Ng, V. (2013). Classifying temporal relations with rich linguistic knowledge. In *Proceedings of NAACL HLT* (pp. 918–927).
41. Laokulrat, N., Miwa, M., Tsuruoka, Y., & Chikayama, T. (2013). Uttime: Temporal relation classification using deep syntactic features. In *Proceedings of the Seventh International Workshop on Semantic Evaluation* (pp. 88–92).
42. Wu, G., & Zhu, Z. (2014). An enhanced discriminability recurrent fuzzy neural network for temporal classification problems. *Fuzzy Sets and Systems, 237,* 47–62. https://doi.org/10.1016/j.fss.2013.05.007.
43. Campos, R., Dias, G., Jorge, A. M., & Nunes, C. (2014). C GTE-Cluster: A temporal search interface for implicit. In *Proceedings of the European Conference on IR Research* (pp. 775–779).
44. Uma, V., Nikhila, L., & Aghila, G. (2016). RaTeR. In *Proceedings of the International Conference on Informatics and Analytics—ICIA-16*. https://doi.org/10.1145/2980258.2980288.
45. Xu, T., Mcnamee, P., & Oard, D. W. (2013). HLTCOE at TREC 2013: Temporal summarization. In *Proceedings of the 22nd Text Retrieval Conference (TREC'13)*.
46. McCreadie, R., Macdonald, C., & Ounis, I. (2014). Incremental update summarization. In *Proceedings of the 23rd ACM International Conference on Conference on Information and Knowledge Management—CIKM'14* (pp. 301–310). https://doi.org/10.1145/2661829.2661951.
47. Wang, B., Liakata, M., Tsakalidis, A., Kolaitis, S. G., Papadopoulos, S., Apostolidis, L., et al. (2017). TOTEMSS: Topic-based, temporal sentiment summarisation for Twitter. In *Proceedings of the IJCNLP, System Demonstrations* (pp. 21–24).
48. McCreadie, R., Santos, R., Macdonald, C., & Ounis, I. (2018). Explicit Diversification of Event Aspects for Temporal Summarization. *ACM Transactions on Information Systems, 36,* 1–31. https://doi.org/10.1145/3158671.
49. Radinsky, K., & Horvitz, E. (2013). Mining the web to predict future events. In *Proceedings of the Sixth ACM International Conference on Web Search and Data Mining—WSDM'13* (pp. 255–264). https://doi.org/10.1145/2433396.2433431.
50. Zhang, S., Bahrampour, S., Ramakrishnan, N., Schott, L., & Shah, M. (2017). Deep learning on symbolic representations for large-scale heterogeneous time-series event prediction. In *IEEE Conference on Acoustics, Speech and Signal Processing (ICASSP)* (pp. 5970–5974).

A Survey on Multi-resolution Methods for De-noising Medical Images

G. Bharath, A. E. Manjunath and K. S. Swarnalatha

Abstract Processing of medical images is important to improve their visibility and quality to facilitate computer-aided analysis and diagnosis in medical science. Such images are usually tainted by noise due to impediments in image capturing devices, unsupportive environment or during transmission over the network. Multi-resolution is a profound technique for decomposing the images into multiple scales and is widely used for image analysis in detail. This paper describes various multi-resolution techniques such as discrete wavelet transform, multi-wavelet transform, and Laplacian pyramid to reduce a wide variety of noise in images. Also, an image de-noising algorithm based on multi-resolution analysis for noise reduction has been described.

Keywords Multi-resolution · Image de-noising · Wavelet transforms · Laplacian pyramid · Threshold

1 Introduction

Diagnosis through medical imaging is massively employed in the field of medical sciences for its reputation in producing visual representations of internal structures of a body and functioning of internal organs or tissues. Medical imaging is the method of capturing instances of a physiological structure to extract information about structures that are hidden by skin and bones, to diagnose and treat diseases. Sources of medical images are X-rays imaging, computed tomography (CT), magnetic resonance imaging (MRI), mammographic scanning, ultrasound imaging or positron emission tomography (PET) which yield images by using noninvasive techniques. In real life, a primary concern is to validate if such images are worthy and useful to the precision required. According to researches, the fundamental stages during which

---

G. Bharath · A. E. Manjunath
Department of CSE, R V College of Engineering, Bangalore 560059, India

K. S. Swarnalatha (✉)
Department of Information Science and Engineering,
NITTE Meenakshi Institute of Technology, Bangalore 560064, India
e-mail: swarnalatha.ks@nmit.ac.in

noise is imparted into images are image acquisition and transmission. Some of the significant variants of noise that predominantly affect medical images are Gaussian noise, impulsive noise, speckle noise, etc. The effect of such noise is reflected in the efforts applied by scientists to invent different techniques for removal. Any denoising technique is called the efficient method if it reduces the impact of noise without compromising with the actual characteristics of original images.

However, several de-noising methods have been proposed and established for removal of noise from biomedical images. Although significant results have been witnessed through linear methods of de-noising, it is observed that the quality of outcome is not sufficient for medical images as diagnosis seeks high reliability. Multi-resolution methods of noise removal analyse an image at different levels of pixel resolution which is implemented by decomposing an image into varying scales (resolution) initiating from the actual level. It has been proved as a powerful means of noise removal and analysis which retains the original quality of the image. Smaller and minute details can be well extracted at higher resolution, whereas bigger components can be looked into in lower-resolution samples. Multi-resolution algorithms are based on two types of transforms, namely wavelet transform and contourlet transform.

Wavelet transform renders image's time–frequency representation with its capability to simultaneously extract information about time and frequency of the signals. It is efficient in identifying discontinuities with zero dimensions (points). This paper elaborates on the techniques based on wavelet transform to convert an image into its equivalent coefficients which are utilized for image analysis and noise removal.

The remnant of the paper is organized into different sections as: Sect. 2 describes the different types of noise a medical image could be affected, and Sect. 3 analyses the wavelet-based multi-resolution techniques for de-noising medical image. The methodology of de-noising is explained in Sect. 4, and finally, Sect. 5 draws the conclusion of the paper.

2 Types of Noise in Medical Images

Noises are unwanted signals that are multiplexed with original signal that corrupt an image. Most commonly found noise types are discussed below.

2.1 Gaussian Noise

Gaussian noise is an additive type of noise that has the same distribution function as normal distribution, which is also termed as Gaussian distribution. A random variable for Gaussian probability distribution function is defined as

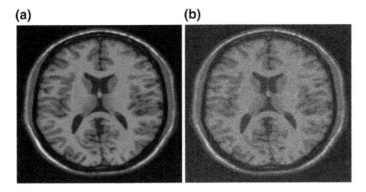

Fig. 1 a Primary image. b Image tainted by Gaussian noise

$$p_G(z) = \frac{1}{\sigma\sqrt{2\pi}} e^{-\frac{(z-\mu)^2}{2\sigma^2}}$$

where z denotes the grey level, μ is the mean value, and σ is the standard deviation. Commonly, computed tomography (CT) images are tainted by Gaussian noise. It has the same effect on each pixel of an image resulting in zero mean collectively for complete image noise (Fig. 1).

2.2 Speckle Noise

Speckle noise is both multiplicative and additive in nature. It is randomly distributed in an image and found inevitable in an image [1]. Impact of speckle noise is generally found in ultrasound images, and it can be mathematically represented as

$$F(i, j) = g(i, j) * m(i, j) + n(i, j) \quad (1)$$

where $F(i,j)$ is the observed image, $g(i,j)$ is the actual noise-free image, $m(i,j)$ is the multiplicative component of the noise, $n(i,j)$ is the linear component of the noise, and i, j represent the horizontal and vertical indices of the image. As speckle noise is granular in nature, it is also termed as texture of images [1]. There is no statistical dependence of this noise on the image signal.

2.3 Impulsive Noise

Impulsive noise also termed as salt-and-pepper noise is a short-term noise. It has its name because of the black and white spots caused in the image during image

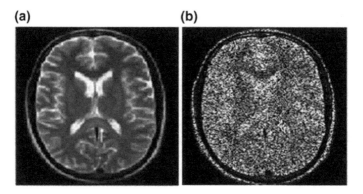

Fig. 2 **a** Primary image. **b** Image with speckle noise

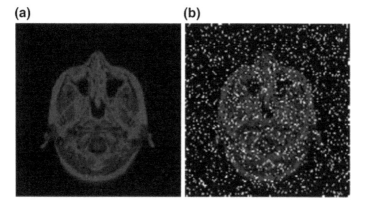

Fig. 3 **a** Primary image. **b** Image tainted by impulsive noise

acquisition or atmospheric disturbance during transmission. It is randomly distributed and predominantly is independent and has no correlation with the image pixels [1]. Also, it is seen that not all pixels are affected by impulsive noise rather it is distributed unevenly. In an image, it is observed that salt-and-pepper noise density is usually half of the total noise density which can be mathematically represented as below

$$y_{ij} = \begin{cases} 0 \text{ or } 255 \text{ with probability } p \\ x_{ij} \text{ with probability } 1-p \end{cases}$$

where y_{ij} represents pixel in an image, p is the probability of salt-and-pepper noise, and x_{ij} represents the pixel not affected by noise (Figs. 2 and 3).

3 Multi-resolution Image Processing Method

Multi-resolution techniques are mathematical approaches to decompose an image into multiple levels of scales. In such methods, an image is sampled into various pixel resolutions starting from the actual resolution level of the image till the required scale. As multi-resolution analysis and human visual system are in accordance with each other, this method produces better images with fewer artefacts [2]. This technique provides deeper insight into understanding characteristics of an image. Amongst all the multi-resolution techniques, discrete wavelet transform (DWT), multi-wavelet transform and Laplacian pyramid decomposition methods are quite commonly used to divide an image into a hierarchy of scales that ranges from most gritty scale to the finest one. Wavelet transform techniques are comparatively new methods that are used to seek more information about the signals which are not readily obtained from the signal observations and analysis. The most popular transform used in this method is Fourier transform. Thresholding through a nonlinear technique is very simple way of suppressing noise signals by comparing the amplitude of the image signals with the thresholds defined using the wavelet transforms. The idea here is to compare the wavelet coefficients with the threshold value; if the signal value is greater than the threshold, then it is left unaltered or reduced towards zero else if it is lesser than the threshold it is zeroed. Threshold behaves as the decider that resolves weaker coefficients, mostly representing noise signals, and the significant coefficients of the image forming signals.

Discrete Wavelet Transform (DWT)

The first multi-resolution decomposition method for image analysis introduced is the DWT [3]. It is derived from the postulate that amplitude of the signal is distinct enough from the amplitude of the noise signal to perform clipping, threshold comparing and shrinking the strength of the noise signal for noise reduction [4]. It provides adequate information about an image and reduces computation time for analysis [5]. In this method, wavelets have a single wavelet function and single scaling function. All wavelet functions are derived from a mother wavelet function which can mathematically be described as below:

Consider a mother wavelet function denoted by $\psi(t)$, and then other wavelet functions will be

$$\psi a, b(t) = 1/\sqrt{|a|} * \psi((t-b)/a)$$

where a and b are arbitrary constants and represent parameters for dilation and translation, respectively.

DWT properties that are attributed for separating noise signals from original signal are energy compaction, smoothness of noisy signal, regularity, and count of vanishing moments [5]. DWT is appreciable due to its property of compressing real-world image signals with very few large wavelet coefficients and scaling coefficients ruling the representation.

Fig. 4 Sub-band distribution for multi-wavelet transform

L_1L_1	L_1L_2	L_1H_1	L_1H_2
L_2L_1	L_2L_2	L_2H_1	L_2H_2
H_1L_1	H_1L_2	H_1H_1	H_1H_2
H_2L_1	H_2L_2	H_2H_1	H_2H_2

Multi-wavelet Transform

Unlike DWT multi-wavelet transform is a modern and more intriguing concept that has more than one scaling function and mother wavelet function to represent image signals. A multi-dimensional vector is normally used to represent a multi-wavelet function [5]. DWT decomposes a signal into one low-pass coefficient and one high-pass coefficient following it, whereas a multi-wavelet decomposition results in multiple low-pass and high-pass coefficients; that is, a second low-pass coefficient follows first low-pass coefficient, and the same happens for high-pass coefficients. Mathematically, a multi-wavelet can be represented as a set of vectors as

$$\Psi(t) = [\psi_1(t)\psi_2(t), \ldots, \psi_r(t)]^T$$

where r represents the dimension of the vector. If $r = 1$, $\psi(t)$ simply is a wavelet. Multi-wavelet decomposition requires pre-filtering to convert a signal into vector form which is a preprocessing stage also known as multi-wavelet initialization. A stream of vectors is generated by the pre-filters from the scalar input image, thus creating initial coefficients.

Multi-wavelet transform provides better localization of image in both spectral and spatial domains than other methods. Division of wavelet into multiple sub-bands through multi-wavelet decomposition is as shown in Fig. 4.

where L1 and L2 are low-pass sub-bands and H1 and H2 are high-pass sub-bands.

Laplacian Pyramid

Laplacian pyramid method of multi-resolution image analysis is used for decomposing an image into multiple scales (resolutions). An input image is decomposed into different resolutions beginning from the actual resolution to a level where the resolution narrows down to 1×1. The difference between the resolutions also called as the difference in details provides information for analysis. Many multi-resolution applications prefer to use Laplacian pyramid over wavelet techniques attributing to its property of generating unscrambled frequency and only one high- and low-pass sub-bands [6]. Laplacian method is widely used for applications such as compression and harmonization [5]. First, the correlations between the pixels are removed by trimming a low-pass filtered copy of the image from the original image itself [5]. If I be the original image, a low-pass filtering of I yields J with an error of $E = I - J$. After subsequent filtering through the low pass filters of the output image obtained from previous filtering, yields a pyramid structure when the arrays are stacked one above the other as show in the Fig. 5.

Fig. 5 Laplacian pyramid of decomposed image

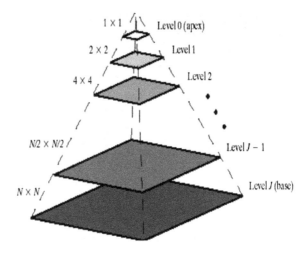

Fig. 6 Algorithm to de-noise a medical image

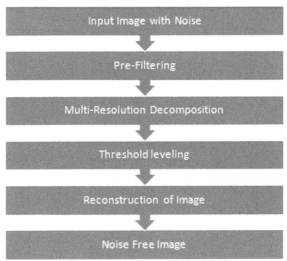

The difference between the convolutions of two equivalent functions with the original image gives the node value in the pyramid.

4 Methodology of De-noising

The algorithm to remove or reduce noise from a medical image using multi-resolution analysis to obtain a noise-free and a loss-less image includes the stages as depicted in Fig. 6.

The algorithm begins with preprocessing of a noisy image where it is minimized to achieve algorithm's computation time reduction before multi-resolution transforms are employed.

Thereafter, multi-resolution decomposition is executed to generate wavelet coefficients and sub-bands at levels depending on the type of method employed. The coefficients of the wavelets cross through the stage where it is determined what action must be taken by comparing their values with the thresholds.

Thresholding, a simple technique, is nonlinear method employed to distinguish noise from signal. It runs on a single coefficient at once. The basic presumption of thresholding is that noise is diffused across all coefficients and amplitude of noise is not high which consequently facilitates noise separation. Threshold selection is very crucial as the result of de-noising algorithm is dependent on the threshold values used. A noise-retained image ensues with a small value of threshold, whereas the selection of large threshold results in smooth image but destroys details.

The ultimate process of the drive chain is image reconstruction to recover the image from its equivalent coefficients. Prior to this stage, most of the applications include edge detection for recovering lost contours due to wavelet or pyramid decomposition. Moreover, a de-noised image is retrieved from the process which could now be significant in the field of its application.

5 Conclusion

In this paper, we present an overview of medical image de-noising techniques based on multi-resolution analysis. The three techniques addressed in this paper are relatively advantageous over other ones. It is observed through research that scalar wavelet decomposition is computationally faster than other methods. Multi-wavelet transform has superior performance than DWT for image processing applications. Ultimately, Laplacian pyramid technique provides much finer performance in de-noising than any other method. The algorithm elucidated here that implements these methods is found successful in rendering clean images from the study.

References

1. Madhura, J., & Ramesh Babu, D. R. (2017). A survey on noise reduction techniques for lung cancer detection. In *International Conference on Innovative Mechanisms for Industry Applications* (pp. 637–640). IEEE.
2. Lukin, A. (2006). A multiresolution approach for improving the quality of image denoising algorithm. In *Proceedings of International Conference Acoustics, Speech and Signal Processing, ICASS-06* (vol. 2, pp. 857–860).
3. Mallat, S. (1989). A theory of multiresolution signal decomposition. The wavelet representation. *IEEE Transactions Pattern Analysis and Machine Intelligence, 11,* 674–693.

4. Hassan, H., & Saparan, A. (2011). Still image denoising based discrete wavelet transform. In *IEEE International Conference on System Engineering and Technology* (pp. 188–191).
5. Vanathe, V., Bhopathy, S., & Manikandan, M. A. (2013). MR image denoising and enhancing using multi-resolution image decomposition techniques. In *International Conference on Signal Processing, Image Processing and Pattern Recognition*. IEEE.
6. Nguyen, H. T., & Linh-Trung, N. (2010). The Laplacian pyramid with rational scaling factors and application on image denoising. In *10th International Conference on Information Science, Signal Processing and their Applications* (pp. 468–471). IEEE.
7. Henkelman, R. M. (1985). Measurement of signal intensities in the presence of noise in MR images. *Medical Physics, 12*(2), 232–233.
8. Malini, S., & Moni, R. S. (2015). Image denoising using multiresolution analysis and nonlinear filtering. In *Fifth International Conference on Advances in Computing and Communications* (pp. 387–390). IEEE.

Performance Analysis of Vedic Multiplier with Different Square Root BK Adders

Ranjith B. Gowda, R. M. Banakar and Basavaprasad

Abstract Multiplication is the main basic operation used by many of the digital signal processor (DSP) and vector processors. DSP application repeatedly performs the operations like signal processing, filtering, processing of discrete signal data, and radar signal processing and use intensive fast Fourier transform (FFT) operations. FFT computation uses butterfly structures, where multiplication is the basic operation. DSPs have to execute a large number of instructions per second, which in turn uses so many FFT computations, and hence, the multiplication operation decides the performance of DSP. Designing a high-performance multiplier improves the overall performance of the processor. Many multiplier architectures have been proposed in the past few decades with the attractive performance, power consumption, delay, area, throughput, etc., and the most acceptable multiplier among them is the Vedic multiplier. When high performance is necessary, Vedic multiplier will be the best choice. Operation of Vedic multiplier is based on ancient Vedic mathematics. This earlier multiplier has been modified to improve the performance. There are 16 sutras for the multiplication operation in this method. These sutras are used to solve large range of multiplication problems in a natural way. This method of multiplication is based on Urdhva Triyagbhyam sutra, which means horizontal and cross-wire technique of multiplication operation. This method uses partial product generation in parallel and eliminates the unwanted steps with zero. Urdhva Triyagbhyam sutra is an efficient sutra which enhances the execution speed of the multiplier by minimizing the delay. This work describes the overall performance of the Vedic multiplier with different high-speed adders like regular square root BK adder (RSRBKA), Modified square root BK adder (MSRBKA) and proposed optimized square root BK adder (OSR-BK-A). The proposed designs are simulated and synthesized in Xilinx ISE 14.7, and the results are tabulated.

R. B. Gowda (✉) · Basavaprasad
Department of ECE, Government Polytechnic, Sorab, Shimoga 577429, India
e-mail: ranjithgowda789@gmail.com

R. M. Banakar
Department of ECE, BVB College of Engineering and Technology, Hubli 580031, India

Keywords Vedic multiplier · Urdhva Triyagbhyam sutra · Regular square root Brent Kung adder (RSRBKA) · Modified square root Brent Kung adder (MSRBKA) · Optimized square root Brent Kung adder (OSRBKA)

1 Introduction

Multiplication is the main basic arithmetic operation used in all the DSPs. DSPs are intended to perform many operations like image processing from satellite, brain signal processing, signal processing for vibration analysis, filtering in communication and ALU of microprocessors where fast multiplication operation is needed [1]. These operations include intensive FFT computations. FFT operation is based on the multiplication of complex input data used by the butterfly structures. DSPs have to execute billions of FFT operations per second and hence the multiplication operation. Therefore, designing a fast multiplier is the basic requirement to improve the performance of DSPs [2, 3]. Currently, instruction cycle time is dependent on the multiplication operation time. As the processor technology growing in faster rate, there is a demand for high-speed processing operations. In most real signal processing and multi-dimensional signal processing applications, like image/video, the execution speed of the arithmetic operations such as addition, multiplication and other operations decides the overall speed of the processor. Designing a faster multiplier is a challenging task to achieve and is the interesting field of research over many decades. Much application requires improved performance by reducing power and delay. The Vedic multiplier is the best choice for minimum power consumption and high-speed applications.

It is possible to minimize the power consumption by optimizing the different levels of architecture. Power expending in a digital circuit depends on the specific technology used, circuit structure, topology and circuit design style and architecture used at higher level of implementation. Digital multiplier must be fast, reliable and high efficient to improve the performance. Depending on the arrangement of the components, different multipliers are available. Choosing a particular type of multiplier is application dependent. In most of the signal processing applications, multiplier, critical path determines the longest path delay and hence the speed. This path length determines the operational speed of a processor and hence the performance.

Earlier multiplication technique involves summation, subtraction and data shifting operations. Various multiplication algorithms were proposed till today and having their own advantages, disadvantages, circuit complexity, throughput, etc. Multiplier needs a larger circuit area for implementation. As the resolution of the result is high, it consumes more area. An n-bit multiplier uses n logic gates. Latency and throughput are the two important parameters of many DSPs. Latency is the time duration between when the inputs are applied to the time and when the output becomes measurable. Throughput is the number of multiplication operation executed in the given amount of time. Multiplier circuit increases not only the area but also the power dissipation.

Hence, many optimization techniques are needed to minimize the power utilization in the multiplier blocks.

Digital multiplier blocks are the most essential operational blocks in DSPs. The performance of DSP is mainly evaluated by the performance of multipliers [4]. The most frequently used multipliers in digital logic circuits are the array multiplier and Booth multiplier. Since the partial products are computed independently in array multiplier, they need less computation time with a penalty of increase in the area. As the length of the input bits increases, array structure area increases which increases the critical path length. Booth algorithms are the another multiplication algorithms which require large number of booth array elements for increased speed of multiplication operation, which in turn needs larger partial sum and carry registers. Since this involves larger propagation delay, there is a decrease in the performance. Since multipliers are the basic building blocks of DSPs, it is still an interesting area of research and many new algorithms for multiplication were presented in the literature [5].

In this work, Urdhva Tiryagbhyam sutra is used to perform the multiplication operation using binary numbers. This method is same as that of array multiplication operation. First, a 2*2 multiplier architecture is designed using Urdhva Tiryagbhyam sutra, and then, a 4*4, 8*8 and 16*16 multiplier structure has been designed. The performance of 16*16 Vedic multiplier is studied using three different square root BK adder architectures. The VLSI design parameters like area, power and delay are calculated for all these structures to analyze the performance of Vedic multiplier.

This paper is organized as follows: Sect. 2 gives the multiplication method using Urdhava Tiryagbhyam sutra, adders for Vedic multipliers are discussed in Sect. 3, Vedic multiplier with different BK adders is discussed in Sects. 3.1–3.4, Sect. 4 gives the simulation and comparison of results with different adders, and Sect. 5 concludes.

2 Urdhava Tiryagbhyam Method of Multiplication

Urdhava Tiryagbhyam [6] is a multiplication method, which uses horizontal and vertical cross-wire method for multiplication and generates partial products in parallel. Figure 1 shows the operation of 4*4 multiplication operation using this method. This method can be extended for any number of input data lengths. The speed of execution can be increased by generating the partial products in parallel. This gives an advantage of using it in high-speed DSP applications. As the partial products are computed in parallel, this multiplier does not dependent on CPU clock frequency of the system [7]. This reduces the operating clock frequency of the CPU and hence less number of switching activities. As there is few number of switching activities, the power consumption of the design is less and can be used for low-power applications. Because of its regular structure [8], the power consumption during processing increases with increase in the width of the data bus. Because of its regular structure, it occupies optimum area on silicon chip. As the length of the input bits increases, the effect of increase in the area and delay is very slow as it is compared to the other architectures.

Figure 1 shows the algorithm needed for the design of 4*4 multiplication operation using Urdhava Tiryagbhyam method. It takes two four-bit binary numbers: a3 a2 a1 a0 and b3 b2 b1 b0. It requires seven steps to obtain final result, and each step involves partial product generation. Initially, in the step 1, LSB bits of both the data are multiplied, and this results in LSB of the final result. In the step 2, product of LSB bit of first data and the second bit of second data are computed, and product of LSB bit of second data and the second bit of first data are computed. This is called cross-wire multiplication. The obtained intermediate products are summed together, the LSB bit of the result is the next higher result bits of final sum, and the bits left are carried to the next step. This process repeats till all the seven steps are completed. Figure 1 shows all the seven steps involved in detail. Each step has the expressions given below:

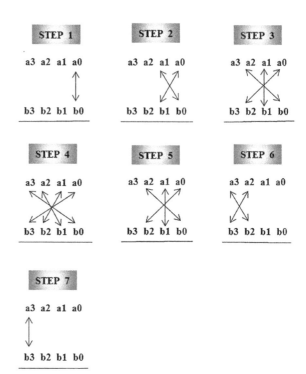

Fig. 1 Urdhava Tiryagbhyam method of multiplication for two 4-bit numbers

$$r0 = a0\,b0$$
$$c1\ r1 = a1\,b0 + a0\,b1$$
$$c2\ r2 = c1 + a2\,b0 + a1\,b1 + a0\,b2$$
$$c3\ r3 = c2 + a3\,b0 + a2\,b1 + a1\,a2 + a0\,b3$$
$$c4\ r4 = c3 + a3\,b1 + a2\,b2 + a1\,b3$$
$$c5\ r5 = c4 + a3\,b2 + a2\,b3$$
$$c6\ r6 = c5 + a3\,b3$$

where r_m represents the sum bits of final result and c_n represents the carry generated from each step. The values c6 r6 r5 r4 r3 r2 r1 r0 give the final product [9]. The equations listed above are the general formula applicable to multiplication of any number input data size. In general, higher bit-size multiplication operation is obtained by first designing this 4*4 multiplier and then using this multiplier repeatedly as many times as required. For example, it requires four 4-bit multipliers for multiplication of two 4-bit input data, two adders for addition of partial products and generated intermediate carry. The 4*4 product results in 8-bit length data and LSB 4-bits are the final result, and the 4-bits left are forwarded to the next step. This procedure repeats for three steps in this example. Similarly, it is possible to design 8*8 and 16*16 multiplier with this procedure. The hardware structure of 4*4 multiplier in gate level is as shown in Fig. 2.

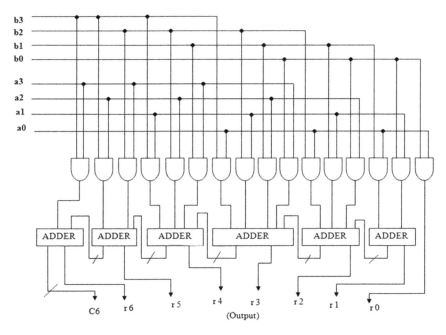

Fig. 2 Logical structure of 4 × 4 Urdhva Tiryagbhyam multiplier

Table 1 Comparative analysis of different multipliers

Multipliers	LUT's	Delay (ns)	Power (W)
Booth	818	89.09	0.267
Wallace tree	670	28.44	0.192
Vedic	514	23.81	0.185

2.1 Comparative Analysis of Multipliers

In this proposed work, different multipliers like Booth multiplier, Wallace tree multiplier and Vedic multiplier are designed and simulated in Xilinx ISE 14.7 tool. VLSI performance parameters like silicon area on chip, delay of data path and power utilization are calculated to select the best multiplier for the proposed design. These multipliers are designed for 16-bit input data, and the simulation results are as shown in Table 1.

These comparative results show that Vedic multiplier is best suited when the application needs low-power and high-speed operation. In the following sections, the performance of Vedic multiplier is studied with three different square root BK adders.

3 Adders for Vedic Multiplier

Adder is combinational circuit used in most of the digital circuits for addition of input data bits. Multiplication operation can be obtained by using repeated addition operation. This requires many adders with fast operation and less delay. In this multiplication method, after multiplication operation in each step, partial products are generated. In 4*4 multiplier, as explained in the previous section, seven steps are involved. Each step generates partial products, and these partial products generated in different steps are summed together to obtain the final resultant bits. The proposed 16*16 Vedic multiplier is designed using 2*2, 4*4 and 8*8 Vedic multipliers. These multipliers generate different length partial products.

Partial products generated in 4*4 Vedic multiplier require various size adders like 6-bit, 12-bit and 24-bit adder. In this multiplier, it is possible to generate partial products from different stages. In order to add these generated partial products from different size multipliers, it is required to have different size adders. These adders add the partial products at various stages as soon as they are generated, and the generated carry will propagate to the next stage. Figure 3 shows the partial products of 8-bit Vedic multiplier. It also shows how to group the partial products for addition using specific size adder.

To obtain the faster multiplication operation, an adder performance should be high. The performance of the adder greatly influences the performance of multiplier. In this work, the performance of the Vedic multiplier is studied with the three different

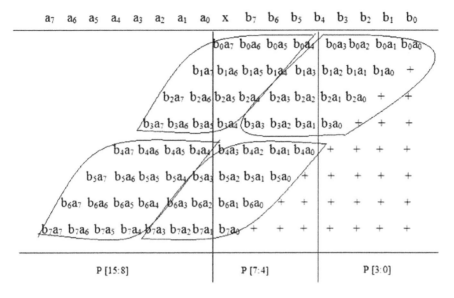

Fig. 3 Grouping of partial product generated in 8*8 Vedic multiplier

square root BK adders like regular square root BK adder, modified square root BK adder and proposed optimized square root BK adder. In this proposed work, a 16-bit Vedic multiplier is built starting from 2*2 Vedic multiplier, different adders are used with this multiplier for partial product addition, and simulation results are obtained.

3.1 Vedic Multiplier with Square Root BK Adders

Brent Kung adder is the most commonly used adder when high-speed operation is required. It gives optimal number of stages by using logarithmic architecture. Because of this logarithmic architecture, there is an asymmetric load on each intermediate stage. BK adder is an example of parallel prefix adder which does the computation based on generate (gi) and propagate (pi) signals. The cost of implementation and routing complexity is minimum in prefix adders. Since the gate/logic-level depth of BK adder is O ($\log_2$ (n)), there is decrease in the speed of operation as compared to CLA adders. For the applications, which require enhanced speed of operation with high throughput, Brent Kung adders are best suited. Since BK adder uses parallel prefix tree structures, they are very faster as compared to other types of adders [10].

3.2 Vedic Multiplier Using Regular Square Root Brent Kung Adder

16-bit regular square root BK adder (RSRBKA) uses five groups of BK adder structures. Among these, four groups contain a BK adder, RCA adder and a multiplexer unit. In the square root structure, the size of BK adder, RCA and multiplexer unit increases from LSB stage to the MSB stage by one bit. For example, in 16-bit RSRBKA the first stage in four groups is of 2-bit size, second stage is 3-bit size, third stage is 4-bit size, and fourth stage is 5-bit size. This structure is called square root structure. Brent Kung adder is a prefix adder used as the first stage of CSA adder. CSA adder uses two separate adders. First-stage adder performs the addition of the input bits by assuming input carry is 0, and second-stage adder adds the input bits by assuming carry input is 1. Finally when the real input carry generated by the previous stage is known, the multiplexer unit simply selects the output of either upper stage or the lower stage. If input carry is zero, then multiplexer selects the output of upper-stage adder; else, it will select the output of lower-stage adder. Here, BK adder performs the addition of the input data bits by assuming carry input is zero and RCA is used to add the input bits by assuming input carry is one. Finally when the actual carry input is known, the output of either BK adder or RCA will be drawn out by the multiplexer. If input carry is 1, select the output of RCA adder; else, select the output of BK adder. The block diagram of regular square root BK adder is shown in Fig. 4.

The design of 16*16 Vedic multiplier requires the design of 2*2, 4*4, 8*8 Vedic multipliers. To perform the summation of partial products generated by these multipliers, different sizes of regular square root Brent Kung adders like 2-bit, 4-bit, 6-bit, 12-bit and 24-bit are required. These different size adders are designed to add the

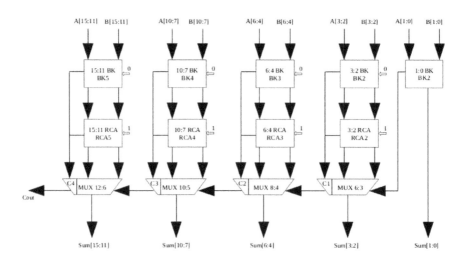

Fig. 4 Structure of 16-bit regular square root BK adder

partial product generated during the multiplication operation. A 16-bit Vedic multiplier is designed using regular square root Brent Kung adder structures to study the performance of multiplier.

3.3 Vedic Multiplier Using Modified Square Root Brent Kung Adder

16-bit modified structure of square root BK adder (MSRBKA) uses five groups of BK adder structures. Among these, four groups contain a BK adder unit, BEC adder unit and a multiplexer unit. In the square root structure, the size of BK adder, BEC and multiplexer unit increases from LSB stage to the MSB stage by one bit. Modified square root BK adder uses BEC unit, rather RCA, as in regular square root BK adder architecture. Here, the advantage of using BEC instead of RCA is that the total logic gates requirement for the implementation can be reduced and hence reduced the overall area of the architecture. Designing BEC is easy as compared to RCA structures. RCA has another disadvantage of carry propagation delay through all the blocks. This rippling of carry through each block of RCA increases the delay. This can be overcome by using BEC structures. BEC will do the same function as that of RCA with much less delay. This shows that using BEC in the CSA adder increases the speed of operation.

Here, BK adder performs the addition of input data bits by assuming zero input carry and BEC performs addition of the input bits by assuming input carry is one. Finally when the actual carry input is known, the output of either BK adder or BEC will be selected by the multiplexer. If input carry is 1, select the output of BEC adder; else, select the output of BK adder. Different sizes of modified square root brent kung adders are designed to add the partial product generated during the multiplication operation. A 16-bit Vedic multiplier is designed using these modified square root BK adder structures to study the performance of multiplier. The block diagram of modified square root Brent Kung adder is shown in Fig. 5.

3.4 Vedic Multiplier Using Proposed Optimized Square Root Brent Kung Adder

In this new approach, the structure of BEC has been changed to reduce the combinational delay of the structure and power consumption. Instead of using XOR gate in BEC, a combination of XNOR gate, OR gate and NOT gates is used. Square root structure of BK adder remains same. A BK adder will be used to add two input data when $c_{in} = 0$ and modified BEC will be used to add 1 to the result of BK adder when $c_{in} = 1$. Including this, a new BEC structure has the advantage of removing

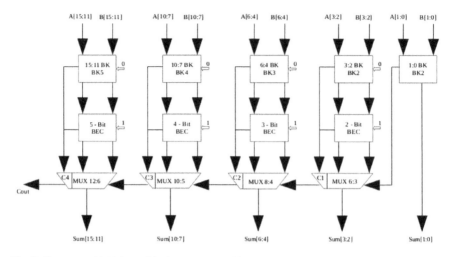

Fig. 5 Structure of 16-bit modified square root adder

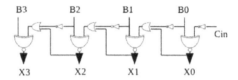

Fig. 6 Modified structure of BEC

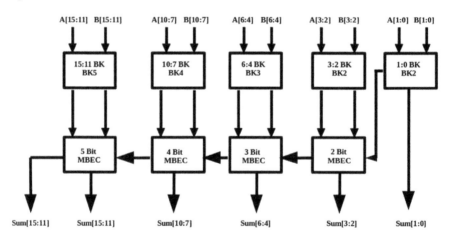

Fig. 7 Block diagram of 16-bit proposed optimized square root BK adder

multiplexer stage and which has the impact in reducing the length of critical path. Structure of modified BEC is as shown in Fig. 6.

The block diagram of the proposed optimized square root BK adder (OSRBKA) is as shown in Fig. 7. This new approach reduces the overall delay and power consumption, with a small increase in the area, as compared to the other architectures. This proposed OSRBKA is used with Vedic multiplier to study the performance of the designed multiplier. As the path length of the critical path of this new design is minimum, as compared to previous square root adders, there is an improvement in the performance of the multiplier.

4 Simulation Results and Comparison

Various adders like regular square root BK adder, modified square root BK adder and proposed optimized square root BK adder are used with the proposed 16-bit Vedic multiplier. These architectures are designed, simulated and synthesized using Xilinx ISE 14.7 tool. The main VLSI design constraints like area, power consumption and delay are calculated for all these architectures to analyze the performance of the multiplier, and the results are tabulated in Table 2. The percentage of improvement in the delay, area and power consumption of the proposed BK adder with Vedic multiplier as compared to RSBK adder and MSBK adder is tabulated in Table 3. The bar chart representation of Tables 2 and 3 is shown in Figs. 8 and 9, respectively. The comparison of these architecture result shows that the proposed BK adder with Vedic multiplier architecture is having higher operational speed, less silicon chip area requirement and lower power consumption as compared to the other square root BK adder architectures.

Table 2 Simulated results of 16-bit Vedic multiplier with different square root BK adders

16-bit Vedic multiplier with following adders	Delay (ns)	Area (LUT's)	Power (mW)
regular square root BK adder	27.223	750	208
modified square root BK adder	29.622	732	201
proposed optimized square root BK adder	26.183	579	169

Table 3 Percentage of improvement in the proposed BK adder with Vedic multiplier as compared to RSBK adder and MSBK adder

16-bit Vedic multiplier with following adders	Delay (ns)	Area (LUT's)	Power (mW)
% Improvement (as compared to RSBKA)	3.82	22.8	18.75
% Improvement (as compared to MSBKA)	11.61	20.9	15.92

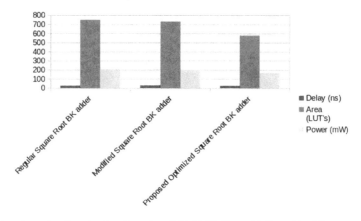

Fig. 8 Bar chart showing simulated results of 16-bit Vedic multiplier with different square root BK adders

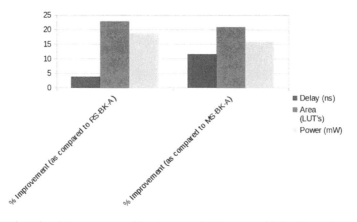

Fig. 9 Bar chart showing percentage of improvement in the proposed BK adder with Vedic multiplier as compared to RSBK adder and MSBK adder

5 Conclusion

In this work, a 16-bit Vedic multiplier is designed using Urdhava Tiryagbhyam sutra. The performance of this multiplier is studied with three different types of square root adders like regular square root BK adder (RSRBKA), modified square root BK adder (MSRBKA) and proposed optimized square root BK adder (OSRBKA). All these architectures are designed and simulated using Xilinx ISE 14.7. Various VLSI design constraints like delay, area and power consumption are calculated for all these architectures. From these results, it is concluded that the proposed optimized square root Brent Kung adder architecture is best suited when an application requires enhanced operational speed with minimum power utilization and less silicon area. Vedic multiplier is one of the fastest multipliers used in the digital design, and in this work, a performance study of this multiplier with the different square root BK adders is carried out.

References

1. Wallace, C. S. (1964). A suggestion for a fast multiplier. *IEEE Transactions on electronic Computers, 1,* 14–17.
2. Booth, A. D. (1951). A signed binary multiplication technique. *The Quarterly Journal of Mechanics and Applied Mathematics, 4*(2), 236–240.
3. Bharath, J. S. S., & Tirathji, K. (1986). *Vedic mathematics or sixteen simple sutras from the vedas*. Varanasi (India): Motilal Banarsidas.
4. Kulkarni, S. (2007). *Discrete fourier transform (DFT) by using vedic mathematics*. Report, vedicmathsindia. blogspot. com.
5. Nicholas, A., Williams, K., & Pickles, J. (1984). *Application of urdhava sutra*. Roorkee (India): Spiritual Study Group.
6. Saokar, S. S., Banakar, R., & Siddamal, S. (2012). High speed signed multiplier for digital signal processing applications. In *2012 IEEE International Conference on Signal Processing, Computing and Control (ISPCC)* (pp. 1–6). IEEE.
7. Naik, R. N., Reddy, P. S. N., & Mohan, K. M. (2013). Design of vedic multiplier for digital signal processing applications. *International Journal of Engineering Trends and Technology, 4*(7).
8. Thapliyal, H., & Srinivas, M. (2005). An efficient method of elliptic curve encryption using ancient Indian vedic mathematics. In 2005 *48th Midwest Symposium on Circuits and Systems* (pp. 826–828). IEEE.
9. Dhillon, H. S., & Mitra, A. (2008). A reduced-bit multiplication algorithm for digital arithmetic. *International Journal of Computational and Mathematical Sciences, 2*(2).
10. Saxena, P. (2015). Design of low power and high speed carry select adder using Brent Kung adder. *International Conference on VLSI Systems, Architecture, Technology and Applications (VLSI-SATA)*. IEEE.

Analysis of Traffic Characteristics of Skype Video Calls Over the Internet

Gulshan Kumar and N. G. Goudru

Abstract Skype is an important application of real-time systems (RTS). It uses Transmission Control Protocol (TCP) for connection establishment and User Datagram Protocol (UDP) port for transfer of audio and video data. In spite of the popularity of Skype, relatively little is known about its traffic characteristics. In this paper, the sender is sending the audio visual data using UDP port at a constant bit rate (CBR). The destination receives the data and checks with its buffer threshold values. The minimum and maximum threshold values are fixed based on the receiver buffer. When the sender data is higher than the threshold value, the destination asks source to reduce the flow rate by sending an explicit control packet. When the sender rate is lower than the threshold value, the destination can ask the source to increase the sending rate. The introduction of feedback system using a control message overcomes the congestion at the receiver, minimizes the data loss, increases optimal utilization of resources, and enhances the quality of expectation. In Skype application, an ordinary node acts one time as a sender and other time as a receiver. The proposed technique is to be built at both the end nodes. Important quality parameters such as packet loss due to congestion, one-way packet delay, effect of queuing delay on sender performance of feedback are analyzed using graphs and statistical data. Mathematical models are used to analyze the Skype performance. MATLAB software is used to simulate the system and for model authentication.

Keywords Skype · Audio · Video · UDP · Feedback · VoIP

G. Kumar · N. G. Goudru (✉)
Department of ISE, Nitte Meenakshi Institute of Technology (Affiliated to Visvesvaraya Technological University), Bangalore 560064, India
e-mail: goudru.ng@nmit.ac.in

G. Kumar
e-mail: gulshankr1nt35@gmail.com

© Springer Nature Singapore Pte Ltd. 2019
N. R. Shetty et al. (Eds.), *Emerging Research in Computing, Information, Communication and Applications*, Advances in Intelligent Systems and Computing 906,
https://doi.org/10.1007/978-981-13-6001-5_22

...uction

... of the most popular real-time applications having more than 50 million ...eal-time communication (RTC) uses UDP port for transport of the data. ... provide good interactivity, the important network parameters such as ... packet losses, and delays are to be controlled. TCP is regarded as an ... protocol for audio and video data transportation in Skype application ... its retransmission policy and in-ordered delivery mechanism that induce ... delay. Hence, in the delivery of real-time audio and video data, no literature ... use of TCP. Real-time video applications employ UDP sockets by using an ...ngestion control algorithm which are implemented at the application layer ...odel. Skype is designed based on a peer-to-peer (P2P) Voice over Internet ... (VoIP) architecture by Kazaa. For the end users, the Skype application ...any services like voice communication, video communication, file transfer, ... In Skype, the communication between users is established by applying a ...nal end-to-end IP paradigm and is multiplexing voice and video blocks. UDP ... nonresponsive to network traffic is continuously sending the originated ...to the network causing congestion at the ingress point of the receiver. The ...tion causes packet losses as well as induces delay. To ease from the congestion, ...pose a control-based feedback system which controls the sending rate at Skype ...ation in both directions.

... queue length measurement and implementation of feedback system are illus-...in Fig. 1. The data arrival rate causes the formation of queue at the receiver. ...stantaneous queue length is measured and compared with the buffer threshold ... such as threshold minimum and threshold maximum and then sends the sig-...) the sender either to increase or decrease the sending rate. This is being the ...and innovative work to control the congestion in Skype application and reduce ...

...lated Work

...lications with real-time communication between the Internet peers are ... popular. Real-time communication still needs the enhancement in the qual-...ices. Skype is a real-time communication system which requires improve-... in the transmission of voice and video data by minimizing the loss and ...quate research work is not carried out to study the impact of network ...lls and user quality of experience. Transmission Control Protocol (TCP) ...des reliable transmission is unsuitable for Skype as it introduces sig-...ncy because of its retransmission policy and in-order delivery of data ... researchers Gaetano Carlucci and others discuss congestion control in ...lications by introducing explicit type of feedback system at the receiver ... sending rate by the source. This results in the reduction of queuing delay and packet loss due to buffer overflow. The authors focus on the application of noise filtering using Kalman filter and able to reduce delay causing due to noise [1]. The authors Kumara W. G. C. W and others analyzed the Skype video calls carried out over the Internet between two long distance nodes using the objective video quality assessment-based tools and found that the video call is not fully satisfactory.

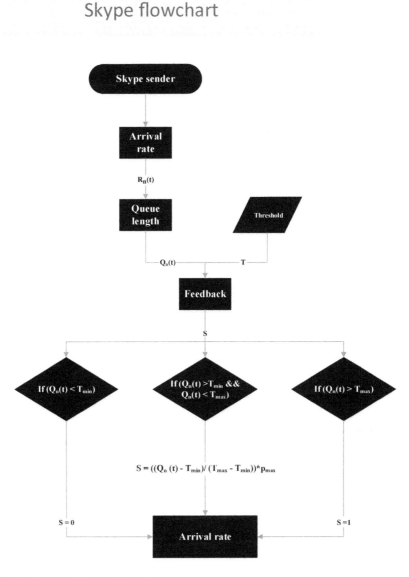

Fig. 1 Feedback congestion control system

Video quality is low in the real environment compared to the values obtained during simulated results. The researchers tried to find a video quality assessment of the WAN calls using the Internet for Skype. The authors continue to work on the quality of services in the Skype [2]. The authors measure the characteristics of Skype video calls on a controlled network testbed. By changing the packet loss rate, propagation delay and bandwidth studied the performance of Skype by adjusting its sending rate. The authors found that Skype works fine for mild packet losses and propagation delays and utilizes efficiently the available network bandwidth. Model-based extrapolation is made and performed a numerical analysis to understand the network impact on Skype [3]. P. M. Santiago del Rio and others tried to reduce the process cost to detect Skype traffic. They used real traces from a public operator. They proposed application for Skype traffic classification at 1 Gbps and 3.7 Gbps which read, respectively, from network interface card and from memory using 8 cores CPUs platform. They show that Skype traffic identification is worth at the high-speed networks typically ranging over 10–40 Gbps. They also measure throughput at these two speeds [4]. The authors Molnár and others use strong encryption to provide secured communication in the Skype network. They introduced algorithms to detect types of communication initiated by Skype client for detection of the Skype servers using packet header and without the information in payload. The traffic analysis was made in both fixed network environment and mobile network environment. The algorithm is also validated based on active test measurements [5]. The authors Dario Bonfiglio and others study the traffic stream characteristics generated by voice and video communication and signaling traffic by Skype application. They apply both active and passive measurement schemes for a deep understanding of the Skype traffic. The authors develop a testbed for Skype service type identification, improve the performance of transportation protocol especially UDP, minimize path losses, and avoid congestion [6]. The authors Haiyong Xie and others study the Voice over Internet Protocol (VoIP) quality of Skype by using large-scale end-to-end measurements. They investigate the impacts of the access capacity constraint and the AS policy constraint on the VoIP quality of Skype. The authors reveal that overall VoIP quality of Skype degrades significantly, and a large percentage of VoIP sessions will have unacceptable quality. The result clearly demonstrates the potential danger of building VoIP applications based on P2P networks without taking into account operational models of the Internet [7]. The authors Saikat Guha and others tried to understand how P2P VoIP traffic in Skype differs from traffic in P2P networks and from traffic in traditional voice communication networks [8]. The researcher Salman A. Baset analyzes key Skype functions such as login, NAT and firewall traversal, call establishment, media transfer, codecs, and conferencing under three different network setups. Performance analysis is performed by careful study of the Skype network traffic and by intercepting the shared library and system calls of Skype [9]. A model-based UDP traffic analysis for performance analysis of Skype has not been studied yet by any researchers. An innovative approach is carried out in the work.

3 Mathematical Models in Skype Network

The analysis of a system using mathematical models is regarded as one of the best techniques. With reference to [1, 10], mathematical model representing CBR flow is given by:

$$\dot{R}(t) = 0 \tag{1}$$

Using Eq. (1), the arrival rate of data packets in iterative form is given by:

$$R_n(t) = R_{n-1}(t) + C * t - \frac{R_{n-1}(t)}{\text{OWD}} S(t) \tag{2}$$

where $R_n(t)$ is the packet arrival rate, C is sending data rate by Skype, OWD is one-way packet delay, and S is the packet dropping probability due to congestion. The queue length dynamics at the receiver is given by:

$$Q_n(t) = Q_{n-1}(t) + (R_{n-1}(t) - C_b) * t \tag{3}$$

where $Q_n(t)$ is the current queue length, $Q_{n-1}(t)$ is previous queue length, R_{n-1} is the packets arrival in the previous time interval, C_b is the bottleneck link capacity, and t is the time in seconds. The queuing delay at the receiver is given by:

$$Qd_n(t) = \frac{Q_n(t)}{C_b} \tag{4}$$

where $Qd_n(t)$ is queuing delay, $Q_n(t)$ is the instantaneous queue length, and C_b is the Bottleneck link capacity. The transmission delay in second is given by:

$$Td_n(t) = \frac{R_n(t)}{C_b} * 0.001 \tag{5}$$

where $Td_n(t)$ gives transmission delay, $R_n(t)$ is the packet arrival rate, and C_b is the bottleneck link capacity. One-way delay is given by:

$$\text{OWD} = Td_n(t) + Qd_n(t) \tag{6}$$

where $Td_n(t)$ is the transmission delay and $Qd_n(t)$ is the queuing delay. The probability of packet dropping to control congestion and sending feedback control message to source is given by:

$$S(t) = \begin{cases} 0, & q(t) \in [0, t_{\min}] \\ \frac{Q_n(t) - t_{\min}}{t_{\max} - t_{\min}} P_{\max}, & q(t) \in [t_{\min}, t_{\max}] \\ 1, & q(t) \geq t_{\max} \end{cases} \tag{7}$$

where $Q_n(t)$ is the instantaneous queue length measured at time t, t_{min} and t_{max} are minimum and maximum threshold values fixed depending on the buffer size, and P_{max} is the maximum packet loss probability fixed based on the link condition.

4 Simulation and Performance Analysis

A number of experiments are conducted using simulation to study the performance of Skype. The proposed Skype network model is illustrated as shown in Fig. 2. The Skype communication takes place in both ways—meaning person 1 speaks, then person 2 listens; similarly, when person 2 speak, person 1 listens. Thus, the model designed is to be built in nodes at both person 1 and person 2. In the experiment, Packet size is 1000 bytes, $P_{max} = 0.001$ s, queue buffer at the nodes has a minimum threshold value, $t_{min} = 1000$ packets, maximum threshold value, CBR flow rate, $C = 2200$ packets, channel capacity, $C_b = 1500$ packets, initial one-way delay, Owd $= 0.001$ s, statistical data of simulation is captured with an interval $t = 0.1$ s.

The graph in Fig. 3 represents the sending rate of UDP at the source. The sending rate varies over the range [470, 2656] packets with an average of 1483 packets. When the sender transmits 2656 packets, queue overflows and the arrival rate exceeds the threshold value. Then, the feedback system asks the source to reduce the flow rate which falls to 470 packets.

The graph in Fig. 4 represents the queue length at the receiver. The queue length varies over the range [1700, 2080] packets with an average of 1851 packets. When the queue length 2080 packets, he buffer overflow and also exceed the threshold value. Then, the feedback system asks the source to reduce the flow rate; thus, it falls to 1700 packets.

The graph in Fig. 5 represents queuing delay which can be defined as the amount of time the packet has to wait in the queue for further process. The queuing delay varies over the range [1.1732, 1.4003] s, with an average value of 1.2272 s.

The graph in Fig. 6 represents the variation of queuing delay on the sender Skype. When the communication starts, gradually delay starts increasing till queue length exceeds threshold. Then, feedback system asks sender to reduce the flow rate, and

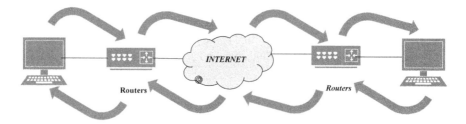

Fig. 2 Skype network model

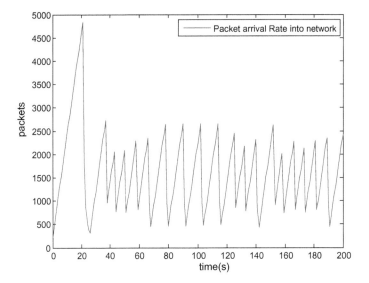

Fig. 3 Variation of packet sending rate by UDP

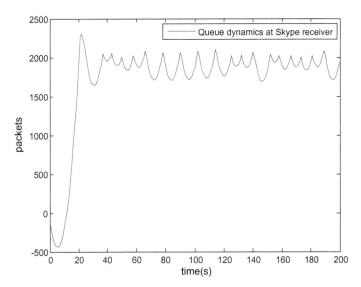

Fig. 4 Queue dynamics at the receiver

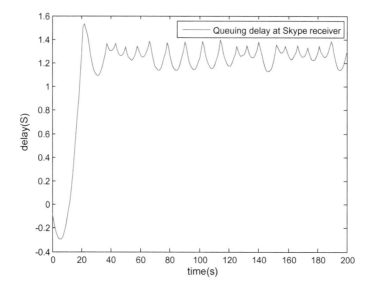

Fig. 5 Queuing delay due to burst traffic

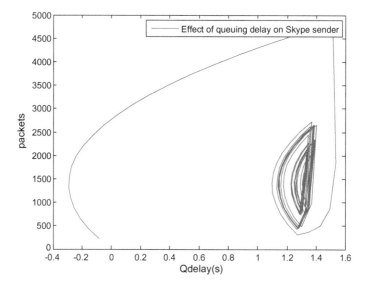

Fig. 6 Variation in sending rate by UDP due to queuing delay

queuing delay starts dropping. Most of the packets are transmitted with a delay ranging between 1.1 and 1.3 s with an average delay of 1.2 s.

The graph in Fig. 7 represents the transmission delay which can be defined as time taken by sender to transfer the packets to the receiver. The time varies over the range [0.0003 ms, 0.0018 ms] with an average of 0.001063 ms.

Analysis of Traffic Characteristics of Skype Video … 295

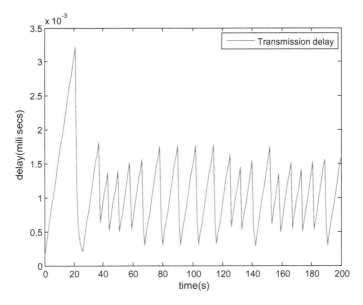

Fig. 7 Variation in transmission delay

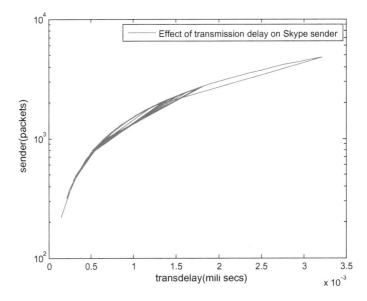

Fig. 8 Variation in sending rate due to transmission delay

The graph in Fig. 8 represents the variation of sending rate with respect to transmission delay. The simulation graph says that large number of packets are transmitted with a delay variation over the range [0.0003, 0.0018] ms.

The graph in Fig. 9 gives the one-way delay in milliseconds in which the time is taken by the packet to move from sender to receiver. Time varies over the range [1.1443, 1.3397] milliseconds, with an average value of 1.228332 ms.

The graph in Fig. 10 represents the feedback sent from receiver to sender, which informs congestion status so that sender can reduce the sending rate or can increase

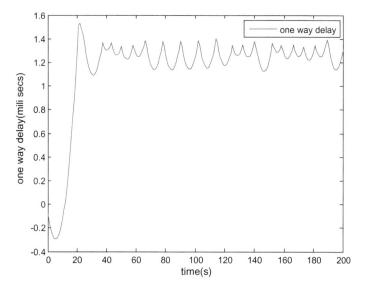

Fig. 9 One-way packet sending delay

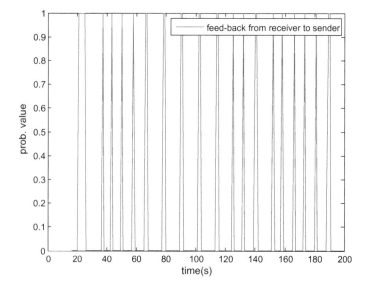

Fig. 10 Feedback received by sender

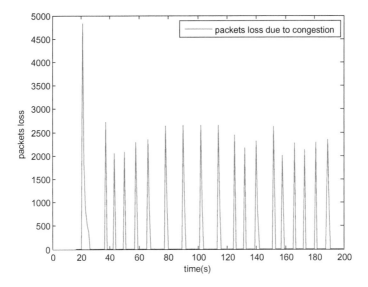

Fig. 11 Packet loss due to queue overflow

the sending rate. It varies from 0 to 1. When the value is 0, there is no congestion, and when the value is 1, there is severe congestion.

The graph in Fig. 11 represents the number of packets lost due to congestion. If the value of feedback is 1, it means all the transported packets are dropped; if feedback value is 0, it means all the transported packets are received without loss, when the value lays between 0 and 1 some packets are lost.

5 Conclusion

Skype is beyond any doubt that P2P VoIP application in the current Internet application. Its amazing popularity drawn the attention of Internet service providers and the researchers a curiosity in knowing and understanding its internal mechanisms and traffic characteristics. A stochastic models-based analysis has been made in this work. The benefit of using mathematical models in the analysis of a system is to capture the performance of all the entitics and making the complexities would be intractable. The objective has been met in this work. There is a lot of scopes to enhance this work for multiuser environment with heterogeneous traffic.

References

1. Carlucci, G., De Cicco, L., Holmer, S., & Mascolo, S. (2017). Congestion control for web real-time communication. *IEEE/ACM Transactions on Networking*, pp. 1063–6692 © 2017, IEEE.
2. Kumara, W. G. C. W., Jayasekara, J. M. N. D. B., & Gunarathne, T. (2013). Video quality and traffic characteristics of Skype video calls in WAN. In *SAITM Research Symposium on Engineering Advancements*.
3. Zhang, X., Xu, Y., Hu, H., Liu, Y., Guo, Z., & Wang, Y. (2012). *Profiling Skype video calls: Rate control and video quality* (p. 100871). Institute of Computer Science & Technology: Peking University, Beijing.
4. Santiago del Río, P. M., Ramos, J., García-Dorado, J. L., & Aracil, J. (2011). *On the processing time for detection of Skype traffic*. 978-1-4244-9538-2/11/$26.00 c, IEEE.
5. Molnár, S., & Perényi, M. (2011). On the identification and analysis of Skype traffic. *International Journal of communication systems, 24*(1), 94–117. Wiley & Sons, Ltd.
6. Bonfiglio, D., Mellia, M., Meo, M., & Rossi, D. (2009). Detailed analysis of Skype traffic. *IEEE Transactions on Multimedia, 11*(1), 117–127.
7. Xie, H., & Yang, Y. R. (2007). *A measurement based study of the Skype peer to peer VoIP performance*. Computer Science Department, Yale University.
8. Guha, S., Daswani, N., & Jain, R. (2006). *An experimental study of the Skype peer-to-peer VoIP system*. USA: Cornell University.
9. Baset, S. A., & Schulzrinne, H. *An analysis of the Skype peer-to-peer internet telephony protocol*. Department of Computer Science, Columbia University, New York NY 10027.
10. Goudru, N. G., & Vijaya Kumar, B. P. (2016). Enhancement of performance of TCP using normalized throughput gradient in wireless networks. *Journal of Information and Computing Science (JIC)* (published by World Academic Press, World Academic Union, England, UK), *11*(3), 214–234.

Smart Tourist Guide (Touristo)

M. R. Sowmya, Shashi Prakash, Shubham K. Singh, Sushent Maloo and Sachindra Yadav

Abstract Individuals travelling, frequently think that it's hard to seek places and find nearby amenities, and this issue even looks greater when we cannot talk the neighbourhood dialect. Additionally while travelling in groups, individuals like exploring different places and may get lost, which again wind up troublesome for their companions individuals to find different individuals from the group. Touristo is a project about building up an android-based application in the field of travel and tourism. Android is a Google-developed programming language for mobiles and tablets. Additionally, Firebase is a real-time database which is utilized for information stockpiling and handling. Hence using the features provided, we intend to develop the application. By examining the above issues and different others, we are taking this venture with the goal to develop such application to overcome above issues and serve clients better.

Keywords Android · Firebase · Location-based services · Real-time location · GPS

1 Introduction

Flying out to different spots for different reasons may it be for relaxation or religious or work and has been in extraordinary practice by individuals nowadays. After soliciting a vast number from individuals who had come to visit the city from various parts of the world, it has been discovered that the significant inconvenience they have is in finding the eateries, place to stay, spots to visit in the city and travellers spot, they face such trouble as the visitor may not know the dialect and culture of the specific place or may experience issues having a discussion, even in the wake of getting the points of interest of the city is troublesome as each individual has their own diverse supposition, and therefore, it gets hard for sightseers to get valuable data from the same. It has additionally been discovered that when individuals go in

---

M. R. Sowmya (✉) · S. Prakash · S. K. Singh · S. Maloo · S. Yadav
Department of CS&E, Nitte Meenakshi Institute of Technology, Bengaluru 560064, India
e-mail: sowmyamr@gmail.com

© Springer Nature Singapore Pte Ltd. 2019
N. R. Shetty et al. (eds.), *Emerging Research in Computing, Information, Communication and Applications*, Advances in Intelligent Systems and Computing 906,
https://doi.org/10.1007/978-981-13-6001-5_23

gatherings, some may lose all sense of direction in the group or go elsewhere, and if the individual's portable goes down, it turns into a problem to find each other.

Henceforth, this inspired us to do some exploration on the past criticisms and audits by the clients and furthermore to adjust this undertaking to attempt and wipe out these issues.

In this paper, we have discussed the design, working and development of Smart Tourist Guide and conclusion. The rest of this paper is outlined as follows: Sect. 2 gives the literature survey, Sect. 3 displays the proposed strategy used to complete the endeavour, Sect. 4 clarifies the usage and the data flow of the application and Sect. 5 gives the conclusion and blueprints future work.

2 Literature Survey

2.1 Related Work

There have been many researches done and papers published in this field, and it includes certain areas of expertise, issue proclamation and their solutions. We had the opportunity to go through some research work and to comprehend the thought and know the solutions for achieving the final output. Hence, here we have discussed some ideas that we encountered and those which helped us to comprehend and execute our idea.

Study of Google Firebase API for Android

Firebase is an API provided by Google for storing and synchronizing the database in your android, IOS or Web application. A real-time database stores information in the database and gets data from it very expeditiously; however, Firebase isn't only a real-time database [1].

Real-Time Location Tracking Application based on Location Alarm

The primary motivation behind this application is to provide users with a convenient route to access information about places around that location. The user can get nearby places like hotels, restaurants and event notifications that happen. The user can join the group of people who have the have a similar enthusiasm through this application. The user can also obtain the address of the destination he wishes to travel. It also provides transportation and costs along with the route. The application provides the real-time location of the client using the GPS system. The position is obtained through the latitude and longitude of the device. The application also contains special features, such as the restricted area, which is limited by the administrator, where users are not authorized. If the user accesses the reserved area, the administrator gets the alarm notification [2].

Smart Tourist Guide (Touristo)   301

Android Application: Travel Guide

The quantity of individuals interested in tourism increases every day. They travel to different countries or cities for different purposes, such as visits, religion, work, and business. This application provides a basic communication between residents and foreigners. It has two buttons on the main page where the first is for sentences containing basic phrases and frequently used in the particular region and the second button is to travel where it shows the distance between the source and destination, the shortest path, the half from which user can reach the destination such as cars, buses, cars, and taxis and also shows the possible cost of travelling to the desired location. The application additionally has a few constraints, i.e. the application is customized just for the particular city and does not identify the situation progressively. User should physically enter the location they are in and the destination they want to travel to. What's more, the primary disadvantage is that the client must know the name of the place, and the sentence and phrases in the local language are limited [3].

The Services Based on Location

Location-based services provide a custom option for customer located in a situation of geographic location data given by the database. This property utilizes geographic data frameworks, this sort of data can be acquired from customer side, such as GPS or server side, as a positioning service arranged for the cellular network operator and communication system technology to transmit information to an application that can process and react to the requested service [4].

Smart Tourist Guide

Smart Tourist Guide provided the plan of the guide and implemented the guide as a mobile application, that is intelligent tourist guide, mobile phone users can seek tourism-related orientation based on your needs using smart travel guide, travellers can get well-said (with few words) information on an attraction, as (word-based), picture-based and video. Above all, the intelligent journey guide can give travellers information about a location, which can be got using a map. The traveller can find the fence near the places and visit the place after knowing the distance between the position of the individual and the tourist place. As you move, the application updates the file location on the map [5].

2.2 Existing System

While travelling to new cities, people have to either keep asking localities or search the Web for places and amenities and needs to perform to much work and sometimes multitasking too. It would have been much easier if all this work can be done using a single application, because of inaccessibility of such application travellers are confronting issues while travelling. They need to pay great looking measure of heading out spending plan to nearby aides and operators to get data. The correct visitor control isn't accessible which could identify a present area, ascertain remove and give appropriate rules.

Different issues emerge while a visitor goes to a traveller spot which we can address by improvement of the framework which will give precise worldwide situating framework (GPS) organizes, give legitimate ongoing bearing, points of interest and pictorial data about the area.

Disadvantages of the existing system

- Locating nearby amenities
- Getting directions
- Information about places requires searching
- Locating Group members
- Renting local guide gets costlier.

2.3 Proposed System

We intend to propose and develop such an application for the Android users to work as a travel application. The proposed application will allow users to select a location or even they can auto-detect. Further they can know the nearby amenities and places of interest. They can also know the directions to the destination. While selecting the nearby place will provide users with the pictorial information of the place and ratings.

Another important feature of the application is, in the chat box they can create group while travelling to different places, other users can join the group and hence chat, send information or images. Also they can even know the last known location of the other users and the directions to that location using maps.

Advantages of the proposed system

- Organized search
- Basic information about the place is available
- Locating places
- Locating users
- Chatting helps to know more information about any queries
- Additionally, application is free to operate and easy to use.

3 Methodology

3.1 Methodology Adopted

1. **Initiation**
   For initiation of project, first of all literature survey was done from which we came to know how the app can be made and what all things can be implemented in the application.
   Market study was also done to know that what type of application should be made so that user can be attracted to the app and should be useful to users.
2. **Analysis**
   Analysis of the application development is the process where the problem is identified that can be solved by this app. The cost required for the development of the application was analysed and how much benefits can be gained from that project.
3. **Design**
   Characterize Project Goal Determine Outcomes, Objectives, and also Deliverables Identify Risks, Constraints, and Assumptions Prepare a Visual Aid.
4. **Construction**
   Amid the construction phase, every single components and application features are created and incorporated into the product, and the construction phase is in some sense an assembling procedure, where accentuation is put on overseeing assets and controlling tasks to advance costs, timetables and quality.
   But if the user is new then they have to register using email id and have to provide details like full name and the password.
   If the users are existing users then the user needs to provide password to enter the app.
   If the user forgets the password then can reset their using their registered email id.
5. **Testing**
   Identified Types of Testing.
   It was an obligatory prerequisite that the application needs to work in all android gadgets as the end customers can have differing gadgets.
6. Be that as it may, utilizing representations can be valuable while dealing with an undertaking since they give colleagues and partners an effortlessly reasonable preview of the venture's objectives, results, expectations, items, administrations or potentially usefulness.
7. Legal requirements for the app was also examined what all the paperwork and documentations are needed for the development of the application.
8. To guarantee that the application worked in every one of the android gadgets, we chose least form of android platform, i.e. Kit Kat version 4 with the goal that it can cover around 97% of the android clients.
9. The possibilities of the risks and the failure of the project were analysed so that it can be overcome by developing this application.

3.2 Use Case Diagram

The user can register using various authentications available. Once registered, will be automatically identified while next login. While they can search for different places, they can also use the current location service and hence get to know the nearby places. Also they can know the distance and directions between the two locations. Another act they can perform is chatting with their group members (Figs. 1 and 2).

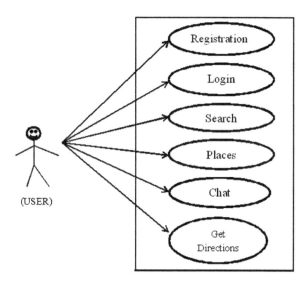

Fig. 1 Use case diagram for a user

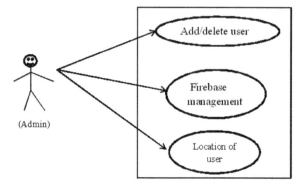

Fig. 2 Use case diagram for Admin

Admin can add and delete users. Manage firebase. Can know the last known location of the users.

4 Implementation

4.1 Data Flow Diagram

A flow diagram of data is a graphical representation of the flow of data through a system, modelling its process aspects. It is often used as a preliminary step to create a general description of the system without going into details. The following flow chart shows the flow of application data and the interaction between the various application interfaces (Fig. 3).

- First activity will be the authentication (SIGNIN) of users. Here the user will get three authentication type, i.e. Facebook, G-mail, Email.
- Sign in to connect to server, i.e. it will access FIREBASE and write (for new users) or check for the users availability (existing user).
- After authenticating the user, MAIN ACTIVITY starts (welcome user).
- Next we can see three activity tabs, i.e. HOME, CHAT and INFO, respectively (Fig. 4).
- Main activity: It will show the location detection or searching option.
- Pin location: Auto-detect the current location of the user.
- Search place: Allows user to search for a place.

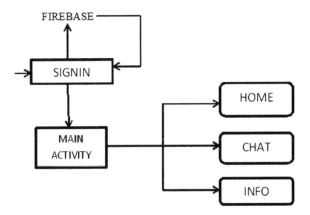

Fig. 3 Data flow diagram of the application

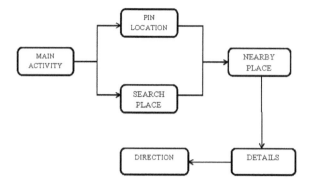

Fig. 4 Main activity

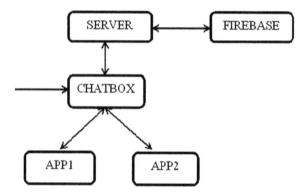

Fig. 5 Chat box

- Nearby place: Will show all the nearby places by that location.
- Details: Details of that place like rating, image, location.
- Direction: Direction and distance of the destination selected from the current place (Fig. 5).

FIREBASE: The Fire-store is used to store the messages sent and received like text and images.

SERVER: Used for data transfer and receiving will be connected to firebase and app.

CHAT BOX: Hence, the apps use chat box to send and receive messages via server which gets stored in the firebase storage.

5 Results

See Figs. 6, 7, 8, 9, 10 and 11.

Fig. 6 Sign up

Fig. 7 Pick location/auto-detect

6 Conclusion and Future Work

On completion of the project, we can conclude that we were able to design such application which can serve users while travelling to newer places. The users of the application, frequent travellers, explorers, people travelling in groups or any casual user can chat in the group or create one. The application is helpful to each of these people. For frequent travellers, the application comes in help to determine location and nearby places when they are travelling to some far away city. The application allows clients to discover adjacent places at any location and in addition, pictorial information and rating for a specific place. For people travelling in groups, the application's main purpose is to track the location of the users so that other users can see.

Smart Tourist Guide (Touristo)

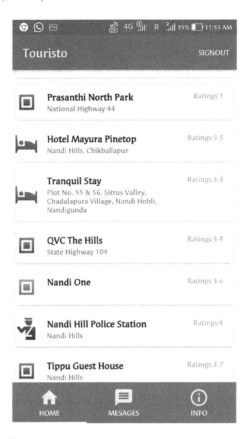

Fig. 8 List of nearby places

Overall, the application which we developed is going to help people with their trips regardless of whether it is work tour or leisure and it makes easier for them while travelling.

Further for future developments of the application, we can work on live tracking of group members and their status too. Each member should be able to live track down every other member of the group and also every user can update their current status. The idea to make this app a source to plan people's trip can be worked upon. With maintaining database for different places and real-time data updation can help to guide travellers with their tour.

Fig. 9 Details of the place along with get direction

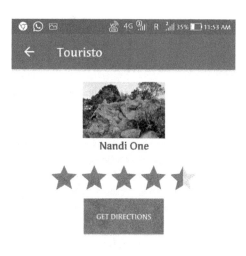

Fig. 10 Chat groups

Fig. 11 Get location of the user

References

1. Singh, N. (2016). Study of google firebase API for android. *International Journal of Innovative Research in Computer and Communication Engineering, 4*(9), 16738–16743.
2. Ghadiyali, A., Tiku, A., Bandevar, S., & Tengale, R. (2015, April). Real time location tracking application based on location alarm. *International Journal of Engineering and Computer Science, 4*(4), 11352–11355. ISSN: 2319–7242.
3. Ferdaus, J. (2015). *Android application: Travel guide*. Research September 2015. https://doi.org/10.13140/rg.2.1.4865.4569.
4. Muzumil, S. (2014). *Fast finder android application* (Vol. 2, No. 2). A Thesis Presented to the Department of Software Engineering Mehran University of Engineering & Technology, Jamshore.
5. Jinendra, D. R., Bhagyashri, J. R., & Pranav, A. (2012). Smart travel guide. In *1st International Conference on Recent Trends in Engineering &Technology*.
6. Watkar, P. D., & Shahade, M. R. Smart travel guide: Application for mobile phone. *International Journal of Research in Science & Engineering, 1*(1). e-ISSN: 2394–8799.

7. Badekar, N., Parakh, T., Lokhande, P., Shelake, K., & Dhawas, N. A. (2015). Smart travel guide for android-enabled devices. *International Journal of Advanced Research in Computer and Communication Engineering, 4*(2).
8. Alshattnawi, S. (May, 2013). Building mobile tourist guide applications using different development mobile platforms. *International Journal of Advanced Science and Technology, 54.*

Thefted Vehicle Identification System and Smart Ambulance System in VANETs

S. R. Nagaraja, N. Nalini, B. A. Mohan and Afroz Pasha

Abstract VANETs act as crucial component of intelligent transportation system (ITS). VANETs are capable of providing connectionless communication between mobile nodes (vehicles) and static nodes such as RSU and BSU to enhance safety and comfort of vehicles on the highways or in urban environments. There is no system in place for finding stolen vehicles and providing faster movement of ambulances on heavy traffic lanes. In this paper, we discuss about automated system for traffic management such as for finding stolen vehicles and providing faster movement of ambulances on heavy traffic lanes, and we are coming up with two systems namely thefted vehicle identification system (TVIS) and smart ambulance system (SAS), respectively.

Keywords Vehicular ad hoc networks (VANETs) · On-board units (OBU) · Road side units (RSU) · Base stations' units (BSU) · Dedicated short-range communication (DSRC)

1 Introduction

VANET is a type of mobile ad hoc networking technology, created by establishing a network of vehicles with road side units and base station unit, used for communication purpose on highways or in urban environments as shown in Fig. 1.

VANETs are the crucial components of intelligent transportation system [ITS]. DSRC is one of the communication standards in VANETs. DSRC band is distributed into seven conduits. The frequency range of the channel is from 5.850 to 5.925 GHz.

S. R. Nagaraja (✉) · N. Nalini · B. A. Mohan · A. Pasha
Department of Computer Science Engineering, Nitte Meenakshi Institute of Technology, Bangalore 560064, India
e-mail: nagaraj.sr@nmit.ac.in

© Springer Nature Singapore Pte Ltd. 2019
N. R. Shetty et al. (eds.), *Emerging Research in Computing, Information, Communication and Applications*, Advances in Intelligent Systems and Computing 906,
https://doi.org/10.1007/978-981-13-6001-5_24

Fig. 1 Vehicular network communication scenario [1]

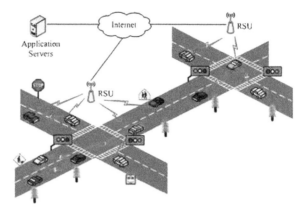

VANETs are an upcoming high-tech technology, combination of ad hoc network, wireless LAN, and cellular technology to attain intelligent vehicle communication system [1].

VANETs have been receiving and acquiring a lot of focus due to the enormous variations in support they can offer. The VANET technology is mainly used to provide defense and amenity to the commuters and also to prevent them from unfortunate incidents.

Vehicular traffic will occur when the size of vehicles' flow requires more space than the existing road capacity. On that time, there is no system in place to providing faster movement of ambulances on heavy traffic lanes, and also, there is no system in place for finding stolen vehicles. In this paper, we discuss on system architecture, complex problems, and technical challenges, related work based on thefted vehicle identification system (TVIS), and smart ambulance system (SAS) [2–12].

2 System Architecture

System architectures can be classified into dissimilar forms based on different view points in VANETs. With regard to the vehicular communication view point, it can be classified into three layers. Infrastructure domain is the upper layer comprising of base station, servers, gateway, and road side unit (RSU). Ad hoc domain is the middle layer; finally, in-vehicle domain is the lower layer. There are five types of communications which are possible in VANETs, and they are V2V, V2RSU, RSU2RSU, RSU2 BS, and BS2BS. The on-board units [OBU] are fixed in vehicles. These vehicles are interacting with RSU. Road side units [RSUs] are fixed near the junctions. These units collect the information from OBU and sends to BSU. Base station units [BSUs] are centralized units. The computation will take place, and the result will send to OBU through RSU as shown in Fig. 2 [13, 14].

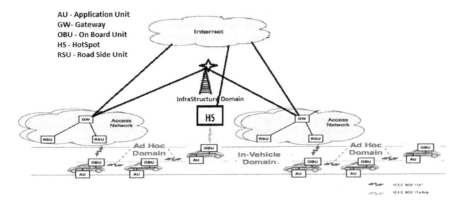

Fig. 2 VANET system architecture

3 Complex Problems and Technical Challenges

Networks usually instill quite a lot technical issues regarding data processing and sharing on network involving network management. Along with the above, networks also throw additional technical challenges in discovering of new networks, congestion control challenges, network control and routing issues, collective information processing, and querying. Following are the list of additional challenges involved in VANETs [14, 15], and they are:

- Vehicular network discovery;
- Network control and routing;
- Congestion control;
- Collaborative signal and information processing;
- Tasking and querying;
- Security.

4 Related Work

According to the recent survey, lakhs of stolen vehicles have gone untraced because there is no system in place to identify and track the stolen vehicles. This invention primarily focuses on this issue which consists of inbuilt sensors in vehicles, and when vehicle is stolen, a complaint has to be registered and saved in a database. The stolen vehicle sends vehicle information via sensors at every junction which consists of a large database which registered complaints. When vehicle information sent by vehicle via sensors is matched with vehicle id in the registered complaint, appropriate message in the form of voice or text message will go to the concerned authorities such as police force and to the owner of the vehicle, thereby stolen vehicles can

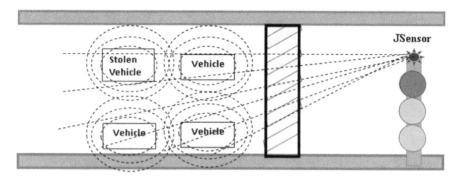

Fig. 3 Thefted vehicle identification system (TVIS)

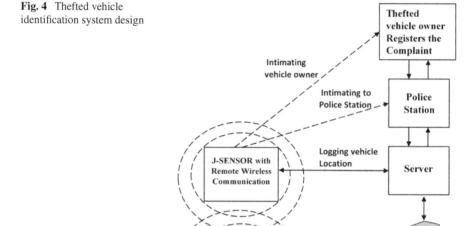

Fig. 4 Thefted vehicle identification system design

be identified and caught with the help of smart thefted vehicle identification system (TVIS) as shown in Fig. 3, and design of this system is shown in Fig. 4. This system can be deployed across various junctions across cities to get hold of stolen vehicles.

Algorithm: Smart Vehicle Theft Identification (SVTI)

Procedure or Description of Method: Thefted Vehicle Identification System (TVIS)
Input: Owner of vehicle registers complaint in Database

Assumptions:

1. All vehicles are equipped with Vsensors to dispatch vehicle info.
2. All Jsensors across junctions deployed to receive vehicle's info.

Output: Detecting and recovering the stolen vehicle.
Step 1: Start vehicle's owner registers complaint with vehicle_Id.
Step 2: Deploying Jsensors across all traffic junctions.
Step 3: Vehicle upon reaching in any junction dispatches info from Vsensors to Jsensors.
Step 4: If vehicle_id matches with that in database, then information is passed to traffic control room and owner's stolen vehicle is traced.
Step 5: Stop.

In real-time scenarios, due to heavy traffic on roads nearby signals, ambulances reaching hospitals faster are difficult which lead to loss of life, so providing timely medical assistance to patients in ambulance is a challenging task. One of the possible solutions for clearing the way for ambulances on heavy traffic signals is to have a smart ambulance system in place. In our research work referring to the module, which dealt with dynamic signaling system, can further be enhanced with smart ambulance system consisting of RF receiver which sends signals. Existing traffic signaling system must comprise of a sensor to receive RF signals sent by ambulance.

Upon receiving the RF signal from ambulance, the traffic signaling system puts the others lanes onto hold by turning the indicator light to red and by turning ambulance stationed lane to green, thereby allowing ambulances to move freely, and the indicator will remain green till all the ambulances in that lane cross the junction. In case the vehicles stationed in front of ambulances do not provide way for its free movement, the ambulance sensor in SAS then captures the vehicle id and sends to nearest traffic control room; for such vehicles, further action can be initiated.

This system is designed for the cities with dense traffic. Most of the time, the traffic will be at least for 100 m. In this distance, the traffic police cannot hear the siren from the ambulance. Then, the ambulance has to wait till the traffic is left. Sometimes to free the traffic, it takes at least 30 min. So by this time, anything can happen to the patient. So, this application module avoids these disadvantages. The second feature is the information system in the ambulance. The system will inform the status of the patient to the hospital as the command giving to the system in the ambulance as shown in Figs. 5 and 6.

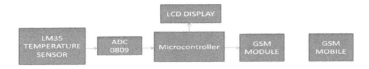

Fig. 5 Ambulance block diagram

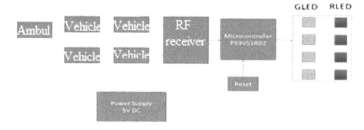

Fig. 6 Smart ambulance system (SAS)

Algorithm: Smart Ambulance Systems (SAS)
#Algorithm for Smart Ambulance systems
Inputs: Ambulance arrival at traffic junction
Output: The signal will change green
Step 1: Start
Initialize the variables flag=0; [Ambulance has not arrived at Traffic junction]
Step 2: If ambulance arrives at traffic junction then

Flag=1;

Step 3: if (flag==1) then

Void Indicator (int green, int red);
Ambulance lane light indicator is turned to green
Green=1; [green variable is set to 1 and red=0]
Other lanes light indicator is turned red
Red=1; [red variable is set to 1 and green=0]

Step 4: if vehicles in-front of ambulance do not provide way then

Vehicle's info is sent to concerned authorities

Step 4: repeat step 3 until flag=0 [Ambulance leaving traffic lane]
Step 5: stop

Description:
When the ambulance comes to any traffic junction, it sends the information through RF transmitter, and the dynamic signaling system senses the information through RF receiver sensor and automatically gives green signal for this ambulance till it leaves the junction. So whenever the ambulance comes near the traffic, the ambulance will transmit a code; say "emergency"; the receiver will receive this signal. Then, it immediately makes all the signals red in other lanes.

Whenever the driver wants to send the information about the status of the patient to the hospital, by a switch press the system will send the information through SMS.

We developed a smart street light system (SSLS) as shown in Fig. 7. This system will help to save the power in such a way that if any vehicle or object will come near the street light, then only that street light will remain on; otherwise, lights remain

Fig. 7 Smart street light system in real-time scenario

off and if no vehicles or objects in its range street light remain in off state. This system embedded with light sensors will sense the light and make the decision either street light on or off state. In morning time, street lights are always off state, and in nighttime, street lights are on state based on vehicle or object in its range, and also, no manual operations are needed. Through this system, we can save maximum power, and also, it reduces the manual operations. In this system, each street light embedded with light sensor, IR sensor, and object detection sensor.

We developed a smart street light system (SSLS) as shown in Fig. 7.

5 Results

When this system is deployed in traffic signals, it will give way to ambulance by changing the traffic signal lights and stolen vehicles can be identified and catch with the help of Smart Vehicle Theft Identification as shown in Figs. 8, 9, 10 and 11.

Figure 8 shows the traffic flow in a junction with different kinds of vehicles in real-time scenario.

Fig. 8 Traffic flow in real-time scenario

Fig. 9 Ambulance in real-time scenario

Fig. 10 Giving green signal to ambulance through smart ambulance systems

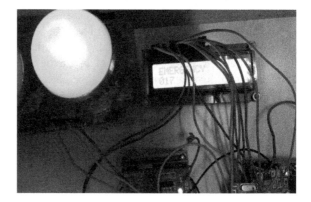

Fig. 11 Thefted vehicle identification system (TVIS)

Figure 9 shows that ambulance comes to near the junction with vehicular traffic.

Figure 10 shows that emergence message in its display and SAS will makes the signal green till ambulance leaves the junction.

TVIS embedded with ARM-7 processor displays other components as shown in Fig. 11.

6 Conclusion

This research work presents developing a thefted vehicle identification system (TVIS) and smart ambulance system (SAS). These kinds of systems are very useful to people. Smart ambulance system (SAS) helps to providing faster movement of ambulances on heavy traffic lanes and stolen vehicles can be identified and catch with the help of thefted vehicle identification system (TVIS), and also, we developed a smart street light system (SSLS) will help to save the power in such a way that if any vehicle or object will come near the street light, then only that street light will remain on state; otherwise, lights remain off state and if no vehicles or objects in its range street light remain in off state. SSLS system also embedded with light sensors will sense the light and make the decision either street light on or off state. In morning time, street lights are always off state, and in nighttime, street lights are on state based on vehicle or object in its range, and also, no manual operations are needed. Through this system, we can save maximum power and also it reduces the manual operations.

References

1. Nagaraja, S. R., Nalini, N., Rama Krishna, K., Satish, E. G. (2015). Alternative path selection through density based approach to controlling the vehicular traffic in VANETs. *International Journal of Advanced Research in Computer and Communication Engineering, 4*(6). https://doi.org/10.17148/IJARCCE.2015.41059. ISSN 2278-1021.
2. Sattari, M.R.J., Noor, R. M., Ghahremani, S. (2013). Dynamic congestion control algorithm for vehicular ad-hoc networks. *International Journal of Software Engineering and Its Applications, 7*(3).
3. Le, L., Baldessari, R., Salvador, P., Festag, A., & Zhang, W. (2011). Performance evaluation of beacon congestion control algorithms for VANETs. Publication in the IEEE Globecom.
4. Konur, S., & Fisher M. (2011). Formal analysis of a VANET congestion control protocol through probabilistic verification. ©2011 IEEE.
5. Sepulcre, M., Gozalvez, J., Harri, J., & Hartenstein, H. (2010). Application-based congestion control policy for the communication channel in VANETs. *IEEE Communications Letters, 14*(10).
6. Piran, M. J., Rama Murthy, G., & Praveen Babu, G. (2011). Vehicular ad hoc and sensor networks; principles and challenges. *International Journal of Ad hoc, Sensor & Ubiquitous Computing (IJASUC), 2*(2).
7. Abdalla, G. M. T., AbuRgheff, M. A., & Senouci, S. M. Current trends in vehicular ad hoc networks. *Ubiquitous Computing and Communication Journal.*
8. Konur, S., & Fisher, M. (2011). Formal analysis of a VANET congestion control protocol through probabilistic verification. In *Proceedings of 73rd IEEE Vehicular Technology Conference (VTC2011-Spring), Budapest, Hungary, May 2011.*
9. Manvi, S. S., & Kakkasageri, M. S. (2008). Issues in mobile ad hoc networks for vehicular communication. *IETE Technical Review, 25*(2), 59–72, March–April, 2008.
10. Darus, M. Y. B., & Bakar, K. A. (2011). Congestion control framework for disseminating safety messages in vehicular ad-hoc networks (VANETs). *International Journal of Digital Content Technology and Its Applications, 5*(2).
11. Liu, Y., Bi, J., & Yang, J. (2009). Research on vehicular ad hoc networks. IEEE.
12. Baumann, R., Heimlicher, S., May, M. (2008). Towards realistic mobility models for vehicular ad-hoc networks.
13. Nagaraja, S. R., & Dr. Nalini, A. G. (2015). Alternate path selection algorithm by virtue of proactive congestion control technique for VANETS. *International Journal of Computer Science Trends and Technology (IJCST), 3*(2). ISSN 2347-8578.
14. Nagaraja, S. R., & Dr. Nalini, N. (2016). Performance analysis of proactive congestion control techniques for VANETs. In *IEEE—Wispnet*. IEEE, 23–24 Mar 2016. 978-1-4673-9338-6/16/2016.
15. Nagaraja, S. R., & Dr. Nalini, A. G. (2016). Congestion control in VANETs using re-routing algorithm. In *IEEE-Wispnet*. IEEE, 23–24 Mar 2016. 978-1-4673-9338-6/16/2016.

Performance Study of OpenMP and Hybrid Programming Models on CPU–GPU Cluster

B. N. Chandrashekhar and H. A. Sanjay

Abstract Optimizing complex code of scientific and engineering applications is a challenging area of research. There are many parallel and distributed programming frameworks which efficiently optimize the code for the performance. In this study, we did a comparison study of the performance of parallel computing models. We have used irregular graph algorithms such as Floyd's algorithm (shortest path problems) and Kruskal's algorithm (minimum spanning tree problems). We have considered OpenMP and hybrid [OpenMP + MPI] on CPU cluster and MPI + CUDA programming strategies on the GPU cluster to improve the performance on shared–distributed memory architecture by minimizing communication and computation overlap overhead between individual nodes. A single MPI process per node is used to launch small chunks of large irregular graph algorithm on various nodes on the cluster. CUDA is used to distribute the work between the different GPU cores within a cluster node. Results show that from the performance perspective GPU, implementation of graph algorithms is effective than the CPU implementation. Results also show that hybrid [MPI + CUDA] parallel programming framework for Floyd's algorithm on GPU cluster yields an average speedup of 19.03 when compared to the OpenMP and a speedup of 15.96 is observed against CPU cluster with hybrid [MPI + OpenMP] frameworks. For Kruskal's algorithm, average speedup of 27.26 is observed when compared against OpenMP and a speedup of 20.74 is observed against CPU's cluster with hybrid [MPI + OpenMP] frameworks.

Keywords CPU · GPU · CUDA · MPI · OpenMP

1 Introduction

Parallel and distributed computing technologies have been used for fast and scalable processing of large irregular graph algorithms in various applications such as

---

B. N. Chandrashekhar (✉) · H. A. Sanjay
Department of Information Science and Engineering, Nitte Meenakshi Institute of Technology, Bengaluru 560064, India
e-mail: chandrashekar.bn@nmit.ac.in

social networks and system biology. Electronic design automation (EDA), computing canonical form of different bounded matrices, etc., which are solved using the graphs and algorithms have millions to trillions of vertices and edges. Most prominently for various research and scientific applications CPU has become commonplace. The idea of using GPU for irregular graph computations is popular because of its high computing capacity due to that there will be a tremendous performance improvement at low cost when applied to suitable problems.

Floyd's and Kruskal's algorithm are the most popular graph algorithm for optimal routing in irregular graph algorithm. In sequential execution of Floyd's algorithm, it requires a long time to find the shortest path between all pairs of vertices in a graph and similarly in a Kruskal's algorithm requires more time to find a minimum spanning tree (MST) [1] in which sum of edge weights is as small as possible. So, it is a difficult task to solve these issues during large irregular graph using serial programming model. And therefore, in our study we use the parallel programming models OpenMP and hybrid [OpenMP + MPI] on CPU and [MPI + CUDA] on GPU. These frameworks take less execution time than the serial execution. And then, it is easy to compute the shortest path in a graph or many other applications. Parallelization of Floyd's serial algorithm, in which each process is allotted as a subsection of vertices and in each stage of calculation, all process creates a set of threads on shared memory architecture. The set of threads per core runs concurrently; i.e., each small chunks of an algorithm are run parallel to solve large irregular Floyd's graph algorithms. Minimum spanning tree algorithm is built on Kruskal's method. Each process gets a subsection of the irregular graph and then finds a native minimum spanning tree (forest). Following, processes combine their MST edges till only single process leftovers, which embrace edges that form a minimum spanning tree of a graph [2].

In this study, even we have demonstrated the performance exploration of different frameworks: OpenMP and hybrid implementation of [MPI +OpenMP and MPI + CUDA]. The comparative study has been conducted on two irregular graph algorithms such as Floyd's algorithm and Kruskal's algorithm for a large number of vertices, where OpenMP is for shared memory architecture [3]. In the OpenMP, the compiler creates numerous threads that stake a common data space. After identification of which loops to be parallelized, apply various compiler pragmas and the compiler handles all that horrible parallel junk. Then, compilers take care of parallelizing irregular graph applications automatically and boost synchronization and communication directives when needed. And hybrid [MPI + OpenMP] is for distributed memory architecture implementation in which MPI is used for communication and OpenMP is for computation to provide communication and computation overlap: that is, while a few threads are interconnected with each other, and a few threads are implementing an application. Here, each thread handles its own communication demands. Therefore, the hybrid framework is to improve the parallel throughput of applications running on clusters of multi-core nodes on CPU and a hybrid [MPI + CUDA] framework with multi-core architecture for acceleration of an application on GPUs. In which host (CPU) processor spawns multithreaded processes (kernels) on to the GPU device. The GPU has a scheduler that will then assign the kernels to different computing capability GPUs hardware is present. For each irregular graph algorithms, parallel programs have been developed in terms of three frameworks on CPU and GPU cluster, and their performance is observed for the large graph. All the parallel programming model are compared using applications performance in terms of speedup.

The details of the study are organized as follows: Sect. 2 briefly reviews related work of the three parallel computing models for irregular algorithms. Section 3 presents selected three frameworks implementation for irregular graph algorithms. Section 4 presents experimental results and performance analyses, and finally, Sect. 5 draws some conclusions and future work.

2 Related Work

This section discusses related work on performance comparison of programming models such as OpenMP, MPI + OpenMP, and MPI + CUDA for irregular graph applications on the CPU and GPU. Most of the earlier research is the subset of work what we have proposed.

Huang and Guo [4] present the execution of Prim's algorithm using the smallest data equivalent primitive under CUDA design on GPU to solve most of the graph problems. The result demonstrates that the algorithm efficiently increases the performance as compared to the CPU. The reason is the case that since the number of cores in the CPU is less compared to that of GPU, it would take longer time as well as the cost of communication is more. In this paper author not considered CPU based programming models OpenMP, and hybrid [MPI + OpenMP] for their comparison.

Kang et al. [5] proposed performance comparison of MPI, OpenMP, and MapReduce research work, and they have demonstrated that if a problem is small enough to be accommodated and the computing resources such as cores and memory are sufficient enough, then OpenMP is a better choice. But MPI is preferred when the input is large, and computation is quite complex. When the input size is huge, and when the computational work is out of iterative processing technique, MapReduce can be a very useful framework. OpenMP is the easiest to use among all of them because there is no special attention to be paid because we need to place some directives in the sequential code. In this paper author not considered GPU based MPI + CUDA programming models for their comparison

Qingshuang [6] has implemented all pair shortest path algorithm based on MPI + CUDA distributed programming model. We can notice that this combination gives the powerful programming environment and instruction set to build the kernel functions by making direct use of C/C++ languages. This idea highly lightens and stresses that the programmer can focus on the parallel algorithm designing only, without having to know the other GPU hardware programming details. All these noteworthy advantages of CUDA reduce the GPU programming difficulty and also speed up the popularity and use of GPU architecture for HPC applications. To use the GPU cluster's powerful parallel computing performance, the parallel algorithm can combine the strength of both MPI and CUDA technologies together and henceforth provide two-level parallel computing. The experimental results show that the MPI + CUDA-based parallel algorithm provides full, powerful computing capability of the GPU cluster, and can, therefore, a time speed of about 100 times faster than a single technology usage can be achived. In this paper, author not considered CPU-based programming models OpenMP and hybrid [MPI + OpenMP] for their comparison.

Loncar [7] proposed the research work authors which were discussed how to parallelize minimum spanning tree algorithms using MPI. Implementations of tree traversal algorithms of this research work were tested on a cluster containing 32 computing nodes where each of this computer in the cluster had two Intel Xeon E5345 2.33-GHz quad-core CPUs and 8 GB of memory space along with installation of Scientific Linux OS installed. They used OpenMPI v1.6 implementation of the MPI standard. The cluster nodes were connected to the network with a throughput of 1 Gbit/s. Both of these procedural implementations were compiled using GCC 4.4 compiler. This cluster has also proved its effectiveness in testing algorithms through the use of more than 256 processes. And also, by testing graphs with densities ranging up to 20% on varying number of vertices and number of edges. Dissemination of edges in graphs was consistently random, and all edge weights were distinctive. But the major issues of this research work are also possible that due to the heavy memory necessities for large graphs, not possible to partition all input into small chunks to cluster nodes. However, in our work divided large applications into small chunks of applications and executed using different parallel programming models OpenMP and hybrid [MPI + OpenMP and MPI + CUDA] on CPU–GPU cluster and compared their results.

Rostrup et al. [8] projected a parallel algorithm for computing the minimum spanning tree of very large graphs with minimum utilization of memory by presenting Kruskal's algorithm. The paper presents a data-parallel Kruskal's (DPK) algorithm that solves the limitations of the previous work that could not map well to the fine-grained parallelism of the GPU. In order to keep low memory utilization, each edge is stored once. The full problem is divided into sub-problems so that optimal parallelism can be exploited while keeping low provisional storage. The algorithm was implemented on an NVIDIA Tesla T10 GPU. In this paper, author has not utilized CPU computational power by using a framework like OpenMP, hybrid [MPI + OpenMP and MPI + CUDA].

3 Proposed Parallel Computing Frameworks

In this framework, we have proposed three parallel programs using OpenMP, hybrid [MPI + OpenMP], and [MPI + CUDA]. Parallel processing is employed to split a large irregular graph application into small chunks of applications which will be solved concurrently, utilizing multi-core architecture. However, it is important to note that while computing concurrently results in one small chunk should not affect the other chunks. Thus, each small chunk should be modeled independently.

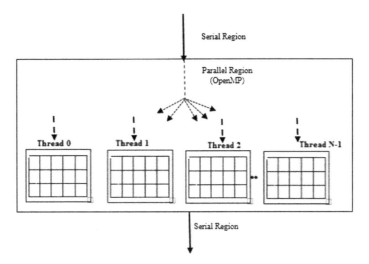

Fig. 1 OpenMP framework for implementation of irregular Floyd's graph algorithms for shortest path problems

3.1 Parallelization with OpenMP Framework

As we can observe in Fig. 1 OpenMP framework for implementation of irregular Floyd's graph algorithms for shortest path problems. OpenMP in an API comprises multiple compiler directives for creation and manipulations of threads. Threads created are shredded same address space; hence, communication between threads is very efficient. In our proposed frameworks, as soon as irregular Floyd's graph algorithms for shortest path problems launched on different nodes of a cluster using one MPI process, each node of a CPU creates a set of threads on shared memory architecture. The set of threads per core runs concurrently. A specified small chunk of as irregular Floyd's graph algorithms is divided among threads, and each thread will access the same information, i.e., variable, objects, and parameters in shared memory. OpenMP uses fork/join model to switch execution serial to parallel as we see Fig. 1 at the entrance of parallel chunk, a single thread is split into several sets of threads once it finished, and the new serial thread will start.

In this work, we made use of OpenMP directives that offer provision for synchronization, concurrency, and data handling while avoiding the need for clearly setting mutexes, condition variables, data scope, and initialization. Floyd algorithm is one of the extremely known algorithms to this problem, which iteratively searches the shortest paths by considering the transitional nodes (vertices) one by one [9]. The recursive Floyd's algorithm may be parallelized by applying the parallel directive: #pragma omp directive [clause list] implements sequentially till they compiler encounter the parallel directive, the clause if (scalar expression) defines whether parallel models consequences in creation of thread, amount of parallelization: The clause num_threads (integer expression) stipulates the amount of threads that are produced by the parallel directive, data handling: The clause private (variable list) specifies that the established variables identified in local to each thread that is individually thread have its private copy of each variable in the list such as firstprivate (variable list): This is equivalent to the private clause, but the value of variables on entering the threads is adjusted to parallel values before the parallel directive shared (variable list): It shows that all variables in the list are pooled across all the threads. In the classic Floyd's version, all repetitions of the ith loop can be performed totally in parallel, so here we considered for directive: #pragma omp for [clause list]: Then for omp, directive is used to divide parallel repetition places through threads. Here, clause list specifies the lastprivate: This clause deals with how multiple local copies of a variable are written back into a single copy at the end of the parallel for loop finished. A loop can be parallelized by accessing subroutine calls from OpenMP thread libraries and inserting the OpenMP compiler directives. In this style, the threads can attain new tasks, unprocessed loop iterations, directly from local shared memory. The section directives: #pragma omp section [clause list]: the section directive supports non-iterative parallel task assignment. It allocates the struck block parallel to each section of one thread. The clause list includes private, firstprivate, lastprivate, reduction, and nowait. Clause list nowait: It stipulates that no implicit synchronization among all threads at the end of the section directive. The other directives are that we are considered are a barrier, single, master, critical, atomic, ordered, and flushed directives.

We understood that way of OpenMP performed at a minor level. It was somewhat easy to increase the speedup. Subsequently, it is clear that OpenMP is the best framework for quick, easy, and safe parallelizing. It is easy to learn, quickly to use and supported by most of the existing compilers.

3.2 Parallelization with Hybrid [OpenMP + MPI] Framework

Figure 2 shows hybrid [OpenMP + MPI] framework on CPU cluster environment; here, hybrid framework uses both the OpenMP + MPI models together for parallelism. A hybrid model needs a more advanced programming model than either type of the components so as to accomplish shared and distributed memory allocations with multiple cores. We are aimed to use a higher number of cores in the irregular application system and reduce the computation time. In this study, we are basically aiming to provide information on basic programming models and compare their performances in terms of computation time. With the hybrid model, Y task is created, and X threads are employed to each task to distribute the small chunks among threads. In this way, $X \times Y$ cores are used with hybrid parallelization. In Fig. 2, we can notice that MPI is used to launch large irregular graph algorithm on various nodes in the cluster and OpenMP to distribute the work between the cores within a cluster node. Further, OpenMP parallel subdivisions are retained separate from MPI communication, since this streamlines synchronization equally for the user code and the MPI implementation. In order to utilize existing resources, avoid internode communica-

Performance Study of OpenMP and Hybrid Programming Models ... 329

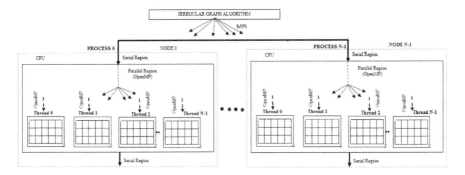

Fig. 2 An OpenMP + MPI framework for implementation of irregular Floyd's and Kruskal's graph algorithms

tion, and increase the intranode computation in our study, we have considered only one MPI process on each node of a cluster. By using MPI, various tasks run concurrently on distinct cores defined by the user and each core has its private memory. MPI offers numerous library routines to synchronize message passing in several modes like blocked and unblocked message passing. MPI programming does not support a shared memory architecture; thus, cores are not able to read the same information stored in the memory. Therefore, cores are required for a messaging protocol and standardized by the MPI, if information from other cores is needed. Hybrid MPI + OpenMP implementation can also take advantage of possibly non-blocking communication which are overlapped with computations The simple difference among OpenMP and OpenMP + MPI is the code is run for a defined number of MPI task time concurrently while OpenMP is solving the optimization problem in parallel according to the user-defined method.

3.3 Parallelization with Hybrid [MPI + CUDA] Framework

Figure 3 shows that each node in a cluster consists of two NVIDIA graphics cards such as NVIDIA QUADRO K2000 and NVIDIA. Quadro 2000 is built on the CUDA architecture. Each GPU consists of several streaming multiprocessors (SMs) and a data transfer from CPU to GPU by the use of the PCI-E bus. The framework contains three sections, library functions, runtime, and CUDA driver. We can invoke the virtual instruction set and memory of the parallel computational elements which are in CUDA GPUs by using CUDA. In order to speed up the performance of large irregular graph applications, we used GPUs with CUDA a general purpose parallel computing framework [10]. It compromises of threads, blocks, and grids shared memory and barrier synchronization. This framework has gained popularity for multithreaded programming on multiple-core GPUs. Scientific domain and academia are already utilizing CUDA to accomplish tremendous speedup on their research. In this

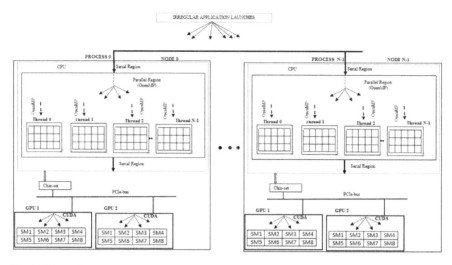

Fig. 3 CUDA framework for implementation of irregular Floyd's and Kruskal's graph algorithms

research work, in order to minimize communication and computation overlap, we have used hybrid [MPI + CUDA] programming strategy. In this strategy, on GPU cluster, we have used a single MPI process per node to launch small chunks of large irregular graph algorithm on various nodes and CUDA to distribute the work between the different capability GPUs cores within a cluster node. Then, it calls the CUDA kernel, in which complete GPU threads run the kernel in concurrently. Kruskal's algorithm is implemented as a thread in a thread block with two dimensions mgrid × ngrid, and a thread block includes mblock × nblock of threads, where each thread block has separate shared memory area in which variables invoked among threads in the block. Thus, the application directed to GPU in the CUDA environment is accessed as threads. The threads are gathered according to the part of a thread block. Therefore, obtaining larger parallelism, a huge number of threads are invoked simultaneously. In these irregular applications on CUDA environment, the threads are designated as a stream-based function written in C called as kernel functions. The irregular application has two small chunks of code directed to CPU and GPUs which is primarily called by CPU (main program) and kernel routines called as a thread on GPU. The kernel routines are defined with the __global__ directive so that it is implemented on GPU. In the kernel functions, gridDim, blockDim, blokIdx, and threadIdx implicitly declared by CUDA runtime are accessible to specify the size of the grid and thread of block. Lastly, transfers the output data from CPU to GPU and runs as multiple threads thus increase productivity and performance of the irregular graph applications.

However, a minimum spanning tree with Kruskal's algorithm and shortest path with Floyd's algorithm has exertion of fine-grained parallelization in a cluster due to which the recursive calculation is performed. Therefore, it is worth to implement

Floyd's and Kruskal's algorithms on GPUs where the massively parallel environment is equipped with a large number of stream processors as shown in Fig. 3.

4 Experimental Setup and Performance Analysis

The experimental environment is a cluster of six nodes which are used each of which has six-core/socket Intel Xeon CPU processor at 2.40 GHz of 31 GB RAM as host CPUs with two CUDA-enabled GPUs (NVIDIA Quadro K2000 and Quadro 2000) as device accelerators and configuration includes the system type that is Dell precision R5500 with 227 cores and 192 cores. Each node is configured with MPICH2-1.2 for MPI library. The compilers used are GCC version 4.4.7 and NVIDIA nvcc version 5.0 with fedora 24 operating systems. OpenMP was installed on a single node since it supports only shared memory model but not distributed memory model.

In this section, the performances of the above-mentioned three parallelization frameworks are compared with different numbers of vertices for Floyd's and Kruskal's algorithm irregular graphs using OpenMP threads, hybrid [OpenMP + MPI and MPI + CUDA]. The results are provided in terms of the total computation time in seconds. The implementation of the proposed hybrid parallel computation [MPI + CUDA] framework should show the superior performance on the OpenMP parallel programming model and hybrid [OpenMP + MPI] model for different data input sizes.

In this section, the speedup is the ratio of time required to execute the applications in Programming Model 1 and time required to execute applications in Programming Model 2 has been shown. The speedup measures the performance gain.

$$\text{Speedup} = \left[\frac{E_{p1}}{E_{p2}}\right] \qquad (1)$$

where

E_{p1} is Elapsed_Time of programming model 1 computation
E_{p2} is Elapsed_Time of programming model 2 computation
Elapsed Time $= \left[\text{Excution\_Time} \times 10^{-6}\right]$

$$\text{Execution\_Time} = [\text{Proces\_End\_Time} - \text{Process\_start\_Time}] \qquad (2)$$

Table 1 Execution time of OpenMP, hybrid [OpenMP + MPI and MPI + CUDA] and speedup of CUDA against OpenMP and hybrid framework on a cluster

Node size	Frameworks execution times in microseconds			Speedup in microseconds	
	OpenMP (μs)	Hybrid [OpenMP + MPI] cluster (μs)	Hybrid [MPI + CUDA] cluster (μs)	Speedup of MPI + CUDA against OpenMP (μs)	Speedup of MPI + CUDA against OpenMP + MPI (μs)
100	10.32423	15.049295	0.986743	10.4629372	15.25148392
300	20.000092	17.99702	0.89769	22.2795085	20.0481458
500	30.000103	21.069183	1.006662	29.801565	20.92974901
700	50.064568	31.816783	3.005478	16.6577722	10.58626382
1000	80.000109	65.000105	5.0034591	15.9889603	12.99103354

4.1 Comparison Results of Parallel Frameworks OpenMP, Hybrid [OpenMP + MPI and MPI + CUDA] for Irregular Floyd's Graph Algorithm

Table 1 shows the computed execution time of OpenMP, hybrid [OpenMP + MPI and MPI + CUDA] and even speedup of hybrid [MPI + CUDA] against OpenMP and hybrid [OpenMP + MPI] framework on a CPU's cluster using Eqs. (1) and (2) for irregular Floyd's graph algorithm for varying number of vertices (nodes), respectively. For this application, the OpenMP gave the best performance for where four threads were used. On the CPU [hybrid OpenMP + MPI] cluster with six machines total 6 processes were considered. And also, the application was executed on NVIDIA Quadro K2000 and Quadro 2000 GPUs for performance acceleration. Due to the computational and communication overhead, the hybrid [OpenMP + MPI] cluster showed less improvement against hybrid [MPI + CUDA].

Figure 4 shows execution time of irregular graph algorithm (Floyd's) on different programming frameworks such as OpenMP, hybrid [MPI + OpenMP] on CPU, and hybrid [MPI + CUDA] on GPU for varying node (vertices) sizes. In the figure, X-axis shows the size of the nodes and Y-axis shows the execution time in microseconds. We observed that during the initial experiment for less number of nodes ($\leq$100), the OpenMP framework with computing resources such as cores and memory is sufficient. OpenMP is a great choice which yields better execution time 10.32423 μs. On the hybrid cluster [MPI + OpenMP] on CPU was 15.049295. And on GPU, cluster with hybrid [MPI + CUDA] programming strategy yields good optimal results 0.986743 μs. As we observed from CPU cluster that as we increase the node size there will be increase in execution time to avoid this problem we executed the applications on GPUs cluster with hybrid [MPI + CUDA] framework we found significant

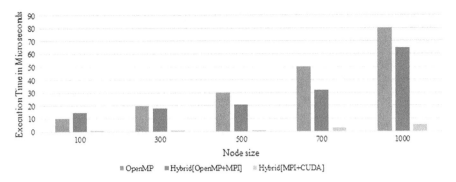

Fig. 4 Execution time of OpenMP, hybrid [OpenMP + MPI] on CPU cluster, and hybrid [MPI + CUDA] on GPU cluster

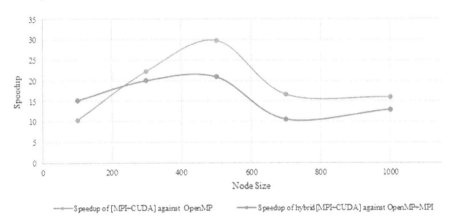

Fig. 5 Speedup of hybrid [MPI +CUDA] on GPU cluster against OpenMP, hybrid [OpenMP + MPI] frameworks on CPU cluster

improvement in the performance due to more number of cores available on GPUs, kernel optimization, improved GPUs memory bandwidth.

Figure 5 shows the speedup comparison irregular graph Floyd on the CPU using parallel OpenMP, on CPU's cluster using hybrid parallel [OpenMP + MPI] and on GPU's cluster using hybrid [MPI + CUDA] programming model where 'x' axis shows the node (vertices) size and 'y' axis shows the speedup in microseconds. As we observed a speedup of hybrid [MPI + CUDA], parallel programming framework on GPU's cluster yields better average speedup of 19.03 when compared against OpenMP and 15.96 against CPU's cluster with hybrid [MPI + OpenMP] programming using Eq. (1).

Table 2 Execution time of OpenMP, hybrid [OpenMP + MPI and MPI + CUDA] and speedup of MPI + CUDA against OpenMP and hybrid framework on a cluster

Node size	Frameworks execution times in microseconds			Speedup in microseconds	
	OpenMP (μs)	Hybrid [OpenMP + MPI] cluster (μs)	Hybrid [MPI + CUDA] cluster (μs)	Speedup of MPI + CUDA against OpenMP (μs)	Speedup of MPI + CUDA against OpenMP + MPI (μs)
100	28.39163	39.12817	1.973486	14.38654	19.826929
300	55.00025	49.49181	1.79538	30.63432	26.06259
500	82.50028	57.94025	2.013324	40.97715	27.208674
700	137.6776	87.49615	6.010956	22.90444	13.762143
1000	220.0003	178.7503	10.00692	21.98482	16.888344

4.2 Comparison Results of Parallel Frameworks OpenMP, Hybrid [OpenMP + MPI and MPI + CUDA] for Minimum Spanning Tree Problems Using Kruskal's Algorithm

Table 2 shows the computed execution time of OpenMP, hybrid [OpenMP + MPI and MPI + CUDA] and even speedup of hybrid [MPI + CUDA] against OpenMP and hybrid [OpenMP + MPI] framework on a cluster using Eqs. (1) and (2) for minimum spanning tree algorithms with varying vertices. On the hybrid [OpenMP + MPI], cluster of six machines is with total six processes. And also, the application was executed on the GPU cluster, with hybrid programming [MPI + CUDA] programming strategy having NVIDIA Quadro K2000 and Quadro 2000 on each node in order to make performance acceleration. Due to the computational and communication overhead, the CPU-based cluster hybrid [OpenMP + MPI] programming strategies showed not as much of improvement against hybrid [MPI + CUDA] programming strategies.

Figure 6 show execution time of a minimum spanning tree problems using Kruskal's graph algorithm for varying number of vertices (nodes) respectively on different programming frameworks such as OpenMP, hybrid [MPI + OpenMP] on CPU, CUDA on GPU for varying node (vertices) sizes. In the figure, X-axis shows the size of the nodes and Y-axis shows the execution time in microseconds. We observed that during the initial experiment for less number of nodes (≤ 100), the OpenMP framework with computing resources such as cores and memory is sufficient. OpenMP is a good choice which yields better execution time 28.39163 μs where on hybrid [OpenMP + MPI] cluster on CPU was 39.12817. And on GPU cluster with hybrid [MPI + CUDA], programming strategy yields good optimal results 1.9734 μs when an application is desirable to be implemented in parallel and dis-

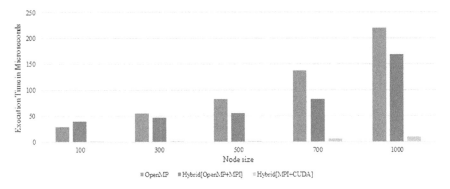

Fig. 6 Execution time of OpenMP, hybrid [OpenMP + MPI] on CPU cluster and hybrid [MPI + CUDA] on GPU cluster

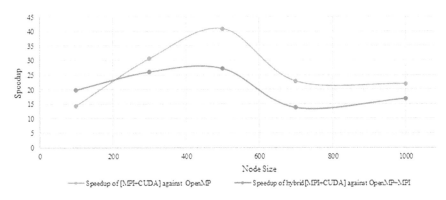

Fig. 7 Speedup of hybrid [MPI + CUDA] on GPU cluster against OpenMP, hybrid [OpenMP + MPI] frameworks on CPU cluster

tributed mode with complex synchronization between processes. As we observed from CPU cluster that as we increase the node size there will be increase in execution time to avoid this problem we executed the applications on GPUs cluster with hybrid [MPI + CUDA] framework we found significant improvement in the performance due to more number of cores available on GPUs, kernel optimization, improved GPUs memory bandwidth.

Figure 7 shows the speedup comparison irregular graph Floyd's algorithm on the CPU using parallel OpenMP, on CPU's cluster using hybrid parallel [OpenMP + MPI] and on GPU's cluster using hybrid [MPI + CUDA] programming model, where 'x' axis shows the node (vertices) size and 'y' axis shows the speedup in microseconds. As we observed, speedup of hybrid[MPI + CUDA] parallel programming framework on GPUs cluster yields better average speedup of 27.265535 when compared against OpenMP and 20.749736 against CPU's cluster with hybrid [MPI + OpenMP] cluster using Eq. (1).

5 Conclusion

In this study, we have considered parallel programming frameworks such as OpenMP, hybrid [OpenMP + MPI and MPI + CUDA] to compare their performance using two irregular applications which are Floyd's algorithm (shortest path problems) and Kruskal's graph algorithms (minimum spanning tree problems). From experimental results, we can conclude that for a minimum number of nodes with sufficient memory and cores, OpenMP yields good performance. If node size is reasonable and the application is compute-intensive, then hybrid [OpenMP + MPI] is optimal for shared–distributed architecture. We have also observed that hybrid [MPI + CUDA] framework on GPU cluster was outperformed other frameworks such as OpenMP, hybrid [OpenMP + MPI] which are implemented on CPU cluster.

Future Work

We will carry out the computation of same graph algorithms by making use of the combination of MPI, OpenMP, CUDA, i.e., hybrid programming model [OpenMP + MPI + CUDA] which has been proven extremely efficient for implementing irregular applications. By combining CUDA with MPI, the libraries can send and receive the GPU buffers in an easy and direct way without having the overhead of transferring them to the host memory. By using OpenMP in a combination, it is possible to share a small portion of GPU computation task to the CPU, so that CPU computing power will be utilized or otherwise CPU will be used just for data transfer.

References

1. Osipov, J. V., Sanders, P., Singler, J. (2009). The filter-kruskal minimum spanning tree algorithm. In *ALENEX'09* (pp. 52–61).
2. Lončar, V., & Škrbić, S. (2010). Parallel implementation of minimum spanning tree algorithms using MPI. Serbia: Faculty of Science, Department for Mathematics an Informatics, University of Novi Sad.
3. Ravela, S. C. (2010). *Comparison of shared memory based parallel programming models* (Technical Report MSC-2010-01). Blekinge Institute of Technology.
4. Huang, Y., & Guo, S. *Design and implementation of parallel Prim's algorithm*. Zhengzhou, China: Zhengzhou Information Science and Technology Institute.
5. Kang, S. J., Lee, S. Y., & Lee, K. M. (2015). Performance comparison with MPI, OpenMP and map reduce in practical problems. *2015*, Article ID 575687. http://dx.doi.org/1001155/2015/575687.
6. Qingshuang, W. All-pairs shortest path algorithm based on MPI + CUDA distributed parallel programming model. Wuhu, Anhui, 241003, China: College of Territorial Resources and Tourism, Anhui Normal University.
7. Loncar, V. (2014). Parallelization of minimum spanning tree algorithms using distributed memory architecture. Serbia: SrdjanSkrbic and AntunBalaz Scientific Computing Laboratory.
8. Rostrup, S., Srivastava, S., & Singhal, K. (2011). Fast and memory-efficient minimum spanning tree on the GPU. In *2nd International Workshop on GPUs and Scientific Applications (GPUScA 2011)*. Geneva: Inderscience.

9. Cormen, T. H., Leiserson, C. E., Rivest, R. L., & Stein, C. (2009). *Introduction to algorithms*. Cambridge: MIT Press.
10. Barney, B. (2007). *Introduction to parallel computing*. Lawrence Livermore National Laboratory. https://computing.llnl.gov/tutorials/parallelcomp/.

Signature Analysis for Forgery Detection

Dinesh Rao Adithya, V. L. Anagha, M. R. Niharika, N. Srilakshmi
and Shastry K. Aditya

Abstract Forgery of signature has become very common, and the need for identification and verification is vital in security and resource access control. There are three types of forgery: random forgery, simple or casual forgery, expert or skilled or simulated forgery. The main aim of signature verification is to extract the characteristics of the signature and determine whether it is genuine or forgery. There are two types of signature verification: static or offline and dynamic or online. In our proposed solution, we use offline signature analysis for forgery detection which is carried out by first acquiring the signature and then using image pre-processing techniques to enhance the image. Feature extraction algorithms are further used to extract the relevant features. These features are used as input parameters to the machine learning algorithm which analyses the signature and detects for forgery. Performance evaluation is then carried out to check the accuracy of the output.

Keywords Signatures · Forgery · Image processing · Neural network · Feature extraction · Authentication

---

D. R. Adithya (✉) · V. L. Anagha · M. R. Niharika · N. Srilakshmi · S. K. Aditya
Department of Information Science and Engineering, Nitte Meenakshi Institute of Technology,
Bangalore, India
e-mail: adithyarao121@gmail.com

V. L. Anagha
e-mail: anagha1996@gmail.com

M. R. Niharika
e-mail: niharikamr03@gmail.com

N. Srilakshmi
e-mail: nsrilakshmi06@gmail.com

S. K. Aditya
e-mail: adityashastry.k@nmit.ac.in

© Springer Nature Singapore Pte Ltd. 2019
N. R. Shetty et al. (eds.), *Emerging Research in Computing, Information, Communication and Applications*, Advances in Intelligent Systems and Computing 906,
https://doi.org/10.1007/978-981-13-6001-5_26

1 Introduction

Handwritten signatures are used to distinctively recognize a person and check whether the person has authorized the contents of the document. It also tells us that the author of the signature was physically present at the time of signing. Handwritten signatures have become a means to authenticate documents in various government sectors and nongovernment sectors. Only when the person in authority signs the document, the document has some value and meaning. Signatures are also used in documents issued by government such as PAN cards, identity cards and driving licence. In the banking sector, it is used on a cheque to withdraw money. In colleges and schools, circulars are authenticated by the principal's or head of the department's signature.

Hence, signature plays a vital role in our day-to-day life. Forging a signature is equal to stealing a person's identity and is considered a very serious crime and is punishable under the law. Manually checking each signature for forgery is very tiring and time-consuming job. The proposed solution makes use of the various geometric features of the signature to check whether it is a forgery. Therefore, it automates the entire process in a cheap and efficient manner. Some of the challenges in identifying the forgery are:

- Signature of a person does not always remain constant. It keeps varying according to the person's mood, posture and time. Therefore, the data set should be collected over a period of time at different mood and posture.
- Highly skilled forgeries are very difficult to detect.
- Since it is offline, we do not have any timing information.

The main objectives of this work are as follows:

- Build a simple graphical user interface to input the signature.
- Pre-process the image to remove unwanted noise and distortion.
- Extract the important features from the signature.
- Use the extracted features to train the machine learning model.
- Identify whether the test signature is genuine or forgery.

The above objectives are achieved using the following modules:

- Pre-processing—This module takes the input of a signature from the user and then removes the unwanted noise in the background, and converts it into greyscale image and then a binary image.
- Feature extraction—This module extracts all the two-dimensional shape features from the signature and then stores it in the database to train the neural network.
- Feature selection—This module chooses a subclass of pertinent features out of all the features that are extracted. Only the chosen features are used for further analysis.

2 Related Work

The authors in [1] use image processing and artificial neural network/s to detect forgery in signatures. They collected 100 signatures from each individual user for three total users. The signatures for the data set were collected in different pens, pencils and different colours. They have used image processing techniques, namely colour inversion, filtering and binarization. Images after being pre-processed undergo feature extraction where area, centroid, eccentricity, kurtosis and skewness of the image are extracted and stored in the database. The neural network is then trained with the features that are extracted. The user provides the input signature which is pre-processed and from which the same five features are extracted. The features of the input are fed into the neural network that has been trained with the features of the data set. The network gives the output verifying the input signature.

Authors in [2] propose a method for a signature verification system that uses Freeman chain code (FCC) as data representation. The images first go through a few pre-processing techniques such as binarization, noise removal, cropping and thinning. The FCC was obtained based on the largest continuous part of the signature. It was then split into 4, 8 and 16 equivalent pieces. The image then undergoes feature extraction where signature width, height, aspect ratio, diagonal distance, centres of mass of all foreground pixels and total shift per horizontal/vertical line are obtained. In the end, verification is done by measuring and matching Euclidean distance in k-nearest neighbours. The database which was used for this paper was the MCYT Bimodal Subcorpus Offline Signature which contained 15 genuine and highly skilled forgery samples of 5 individuals totalling to 2250 signatures.

In [3], authors have used the pixel matching technique to identify a forgery. Here the image is first pre-processed to remove the noise and colour and then normalized and converted to a black and white image. Then its properties are adjusted like the angular rotation and size. The signature is rotated so that it lies on the x-axis and is resized to the size of the image in the database. Finally, pixel matching technique is used to compare each pixel from the two images from left to right. A certain threshold is set, and if the percentage of pixels matching is above the set threshold, then the signature in genuine otherwise it is a forgery.

Authors in [4] propose a method for a signature verification system that uses dynamic time warp (DTW). The objective of DTW is to compare the sequences X and Y. The pre-processing techniques that are used in this paper are of two types, Maximum Length Vertical Projection (MLVP) and Minimum Length Horizontal Projection (MLHP), followed by feature extraction, one-dimentional features like vertical projection (VP), horizontal projection (HP), top contour, bottom contour, envelope width and contour ratio. The dynamic time warping (DTW) algorithm which is founded on dynamic programming obtains an optimum match among two sequences of feature vectors by permitting for stretching and compression of sections of the sequences. This feature of the algorithm makes it appropriate for detecting whether the signature is genuine or forged using the DTW algorithm.

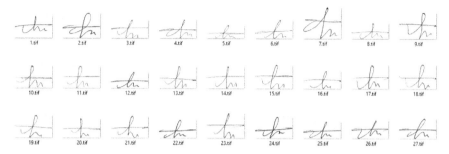

Fig. 1 Signature data set

3 Data set Description

We have used the Persian offline signature data set, UTSig, which comprises 8280 images from 115 classes. Each class contains 27 genuine signatures and 27 forged signatures prepared by 6 forgers [5]. UTSig was selected as the benchmark data set for this work as it contains many samples, classes and forgers [5]. Figure 1 shows the signature data set.

3.1 Genuine Signatures

To obtain genuine signatures, 115 male participants signed 27 times on a form and repeated the action for 3 days that can be used as forgery or disguise. As a result, 3105 genuine signatures were collected.

3.2 Forged Signatures

UTSig consists of 5130 forgeries, signed by 345 participants for three days giving 27 samples each. They were asked to make signatures as similar as possible to genuine samples.

A simple noise removal system where pixels brighter than a threshold was assigned to pure white (255 in greyscale). To estimate the threshold, we scanned five blank papers and found the darkest pixel, which resulted in 237 in greyscale as threshold. The data obtained from each signature is stored in a MySQL database.

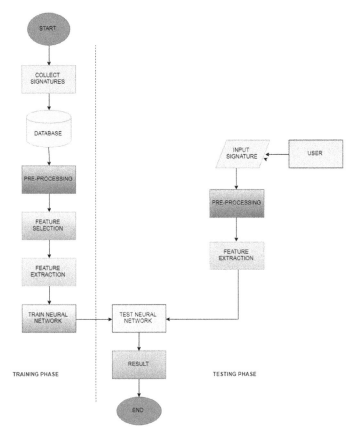

Fig. 2 Flow chart of proposed system

4 Proposed System

Figure 2 describes the proposed system for detecting forgery. It consists of the following phases:

4.1 Training Phase

The training phase of the proposed system consists of the following steps:

- Signature acquisition: Collect genuine and forged signatures of each user and store it in the database of the application. The signatures are collected over a period of time in different moods.

- Pre-processing: All the images in database have to be pre-processed before extracting features. All the images undergo the following processes:
  1. The image is first converted to greyscale from RGB.
  2. This greyscale image is converted to black and white where the entire white portion is numbered as 1 and the black portion as 0.
  3. To obtain accurate values for the features, we then use complement function to turn black into white and vice versa.
- Feature selection: All the chosen features hold different levels of importance to the problem statement. This method is to filter only the crucial features which the verification model depends on.
- Feature extraction: Once the important features are chosen, they must all be extracted from the pre-processed image and stored in a local variable in the system.
- Training the neural network: The mined features are utilized to train the neural network. The network has ten hidden layers and the output of two different classes: one for genuine and one for forged signature.

4.2 Testing Phase

The testing phase consists of user input and verification. Here, the user gives a picture of the signature as input to the verification system. The input image is pre-processed, and all the selected features are extracted. These features are fed into the above trained neural network. The neural network will now detect whether the user's input was a forged signature or a genuine signature.

5 Signature Pre-processing

Signature pre-processing is an essential step to enhance the accuracy of the algorithm and to diminish their computational requirements. Following pre-processing steps are taken into consideration:

- Transformation from colour to greyscale, and finally to black and white.
- Image is resized in order to make them normalized.
- Diminishing the black and white image leads to enormous information loss. It is vital to choose a thinning algorithm which provides a reasonable abstraction of the original signature, with a small noise level.

6 Feature Extraction

The following features are extracted from the signature for forgery detection [6, 7]:

- Area: The definite number of pixels in the area is calculated, and it yields a scalar value.
- Major axis length: The length of the main axis of the ellipse that has the similar standardized subsequent essential instants as the region.
- Minor axis length: The length of the minor axis of the ellipse that has the similar standardized subsequent essential instants as the region.
- EquivDiameter: The diameter of a circle with the equivalent area as the region is calculated and it yields a scalar value and it is computed using the formula: sqrt (4*Area/pi)
- Extent: The fraction of pixels in the region to pixels in the total bounding box is calculated, and it returns a scalar value. It is calculated as the area divided by the area of the bounding box.
- Orientation: Angle between the x-axis and the major axis of the ellipse that has the similar subsequent instants as the region and yields a scalar value. This value is in degrees, ranging from $-90°$ to $90°$.
- Solidity: It is the proportion of pixels in the region to pixels of the convex hull image.
- Perimeter: It calculates the internal space about the border of the region. The regionprops function calculates the perimeter by calculating the distance among every neighbouring pixel around the border of the region. Perimeter is computed as the number of pixels in the external contour of the object.

7 Artificial Neural Network for Forgery Detection

It is an approach made to recreate network to work like the neurons in our brain. The network is trained with the existing data using which it learns and is able to provide output in terms of detection or identification. In an ANN model, there are input nodes that collect the data which is then sent to the hidden layers. These hidden layers are the main part of the model. They process the data given by the input nodes and provide output which is shown by the output nodes. Figure 3 describes the ANN model for signature forgery detection.

We first launch the neural network toolbox in MATLAB. Using all the features that were selected and extracted for each person in the previous process, we train the neural network to differentiate between forged and genuine signatures.

- INPUT: The input for the toolbox is a fixed set of features for all the signatures in the data set.
- TARGET: The target for the toolbox is the matrix of the class it belongs to.

Class 1: Genuine
Class 2: Forged

To test the model, we make the user input a signature from the test set that then undergoes pre-processing and feature extraction. These features are now fed to the neural network.

- TEST: User input signature either genuine or forged.
  After the neural network is given, the test signature it now provides output saying which class the test signature belongs to.
- OUTPUT: Tells the user whether the signature is fake or genuine (Fig. 4).

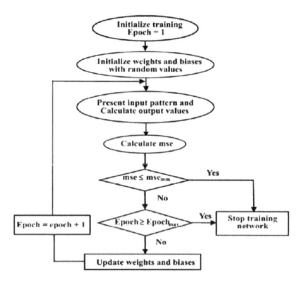

Fig. 3 A training process flow chart

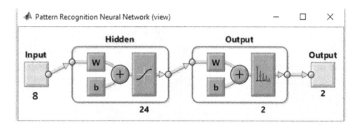

Fig. 4 Pattern recognition neural network

Signature Analysis for Forgery Detection 347

8 Results

After training the neural network using the data collected from five individuals over a period of time, we achieved 90–95% efficiency for the various test data between them. Figure 5 shows how signature is given as input to the system. Figure 6 illustrates how the signature image is pre-processed. Figure 7 depicts the output after testing.

Figure 8 shows the confusion matrix. Here, the diagonal elements indicate perfect match of signatures where signatures detected by the system are correctly matched with the actual signature, while other matrix elements indicate the false detection

Fig. 5 Input of image

Fig. 6 Pre-processing of image

Fig. 7 Output after testing

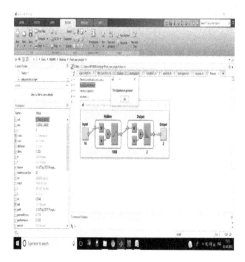

Fig. 8 Confusion matrix

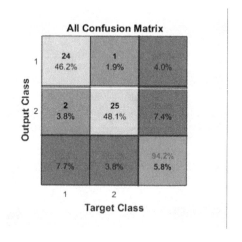

9 Conclusion

In conclusion, we would like to state that the project signature analysis for forgery detection is developed with the intention to automate the signature verification process. Our project uses image processing along with artificial neural network to detect a forgery. Artificial neural network has been used in many applications due to their ability to solve problems with diverse ease of use and the model-free property. Main feature of neural network is its ability to comprehend nonlinear offline problem with selective training, which can lead to sufficiently accurate response. Our verification system based on ANN is able to learn different kinds of signature data sets, by using geometrical offline features and is able to detect whether the signature is genuine or forged. In future, this work can be extended by using a larger data set and extracting more number of features to provide higher level of accuracy. A mobile application can also be built for the same.

REFERENCES

1. Karounia*, A., Dayab, B., & Bahlakb, S. (2010). *Offline signature recognition using neural networks approach*. WCIT.
2. Shankar, A. P., & Rajagopalan, A. N. (2007). Offline signature verification using DTW. *Pattern Recognition Letters, 28*.
3. Bhattacharyaa*, I., Ghoshb, P., & Biswasb, S. (2013). Offline signature verification using pixel matching technique. In *International Conference on Computational Intelligence: Modeling Techniques and Applications (CIMTA)*.
4. Anand, H., & Bhombe, D. L. (2014, May). Enhanced signature verification and recognition using matlab. *International Journal of Innovative Research in Advanced Engineering (IJIRAE), 1*(4). ISSN 2349-2163.
5. Blei, D. M., et al. (2003). Latent dirichlet allocation. *Journal of Machine Learning Research, 3*, 993–1022.
6. Fanga, B., Leungb, C. H*., Tangc, Y. Y., Tseb, K. W., Kwokd, P. C. K., & Wonge, Y. K. (2003). Offline signature verification by the tracking of feature and stroke positions. *Pattern Recognition, 36*, 91–101.
7. Lewis, D. Naive (Bayes) at forty: The independence assumption in information retrieval.

Optimal Sensor Deployment and Battery Life Enhancement Strategies to Employ Smart Irrigation Solutions for Indian Agricultural Sector

M. K. Ajay, H. A. Sanjay, Sai Jeevan, T. K. Harshitha, M. Farhana mobin and K. Aditya Shastry

Abstract In this fast-growing technology, agriculture sector in India is one of the domains where we have observed slow adoption of Internet of Things (IoT) solutions. Unavailability, expensive seasonal labour and inadequate water resources are one of the major problems faced by Indian agriculture. Use of costly industrial standard sensors, energy utilization and placement of sensors also poses a greater problem for adoption of IoT solutions. This paper proposes an IoT-based framework and a strategy for placement of the optimal number of sensors and optimal utilization of battery to address the mentioned issues. Framework adopts NodeMCU and moisture sensor that addresses communication and water scarcity problems. We are using ThingSpeak cloud to store and process the sensor data. Results demonstrate the effectiveness of our strategies and proposed IoT framework.

Keywords Internet of things (IoT) · Sensors · Strategy · Power consumption/utilization

1 Introduction

Agriculture has a vital influence on India's economy. In today's world, as we have a tendency to see the rising in population, agriculture becomes necessary to fulfil the needs of humanity. With every passing year, the water consumption is increasing, especially for agriculture. Advancements in the field of IOT have a wide area of applications such as smart home, enterprise IOT, health care and agriculture.

Developments in field include integration of sensors with cloud platforms which helps in collecting vital information about environmental conditions. The sensors that are being used are high-level industrial sensors which cost around Rs. 2500 per sensor. Using such sensors for acres of land requires lakhs of rupees as investment for

---

M. K. Ajay (✉) · H. A. Sanjay · S. Jeevan · T. K. Harshitha · M. F. mobin · K. A. Shastry
Department of Information Science and Engineering, Nitte Meenakshi Institute of Technology, Bangalore, India
e-mail: ajay.crispy@gmail.com

© Springer Nature Singapore Pte Ltd. 2019
N. R. Shetty et al. (eds.), *Emerging Research in Computing, Information, Communication and Applications*, Advances in Intelligent Systems and Computing 906,
https://doi.org/10.1007/978-981-13-6001-5_27

installing. This investment will be more than the investment required for agriculture. Therefore, we use cost-effective sensors to conduct this experiment.

When you think about real-time application in the agriculture field, number of sensors that are required to cover the crop area comes to picture. So the challenge is to cover the whole crop with minimum number of sensors. For this, we have strategies (triangular, square) which are detailed in the following sections. Once sensors are deployed according to a strategy, the next question is at what intervals the data should be retrieved so that battery is utilized effectively for longer periods. For this challenge, we should learn about soil types, composition of soils and crop that is planted. Different soils have different capacity to hold water.

The IoT framework allows the objects to be controlled remotely across existing network and monitors the weather parameters. Therefore, this project deals with building strategy to compute minimum/optimal number of sensors for a given agricultural space, building strategy to compute frequency of pushing the data to cloud and building a IoT-based framework for smart irrigation. In this paper, Sect. 2 explains the information gathered from the literature survey, Sect. 3 explains the different strategies of sensor deployment, Sect. 4 explains power aware data retrieval, Sect. 5 describes the framework for smart agriculture with experimented results.

2 Related Work

In this survey, literature survey about wireless underground sensor for data transmission, deploying the sensors based on the type of the soil and its parameters, path loss during data transmission and minimizing the number of sensors are discussed.

The work by Yu et al. [1] considers WSN to handle and manage the water resources for agricultural irrigation. The author differentiates the communication of WUSN and the terrestrial WSN. Wireless underground sensor networks can be utilized as intelligent transportation system. The performance of WUSN rests on the spread of electromagnetic waves in the soil, underground channel model, electrical features of soil and positioned solutions of WUSN nodes. The authors Wu et al. [2] have conducted experiments in the laboratory of Bogena wireless signal lessening of ZigBee wireless transceiver unit having 2.44 GHz frequency. Experimental results specified that increase of soil column depth and volumetric water content of the soil could lead to surge of signal attenuation. The work by Akyildiz and Stuntebeck [3] utilizes present machinery for underground sensing by deploying a buried sensor. Here, the depth at which devices are deployed will depend upon the application of the network. Pressure sensors must be placed close to the surface, while soil water sensors should be located deeper near the roots of the plants.

The work by Tejal et al. [4] states that the production processes must be organized conferring to the demands of plants and environment in a location-specific manner. There are four steps of precision agriculture which are data collection, analysis, management decision and farming. According to the existing system and our proposed work, we get some deployment strategies such as triangular, square and hexagonal for sensor nodes in the greenhouse shed using ZigBee technology. The work by Majone et al. [5] mainly deals with how many number of sensors need to be deployed to monitor the field. Using huge number of low-cost sensors may cause problems to the calibration curve which cannot handle huge amount of data.

3 Strategies of Sensor Deployment

The main strategy deals with minimizing the number of sensors and making it cost-effective. As we are considering only the topsoil region, the quantity of water used for the crop growth will be lesser when compared to the water required for the mid-soil region. Most of the moisture level varies constantly in the topsoil area, so concentrating on topsoil gives good yield (Fig. 1).

There are three main strategies of sensor deployment. We have experimented square and triangular method of deployment for our 10 square feet land. Since there is no range of the soil moisture sensor used in the experiment, we deploy the sensors according to the water coverage capacity of the soil. This water holding capacity will vary with respect to type of the soil and its parameters. In square method of deployment, the readings obtained at the corners were the same for our experimented area. Hence, here only one sensor is enough to push the data to cloud. If crop is grown in shape of blocks and each block with one sensor, then square method of deployment is feasible. In triangular method of deployment, we maintained a gradient to differentiate the dry soil and wet soil. The gradient is the point where there is fluctuation in the reading for soil moisture sensor. For our project, gradient is 50.

For 330 ml of water, the coverage area was 14 cm wide and 7 cm depth, for 500 ml of water, the coverage area was 19 cm wide and 12 cm depth. Accordingly, for 10 square feet (experimented land area) and 330 ml water, 10 lines are enough and each with one sensor and for 500 ml water, 8 lines are enough and each with one sensor.

Hence, if the land is divided into multiple lines of the crop, then each individual line can hold one sensor. Drip irrigation is designed in such a way that it will cover the water coverage area of soil and each pipe outlet will pump out the same amount of water. Therefore, this will maintain equal amount of moisture in the land. In one acre of land, this could be arranged in triangular fashion as shown in Fig. 2.

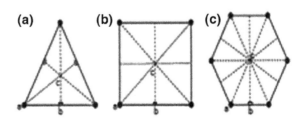

Fig. 1 Strategies of deployment

Fig. 2 Triangular deployment module

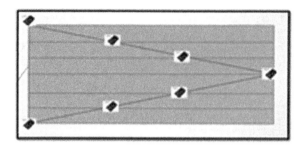

4 Battery Aware

Sensors and raspberry pi are powered with a power supply (normally batteries) to function. Using the batteries for longer periods is a challenge. More often you power them, more battery is drained which is not preferable. Raspberry pi 3 which we are using is powered by 5.1 micro USB supply. The amount of current that a raspberry pi consumes is dependent on what is connected to it. In our architecture, we connect Arduino and few sensors which take 3 A current. Continuously, supplying this current will shorten the battery life. Considering this a challenge, we try to minimize the activity time of raspberry pi. So the key is to identify the longest interval of time, when the change in moisture level will normally be observed. Here the concept of types of soils comes into the picture.

Capacity of soil to hold moisture depends on the size of soil particles. If the soil particles are smaller, it allows more amount of water to hold and vice versa. The theory behind this is as follows. Smaller soil particles give more space between particles. This space is occupied by water that is what we call soil moisture. According to the capacity of soil that can hold moisture, soil is classified into three types namely sand soil, clay soil and silt soil.

Sandy soil has larger soil particles. So, this type of soil will hold water for less time. Clay soil has small particles which is why it holds water for longer periods of time. Sandy soil and clay soil both are not effective for agricultural purposes. Silt soil comes in between sandy soil and clay soil which holds moderate amount of water. Soil that is available is combination of the above-mentioned types of soils. If the soil has more amount of clay, the soil holds water for longer time. So this means that the amount of clay that is present in soil is directly proportional to interval of time to power the sensors.

Different crops have different requirements of water. Few need more water for proper growth, while few grow with less water. Some crops absorb more water quickly and few slowly. For the crops which absorb quickly, the frequency to retrieve should be high. Similarly for crops which absorb slowly, the frequency will be low. When more number of sensors is deployed in the same region of land, it consumes more battery power.

5 Framework for Smart Agriculture

This Framework contains two DHT22 sensors and four soil moisture sensors. The data collected by these sensors will be pushed to ThingSpeak cloud. ThingSpeak is an IoT platform which collects, stores the sensor data and analyses the data. The data collected can be visualized in an app. ThingSpeak provides a graphical user interface to analyse the data and take particular decisions accordingly.

Virtual network computing (VNC) is used for our experiment to make our system portable for outdoors. VNC is user friendly. VNC server is installed in raspberry pi, and VNC viewer is installed in host machine or a mobile phone. The connection is established in a common network through an IP address of raspberry pi. Host connected to particular IP address can access the computer within the network.

5.1 Square Method of Deployment

Square method of deployment is feasible to the crops which are grown in blocks and which are grown in bundle, for example paddy and ragi. This method uses many sensors if we grow the crop in lines and decreases the number of sensors if the crop is grown in blocks. Normally, the crops in India are not grown in blocks. Hence, we use triangular method of deployment for sensor placement. Triangular deployment minimizes the number of sensors and is suitable for all kind of crops. Figure 3 illustrates the square method of deployment.

5.2 Triangular Method of Deployment

Humidity and temperature change according to the condition of the soil and the surrounding area. Soil moisture sensor does not have any range; therefore, we used a gradient concept to place the sensors in the field. This gradient is considered according to the water coverage capacity of the soil and the varying readings from the sensors. Gradient is a point where the threshold value changes for a particular sensor. For our project, the gradient we used is 50 m^3 for soil moisture sensor. Usually with a power

Fig. 3 Square deployment

Fig. 4 Triangular deployment

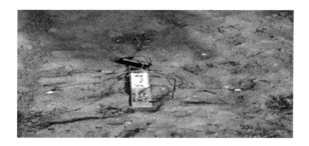

of 3.3 V, wet soil reading is below 300 m^3, and dry soil is above 600 m^3 and ranges to 750 m^3 for completely dry soil. Triangular module will minimize the number of sensors and is suitable for any shape of the land as depicted in Fig. 4.

6 Experimental Setup and Results

The readings were retrieved in the summer season, and the experiment was conducted for both dry soil and wet soil. We have tested this under two types of soil, namely red and black soil. The result for red soil is shown below. We used two DHT sensors and four soil moisture sensors to experiment on 10 square feet. The result is stored in ThingSpeak, where it stimulates a graph based on the input fields. Here x-axis indicates time and y-axis gives the output reading.

6.1 DHT for Dry Soil

When the experiment was conducted under dry soil in the summer season, the temperature obtained is 43.9 °C at particular time as shown in Fig. 5.

When the experiment was conducted under dry soil in the summer season, the humidity obtained is 28.5 g/m^3 at particular time as shown in Fig. 6.

6.2 DHT for Wet Soil

When the experiment was conducted under wet soil in the summer season, the temperature 1 obtained is 29.1 °C and humidity 1 obtained is 59.7 g/m^3 at particular time as shown in Fig. 7.

When the experiment was conducted under wet soil in the summer season, the temperature 2 obtained is 29.9 °C and humidity 2 obtained is 49.0 g/m^3 at particular time as shown in Fig. 8.

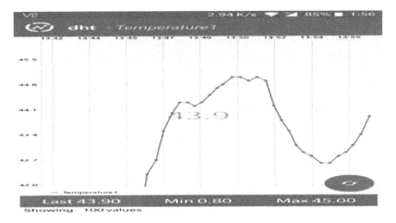

Fig. 5 Temperature reading

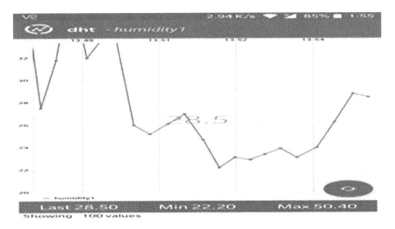

Fig. 6 Humidity reading

6.3 Soil Moisture for Dry Soil

When the experiment was conducted under dry soil in the summer season, the readings obtained for soil moisture 1 is 757.0 m^3 and soil moisture 2 is 757.0 m^3 at particular time as shown in Fig. 9.

When the experiment was conducted under dry soil in the summer season, the readings obtained for soil moisture 3 is 756.0 m^3 and soil moisture 4 is 757.0 m^3 at particular time as shown in Fig. 10.

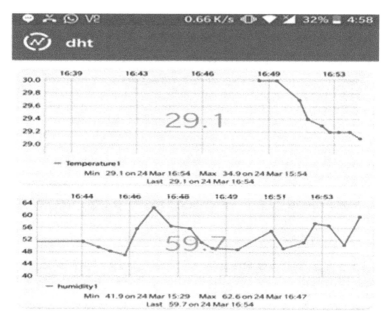

Fig. 7 Temperature 1 and humidity 1 readings

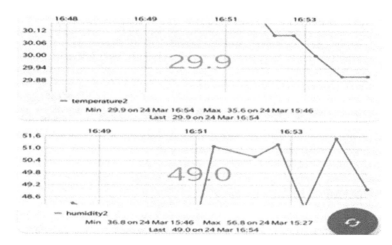

Fig. 8 Temperature 2 and humidity 2 readings

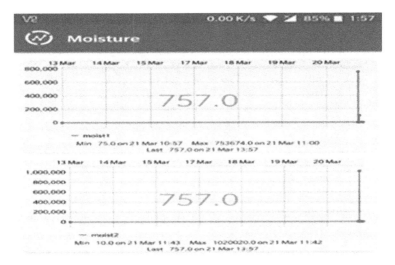

Fig. 9 Readings for soil moisture 1 and 2 for dry soil

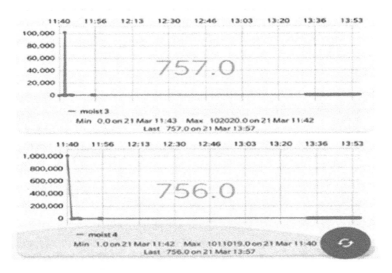

Fig. 10 Readings for soil moisture 3 and 4 for dry soil

6.4 Soil Moisture for Wet Soil

When the experiment was conducted under wet soil in the summer season, the readings obtained for soil moisture 1 is 371.0 m^3 and soil moisture 2 is 354.0 m^3 at particular time as depicted in Fig. 11.

Fig. 11 Readings for soil moisture 1 and 2 for wet soil

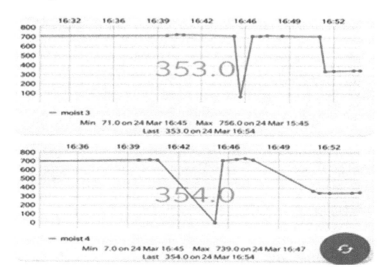

Fig. 12 Readings for soil moisture 3 and 4 for wet soil

When the experiment was conducted under wet soil in the summer season, the readings obtained for soil moisture 3 is 353.0 m^3 and soil moisture 4 is 354.0 m^3 at particular time as illustrated in Fig. 12.

7 Conclusion

The proposed system accomplishes the irrigation system in a better way. Sensor deployment helps farmers and industrial people in many ways. Our work mainly helps the farmers to grow healthy crops by using minimum amount of water. This effective system serves majorly for water management through automatic actuation by pumping water. Energy conservation can be increased by power aware architecture. Data analysis provides the proper amount of water used by the crops, and it educates the farmer to grow different varieties of crops by managing the water resources. This proposed system is cost effective and enriches the agriculture productivity.

In future, the same work can be accomplished using wireless transmission technology like Lora, which communicates in rural areas up to 15 km and has high bandwidth for communication. It can be set up as the whole device, so that there is no human intervention. This technology can also be semi-wired, where the transmitter lies on the field and the receiver is placed above ground to receive multiple signals at the same time.

References

1. Yu, X., Wu, P., Han, W., & Zhang, Z. (2012). Overview of wireless underground sensor networks for agriculture. *African Journal of Biotechnology, 11*(17), 3942–3948.
2. Yu, X., Wu, P., Han, W., & Zhang, Z. (2012). The research of an advanced wireless sensor networks for agriculture. *African Journal of Agricultural Research, 7,* 851–858.
3. Akyildiz, I., & Stuntebeck, E. (2006). Wireless underground sensor networks: Research challenges. *Ad Hoc Networks, 4,* 669–686. https://doi.org/10.1016/j.adhoc.2006.04.00.
4. Le, T. D., & Tan, D. H. (2015). Design and deploy a wireless sensor network for precision agriculture. In *2015 2nd National Foundation for Science and Technology Development Conference on Information and Computer Science (NICS)*, Ho Chi Minh City (pp. 294–299). https://doi.org/10.1109/nics.2015.7302210.
5. Majone, B., Bellin, A., Filippi, E., Ioriatti, L., Martinelli, M., Massa, A., et al. (2013). Wireless sensor network deployment for monitoring soil moisture dynamics at the field scale. *Procedia Environmental Sciences., 19,* 426–435. https://doi.org/10.1016/j.proenv.2013.06.049.

Smart Waste Monitoring Using Wireless Sensor Networks

T. V. Chandan, R. Chaitra Kumari, Renu Tekam and B. V. Shruti

Abstract In today's technology, waste disposal and management is becoming a very big issue for the people, as it is the main cause for the unhygienic environment. This leads to various diseases and human illness. To avoid this situation, we are going to implement a system with the help of Python and IoT. The concept is based on smart waste monitoring system. This will help us to maintain a clean environment in our city. We can manage the waste disposal in various areas of the city. IoT is a concept in which we can operate various devices without any user intervention. We can able to manage all the devices with the help of IoT. Sitting at one place we can able to monitor our system and keep an eye over the city. We are making use of Raspberry pi to operate Raspbian OS to enable device connectivity. Different IR sensors are used to detect the level of the dustbin camera which is also fixed in the area to capture the image of the dust in the area. The information is send to the authorized person and we can take the immediate action related to that.

Keywords Python · IoT · Raspberry pi · IR sensors · Camera

1 Introduction

As the technology is growing, we are finding different ways and remedies for a future development. Building a smart city is one of the parts of our advanced technology. One of the major issues in building a smart city is the waste management and waste disposal. Waste disposal is a very challenging task for the current technology. It is related to the problem of cleanliness and hygiene of our environment as we can keep an eye over the city with the help of raspberry OS. As the technology is growing, people are ignoring such kind of issues that will become a major problem in the future. Monitoring the dust in different areas of locality will be possible with the help of this system. We are going to implement this project with the help of IoT and Python. The IoT is a kind of concept in which we can connect the object and

T. V. Chandan · R. C. Kumari · R. Tekam · B. V. Shruti (✉)
Department of Computer Science and Engineering, NMIT, Bangalore, India
e-mail: shrutibv@gmail.com

monitor without any user intervention. Waste management is all the activities and tasks that will help to manage the waste disposal. From inception to its final disposal we are able to monitor with the help of this device. This system will also provide awareness to the customer towards the cleanliness and hygiene of our environment. In this paper, we will discuss design, implementation, limitation and use case of this system.

2 Literature Survey

In the proposed system, we have found different studies that are related to the need of managing the waste disposal and to create awareness among the people of our country and the measure steps that have to be followed to ensure this. We have seen the uses of Raspberry and Python to detect the level of the dustbin and handling the information related to the waste disposal in the different locations. We have also seen that how IoT is used to communicate with different devices to get the information.

From [1, 2] we have to find the amount of garbage that has been produced everyday in different areas in different ways. In this technique, we can get to know the number of people aware of the waste disposal. Then after this, we have to find proper solution related to other waste disposal.

In [3, 4], we observed for how to fix different sensors in every dustbin at a correct position that will help to detect the level of the dustbin and to keep track of the waste disposal. The sensor will provide the enough information to handle things and to take immediate actions. In [5], we will provide how these sensors will be cost effective and efficient to use. We will provide proper information related to the sensors.

In [6], the Raspberry operating system will provide the communication between the user and the devices and therefore provides remedies to this problem. Our literature survey will provide solution to the problem of waste disposal in the different areas of our city.

3 Wireless Sensor Networks

Wireless sensor network is a collection of different kinds of sensors for monitoring the condition of the environment and storing the information and data at a central site location. It will monitor all the physical conditions of the environment like temperature, wind, weather and humidity. This network will provide wireless connectivity between the devices. This network is also referred to as dust networks because it will provide information as small as the dust. It will work as minute sensors. The dust network is defined by UC Berkley project in DARPA. They are the autonomous network that will pass the data into the main locations. Today, this network is used for various industrial and commercial networks. The major applications of these networks are for military applications. The wireless sensor network is made up of nodes, we can

Fig. 1 Wireless sensor network

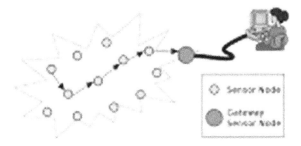

say it is a connection of thousand nodes and these nodes are connected to different kind of sensors. A wireless sensor network consists of a radio transceiver, antenna, electronic circuit and an interface connection between the devices. In wireless sensor networks, sensor nodes can be of variable sizes depending on the network. The cost also depends on the size, connectivity, speed and energy. There will be different types of connectivity between the devices such as star network and multi-hop network. The way of propagating of these sensor nodes can be routing and flooding. Ipsn, ESWN and Senses are various research areas and project workshop conferences where wireless sensor network have been studied (Fig. 1).

4 System Design

In the proposed system, the different kinds of sensors have been placed over the dustbins; sense the level of the dustbin. The IR sensor senses the level of the dustbin whether it is full or not. If the dustbin is full send the mail to the particular authorized person. Another sensor is gas sensor that senses the bad odour smell; if the bad smell is coming from the dustbin, it will send the mail to the particular authorized person. The sensor will immediately detect the area and produce a beep signal in the office to produce an alarm that somebody is trying to pollute the environment. A+++ camera is also fixed in the area to capture the image of the person who is trying to escape from the area after polluting the environment. The image will be immediately sent to the authorized person to take the immediate action (Fig. 2).

5 Implementation

As shown in Fig. 3, the different kinds of sensors have been placed over the dustbins. First will have checking all the components simultaneously, the Raspberry Pi is checking all the conditions and it will be sent through mail to the particular authorized person.

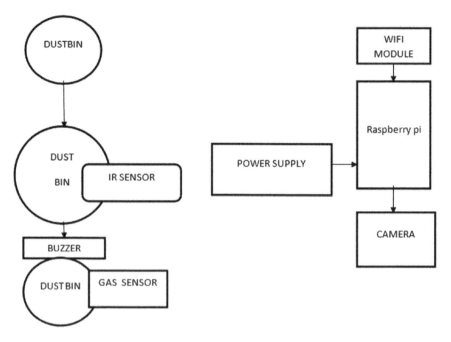

Fig. 2 Architecture design

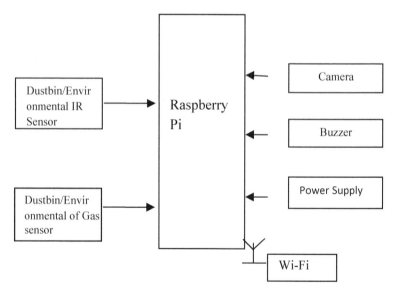

Fig. 3 Block diagram

The Raspberry pi checking the condition senses the level of the dustbin. If it's Dustbin is full, send to the message on the particular authorized person.

A buzzer sensor is also used whenever somebody tries to throw the garbage in the particular area or try to pollute the area with the garbage disposal.

The sensor will immediately detect the area and produce a beep signal in the office to produce an alarm that somebody is trying to pollute the environment.

A camera is also fixed in the area to capture the image of the person who is trying to escape from the area after polluting the environment. The image will be immediately sent to the authorized person to take the immediate action.

Gas sensor will be sensing the bad odour and the smell coming from the dustbin. Whenever the threshold value of the garbage smell reaches its maximum level the gas sensor will quickly identify and send an appropriate message 'bad odour' to the authorized person at that moment to take the immediate action.

6 Result

See Figs. 4, 5 and 6.

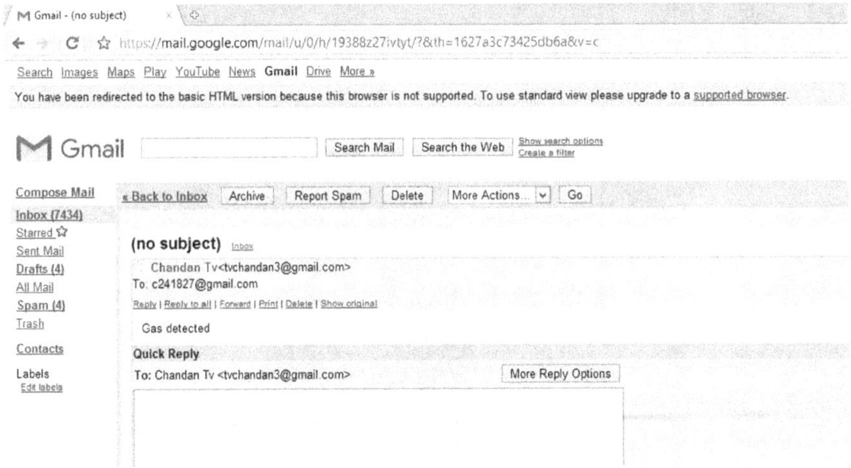

Fig. 4 Gas detected

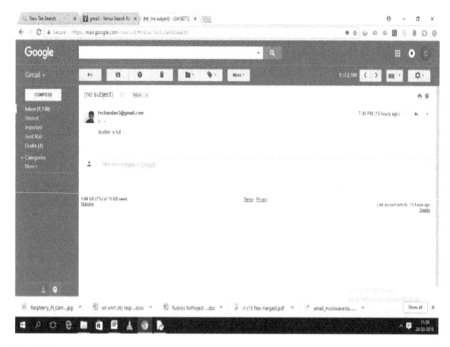

Fig. 5 Dustbin is full

Fig. 6 Capture the image

7 Conclusion and Future Work

The technology has been changed past few decades and there is a great advancement of various experiments. But we are not aware of global cause because of the waste in the city. We are developing a system that will increase the awareness among the people to create the maintenance for the smart city. There will also be encouragement by the government for kind of this project as it will provide a great smart waste monitoring system for the garbage in the city.

As for the future scope of the project, the smart waste monitoring systems will be segregating the garbage in the dustbin. If it segregates the garbage in the dustbin, it will be easy to recycle. The ultrasonic sensor senses the level of the dustbin and it will be showing on android application. If its dustbin is full, ROBO will come and collect the garbage.

References

1. Saji, R. M., Gopakumar, D., Kumar, S. H., Sayed, K. N. M., & Lakshmi, S. (2016). A survey on smart garbage management in cities using IoT. *International Journal of Engineering and Computer Science, 5*(11), 1874–1875, ISSN: 2319-7242.
2. Parkash, P. V. (2016). IoT based waste management for smart city. *International Journal of Innovative Research in Computer and Communication Engineering (An ISO 3297: 2007 Certified Organization), 4*(2), February 2016 copyright.
3. Vishesh, K. K. (2016, May). Smart garbage collection bin overflows indicator using IOT. *International Research Journal of Engineering and Technology (IRJET), 3*(5).
4. Tae-Gue, O., Chung-Hyuk, Y., & Gyu-Sik, K. (2017). Esp8266 wi-fi module for monitoring system application. *Global Journal of Engineering Science and Researches*.
5. Monika, K. A., Rao, N., Prapulla, S. B., & Shobha, G. (2016). Smart dustbin-an efficient garbage monitoring system.
6. Hong, I., Park, S., Lee, B., Lee, J., Jeong, D., & Park, S. (2014). IoT-based smart garbage system for efficient food waste management. In *School of Electrical and Electronics Engineering*, Chung-Ang University, Seoul 151-756, Republic of Korea, Published August 28, 2014.
7. Pandey, N., Bal, S., Bharti, S., & Sharma, A. Garbage monitoring and management using sensors, RF-ID and GSM. Government Engineering Collage, Bharuch, Gujarat Technological University, Ahmadabad, India.
8. Navghane, S. S., Killedar, M. S., & Rohokale, V. M. (2016, May). IoT based smart garbage and waste collection bin. *International Journal of Advanced Research in Electronics and Communication Engineering, 5*(5).

Digital Filter Technique Used in Signal Processing for Analysing of ECG Signal

A. E. Manjunath, M. V. Vijay Kumar and K. S. Swarnalatha

Abstract The area of signal processing holds high significance in biomedical engineering, acoustics and sonar fields. The main finding of coronary heart illnesses is done utilizing ECG. It demonstrates the bio-physiology of cardiac muscles and modifications like arrhythmia and also conduction surrenders. ECG in flag handling is a prime zone of study in bio-signal processing. Present-day advancements in personal computer equipment and computerized channel approach in flag preparing have made correspondence with personal computers through ECG signals suitable. Efficient determination of ECG is a mechanical test. This study exhibits a far-reaching review of computerized sifting strategies to adapt to the clamour curios in ECG flag. The goal of this paper is to separate noteworthy components of ECG utilizing signal preparing methods. Methodologies of different computerized channels for ECG in flag preparing are analysed. Noteworthiness of flag preparing gives off an impression of being with no noticeable indication of immersion in today's world.

Keywords Finite impulse response (FIR) · Infinite impulse response (IIR) · Signal processing · Low frequency filtering · High frequency filtering

1 Introduction

Examining the flag from cathodes on physical surface is at first required in handling of ECG flag utilizing computerized channel. Hence, low recurrence clamour coming about because of baseline wander, movement and breath and high recurrence commotion coming about because of muscle curio and electrical cable or emanated electromagnetic obstruction must be suppressed or killed by the computerized ECG.

A. E. Manjunath
Department of CSE, R.V. College of Engineering, Bengaluru 560056, India

M. V. V. Kumar
Department of CSE, Dr. AIT, Bengaluru 560072, India

K. S. Swarnalatha (✉)
Department of ISE, NMIT, Bengaluru 560064, India
e-mail: swarnalatha.ks@nmit.ac.in

Henceforth, ECG motion at physical surface must experience filtration and enhancement by the electro-cardiograph.

1.1 Signal Processing in ECG

Advanced channels can be intended to have straight stage qualities which keep away from a portion of the twisting presented by exemplary simple channels. Singular layouts are made for each lead from information tested from controlling edifices, from which adequacy and term estimations are made, in the wake of sifting. Blunder of measurement assumes an imperative part on the exactness of ECG analytic explanations. Much research has been done on the issues related to energy line interference (ELI) and baseline wander in the analysis of ECG flag. Various techniques have been proposed for expelling impedances. Various researchers have presented worldwide sifting of AC impedance in digitized ECG as another idea. Some recommended two computerized channels. One depends on summation strategy and different uses slightest squares approach. Genuine ECGs and fake signs utilized as a part of examination by applying each prescient channel and analysed two of the methods (1987). Others have built up a computerized channel using shaft zero strategies. The Chebyshev and Butterworth channels were created. It was concentrated that both sorts work adequately in the ECG image signal. Scientists evaluated two computerized channels and found them to be extremely profitable in limiting sign adulteration. Researchers established a straightforward yet exceptionally coordinated computerized flag handling framework for continuous separating of biomedical filters. Others characterized an advanced channel for concealment of benchmark meander. Some suggested he straight sifting technique for gauge meander reduction. ECG has been engaged as one of the biomedical flag in extensive measure of research. Despite the changes that have been accomplished here, separating of ECG still confronts a few difficulties. The strategies for various computerized channels for pre-processing of ECG signs in regard to their applications have been examined in this paper.

2 Literature Survey

The process of real-time image detection based on two general-purpose processors and fast digital signal processing released over 10 years ago. It is considered as one of the most rapidly growing research areas in the arena of digital signal processing.

2.1 Types of Filtering Approaches

Computerized channel have a wide variety of application such as biomedical flag examination, picture investigation, picture coding and translating strategies. For the most part to filter, digitization of the simple waveform is right off the bat done by a simple to advanced convertor (ADC) and paired qualities are transmitted to DSP gadget which does a continuous convolution operation in non-analogue space utilizing either limited finite impulse response or infinite impulse response calculation. The information prepared is transferred to simple converter (DAC) that creates a separated simple flag. A hostile to associating channel is incorporated before the ADC, and a recreation channel is incorporated after the DAC to fulfil the necessities of sampling theorem with respect to the info waveform and to expel quantization clamour in handled flag. Channels fabricated applying DSP innovation give many advantages over regular simple systems. Most altogether, they are inherently adaptable, as modifying the elements of the channel simply includes changing the program code or channel coefficients. Be that as it may, physical remaking is required with a simple channel. Also, they are impervious to impacts of maturing and natural conditions, as separating procedure relies on upon numerical computations, not mechanical properties of the segments. This renders them perfect for low recurrence signals. For this correct reason, the execution of computerized channels can be characterized with outrageous exactness, inverse to simple channels where a 5% figure is viewed as incredible.

Digital Signal Processing Filter Theory

The basic filtering equation is given as follows:

$$o(t) = \int_{-\infty}^{\infty} ir(\tau)i(t-\tau)d\tau \tag{1}$$

$$o[n] = \sum_{k=0}^{M} ir[k]i[n-k] \tag{2}$$

Using Z-transform, then transfer function is represented by the following equation,

$$H(f) = \frac{(Y(z))}{(X(z))} = \sum_{n=0}^{\infty} h[n] \bigg/ z^{(-n)} \tag{3}$$

The output signal $o[n]$ is represented by the following equation for infinite impulse response,

$$o[n] = \sum_{k=0}^{N} a[k]i[n-k] - \sum_{k=1}^{M} b[k]o[n-k] \tag{4}$$

Table 1 Characteristics of IIR and FIR filters

Sl. No	Characteristics	IIR	FIR
1	Design case	Labour-intensive	Straightforward
2	Stability	No	Yes
3	Direct analogue equivalent	Yes	No
4	Computational load	Low	High

$$H(f) = \frac{\left(\sum_{m=0}^{M} a[m]z^{(-m)}\right)}{\left(1 + \sum_{n=1}^{N} b[n]z^{-n}\right)} \quad (5)$$

where,

$o(t)$ output signal,
$i(t)$ input signal,
t time shift operator,
$ir(\tau)$ impulse response of the filter,
$H(f)$ transfer function,
a and b are constants.

There are real outcomes and practices connected with these two methods to advanced separating, which are outlined in Table 1. Stability is one of the prime criteria in the evaluation of execution of a channel. As appeared in conditions 1 and 2, no recursion or criticism in convolution prepares display FIR channels to be genuinely steady. Interestingly, a small amount of yield flag is constantly encouraged back in IIR channels, commanding careful thoughtfulness regarding configuration to guarantee dependability. This might be seen in an unexpected way: Eq. 5 presents exchange work as the proportion of two polynomials in climbing negative forces of z. In this manner, higher request polynomials are connected with little denominator terms and thus the danger of not well-adapted division. IIR channels are touchy to word length of DSP gadget therefore. By and large, more prominent the request of channel, more noteworthy the danger of flimsiness, so high request IIR channels are normally outlined by falling together a few low request areas.

Direct stage is alluring on account of superior sound framework channels. Consequently, this property is attractive in the preparing of biomedical signs. Direct stage infers that at whatever time postpone felt by one recurrence segment is felt by them all in equivalent measure; along these lines the state of the sifted flag is safeguarded. Direct stage is guaranteed if the motivation reaction is symmetrical as represented in the below equation:

$$h(n) = h(N - n - 1), n = 0, 1, \frac{(N-1)}{2} \quad (6)$$

Its impractical to achieve unadulterated direct stage, particularly experiencing significant change groups, with IIR channels. Different properties likewise settle on FIR channels the perfect decision for different objectives; for instance, they can be provided subjective recurrence reactions by suggesting this in the Fourier region and utilizing the inverse Fourier transform to get the drive reaction. This method is called frequency sampling. Despite the fact that hypothetically it is conceivable to create self-assertive IIR channels, the computational weight in figuring the channel coefficients renders it totally unreasonable by and by. Outline of IIR channels is by and large done by figuring the shafts and zeros for a specific channel and taking in those which exist in the unit hover of the z plane. This method can be tedious, thus there are set-up conditions to get the shafts and zeros of normally utilized channels. For example, Butterworth and Chebyshev sorts. From this appraisal, it may create the impression that FIR channels are better than anything IIR channels. Most cases, they are both utilized. The vital favourable position that IIR sort has is computational proficiency. IIR utilizes way less terms than FIR to develop a channel with a sharp cut-off. Subsequently, IIR channels are more profitable and use less memory asset for a processor with given power. In addition, simple channels can be promptly changed over into proportional IIR advanced channels, with comparative exhibitions. For the most part, this does not remain constant for FIR channels. The framework portrayed underneath utilizations FIR approaches, however, it can be balanced, without any adjustments in equipment, to implement both finite impulse response and infinite impulse response sorts.

Performance Measurement

Signal-to-noise ratio and mean square error which are defined in Eqs. 7 and 8 as shown below.

$$\text{SNR} = 10 \log 10 \frac{\left(\sum_{n=0}^{B} x[n]^2\right)}{\left(\sum_{n=0}^{N}(x_{dn}[n] - x[n])^2\right)} \qquad (7)$$

$$\text{MSE} = \frac{1}{N}\left(\sum_{n=0}^{N}(x_{an}[n] - x[n])^2\right) \qquad (8)$$

3 Filtering Methods

Low Frequency Filtering: Critical twisting into the ECG is presented by traditional simple sifting, a 0.5-Hz low recurrence cut-off, in comparison with the ST portion level. It is from the stage nonlinearities that this mutilation starts. The stage nonlinearities considered are those that happen in zones of ECG flag where there is a sudden change in recurrence substance and wave abundance, as happens where the finish of QRS complex meets the ST section. Strategies to build the low recurrence cut-off without the presentation of stage contortion are given by digital filtering. This can be achieved with a two-directional channel with a moment-sifting pass which is connected backward time that is from end of T wave to onset of P wave. This procedure can be connected to ECG flags that are put away in PC memory that is from ending of T wave to beginning of P wave. ECG signals kept away in memory can be connected to this technique. Ceaseless continuous observations are difficult to obtain. Then again, level stride reaction channel can be utilized to accomplish a zero-stage move, permitting minimization of standard float without recurrence bending.

High Frequency Filtering: The maximum furthest reaches of flag recurrence that can be rendered is controlled by the advanced examining rate (tests every second). As indicated by Nyquist hypothesis, advanced examining must be tested at double the rate of wanted high recurrence cut-off. Since this hypothesis remains constant just for a vast inspecting interim, the AHA report published in 1990 suggested examining rates at three or two times the hypothetical least. Different reviews have demonstrated that information at 600 examples for every 1000 ms are expected to allow the 150-Hz high recurrence computerized channel threshold that is expected to limit plentifulness estimations to 1% in grown-ups. Higher bandwidth might be required for exact assurance of magnitude in newborn children.

3.1 Pre-processing of Signal

The point of separating is to expand the straightforwardness of succeeding preparing operations without the loss of appropriate data. Enhancing signal-to-noise ratio (SNR) to enhance flag quality is a critical objective of pre-processing. A low or terrible SNR suggests that there in impedance in unique flag rendering applicable data hard to distinguish. For the most part, a high or great SNR makes the identification and grouping errand simple. Everyone of the signs do not originate from the electronic action of the heart when measuring the ECG. Numerous potential varieties observed in the ECG might arise due to different reasons. Most of these varieties in ECG that are of cerebral beginning are known as curios, and their reasons might be hardware or subject. Acceptance of 50–60-Hz part in the flag might be finished by encompassing electrical flag. It is expelled by separating by step channel. Figure 1 describes the few techniques for analysis of ECG Signal.

4 Results Analysis

ECG filtration was done utilizing low-pass, high-pass and notch filters independently. The channels are felled towards the end to get the consolidated impact of these channels. Figure 2 portrays crude ECG motion earlier filtration. Figure 3 delineates ECG motion after work of fell channels. At long last, the channels are felled to get the total impact of these channels.

Digital Filter Technique Used in Signal Processing ... 377

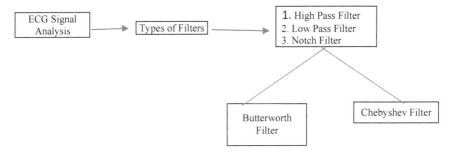

Fig. 1 Processing technique in ECG

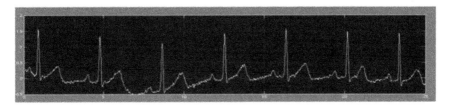

Fig. 2 Raw signal

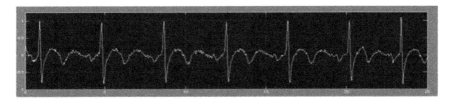

Fig. 3 Filtered signal

5 Conclusion

Separating is a vital strategy in preparing of surface ECG flag. This examination assessed a portion of the different components influencing computerized sifting strategy in the sifting of ECG and analysed the presentations of all IIR channels. At low scope, the IIR channels dispensed with commotion superior to different channels QRS complex is altered and benchmark meander is essentially limited. This work prompts ideal arrangement without the loss of appropriate data in ECG flag. At long last, the channels are felled to get the combined impact of these channels.

References

1. Fedotov, A. A., & Akulov, S. A. (2017). *Mathematical modeling and analysis of errors of measuring transducers of biomedical signals*. Moscow: Fizmatlit (in Russian).
2. Variability, H. R. (2017). Task force of the European society of cardiology and the North American Society of pacing and electrophysiology. *Circulation, 93*(5), 1043–1065.
3. Friesen, G. M., et al. (2017). *IEEE Transactions on Biomedical Engineering, 27*(1), 85–98.
4. Tompkins, W. J. (Ed.). (2016). *Biomedical digital signal processing: C language examples and laboratory experiments for the IBM PC*. New Jersey: Prentice Hall.
5. Theis, F. J., & Meyer-Bäse, A. (2015). *Biomedical signal analysis: Contemporary methods and applications*. Cambridge: MIT Press.
6. Mohamed, E., et al. (2015). In *Proceedings of the 3rd International Conference on Bio-Inspired Systems and Signal Processing*, pp. 428–431.
7. Fedotov, A. A., Akulova, A. S., Akulov, S. A. (2014). Izmer. Tekh., No. 11, pp. 65–68.
8. Benitez, D. et al. (2001). *Computers in Biology and Medicine, 31*, 399–406.
9. Rangayyan, R. M. (2014). *Biomedical signal analysis* (Russian translation). Moscow: Fizmatlit.
10. Moody, G. B., & Mark, R. G. (2014). *IEEE Engineering in Medicine and Biology, 20*(3), 45–50.
11. Pan, J., & Tompkins, W. J. (2014). *IEEE Transactions on Biomedical Engineering, 32*, 230–236.
12. Ruha, A., Sallinen, S., & Nissila, S. (2014). *IEEE Transactions on Biomedical Engineering, 44*(3), 159–167.
13. Kadambe, S., Murray, R., et al. (2014). *IEEE Transactions on Biomedical Engineering, 46*(7), 838–848.

GeoFencing-Based Accident Avoidance Notification for Road Safety

Bhavyashree Nayak, Priyanka S. Mugali, B. Raksha Rao, Saloni Sindhava, D. N. Disha and K. S. Swarnalatha

Abstract The aim of the project is to attempt the reduction of occurrence of accidents by providing effective and precautionary notifications to the user on the progressive Web app, if the user is approaching the accident-prone zone. In the past, due to road accidents, over ten lakhs people had been killed and 50 lakhs had been severely injured, in India. The project provides notifications about approaching vehicles in and around accident-prone areas. Notifications are voice-based so that user could focus on driving and need not constantly view the phone for the precautionary warnings. The project uses global positioning system (GPS) to specify a virtual boundary called the geofence and open-source Google APIs for setting the accident-prone areas on Google Maps. Firebase is used as a backend to store the data on real-time databases for providing a warning on approaching vehicles. The geofenced areas are marked at a sufficient distance from the actual accident-prone zones so as to provide notifications in advance. Necessary notifications are provided on the app, only if the user is in the geofenced area. The prototype also demonstrates temporary geofencing, to provide warnings about the conditions like a roadblock.

Keywords Geofencing · Accident prone · GPS · 000webhost

1 Introduction

Roads and vehicles are a critical part of people's daily routine. The tremendous growth and increase in the number of vehicles on roads leads to traffic congestions which requires to be monitored in an effective way. These people are in different states of mind, most of the time in a hurry or preoccupied with thoughts; unfamiliar and unexperienced with roads and whose behavior can be quite uncertain. Several reasons such as bad roads, over speeding, overloaded vehicles, drunken driving, and failure of sticking onto traffic rules contribute to accidents which maybe fatal or nonfatal.

B. Nayak · P. S. Mugali · B. R. Rao · S. Sindhava · D. N. Disha · K. S. Swarnalatha (✉)
Department of Information Science and Engineering, NITTE Meenakshi Institute of Technology, Bengaluru 560064, India
e-mail: swarnalatha.ks@nmit.ac.in

Hence, there is an immediate need to address this problem by designing efficient systems for accident avoidance. The accident prevention system implemented in this project is based on creating geofences and monitoring the entry of users into these created geofence. Accident-prone areas are geofenced and when users enter such defined areas, they are provided with notifications well in advance so that accidents can be prevented.

Two types of notifications are required to be provided on the progressive Web app: (1) Passive notification—notify driver when in geofenced accident-prone areas and (2) Active notification—notify driver of any approaching vehicles from the opposite direction. Current location of users is fetched through GPS receiver in the smartphone. The real-time database in firebase is used to store the geofence variables. Additionally, geofences can be created at appropriate areas by specifying a set of coordinates to provide other time- and location-specific information such as blocked roads and fallen tree.

2 Related Work

Faiz et al. [1], A great many vehicles are being created every year. A large portion of the mischances happen because of human carelessness, for example, rash driving, absence of good framework for accident prevention and detection and so forth. A prompt protection and rescue process following a casualty can be considered as a tightrope stroll among life and demise.

As smartphones turn out to be such a critical piece of our life in today's world, it is achievable to utilize them in reducing and controlling the severity of a post-mischance casualty.

The android application utilizes the in-built GPS receiver in smart mobile phones to recognize the change in deceleration that happened at the time of accident. Additionally, it also measures the change of pressure detected by the pressure sensor and the change of tilt detected by the accelerometer sensor. These three parameters help detect the occurrence of an accident. The android application is programmed to send the location of occurrence of accident to provide help. Japan experiences enormous harm and disastrous events consistently. The reason for this is lack of timely supply of the right data to the general population who require it. Hence, this comes up as a motivation for the need of such a scheme that effectively delivers time- and location-specific risk information to users. Suyama and Inoue [2] proposed a disaster information system that uses the geofencing technique. The system monitors the movement of users and accordingly supplies risk information to them. Geofencing is a mechanism that uses global positioning system (GPS) or radio frequency identification (RFID) to create a virtual fence in a specific area. The coordinates (pair of latitude and longitude) are required to create a geofence. This application creates a circular geofence at a critical area. This allows the administrator to monitor entry and exit of users into and out of the geofence.

3 Goal

- Using software approach to develop an accident prevention system and developing a mobile application.
- The accident-prone areas are geofenced and notifications are provided to the users through the progressive Web app on the smartphone.
- With the use of software techniques like geofencing, users are notified both passively and actively.
- Our proposed system will reduce the dependency on sensors and other hardware components and the associated problems of deployment and monitoring.

4 Proposed Method

4.1 Geofencing

A geofence is a boundary or perimeter that is virtual on a real-world geographic area. It could be generated by predefining a set of points as boundary such as setting a radius around the selected location point. The geofence can be programmed accordingly to send mobile push notifications, allow tracking of vehicle traffic, trigger alerts or text messages, provide customer targeted advertisements on social media, prevent the use of certain technology, or prompt location-based marketing or safety information.

4.2 Use of Geofencing

The developer or administrator who wishes to use geofencing should set a virtual boundary around the specified location in the software that supports GPS. This could be simply done by drawing a circle of 100 feet around the chosen point of location on Google Maps using Google APIs, during the mobile app development. It could also be done by specifying coordinates of the boundary. Then depending on the program specified by the administrator or developer, this virtual perimeter could trigger a response or an alert when an authorized device enters or exits that boundary.

For the project, geofence technique has been used to mark virtual boundary areas at a sufficient distance from the actual accident-prone areas on Google maps. These areas are used for issuing appropriate precautionary warnings to the users about the nearing accident-prone zones. For every such prone area on bidirectional roads, at least two bounded regions need to be marked on the Google maps, to supply notifications to the vehicles on either side of the road.

5 Flow of Process

An HTML code should be written for the progressive Web app. This should be then hosted using 000Webhost service. Firebase needs to be initialized to store variables that would support the issuing of active notifications.

5.1 Modules Involved

(1) Passive Notification

The components of this module are—Location detection—Marking the user's current location on the map if the user has allowed the app to use the location services (i.e., GPS).

Issuing notifications—Checking if the vehicle enters or is in the geofenced area, and supplying appropriate voice notifications on the app.

Updating the database—Dynamically changing values of the variables associated with a geofence in the database if a vehicle enters or exits from it.

(2) Active Notification

Retrieving data from database—Obtaining variable values associated with the geofence area. Issuing notifications—Checking if there is at least one vehicle each in the two different geofenced areas pertaining to an accident-prone area, and issuing appropriate voice notifications for these vehicles, on the app.

For every accident-prone area, at least two geofences need to be marked to warn the vehicles traveling in either direction of the road. These areas should be at an appropriate distance from the actual accident-prone zone to issue precautionary warning.

5.2 Flowchart

The description of the workflow of the system is depicted in the form of a flowchart.

On launching the app, Google map showing current location appears. If the vehicle enters or is in the geofenced area, a voice notification warning that "the vehicle has entered accident-prone area" needs to be provided. The value of the variable associated with that geofenced area needs to be updated in the firebase. The values of variables of two or more geofenced areas associated with one accident-prone zone are then retrieved from the database If the comparison of the values of the variables, implies that there are two vehicles simultaneously existing in the two opposite geofenced areas marked for an accident-prone area, then a voice notification warning that "there is an approaching vehicle" needs to be provided. If the vehicle exits or is not in the geofenced area, the values of the variables associated with that geofenced area needs to be updated in the firebase. The voice notifications are provided only if the vehicle is in the geofenced area, i.e., the virtual boundary (Fig. 1).

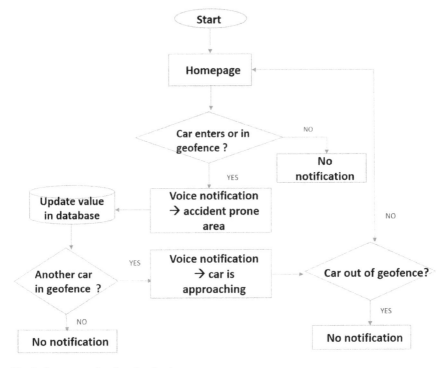

Fig. 1 Sequence of actions involved

6 Experiment and Result

The users must allow the app to access the location services so that it could utilize it for efficient functioning. Also, the cellular data should be switched on, in the user's device. A minimum of 2G network is required. The users who wish to utilize the app must enter the URL of the app in the Web browser that supports adding Websites to the home screen of their mobile. Users must then save this Website to their home screen. They also get an opportunity to save the app with the name of their choice

To demonstrate the effectiveness of the prototype designed, it was deployed in the college premises.

The geofenced areas (covering around 50 m of the road stretch) were marked on the Google Maps using a set of points or coordinates. These boundaries which looked similar to rectangle shape were marked at a distance of about 100 m from the actual accident-prone zone so as to provide precautionary warning to the driver who would be using the app.

In Fig. 2, there are two geofenced areas (50 m in length) marked about 100 m away from the actual accident-prone area.

For simplicity of understanding consider the geofenced area between the Maggi Station and the ATM as geofence area 'GA1' and the one between the parking

Fig. 2 Screenshot showing geofences to notify about accident-prone area

and Potenza Cafe as 'GA2' and the values associated with them be GA_val1 and GA_val2, respectively.

If a vehicle (i.e., user device) enters or is in GA1

- Voice notification that "the vehicle is in the accident-prone area" was provided on app.
- Value of GA_val1 was changed and updated in the database of firebase.

This procedure shall be similar for GA2.

Values of GA_val1 and GA_val2 were obtained from the database. They were compared to check if there were vehicles present in GA1 and GA2 at the same point of time. If so voice notification about "approaching vehicle" was provided on the app.

If a vehicle (i.e., user device) exits or is out of GA1, value of GA_val1 was changed and updated in the database. This procedure shall be similar for GA2 (Fig. 3).

Fig. 3 Screenshot showing geofences to notify about blocked road ahead

This project also demonstrates the use of geofencing for temporary purposes like for providing notifications about road block which has been caused due to fallen tree on the road.

For the screenshot (number) mentioned above, a geofenced area was marked assuming that some tree has fallen around the temple. There would be a road block because of that fallen tree. Therefore, this geofenced area could provide efficient warnings, so that the user of the device (vehicle) can decide on taking an alternate route in advance, as opposed to the user traveling a distance and later finding out about the roadblock situation and then deciding on alternate path to reach the destination.

7 Conclusion and Future Work

A system to avoid accident occurrence by providing precautionary notifications was proposed. The prototype was implemented with the use of geofencing method and was evaluated in the college premises. Users of the app had to add the Website on their smartphone's home screen. It was confirmed that the app provided relevant notifications to the users if they entered or were present in the geofenced areas, which were marked at adequate distance from the real accident-prone zone. Appropriate warnings were also received if there were two or more vehicles present in the two opposite geofenced areas (associated with an accident-prone zone) at the same point of time. At least 2G cellular data and allowance to use the location services were necessary for efficient functioning of the prototype. The values of the variables associated with the geofenced areas were successfully updated and retrieved from the firebase database.

For fatal accident-prone areas, the area of fencing should be larger than the non-fatal ones. Google APIs need to be bought for real-time location monitoring and geofencing. It is important to improve the location accuracy to provide timely warnings. Information sources like the government transport department needs to be added to our system to obtain genuine accident-prone areas information, and this could also be stored and maintained in the firebase database.

References

1. Faiz, A. B., Imteaj, A., & Chowdhury, M. (2015). Smart vehicle accident detection and alarming system using a smartphone. In *1st International Conference on Computer & Information Engineering*, November 26–27, 2015.
2. Suyama, A., & Inoue, U. (2016). Using geofencing for a disaster information system. In *Copyright IEEE ICIS 2016*, Okayama, Japan, June 26–29, 2016.
3. https://developers.google.com.
4. https://www.000webhost.com/.
5. https://en.wikipedia.org/wiki/Progressive_Web_Apps.
6. https://developers.google.com/web/progressive-web-apps/.
7. https://www.mapsofindia.com/my-india/india/road-accidents.
8. https://www.road-safety.co.in.
9. https://firebase.google.com.
10. https://en.wikipedia.org/wiki/Geo-fence.

An Innovative IoT- and Middleware-Based Architecture for Real-Time Patient Health Monitoring

B. A. Mohan and H. Sarojadevi

Abstract WSN is deployed in every sector of human life, an instance of which is in medical device network. With the increasing requirement of handling large patient data and faster data conversion in the hospital and medical diagnostic centres, there is a need for innovative and low-cost ways of interconnecting medical equipment, middleware and network support. Hospitals and medical diagnostic centres require cost-effective sensor network for handling large patient data and fast conversion. The high cost of medical devices, equipment and lack of interoperability and portability necessitates new approaches. This paper presents low-cost and portable approach for medical data transmission from devices to middleware using IoT support and conversion to universal standard called Health Level-7 (HL7) and a method to store in cloud.

Keywords Medical health record · Health monitoring system · IoT · Middleware · Health level-7

1 Introduction

Sensors are used in every part of our life. Medical devices that acquire medical data invariably incorporate sensors communicating wirelessly which forms domain-specific wireless sensor network (WSN). The data collected in this way is transferred to the middleware and converted into HL7 format for portability. The present-day technology in medical domain is of high cost, lacks interoperability in some cases and slows in data conversion. In this paper, we propose middleware with the support of WSN in conjunction with microcontroller for collecting and processing medical data which results in fast and low-cost conversion into HL7 format.

B. A. Mohan (✉) · H. Sarojadevi
Faculty, Department of CSE, Nitte Meenakshi Institute of Technology,
Bangalore 560064, India
e-mail: mohan.ba@nmit.ac.in

Patient data in standardized and convertible format is essential for hospitals, diagnostic centres, research centres and end-users, which are produced by versatile medical equipment from various end points connected through network. Medical device integration is a process of interconnection of medical equipment through which the data is transferred between devices, middleware and servers. Data is in the form of MRI, ECG, surgical data and other data required for electronic medical record. Integrating medical devices will eliminate the need for manual working on data and benefits in faster and more accurate data exchange with improved workflow efficiency. To achieve such integration, IoT-based middleware solutions proposed here will be useful. While there exist efforts towards middleware development, which are proprietary and expensive, this proposed approach reduces cost, besides making it more interoperable.

Many vendors manufacture different medical devices resulting in unstandardized data format. So, a middleware is proposed to convert data into HL7 compliant format, which is an international standard for data exchange between medical equipment and suitable for storage and processing.

The proposed approach consists of following modules.

- Medical equipment interface to IoT (Internet of Things) devices.
- Communication modules (wireless) to connect IoT sensors to Middleware.
- Middleware for information gathering and converting it to HL7 format.
- Method for storing Converted data in the cloud for analysis.

2 Related Work

The hospital management system (HMS) involves data gathering and processing from IoT devices [1, 2]. The data model selected can be structured or unstructured depending on the back-end technology used for data storage in database. The network integration is based on the data exchange format. Researchers are focusing on the infrastructure integration, process automation and application framework. But, a very few are targeting the data model, processing data and linking data across design process.

Development of biomedical interface using middleware [3] for facilitating medical device connectivity was a leading research work done. As many vendors manufacture variety of medical devices resulting in different data formats, middleware is suggested for converting different data formats to international standard format called HL7 (IEEE 1073 Communication Standard).

Many have proposed generalized system for monitoring the health status with the following functions [4]: electrocardiogram, pulse oximeter, thermometer, accelerometer, etc. The centrepiece being the platform TWR-K53N512 Tower System from Free Scale Semiconductor Inc. can collect data from connected sensors and transfer them via USB, Bluetooth or Wi-fi to a local computer or cloud server [5].

2.1 Research in Industry

Many companies like TCS [6, 7], IBM, Infosys and Philips are into research providing the smart hospitals to improve the healthcare facility. To face the challenges posed by different equipment manufacturers for interconnecting them, Internet of Things for Medical Devices (IoT-MD) [7–10] was considered as suitable solution and some working models were also released. These models are proprietary and costlier to implement by all the hospitals [11–13]. So, cost-effective solution is needed for countries like India.

2.2 Python's Support for HL7

Python is a most famous scripting language which includes library module called hl7APIs. HL7 APIs help parsing messages of HL7 into Python objects. It also includes a simple client engine that can send HL7 messages to a minimal lower-level protocol (MLLP) server. The current widely accepted version of HL7 is 2.x series. Python API's at present parse HL7 version 2.x message and has ability to create HL7 v2.x messages.

Medical health record data (HL7 data) is parsed into objects of specific subclasses of container. The container message itself is a subclass of a Python list which can easily access the HL7 message as an n-dimensional list. The subclasses of container class include message, segment, field, repetition and component. However, Python is not very platform generic. There may be compatibility issues.

2.3 Cloud for Storage

Applications today need the ability to simply and securely collect, store and analyse data at massive scale. Nowadays, we find many cloud storage service providers offering storage space for any amount of data from anywhere—websites and mobile apps [14–17]. Cloud is designed to deliver 99% durability and stores data for millions of applications in every industry. It gives customers flexibility in the way they manage data for cost optimization, access control and compliance. Cloud allows user to run powerful analytics directly on their data, with integration from the largest community of third-party solutions and services.

3 Proposed Technique

The architecture shown in Fig. 1 indicates organizations of the main modules for IoT-based medical data transmission and conversion to HL7 at the middleware. The IoT transceivers such as Raspberry Pi/ARM-7 help in connecting medical devices to procure data and propagate to middleware integration. Converted data can be further stored in the centralized servers or in cloud.

The middleware solution proposed here ensures accurate analysis of patient and medical device data, documentation and workflow automation to support improvement in healthcare quality and medical records.

3.1 ARM in IoT

IoT interconnects all the dumb/electronics devices in home to automate its function, which can be further connected to the Internet for remote control. Devices like toasters, washing machines and lights will be able to communicate each other to solve day-to-day problems. It also includes the use of sensors and smart devices in the wider environment to automate processes and give information to human operators, such as road-side units that alerting drivers about potholes or recycling bins alerting authorities when it is full.

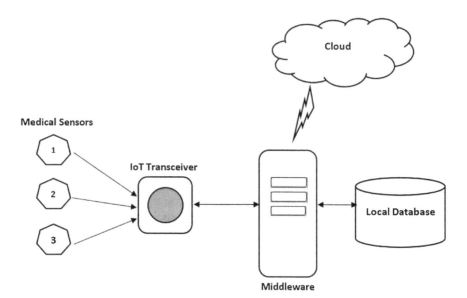

Fig. 1 Proposed middleware architecture

IoT is still in its early stages. However, there are small collection of consumer devices and a few commercial devices that actually link into the wider world. Some of the best-known smart devices are smart thermostats, such as (a) Nest's Learning Thermostat (b) lighting systems like Philips Hue and Light-wave RF devices (c) smart devices such as electronic locks and cameras. Various smart electronic devices are available that allow users to control remotely with various interconnected systems. Further advanced set-ups can be established with specialized equipment and protocols such as Z-wave or ZigBee, which allow dissimilar devices to communicate wirelessly without Wi-fi or the Internet.

Advanced RISC Machine [18] holdings is a designer of low-power chips and processors that can be found in devices such as mobile phones, smart-TVs, smart-watches, tablets and other various computing devices.

ARM is a family of processor architecture for computer processors based on a reduced instruction set computing architecture. ARM is a 32-bit architecture used in many processors and microcontrollers available from various manufacturers. ARM-7 series is largest success of ARM architecture embedded in products ranging from mobile phones to anti-lock braking systems (ABS) in cars.

ARM does not fabricate the processors, but its designs are used by the companies all over the globe, including companies like Apple, Samsung and Qualcomm and used to drive devices all over the home and in the wider world. As the IoT expands, the demand for low-power chips will increase.

3.2 Conversion Details

HL7 is a standard for transferring medical data between medical applications used by various healthcare providers. This standard focuses on application layer of OSI Model called layer 7 or Higher Layer 7. The HL7 standard is given by the Health Level Seven International, an international standards organization and is adopted by other standards issuing bodies such as American National Standards Institute and International Organization for Standardization (ANSI).

HL7 International identifies a number of flexible standards, guidelines and methodologies to follow for successful communication within healthcare system and outside system. Such standards permit information to be processed and shared in a uniform and consistent manner. These data standards are meant to allow healthcare organizations to easily share clinical information.

HL7 International standard defines the following documents format.

- Version 2.x Messaging Standard.
- Continuity of Care Document (CCD).
- Clinical Document Architecture (CDA).
- Clinical Context Object Workgroup (CCOW).
- Structured Product Labelling (SPL).

3.3 Storing Converted Data in Cloud

For the purpose of increased availability and portability, the medical health record (MHR) data is stored in the cloud for easy analytics as proposed in our approach. The MHR data in the cloud is also made available in HL7 format for any of the client purposes. There is a real-time implication also with our proposed method, which is achieved by means of Google's firebase real-time database. Client applications in the form of doctors and patient attendants connected to a single real-time database instance will instantaneously receive the updates as soon as patient parameters are written into database.

4 Implementation

The implementation of proposed architecture is done using the following hardware components.

i. Microcontroller (ARM-7 LPC-2148)—for interconnecting different sensors for sensing temperature, ECG, Blood Pressure, Heart Rate and Eye Blink.
ii. Sensors

    a. Temperature Sensor (LM-35 Thermostat)
    b. Heart Rate Sensors (TCRT-1000)
    c. Blood Pressure Sensors (SEN244)
    d. Drowse Sensors/Eye Blink Sensor (LM-324 Power Operational Amplifier).

iii. GSM Module (SIM-300) for interconnecting microcontroller to the middleware server.

The ARM microcontroller is connected with the above-mentioned sensors as shown in Fig. 2, for sensing the patients' health status, which is then sent to the middleware server [19, 20] for converting it into HL7 format and before storing in cloud. Global system for mobile communication (GSM) is used as wireless technology for interconnecting microcontroller to the middleware.

The patient data received from microcontroller is first stored in MySql database, and then Python parser is used for converting the sensed data to the HL7 format, which is later written into the cloud.

Google's firebase cloud platform is used for storing patient data. It has a real-time database hosted on it for storing data in a JSON format and synchronizes in real time to every connected client. The connected clients share single database instance for receiving instantaneous update. Multiple receivers like doctors and patient attendants can be connected to cloud to get the real-time updates on patient health parameters.

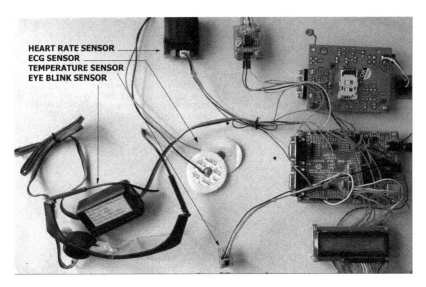

Fig. 2 ARM-7 with sensors

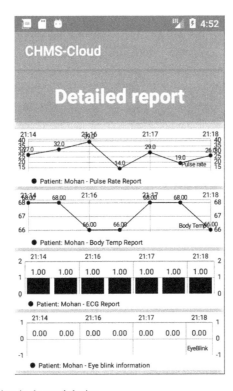

Fig. 3 Patients' live data in doctors' device

Fig. 4 Patients' HL7 data

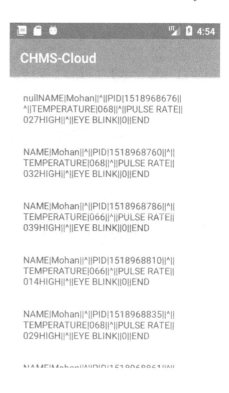

5 Results

This section shows results of creating sample HL7 message from the cloud. Figure 3 shows doctors' interface for monitoring live patient critical health parameters using a hand-held android mobile device. The data displayed corresponds to single patient. The data shown in Fig. 4 represents multiple parameters in HL7 format acquired from the sensors through IoT devices via GSM wireless connections. At present, four fields of patient data, viz. temperature, heart rate, ECG and eye blink status are acquired.

The results presented here represent mere experimental attempts. Real cloud deployment of health data is not recommended as per the standards.

6 Conclusion

A WSN-based solution architecture approach for the conversion of medical data in the middleware using IoT is presented in this paper. The method proposed is observed to be of low cost and interoperable. The cost of the complete set-up accomplished with our approach is less, i.e. about Rs. 40,000 compared to lakhs of rupees of the similar units used at present in the hospital, which are not IoT-driven and on wired network which has strict distance limitations. Integration with cloud is enabled with its seamless storage capacity, real-time implication and instant data conversion. Hospitals and medical centres can now handle the patient data in large quantities and perform the conversion of that data efficiently.

Acknowledgements We sincerely thank the Director, Principal, Dean and HOD of CSE department of our college, NMIT for supporting us in the funding of seed money to carry out this project successfully. We are thankful for their encouragement in publishing this work in ERCICA 2018.

References

1. Mishra, P. A., & Roy, B. (2017). A framework for health-care applications using internet of things. In *International Conference on Computing, Communication and Automation (ICCCA)* (pp. 1318–1323).
2. Thangaraj, M., Ponmalar, P. P., & Anuradha, S. (2016, March). Internet of Things (IoT) enabled smart autonomous hospital management system—A real world health care use case with the technology drivers. In *IEEE International Conference on Computational Intelligence and Computing Research*.
3. Sallabi, F., & Shuaib, K. (2016). Internet of things network management system architecture for smart healthcare. In *Sixth International Conference on Digital Information and Communication Technology and its Applications (DICTAP)* (pp. 165–170).
4. Gomez, J., Oviedo, B., & Zhuma, E. (2016). Patient monitoring system based on internet of things. In *Proceeding of Computer Science* (pp. 90–97). Amsterdam: Elsevier.
5. Hassanalieragh, M., et al. (2015). Health monitoring and management using IoT sensing with cloud-based processing: Opportunities and challenges. In *IEEE International Conference on Services Computing* (pp. 285–292).
6. Laplante, P. A., & Laplante, N. (2016, June). The internet of things in healthcare: Potential applications and challenges. *IT Professional, 18*(3), 2–4, ISSN: 1520-9202.
7. Chhatlani, A., Dadlani, A., Gidwani, M., Keswani, M., & Kanade, P. (2016, December). Portable medical records using internet of things for medical devices. In *International Conference on Computational Intelligence and Communication Networks*.
8. Dridi, A., Sassi, S., & Faiz, S. (2017). Towards a semantic medical internet of things. In *IEEE/ACS 14th International Conference on Computer Systems and Applications (AICCSA)* (pp. 1421–1428).
9. Catarinucci, L., et al. (2015). An IoT-aware architecture for smart healthcare systems. *IEEE Internet of Things Journal, 2*(6), 515–526.
10. Razzaque, M. A., Milojevic-Jevric, M., Palade, A., & Clarke, S. (2015, November). Middleware for internet of things: A survey. *IEEE Internet of Things Journal*, 70–95.
11. Maia, P., Baffa, A., Cavalcante, E., & Delicato, F. C., Batista, T., & Pires, P. F. (2015, November). A middleware platform for integrating devices and developing applications in E-health. In *Brazilian Symposium on Computer Networks and Distributed Systems (SBRC)*.
12. Al-Fuqaha, A., Guizani, M., Mohammadi, M., Aledhari, M., & Ayyash, M. (2015). Internet of things: A survey on enabling technologies, protocols, and applications. *IEEE Communication Surveys Tutorials, 17*(4), 2347–2376.
13. Stankovic, J. A. (2014). Research directions for the internet of things. *IEEE Internet of Things Journal, 1*(1), 3–9.
14. Maia, P., et al. (2014). A web platform for interconnecting body sensors and improving health care. In *Proceedings of Computer Science* (Vol. 40, pp. 135–142). Elsevier.

15. Jara, A. J., Zamora-Izquierdo, M. A., & Skarmeta, A. F. (2013, September). Interconnection framework for mHealth and remote monitoring based on the internet of things. *IEEE Journal on Selected Areas in Communications, 31*(9).
16. Bazzani, M., Conzon, D., Scalera, A., Spirito, M. A., & Trainito, C. I. (2012, September). Enabling the IoT paradigm in E-health solutions through the VIRTUS middleware. In *IEEE 11th International Conference on Trust, Security and Privacy in Computing and Communication*.
17. Lu, D., & Liu, T. (2011). The application of IoT in medical system. *IEEE International Symposium on IT in Medicine and Education, 1,* 272–275.
18. Nagabhushanam, K., Kumar, S. A., & Varma, J. A. (2015, June). Wireless sensor reconfigurable network over IoT using ARM processor. *4*(17), 3107–3109.
19. Chaczko, Z., Kohli, A. S., Klempous, R., & Nikodem, J. (2010, June). Middleware integration model for smart hospital system using the open group architecture framework (TOGAF). In *14th International Conference on Intelligent Engineering Systems (INES)* (pp. 215–220).
20. Tran, T., Kim, H., & Cho, H. (2007, October). A development of ubiquitous biomedical interface for facilitating medical device connectivity. In *7th IEEE International Conference on Computer and Information Technology* (pp. 1094–1099).

Travelling Salesman Problem: An Empirical Comparison Between ACO, PSO, ABC, FA and GA

Kinjal Chaudhari and Ankit Thakkar

Abstract Travelling salesman problem (TSP) is one of the optimization problems which has been studied with a large number of heuristic and metaheuristic algorithms, wherein swarm and evolutionary algorithms have provided effective solutions to TSP even with a large number of cities. In this paper, our objective is to solve some of the benchmark TSPs using ant colony optimization (ACO), particle swarm optimization (PSO), artificial bee colony (ABC), firefly algorithm (FA) and genetic algorithm (GA). The empirical comparisons of the experimental outcomes show that ACO and GA outperform to ABC, PSO and FA for the given TSP.

Keywords TSP · ACO · PSO · ABC · FA · GA

1 Introduction

The collective behaviour of social insects such as ants, bees and of other animals such as bird flocking, fish schooling have motivated designing of intelligent multi-agent systems [1, 2]. The non-sophisticated individuals of such colonies are self-organized; they are capable of solving a complex problem. Swarm intelligence-based algorithms provide a better solution in reasonable time. ACO [3], PSO [4], ABC [5] and FA [6] can be used to solve NP-hard problems such as TSP [7]. On the other hand, evolutionary algorithms such as GA [8] are inspired by the biological evolution. These are population-based metaheuristic algorithms and are designed for global optimization. These algorithms are designed to efficiently explore the search space in order to find near-optimal solutions [9].

K. Chaudhari (✉)
Institute of Technology, Nirma University, Ahmedabad 382481, Gujarat, India
e-mail: 17ftphde21@nirmauni.ac.in

A. Thakkar
e-mail: ankit.thakkar@nirmauni.ac.in

© Springer Nature Singapore Pte Ltd. 2019
N. R. Shetty et al. (eds.), *Emerging Research in Computing, Information, Communication and Applications*, Advances in Intelligent Systems and Computing 906,
https://doi.org/10.1007/978-981-13-6001-5_32

TSP has been studied thoroughly for optimizing the shortest route. Given a list of cities and the distances between each of them, the objective is to find the shortest possible route for a salesman such that he visits each city exactly once and returns to the original city [7]. Many approaches have been developed for solving TSP that provide either an exact or an approximate solution. Algorithms such as branch-and-bound [10], dynamic programming [11] are devised to give the exact solution of TSP with a small number of cities. However, for a large number of cities, identification of an approximate solution may suffice. Various metaheuristic algorithms based on swarm and evolutionary techniques approximate good solutions for the given TSP.

In this paper, we have applied ACO, PSO, ABC, FA and GA for solving few of the benchmark TSPs named Burma14 [12], Bayg29 [13], and Att48 [14] and shown analysis of the empirical results by carrying out each simulation ten times with the given configuration.

The remaining paper is organized as follows: Sect. 2 briefs about TSP and how ACO, PSO, ABC, FA and GA are applied for solving TSP. In Sect. 3, we have given simulation parameter specifications and experimental result discussions; concluding remarks are given in Sect. 4.

2 Related Work

In this section, we have discussed TSP, followed by the discussion of the identified swarm intelligence and evolutionary algorithms, and their use to solve TSP.

2.1 Travelling Salesman Problem (TSP)

In combinatorial optimization, TSP is an NP-hard problem [7]. One of the representations of TSP is a complete undirected weighted graph, $G = (V, E)$. Here, a total of N cities are the vertices of the graph, $V = \{1, 2, \ldots, N\}$ and distance, d_{ij} between two cities i and j, is the distance of the path, i.e., weight of the edge, $E = \{(i, j) \mid i, j \in V\}$. For the given (x, y) coordinates, distance can be evaluated using the Euclidean distance between cities i and j as given by Eq. 1 [7].

$$d_{ij} = \sqrt{(x_i - x_j)^2 + (y_i - y_j)^2} \tag{1}$$

This heuristic information, d_{ij} has to be preserved before applying any approach for solving TSP. Depending on the approach, the shortest route is to be found by adding up the Euclidean distances of the consecutive edges within the tour.

2.2 Ant Colony Optimization (ACO)

Motivated by the foraging behaviour of ant species, ACO was developed to be a multi-agent approach. To symbolize a favourable route which other members of the colony should follow, ants deposit pheromone on the ground. ACO [3] exploits a similar mechanism with artificial ants, behaving in a similar manner to solve optimization problems iteratively.

In each iteration, an artificial ant constructs a solution by walking from vertex to vertex, provided that the vertex was not previously visited by her in that iteration. Based on the pheromone trail, the next vertex is selected to be visited. From any vertex i, if vertex j has not been previously visited by ant k, j can be chosen with probability proportional to the pheromone on edge (i, j). The pheromone deposit values on all edges are updated after every iteration as given by Eq. 2 [15]. The pheromone, τ_{ij} associated with edge (i, j) is updated by all m ants. ACO can be used to solve TSP; the cities are represented by vertices and path joining them are given by edges.

$$\tau_{ij} \leftarrow (1-\rho) \cdot \tau_{ij} + \sum_{k=1}^{m} \Delta \tau_{ij}^{k} \qquad (2)$$

where, ρ is the evaporation rate of pheromone, and $\Delta \tau_{ij}^{k}$ is the pheromone quantity laid on edge (i, j) by ant k as given by Eq. 3 [15].

$$\Delta \tau_{ij}^{k} = \begin{cases} \frac{Q}{L_k} & \text{if ant } k \text{ used edge}(i, j) \text{ in its tour} \\ 0 & \text{otherwise} \end{cases} \qquad (3)$$

where, Q is a constant, and L_k is the length of the tour built by ant k. When ant k is at city i, the probability of going to city j is given by Eq. 4 [15].

$$p_{ij}^{k} = \begin{cases} \frac{\tau_{ij}^{\alpha} \cdot \eta_{ij}^{\beta}}{\sum_{c_{il} \in N(s^p)} \tau_{il}^{\alpha} \cdot \eta_{il}^{\beta}} & \text{if } c_{ij} \in N(s^p) \\ 0 & \text{otherwise} \end{cases} \qquad (4)$$

where, s^p is the partial solution constructed by ant k; $N(s^p)$ are the edges (i, l), where l is the unvisited city by ant k so far. The parameters α and β control the relative importance of the pheromone and the heuristic information, respectively. This heuristic information, η_{ij} is given by Eq. 5 [15] for distance d_{ij} between cities i and j.

$$\eta_{ij} = \frac{1}{d_{ij}} \qquad (5)$$

2.3 Particle Swarm Optimization (PSO)

For solving problems with continuous quantities, PSO has been widely used [4]. It is inspired by the social behaviour of bird flocking and fish schooling organisms. For a given problem, a number of particles are placed in the search space to evaluate the objective function. Each particle determines its movement through the search space [4]. This procedure is repeated in each iteration by all the particles. PSO aims at designing an optimization technique where these particles move close to an optimum solution eventually.

In PSO, each particle i has a form of three D-dimensional vectors, where D is the dimensionality of the search space [4]. For the ith particle, these components include the current position $X_i = (x_{i1}, x_{i2}, \ldots, x_{iD})$, the previous best position $p_{best_i} = (p_{i1}, p_{i2}, \ldots, p_{iD})$, and the velocity $V_i = (v_{i1}, v_{i2}, \ldots, v_{iD})$. The global best position $g_{best} = (g_1, g_2, \ldots, g_D)$ is known to all m particles [16]. The particle swarm works by handling coordinates of the particles and hence, adjusting the orientation [17]. Velocity and position coordinates are updated by each particle after every iteration as given by Eqs. 6 and 7 [4], respectively.

$$V_{id} \leftarrow \omega \times V_{id} + \eta_1 \times \text{rand}() \times (p_{best_{id}} - X_{id}) + \eta_2 \times \text{rand}() \times (g_{best_d} - X_{id}) \quad (6)$$

$$X_{id} \leftarrow X_{id} + V_{id} \quad (7)$$

where, ω is the inertia weight ($0 < \omega < 1$); η_1 and η_2 are acceleration constants, and rand() generates a random value in the interval [0, 1]. The velocities of particles are constrained within interval [$V_{\min}$, $V_{\max}$].

Solving TSP using PSO includes velocity V_i and position X_i to each particle; here, the position denotes the sequence in which the ith particle has visited the cities [18]. The quality of solutions is evaluated after every iteration and the personal best (p_{best_i}) of each particle and the global best (g_{best}) of overall swarm are updated.

2.4 Artificial Bee Colony (ABC)

Foraging behaviour of a honeybee colony has been a research curiosity for many researchers. A model based on reaction-diffusion equations [5] was developed; it leads to the emergence of collective intelligence of honeybee swarms. Three essential components of this model are food sources, employed foragers and unemployed foragers whereas two leading modes of the behaviour include recruitment to a nectar source and abandonment of a source [5].

In ABC algorithm, bees initially select a set of food sources randomly; they provide their respective nectar information to onlookers and hence recruit them, and return back to their food sources. The recruited onlookers become employed bees and exploit the food sources depending upon the probability value associated with the food source [19]; the probability, p_i is calculated by Eq. 8 [5].

$$p_i = \frac{fit_i}{\sum_{n=1}^{SN} fit_n} \qquad (8)$$

where, fit_i is the fitness value of the ith solution; SN is the number of food sources which is the same as the number of employed bees or onlookers [20]. For producing a candidate food position, i.e., the new solution from the old one in memory, ABC uses Eq. 9 [5].

$$v_{ij} = x_{ij} + \phi_{ij}(x_{ij} - x_{kj}) \qquad (9)$$

where, $k \in \{1, 2, \ldots, SN\}$ and $j \in \{1, 2, \ldots, D\}$ are randomly chosen; ϕ_{ij} is a random number in interval $[-1, 1]$, and x_{ij} is the food position, i.e., the solution. Here, D is the number of optimization parameters.

Each candidate solution v_{ij} is evaluated in every iteration and compared with x_{ij} to select the best solution. The nectar gets abandoned for a food source if a position cannot be improved by a predetermined number of cycles; the scouts then replace such food source with a new one. This predetermined number of cycles is called limit for abandonment [19]. For an abandoned source, x_i, the scout discovers a new food source as given by Eq. 10 [5].

$$x_i^j \leftarrow x_{\min}^j + \text{rand}[0, 1](x_{\max}^j - x_{\min}^j) \qquad (10)$$

where, $j \in \{1, 2, \ldots, D\}$. The same can be applied to solve TSP by providing initial solutions, i.e., a reference path. The artificial bees memorize the best solution. In each iteration, the solutions of the best tours are updated until the termination criterion is not satisfied.

2.5 Firefly Algorithm (FA)

FA was developed based on the idealized behaviour of the flashing characteristics of fireflies [6]. This optimization technique is based on the assumption that solution of an optimization problem can be perceived as an agent, i.e., a firefly, that glows proportionally to its quality in a considered problem setting [21]. The flashing characteristics are idealized with three rules: all fireflies are unisex and can get attracted to any other firefly; attractiveness is proportional to their brightness; and the brightness or light intensity of a firefly is affected or determined by the landscape of the objective function to be optimized [6].

In FA, the search space is explored more efficiently as each brighter firefly attracts its partners, irrespective of their sex. The light intensity, $I(r)$ varies monotonically and exponentially with the distance r as given by Eq. 11 [6].

$$I(r) = I_0 e^{-\gamma r} \qquad (11)$$

where, I_0 is the original light intensity and γ is the light absorption coefficient. A firefly's attractiveness is proportional to the light intensity seen by adjacent fireflies. Hence, this attractiveness $\beta(r)$ can be given by Eq. 12 [6].

$$\beta(r) = \beta_0 e^{-\gamma r^2} \qquad (12)$$

where, β_0 is the attractiveness at $r = 0$. The distance between two fireflies, i and j at x_i and x_j, respectively, can be given by the Cartesian distance or the l_2-norm [6]. The movement of firefly i is attracted by another more attractive, i.e., brighter firefly j, which can be given by Eq. 13 [6].

$$x_i \leftarrow x_i + \beta_0 e^{-\gamma r_{ij}^2}(x_j - x_i) + \alpha \epsilon_i \qquad (13)$$

where, the last term provides randomization with the vector of random variables ϵ_i and $\alpha \in [0, 1]$.

Formerly, FA was designed to solve continuous optimization problems. This approach can also be discretized for solving a permutation problem such as TSP [22]; initial solutions are provided and based on the distance, a firefly moves towards a brighter one and hence, updates the solution in each iteration.

2.6 Genetic Algorithm (GA)

GA first selects individuals for producing the next generation and then manipulates the selected individuals by crossover and mutation techniques to form the next generation [8]. This can also be applied for solving TSP.

The chromosomes in GA are potential solutions and can be encoded by permutation coding to represent a tour of TSP. Here, each gene in a chromosome represents the city. An initial random population of chromosomes is generated and the fitness function is evaluated. Selection, crossover and mutation are the genetic operators applied for generating new solutions in each iteration [23].

3 Simulation Parameters and Result Discussions

The simulation parameters used for experimentation are given in Table 1. The experiments have been carried out on the benchmark problems Burma14 (14 cities) [12], Bayg29 (29 cities) [13], and Att48 (48 cities) [14], and results of experimentation are shown in Table 2. In order to verify robustness of the experimentation, we have repeated each simulation for ten times. We have calculated an average of the resulted routes retrieved from these simulations as well as the best route found in our work.

Table 1 Simulation parameter specifications of various algorithms for TSP

Parameter	ACO [24]	PSO [4]	ABC [20]	FA [6]	GA [23]
Maximum iterations	500	500	500	500	500
Number of agents	40 ants	40 particles	40 bees	40 fireflies	40 chromosomes
Initial pheromone on all edges (τ_0)	0.5	NA[a]	NA[a]	NA[a]	NA[a]
Pheromone exponential weight (α)	1	NA[a]	NA[a]	NA[a]	NA[a]
Heuristic exponential weight (β)	5	NA[a]	NA[a]	NA[a]	NA[a]
Evaporation rate (ρ)	0.5	NA[a]	NA[a]	NA[a]	NA[a]
Pheromone quantity constant (Q)	1	NA[a]	NA[a]	NA[a]	NA[a]
Inertia weight (ω)	NA[a]	0.9	NA[a]	NA[a]	NA[a]
Acceleration constant (η_1)	NA[a]	2	NA[a]	NA[a]	NA[a]
Acceleration constant (η_2)	NA[a]	2	NA[a]	NA[a]	NA[a]
Number of onlookers	NA[a]	NA[a]	40	NA[a]	NA[a]
Number of scouts	NA[a]	NA[a]	1	NA[a]	NA[a]
Acceleration coefficient (ϕ)	NA[a]	NA[a]	[−1, 1]	NA[a]	NA[a]
Initial attractiveness (β_0)	NA[a]	NA[a]	NA[a]	2	NA[a]
Light absorption coefficient (γ)	NA[a]	NA[a]	NA[a]	1	NA[a]
Random variable (ϵ)	NA[a]	NA[a]	NA[a]	[−1, 1]	NA[a]
Randomization variable (α)	NA[a]	NA[a]	NA[a]	0.2	NA[a]
Selection strategy	NA[a]	NA[a]	NA[a]	NA[a]	Elitism selection
Mutation operator	NA[a]	NA[a]	NA[a]	NA[a]	Swap

[a] Not Applicable

Table 2 Comparison of average and best route (in *km*) for solving TSP

Algorithm	Route	Few of the benchmark TSPs		
		Burma14 [12]	Bayg29 [13]	Att48 [14]
ACO	Average	31.05	9274.79	35043.34
	Best	30.88	9195.22	34600.71
PSO	Average	32.34	15047.83	109979.87
	Best	30.88	14036.75	91237.09
ABC	Average	32.36	17404.62	107883.76
	Best	30.88	16658.30	101985.88
FA	Average	31.80	14283.51	81182.32
	Best	30.88	13062.40	78479.69
GA	Average	31.49	11023.70	50753.50
	Best	30.88	10018.10	46362.05

The results show that for a small number of cities, e.g., for 14 cities in our experiment, the results are comparable for identified algorithms. However, as the number of cities is increased, the performance of PSO, ABC and FA have degraded drastically. This is because PSO does not work well on optimizing a scattered problem [25]. Consequently, ABC may provide fast convergence but its search space is limited by the initial solution. FA has a limitation of being trapped into several local minima [26].

Hence, for a scattered, population-based optimization problem such as TSP, ACO provides the near-optimal solution even for a large number of cities.

4 Concluding Remarks

In this paper, we have compared ACO, PSO, ABC, FA and GA for solving TSP. We have considered few of the benchmark TSPs and have experimented with identified algorithms. Our simulation outcomes suggest that two algorithms, ACO and GA, outperform in solving TSP for considerably large number of cities.

References

1. Blum, C., & Li, X. (2008). Swarm intelligence in optimization. In *Swarm intelligence* (pp. 43–85). Berlin: Springer.
2. Beni, G., & Wang, J. (1993). Swarm intelligence in cellular robotic systems. In *Robots and biological systems: Towards a new bionics?* (pp. 703–712). Berlin: Springer.
3. Dorigo, M., & Birattari, M. (2011). Ant colony optimization. In *Encyclopedia of machine learning* (pp. 36–39). Berlin: Springer.
4. Poli, R., Kennedy, J., & Blackwell, T. (2007). Particle swarm optimization. *Swarm Intelligence, 1*(1), 33–57.
5. Tereshko, V., & Loengarov, A. (2005). Collective decision making in honey-bee foraging dynamics. *Computing and Information Systems, 9*(3), 1.
6. Yang, X.-S. (2010). Firefly algorithm, stochastic test functions and design optimisation. *International Journal of Bio-Inspired Computation, 2*(2), 78–84.
7. Hoffman, K. L., Padberg, M., & Rinaldi, G. (2013). Traveling salesman problem. In *Encyclopedia of operations research and management science* (pp. 1573–1578). Berlin: Springer.
8. Holland, J. H. (1975). Adaptation in natural and artificial systems. In *An introductory analysis with application to biology, control, and artificial intelligence* (pp. 439–444). Ann Arbor, MI: University of Michigan Press.
9. Blum, C., & Roli, A. (2003). Metaheuristics in combinatorial optimization: Overview and conceptual comparison. *ACM Computing Surveys (CSUR), 35*(3), 268–308.
10. Balas, E., & Toth, P. (1983). Branch and bound methods for the traveling salesman problem. Technical Report, Carnegie-Mellon Univ, Pittsburgh, PA, Management Sciences Research Group.
11. Held, M., & Karp, R. M. (1971). The traveling-salesman problem and minimum spanning trees: Part II. *Mathematical Programming, 1*(1), 6–25.
12. Win, Z. (1995). 14-state in Burma.
13. Groetschel, R. (1995). Juenger, 29 cities in Bavaria, geographical distances.

14. Padberg, R. (1995). 48 capitals of the US.
15. Dorigo, M., Maniezzo, V., & Colorni, A. (1996). Ant system: Optimization by a colony of cooperating agents. *IEEE Transactions on Systems, Man, and Cybernetics. Part B (Cybernetics)*, *26*(1), 29–41.
16. Fang, L., Chen, P., & Liu, S. (2007). Particle swarm optimization with simulated annealing for tsp. In *Proceedings of the 6th WSEAS International Conference on Artificial Intelligence, Knowledge Engineering and Data Bases* (pp. 206–210).
17. Kennedy, J., & Eberhart, R. C. (1997). A discrete binary version of the particle swarm algorithm. In *1997 IEEE International Conference on Systems, Man, and Cybernetics, 1997. Computational Cybernetics and Simulation* (Vol. 5, pp. 4104–4108). New York: IEEE.
18. Wang, K.-P., Huang, L., Zhou, C.-G., & Pang, W. (2003). Particle swarm optimization for traveling salesman problem. In *2003 International Conference on Machine Learning and Cybernetics* (Vol. 3, pp. 1583–1585). New York: IEEE.
19. Karaboga, D., & Akay, B. (2009). A comparative study of artificial bee colony algorithm. *Applied Mathematics and Computation*, *214*(1), 108–132.
20. Karaboga, D., & Basturk, B. (2007). Artificial bee colony (abc) optimization algorithm for solving constrained optimization problems. In *International Fuzzy Systems Association World Congress* (pp. 789–798). Berlin: Springer.
21. Łukasik, S., & Żak, S. (2009). Firefly algorithm for continuous constrained optimization tasks. In *International Conference on Computational Collective Intelligence* (pp. 97–106). Berlin: Springer.
22. Kumbharana, S. N., & Pandey, G. M. (2013). Solving travelling salesman problem using firefly algorithm. *International Journal for Research in Science & Advanced Technologies*, *2*(2), 53–57.
23. Razali, N. M., Geraghty, J., et al. (2011). Genetic algorithm performance with different selection strategies in solving tsp. In *Proceedings of the World Congress on Engineering* (Vol. 2, pp. 1134–1139). International Association of Engineers, Hong Kong.
24. Dorigo, M., Birattari, M., & Stutzle, T. (2006). Artificial ants as a computational intelligence technique. *IEEE Computational Intelligence Magazine*, *1*, 28–39.
25. Yonggang, C., Fengjie, Y., & Jigui, S. (2006). A new particle swam optimization algorithm. *Journal of Jilin University*, *24*(2), 181–183.
26. Farahani, S. M., Abshouri, A., Nasiri, B., & Meybodi, M. (2011). A Gaussian firefly algorithm. *International Journal of Machine Learning and Computing*, *1*(5), 448.

Twitter Data Sentiment Analysis on a Malayalam Dataset Using Rule-Based Approach

Deepa Mary Mathews and Sajimon Abraham

Abstract Opinion characterization is nowadays a potential and intense research focus because of the hasty growth of social media such as blogs and social networking sites, where individuals put in freely their perspectives on different themes. Researches prove that people find it comfortable to opinionate in their mother tongue, be it verbal or written. Given that now almost all social platforms support most of the popular languages, the requirement to mine the sentiments in various dialects is on the rise. However, not all data may be relevant; some may not have any impact on the end result and some may have similar meanings. A preprocessing phase is hence required to help make the dataset concise. In this paper, the authors focus on finding out the polarity of the words input by various users through their reviews exhibited using the South Indian language, Malayalam. Malayalam like the other languages in the Dravidian family exhibits the characteristics of an agglutinative language. The preprocessing process consists of cleaning the data, tokenization, stopword removal, etc. In this paper, authors are focusing on the document-based polarity calculation of the Malayalam reviews. The overall polarity of the corpus is calculated based on the positivity and negativity values of individual documents. It is found that negativity value is higher for the user reviews in our corpus which shows their negative attitude toward the news thread with the classifier accuracy of 89.33%.

Keywords Opinion mining · Sentiment analysis · Stopword removal · Malayalam · Lexicon based · Naïve Bayes · Machine learning

---

D. M. Mathews (✉)
School of Computer Sciences, Mahatma Gandhi University,
Kottayam, India
e-mail: deepamarymathews@gmail.com

S. Abraham
School of Management and Business Studies,
Mahatma Gandhi University, Kottayam, India

© Springer Nature Singapore Pte Ltd. 2019
N. R. Shetty et al. (eds.), *Emerging Research in Computing, Information, Communication and Applications*, Advances in Intelligent Systems and Computing 906,
https://doi.org/10.1007/978-981-13-6001-5_33

1 Introduction

Since English is a universal language, many research works in sentiment analysis are done in English dialect. Owing to the exponential rise in the number of users expressing their opinions in their native languages, the tools that mine English data is becoming less and less useful. In India, where there are 22 officially recognized languages and more than 132 crore people out of which a significant percentage are active Internet users, the amount of data generated on a daily basis is huge. This data, if processed efficiently, can be routed to dig out many valuable patterns like the customers' buying pattern, product feedback, and so on at micro level. However, there are not many tools available to extract these patterns. Hence, it has become imperative to perform sentiment analyses in more languages.

Our intention was to test whether we could automatically predict users' views on a particular topic from their free-text responses expressed in Malayalam language. The essence of sentiment analysis or standpoint analysis is in identifying and extracting relevant information from a given text by means of various algorithms. It determines or quantifies the attitude, opinion, or feeling toward a person, organization, product, or location. Reading the polarity of the texts and classifying them as positive, negative, or neutral will help us determine whether the underlying mood is happy, sad, angry, or excited.

The most vital indicator of sentiments is sentiment lexis. The set of these vocabularies are usually used to convey the polarity. For instance, 'നന്നായി', 'പ്രശംസിച്ചു', 'നേട്ടങ്ങൾ', 'ആകർഷണീയമായ','ആകർഷകമായ', 'സമ്പന്നമായ', 'മനോഹരമായ', 'അഭിനന്ദിക്കുക', etc., are positive Malayalam sentiment words, and മോശമായ, നിന്ദിക്കുക, അബദ്ധം, ചീത്തപറയുക, നിഷ്ഠൂരമായ, etc., are negative Malayalam sentiment words. A list of such words called as sentiment lexicons is created in-house based on the Olam dataset. The authors survey the feasible ways to analyze the user sentiments using Python programming libraries and classification algorithms.

2 Related Works

Limited works have been accounted in Malayalam so far in the area of sentiment analysis though many works have been proposed for sentiment analysis in English. preprocessing phases like stemming, POS tagging, etc., on the Malayalam reviews requires special attention as Malayalam is an agglutinative language. A lightweight Malayalam stemmer called LALITHA which is developed using suffix stripping method is introduced [1] in the year 2013. In the article, the authors demonstrate that TnT and rule-based suit combination are better for Malayalam [2]. In the year 2014, the authors proposed a rule-based approach for extracting sentiments from Malayalam film reviews [3]. The authors in their paper [4] stated that they face a lot of difficulties while dealing with Malayalam due of the morphological variations of the language. The authors in their paper [5], use PageRank method for ranking the judgment in the document. In [6], the machine learning techniques CRF combined with a rule-based approach have been used to carry out the sentiment analysis of Malayalam film reviews. An effort to accomplish aspect-based sentiment analysis in Malayalam is done by the authors in their paper [7] in the year 2016. In the paper "Domain Specific Sentence Level Mood Extraction from Malayalam Text" [8], the authors focused on a specific domain and the SO-PMI-IR formula classifies an input text into one of the two classes that indicate desirable or not desirable. The other contributions in stemming on the Malayalam reviews are depicted in [9, 10]. A hybrid approach of Max Entropy model and certain rules for handling special cases in Malayalam is used by the authors in their paper [11, 12].

3 Methodology

Machine learning algorithms can be used to find meaningful patterns in the dataset that generate insight and help us in taking better decisions and predictions. The two components of machine learning approach are preprocessing, which cleans the data, and the classification of sentiments. In this section, the process framework is discussed. The various steps to be followed are

1. Extract the reviews and build the corpus.
2. Preprocessing of the corpus.
3. Build the sentiment classification system.
4. Test the Classifier.

A. *Extract the Reviews and Build the Corpus*

The data sources are mainly Malayalam News websites, Malayalam blogs, twitter, and facebook. In order to collect a corpus, we scraped the text messages from reviews of popular Malayalam newspapers and blogs. We queried accounts of four newspapers to collect user comments. Here, the authors extracted user reviews from websites like "www.manoramaonline.com", "www.mathrubhumi.com", "www.deepika.com", etc., linked to the particular news thread "two types of passport!" using beautiful soup. Also the YouTube comments based on the same news thread in Malayalam language is collected using TAPoR 3. The reviews that are acquired are based on the news thread "രണ്ടുതരം പാസ്പോർട്ട് എന്ന തീരുമാനത്തെ....". The YouTube comments are stored in document 1, the reviews from manoramaonline.com are stored in document 2, and the comments from other dailies like mathrubhumi.com and deepika.com etc. are stored in document 3. The percentage of positive, negative, and neutral reviews of the whole corpus is shown in Fig. 1. The statistics of the dataset is shown in Table 1.

B. *Preprocessing the Corpus*

This is a key phase in the whole process where the dataset is thoroughly combed and methodically shrunk so that the size of opinionated text is reduced and performance

Fig. 1 Sentiment distribution in the training set

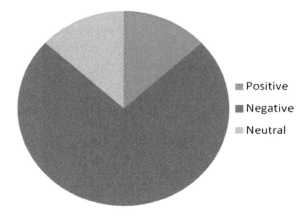

Table 1 Summary statistics of dataset

Number of reviews	136
Percentage of positive reviews	11.76%
Percentage of negative reviews	65.44%
Percentage of neutral reviews	11.76%

is augmented. This step is vital to accelerate the process while dealing with large corpora [13]. Tokenization is performed on the dataset and then the Malayalam stopwords are removed. The stopwords are removed by following a lexicon-based approach with the help of a generic list, which was created in-house, in which more than 100 Malayalam stopwords were listed.

C. *Classification of Reviews/Sentiments*

In document level sentiment analysis, the polarity or the orientation of the reviews are calculated based on the overall opinion expressed in the whole document. Naive Bayes classifiers that make use of training data is used to estimate an experiential likelihood of each class. Later, when the classifier is used on unseen data, it employs the observed probabilities to envisage the majority class for the new features. The probability of an item can be anticipated from observed data by dividing the number of audition in which an item occurred by the total number of auditions. For e.g., while evaluating the passport reviews, if 230 reviews contain the word 'സന്തോഷി' out of 1000 reviews, the probability of 'സന്തോഷി' can be anticipated as 23%. The notation P(A) is used to symbolize the probability of item A, as in P ('സന്തോഷി') = 0.23.

The total probability of all probable outcomes of an examination must always be 100%. Thus, if the test only has two conclusions that cannot happen concurrently then knowing the possibility of either conclusion tells the possibility of the other. For example, if P(negative) = 0.20, then P(positive) can be calculated as $(1 - 0.20) = 0.80$. This works because the events are mutually exclusive and exhaustive. The notation P($\leftarrow$A) can be used to denote the probability of event.

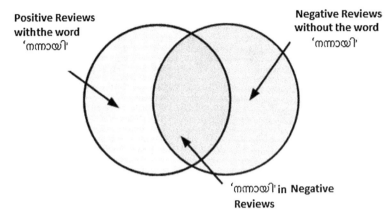

Fig. 2 Probability of സന്തോഷി in positive and negative reviews

If some events occur with the event of interest, then they may be useful for making predictions. Consider, for instance, the review message contains the word 'സന്തോഷി'. For nearly all, this word is expected to be in a positive message; its occurrence in a review is hence a very strong bit of proof that the users are very much satisfied. To enumerate the extent of overlap amid these two proportions, we need to estimate the probability P(positive ∩ സന്തോഷി). Calculating P(positive ∩ സന്തോഷി) depends on the joint probability of the two events, or how the probability of one event depends on the probability of the other. Figure 2 shows the probability of the word 'സന്തോഷി'in positive and negative reviews. If the two events are totally disparate, they are called independent events. If all events were independent, it would be impossible to predict any event using the data obtained by another. On the other hand, dependent events are the basis of predictive modeling. For instance, the appearance of the word 'സന്തോഷി'or 'ഇഷ്ടപെടു' is predictive of a positive review. For independent events A and B, the probability of both happening is $P(A \cap B) = P(A) * P(B)$.

As per Bayes' theorem, if there are two events e_1 and e_2, then the conditional probability of occurrence of event e_1 when e_2 has previously occur is given by the following formula:

$$P(e_1|e_2) = \frac{P(e_2|e_1)P(e_1)}{e_2}$$

This equation is applied to compute the likelihood of a data to be as positive or negative [4]. Table 2 shows the document-based polarity values on the corpus.

The graphical representation of these values is shown in Fig. 3. The overall polarity of the corpus is calculated based on the positivity and negativity values of individual documents. It is found that negativity value is higher for the user reviews in our corpus which shows their negative attitude toward the news thread.

Table 2 Document-based polarity values

Malayalam review document #	Percentage of positivity	Percentage of negativity	Polarity
1	35.8209	64.1791	Negative
2	48.8889	51.1111	Negative
3	42.4242	57.5758	Negative

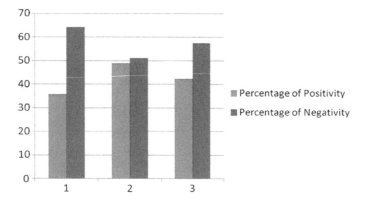

Fig. 3 Graphical representation of polarity in the review documents

4 Evaluation Measures

Table 3 stands for the lessening in the number of words after the preprocessing processes were applied on the dataset. It is apparent that the preprocessing steps performed on the Malayalam review documents lessened the number of words that needs to be analyzed by a great extent. Reduction in the number of words, in turn, helped reduced the time required for the analyses of the corpus. The table shows that nearly 20% reduction of word count is happened on the whole corpus after the preprocessing phases which may considerably reduce the processing time of analysis process. Figure 4 is the graphical representation of the same.

Classifiers are learned or trained on a finite training set. The classifier model generated after the learning phase has to be examined on different test sets. This

Table 3 Reduction in word count after preprocessing phases

Malayalam review document #	No. of unigrams in the document	No. of words after removing punctuations	No. of words after stopwords removal
1	960	950	784
2	1296	1277	1108
3	876	845	717

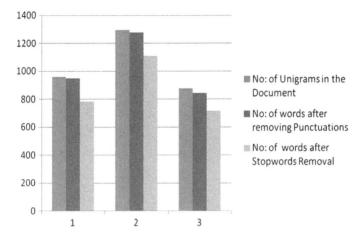

Fig. 4 Graphical representation of reduction preprocessing done on dataset

Table 4 Values of various evaluation measures

Malayalam review document #	Precision	Recall	F Score	Accuracy
1	88.88	75	81.35	85.89
2	88.88	82.75	85.7	89.33
3	73.68	63.63	68.28	71.23

classification produces four outcomes—TruePositive, TrueNegative, FalsePositive, and FalseNegative. The experimental outcomes on the test_data are alternate for the performance on unseen data. The accuracy is taken as the criterion function for assessing the classifier performance experimentally. Various evaluation metrics used in this article are Precision, Recall, F Score, and Accuracy (total number of two correct predictions (TP + TN) divided by the total number of a dataset). The Precision, Recall, F Score, and the Accuracy values calculated document-wise are shown in Table 4 and its graphical representation in Fig. 5.

5 Conclusion and Future Scope

In this paper, the authors proposed document-based and sentence level sentiment analysis on the user comments about "change in passport color" expressed in Malayalam language. This work also demonstrates the various preprocessing phases done on the Malayalam words that were extracted from reviews/opinions of users as expressed through social media platforms, comments sections of news websites, etc. List of Malayalam positive words, negative words and stopwords need to create in-house which takes lots of manual effort. The overall polarity of the corpus is calculated based on the positivity and negativity values of individual documents.

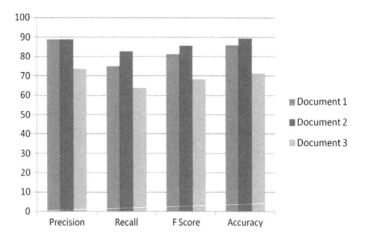

Fig. 5 Graphical representation of various evaluation measures

It is found that negativity value is higher for the user reviews in our corpus which shows their negative attitude toward the news thread with the classifier accuracy of 89.33%. Various other supervised machine learning algorithms can also be used for classification and it is a future scope.

References

1. Prajitha, U., Sreejith, C., & Raj, P. C. R. (2013). LALITHA: A light weight Malayalam stemmer using suffix stripping method. In *2013 International Conference on Control Communication and Computing (ICCC)*. IEEE 2013.
2. Jayan, J. P., Rajeev, R. R., & Sherly, E. (2013). A hybrid statistical approach for named entity recognition for Malayalam language. In *Proceedings of the 11th Workshop on Asian Language Resources*.
3. Nair, D. S., Jayan, J. P., & Sherly, E. (2014). SentiMa-sentiment extraction for Malayalam. In *Proceedings of the 2014 International Conference on Advances in Computing, Communications and Informatics (ICACCI)*. IEEE 2014.
4. Manju, K., David, P. S., & Sumam, M. I. (2017). An extractive multi-document summarization system for Malayalam news documents. https://doi.org/10.4108/eai.27-2-2017.152340.
5. Renjith, S. R., & Sony, P. (2015, June 14). An automatic text summarization for Malayalam using sentence extraction. In *Proceedings of 27th IRF International Conference*.
6. Nair, D. S., Jayan, J. P., & Sherly, E. (2015). Sentiment analysis of Malayalam film review using machine learning techniques. In *2015 International Conference on Advances in Computing, Communications and Informatics (ICACCI)*. IEEE 2015.
7. Anu, P. C., et al. (2016). Aspect based sentiment analysis in Malayalam. *International Journal of Advances in Engineering & Scientific Research*, 3(6), 2–34. ISSN: 2349-3607 (Online), ISSN: 2349-4824 (Print).
8. Mohandas, N., Nair, J. P.S., & Govindaru, V. (2012). Domain specific sentence level mood extraction from Malayalam text. In *2012 International Conference on Advances in Computing and Communications (ICACC)*. IEEE 2012.

9. Balasankar, C., Sobha, T., & Manusankar, C. (2016). Multi-level inflection handling stemmer using iterative suffix-stripping for Malayalam language. In *2016 International Conference on Advances in Computing, Communications and Informatics (ICACCI)*. IEEE 2016.
10. Pragisha, K., & Reghuraj, P. C. (2013). STHREE: Stemmer for Malayalam using three pass algorithm. In *2013 International Conference on Control Communication and Computing (ICCC)*. IEEE 2013.
11. Anagha, M., Kumar, R. R., Sreetha, K., Rajeev, R. R., & ReghuRaj, P. C. (2014). Lexical resource based hybrid approach for cross domain sentiment analysis in Malayalam. *An International Journal of Engineering Sciences, Special Issue iDravadian*.
12. Anagha, M., Kumar, R. R., Sreetha, K., & Raj, P. C. R. (2015). A novel hybrid approach on maximum entropy classifier for sentiment analysis of Malayalam movie reviews. *International Journal of Scientific Research*, Special Issue June 2015. ISSN No 2277-8179.
13. Palmer, D. D. (2000). Tokenisation and sentence segmentation. In *Handbook of natural language processing* (11–35).
14. Thulasi, P. K. (2016). Sentiment analysis in Malayalam. *International Journal of Advanced Research in Computer and Communication Engineering, 5*(1).

An IoT-Based Smart Water Microgrid and Smart Water Tank Management System

Shubham Kumar, Sushmita Yadav, H. M. Yashaswini and Sanket Salvi

Abstract Water is most important resource which needs to be managed smartly. Managing house water supply in a society consisting of water tanks, motors, and pumps automatically is an important task for efficient consumption of water. In this paper, we propose a smart solution for leakage detection in the tank using its dimensions and sensor data. The data from each house is stored on the cloud for analyzing the water consumption of each house in a society and main water supply, through GSM/GPRS 900a module. A hybrid application, Smart Water Grid, is responsible for monitoring the water level in the tank continuously, to control the motor automatically, and it consists of an inspection mode to detect the leakage in the tank and its dimension.

Keywords GSM SIM 900a · Arduino · ThingSpeak cloud · Leakage detection and dimension · Hybrid application

1 Introduction

In recent times, use of the android/hybrid apps and IoT devices have gain greater importance for daily life. In day-to-day activities, water wastage in the houses has increased and proper supply of water from main source is getting wasted. We developed two IoT devices, one for automatic water tank that senses the water tank level and stores it in cloud for analysis using GPRS module and second device, motor with an Arduino and GSM module to get automatically on and off. A hybrid application "Smart Grid" has been developed to monitor the water level, control the motor, detect leakage and gives approximate dimension of the leakage in the tank.

In this paper, Sect. 2 is for the overview of the Smart Microgrid. In this, we will see how the data is getting stored from the device to the cloud and how the stored

S. Kumar (✉) · S. Yadav · H. M. Yashaswini · S. Salvi
Department of Information Science and Engineering,
Nitte Meenakshi Institute of Technology, Bangalore, India
e-mail: shubham.kumar008@gmail.com

© Springer Nature Singapore Pte Ltd. 2019
N. R. Shetty et al. (eds.), *Emerging Research in Computing, Information, Communication and Applications*, Advances in Intelligent Systems and Computing 906,
https://doi.org/10.1007/978-981-13-6001-5_34

data is used for leakage detection and proposed system architecture. In Sect. 3, we discuss about the smart water level device and motor controlled device, how they function and provide data for automatic tank system and leakage detection. In Sect. 4, we discuss the hybrid app architecture, system flow information, and the leakage detection/dimension algorithm of the water tank. In Sect. 5, we discuss the issues faced by the Smart system and finally we will draw conclusion and future works.

2 Related Work

In [1] this project, we are presenting the idea of smart water tank management system which is operated with Arduino microcontroller. By using this microcontroller, we are preventing the manual intervention for continuous water supply.

It can be also used for other industries. The main focus of this project is to provide the optimal water distribution, and it also reduces the man power which is involved in operating the water management manually. We can easily see that many of water resources are wasted because of lack of inefficient and poor water allocation and lack of integrated water management system. Measuring the level of water manually is a big task for government and residential people.

Our project helps us to automatically measure the level of water in the tank and prevent the wastage of water resource available. We all know that water is very essential for each and every living creature in this world so, wasting water is not good for anyone. So monitoring the water management system automatically helps us to reduce the wastage of water.

The system is made using the ultrasonic sensor which will sense the accurate level of water and according to that we can smartly manage our system through the mobile app which is used by each and every person in today's world.

We are dealing with a system, i.e., water tank monitoring [2] system which has only one water tank and one pump. The procedure to implement this system is very easy. But what we have to deal with the monitoring system consists of many valve, tanks, and pumps. Managing this type of system is a challenging task with existing resource and technology available. Also we are living in an era where everything is controlled with mobile application or we can say ubiquitous computing system.

Ubiquitous computing is concept of software engineering where computing can occur at any device in any location and any time not necessarily desktop computing. This type of system is used in ships where there are multiple tanks and valves. Also the main challenge is to control and monitor this system remotely through Web application or mobile application. Here we are using mobile application to control and monitor the system remotely.

The water tank system can be loaded with bunch of sensors on all three components, i.e., tank, valve, and pump. The data from each sensor is stored in form of tables in database. The database consists of three tables. The first table is for sensor value of tank. The second and third tables consist of input and output values of sensors which control signal.

The mobile application fetches the data from database which is stored on cloud. The mobile application fetches the data and does the calculation based on tables. Then it shows the user current status of system. Also it shows the user whether the system is in safe state or the system needs some set of operation to be performed to bring the system in safe state.

The developed system solves various issues of monitoring and controlling the circulation system of ships. The sailor can remotely access the current status of tanks and valves and can perform on or off operation on a particular component in order to control the ship properly.

Leakage at thermal power plants, water distribution system, overhead tanks, boilers, etc., can occur due to aging of infrastructure and environmental conditions of the tanks which should be detected and dimensions should be analyzed [3]. Due to these problems, it is required to detect the leakage system wirelessly using the microcontroller and various sensors through which it can be made portable and can perform nondestructive techniques.

The sensors are placed to detect the humidity, temperature, pressure, and sound detection around the leakage areas of the tanks using Arduino microcontrollers and sensors.

The data is received from the sensors that are located outside the tanks to detect the humidity and temperatures in the environment; the data will be stored though the Zigbee and will be uploaded and processed with the GUI developed in the LabVIEW which will be transferring to the Web page.

The data shown on the Web page can be analyzed for the future processing and visualization. Zigbee is used for connecting the Internet and http requests to the Web pages and will be able to get and put the data through Zigbee.

The leakage with different parameters can be found out with low power source using different types of testing values.

The leakage detection in the water grids can be made automatic, and a tool to provide more realistic leakage is EPANET [4]. More experiments have to be performed with the pressure data sets for realistic leakage detection using EPANET tool.

The combination of flow and pressure features allows more significant improvement in the leakage detection and leakage dimension detection methods. At regular intervals, the temporal improvements should be done to improve the flow and pressure data sets to get the proper leakage detection methods. According to EPA reports, the leaks of average household are responsible of wastage of 10,000 gallons wasted every year, with more than 1 trillion gallons of water wasted annually nationwide. With performing frame-by-frame evaluation, the algorithm allows to perform with more accuracy in the online monitoring of the data sets of flow and pressure data.

3 Proposed Framework

In this paper, we monitor the real-time height of the water tank in each house. Before we develop the whole Microgrid system, the structure needs to be analyzed (Fig. 1).

A. Overall Architecture

The Microgrid system consists of two devices and a hybrid application. The first device, smart level device, measures the height of water tank to control the level and uploads the real-time data to cloud. Graph below shows how the data is stored in the cloud from smart level device.

Another device, motor controlled device, switches on/off the motor automatically by receiving the signal from the smart level device through GSM module. The relay is used to control the motor once they receive the signal.

We developed a hybrid application to monitor, control, and check the leakage dimension through the algorithm developed (Fig. 2).

B. Tank Unit

The Smart level device uses the ultrasonic sensor to sense the tank height and uploads the sensed data to the cloud every minute. The cloud is used for storage and analysis purpose; the GSM/GPRS module has been used configured with Arduino and UR sensor to upload the data to the cloud (Fig. 3).

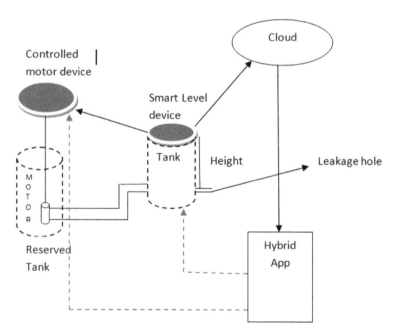

Fig. 1 Proposed smart water tank

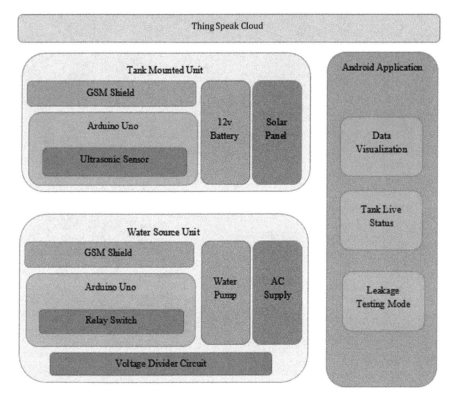

Fig. 2 Complete smart water tank block diagram

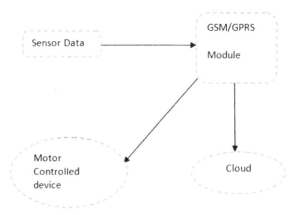

Fig. 3 Tank unit

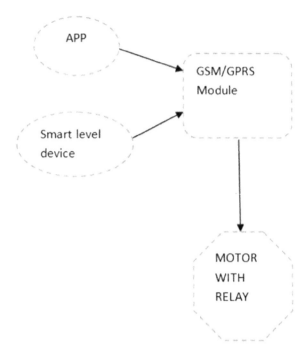

Fig. 4 Water pump unit

The smart level device sends signal to the controlled motor device to ON or OFF the motor at the specified time (when tank height reaches top and it reaches bottom) automatically (Fig. 4).

C. Water Pump Unit

The second controlled motor device uses the relay connected with a motor to get on/off automatically. The motor and relay are configured with Arduino and GSM module to receive the signal and perform the specified operation to make motor on/off.

D. Hybrid App

We decided to develop hybrid app as it combines the feature of both Web and native apps [5]. The storage area is the cloud which is used for analyzing and visualization of the sensed data. The app will be able to monitor the current level of the water tank automatically; it fetches the data from cloud with an interval continuously. The app will be able to force on or force off the motor by signaling the controlled motor device; the priority will be given to the app to control the motor.

The signal will be in the form of text SMS sent from the app to the GSM module in the controlled motor device which will automatically turn on/off the motor using the relay. The app will provide an inspection mode which will be turn on by the user occasionally in the night when they want to check if there is any leakage in the house water tank.

E. Inspection Mode

Once the inspection mode gets on, the user will not be able to use the tank over the inspection time. During the inspection mode, the app will get the start time and end time for further calculations; 6 h is the maximum time for which inspections will be on; if there is no leakage, the inspection gets automatically off. Once the inspection mode gets started, the timer starts and the calculation for leakage detection and dimension (if any found during the specified time) is performed in the app continuously.

F. Leakage Detection Method

Leakage Detection Method: Once the timer starts, the first value app fetches from the cloud it gets stored and later on every 30 s the data will be fetched from cloud and compared with the starting value if any changes occur then the level is decreasing and then the tank has leakage in it somewhere.

Leakage dimension algorithm:

We propose a new algorithm by using concepts of atmospheric pressure, physics, and Bernoulli's principle to determine the dimension of the leakage.

1. We use the physics formula $Q = AV$, where

   Q: Rate of Flow of water from the tank,
   A: Area of the leakage found,
   V: Velocity of the water flowing out from the tank.

2. First we find the Q, the provided values are (a) Difference in the level of the tank from starting the timer to the time leakage is found.
3. (b) Time difference (starting time level when it gets decreasing and time when level gets constant).
4. We will get the Q value suppose it to be X.
5. Next the velocity needs to be calculated.
6. The velocity will be found with the help of Bernoulli's principle and atmospheric pressure, the pressure needs to be calculated at the each level of the tank.
7. The velocity can be calculated by using the Bernoulli principle and atmospheric pressure; we derive at the following equation for velocity:
8. $V = \operatorname{sqrt}(2*g*h1)$.

Next using the above results, the area of the leakage will be found ($A = Q/V$); from the area of the leakage, the radius can be calculated. Using the proposed leakage detection/dimension algorithm, the approx. dimension can be found in the leakage hole (Fig. 5).

The app controls, monitors, and calculates the dimension of leakage using this algorithm and notifies the user with the required details of the leakage hole.

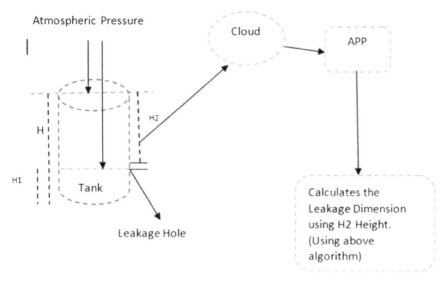

Fig. 5 Leakage detection and approx. dimension

Fig. 6 Arduino UNO

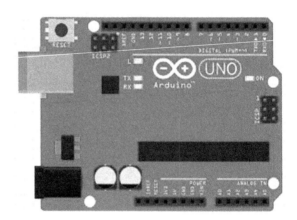

4 Implementation

A. Hardware Requirements

i. *Arduino Uno*

Arduino UNO is a board which supports ATmega328P microcontroller. It consists of 14 digital pins, a USB connection, reset button, 6 analog input, ICSP header, and 16 MHz quartz crystal. It is mainly responsible for supporting microcontroller. It has flash memory of 32 KB, SRAM 2 KB, EEPROM 1 KB. Arduino software is used for programming Arduino UNO (Fig. 6).

Fig. 7 Ultrasonic sensor

ii. *Ultrasonic sensor*

HC-SR04 ultrasonic module has four pins, ground, VCC, Trig, and Echo. The ground and VCC pins should be connected to the Arduino 5 v and ground pins, respectively. The Echo and Trig pins need to be connected to the digital pins of the Arduino to read the sensor values and to print the sensor values on the serial monitor. The distance calculated from ultrasonic sensor will be in meters(m), that can be modulated to any form of units (Fig. 7).

iii. *GSM/GPRS SIM 900A*

Global system for mobile communication has frequencies of 900/1800 MHz. AT commands can be used for setting up the frequency band. The baud rate can also be configured using AT commands from 1200 to 115,200. It provides internal TCP/IP stacks to enable Internet connections. SIM900a is compact and wireless module (Fig. 8).

iv. *Relay Board*

Relay is an automatic electrically operated switch. It is mainly used for a circuit where it is essential to control a device by a low power signal or one signal. Contractor is a type of relay that handles electric motors (Fig. 9).

v. *Submersible Motor*

A submersible pump is a device that contains closed motor attached to the pump body, basically tries to push fluid to the flat area as opposed to other pumps having to pull fluids (Fig. 10).

Fig. 8 GSM/GPRS module

Fig. 9 Relay board

Fig. 10 Submersible motor

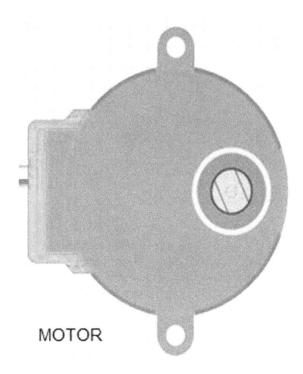

B. **Software Requirements**

i. *Arduino software*

Power up the Arduino by connecting USB cable to the computer; you will see the LED label on light up. Select the required port in the Arduino software and upload the code to the Arduino, with respect to the code the output can be seen on the serial monitor.

ii. *ThingSpeak cloud*

The ThingSpeak cloud is generally used for IoT application for storing data and retrieving data from the cloud directly to the IoT devices which have connectivity to the Internet.

The data sensed by the sensors used in the IoT devices will send to the ThingSpeak cloud using the HTTP protocol request and response methods (GET, PUT, DELETE). The data stored in the cloud is stored in the form of channels; user needs to make private or public channel to store the data from IoT devices; once they make channels, the ThingSpeak will provide API keys to read and write on the cloud.

iii. *Ionic framework*

Ionic framework is open-source framework for hybrid app development. Ionic framework is built on top of angular JS and Cordova. It provides services and tools for

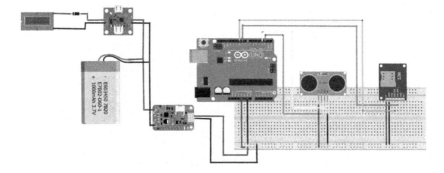

Fig. 11 Schematic diagram of sensor device

developing apps using Web technologies like HTML, CSS, angular JS, JavaScript, and Node JS framework support for server-side programming.

Ionic provides ionic creator to make UI of apps easy by using drag and drop icons which are difficult to program by using any editor.

iv. *Atom*

It is a free and open-source text editor. Atom works on different operating systems like OS X, Linux, and Windows X. A user can build their own packages and can install packages. It consists of many themes which catch user's attention and hold their interest. It makes user to understand and access the code easily by managing multiple files of a project or the whole project.

As shown in Figs. 11 and 12, ultrasonic sensor reads the level of the water in the tank and sends data through GSM/GPRS module to the cloud. For the GSM/GPRS modules, the connection has to be made properly by using the AT commands for uploading the data to the cloud in regular intervals. The issue was in connection with the cloud and to get the data at regular intervals, the TCP connections had to made first with the cloud, and then get the data with GET request in regular intervals.

The synchronizing of the timer and algorithm needs to be implemented properly in the app, proper functions/methods need to address for the checking of constant data receiving from the cloud to make the inspection mode off automatically.

The motor is controlled using relay board. Figures 13 and 14 show the circuit connection between relay and GSM/GPRS module. Figures 12 and 14 are the completed image of the circuits.

5 Result

Input: height and capacity of tank

The proposed algorithm is tested for different test cases as shown in the Table 1. In the Table 1, if there are any changes in the level of the water in the tank, then the

Fig. 12 Actual circuit

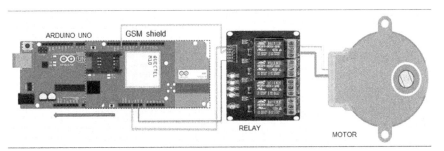

Fig. 13 Schematic diagram of motor device

leakage is detected and the dimensions of leakage are calculated for that time period given in column inspection time.

6 Conclusion and Future Work

This paper proposed and Smart Microgrid system to automatically turn on/off the house motors using the IoT devices according to the tank levels and using the cloud for analyzing the data for efficient and no wastage of water will be done and an app that controls the house motor and finds the leakage dimension in the water tank of each houses.

Fig. 14 Actual circuit

Table 1 Test cases

Test cases	Readings				Calculating leakage	
	Inspection time (minutes)	Initial level (liters)	Final level (liters)	Change in level	Rate of flow of water (l/s)	Radius of leakage (cm)
01	30	7	3.75	Yes	0.00368	0.292
02	20	5	2.95	Yes	0.001708	0.0649
03	60	6.4	6.4	No	No leakage	No leakage
04	45	5.51	3.78	Yes	0.0006407	0.04149
05	20	6.5	6.5	No	No leakage	No leakage

The proposed system includes tank automatic controlling and leakage detection/dimension; in future works, we expect to send more data to cloud and analyze it more and can generate some algorithm to find the lifetime of the tank and can get proper dimensions of the leakage.

References

1. Paul, S., Das, M., Sau, A., & Patra, S. (2015). Android based smart water pump controller with water level detection technique. *International Journal of Advanced Research in Computer and Communication Engineering, 4*(12). ISSN 2278–1021.

2. Adsul, S., Mevekari, R. G., & Sharma, A. K. (2016). Leakage detection system development. In *IEEE International conference on Automatic Control and Dynamic Optimization Technique (ICACDOT)*, 978-1-5090-2080-5/16/31.00. IEEE 2016.
3. Verma, S., & Prachi. (2012). Wireless sensor network application for water quality monitoring in India. In *National Conference on Computing and Communication Systems (NCCCS)*, Publisher IEEE. Date of Conference 21–22 November 2012. (pp. 1–5).
4. Patil, Y., & Singh, R. (2014, June). Smart water tank management system for residential colonies. *International Journal of Scientific and Engineering Research, 5*(6). ISSN 2229–5518.
5. Kwon, D., Tak, H., & Cho, H. G. (2013, February). Water tank monitoring and visualization system using smart phones. *International Journal of Machine Learning and Computing, 3*(1).

Smart Sensing for Vehicular Approach

Mukesh Chandra Sah, Chandan Kumar Sah, Shuhaib Akhter Ansari, Anjit Subedi, A. C. Ramachandra and P. Ushashree

Abstract Every day around the world, a humongous amount of people die from road accident and the subsequent injuries. There are many problems which are largely prevalent in the everyday life of a driver around the globe. Some of the techniques that are available in the market are too expensive to implement on a common vehicle. If we take a look around the common household in an Indian society, most of the people are using average cost vehicles and they are not able to afford the existing techniques which can detect the obstacle to prevent from the road accident. The survey has been conducted on the problems which are being faced by the driver at the time of driving and we have proposed a suitable and less expensive ways to implement the solutions of, not all the problems, but few of them to detect the causes of road accident by using some sensors like ultrasonic sensor, ldr sensor, ir sensor and prevention from collision. As smart-driver assistance system, invisibility problem is our main focus in this project. The concept is that it assists the driver with information and actions. In our proposed work, the smart-driver assistant system will provide the information after analyzing results of various sensors existing in the system and then if the driver is unable with actions necessary to ensure the driver's safety. Invisibility in fog is one of the major reasons of road accidents, various approaches have been made to counter this problem. We have found that ultrasonic sensor can be used to counter this problem. The sensed information is provided to the driver who takes appropriate action depending on the information. However, there are cases where the driver is incapacitated or unable or there are cases where the driver actually needs to drive faster for some urgency. In such cases, the smart-driver assistant system comes in play and slows down the vehicle for the drive, which changes their direction. If unable, the system slows the vehicle itself and if still not stopped, it stops the vehicle at 20 cm away from the obstacle. The proposed work has been tested with four parameters and found to be a better solution.

Keywords Smart-driver assistance system · LDR · IR · Ultrasonic sensor

---

M. C. Sah (✉) · C. K. Sah · S. A. Ansari · A. Subedi · A. C. Ramachandra · P. Ushashree
Department of CSE, Nitte Meenakshi Institute of Technology, Bangalore 560064, India
e-mail: mukeshsah9211@gmail.com

© Springer Nature Singapore Pte Ltd. 2019
N. R. Shetty et al. (eds.), *Emerging Research in Computing, Information, Communication and Applications*, Advances in Intelligent Systems and Computing 906,
https://doi.org/10.1007/978-981-13-6001-5_35

1 Introduction

Transportation has a key influence in the improvement of our civilization. Transportation intends to move individuals, creatures, or merchandise starting with one place then onto the next. Furthermore, it can be accomplished by different means of land (street and rail), water, air, pipeline, link, or space. Be that as it may, the most well-known and the most utilized as a part of our everyday life are street transports like transport, auto, cruiser, and so on. Road accidents are an essential stress for the both made and making the world. Diminishing these setbacks and completing directing factors for the security of drivers and individuals by walking is of most extraordinary hugeness as the decrease and flow examinations exhibit that most of the road mishaps happen in light of roadwork or obstacles or clashing speed. The essential point is that roads mishaps are dangerous and furthermore hurting to both individual lives and properties. These should be avoided. There have been various progresses to control the development prosperity and the starting late impelled driver help structures are to some degree unmistakable. There are various such unique structures to control the speed of the vehicles normally and avoid horrendous repercussions. One such system to control the vehicles speed is cruise control (CC). It keeps up the speed of the vehicle at customer reset capacity. The adaptable excursion control (AEC), which adds to CC the capacity of keeping a shielded division from the principal vehicle. A drawback of these structures is that they are not unreservedly fit for perceiving straight and twisted parts of the road; where the speed must be conveyed down to avoid setbacks [1].

2 Related Work

Smart vehicles are a thing of today, a rave among the vehicles. Regular vehicles while helpful and responsive to human info, while vital does not give the wellbeing and security of the smart vehicles which are involved numerous highlights that require no human information and help when the driver cannot and along these lines maintaining a strategic distance from the workload. Earlier investigations and examines have indicated numerous enhancements in the ideas and outline of such vehicles with keen highlights. In 1983, an entire creation four-wheel electronic hostile to slide control was presented on the Toyota Crown. In 1987, Mercedes-Benz, BMW, and Toyota delivered their first footing control frameworks. In 1990, Mitsubishi discharged the Diamante (Sigma) in Japan which highlighted another electronically controlled active trace and traction control system (the first mix of these two frameworks on the planet) that was created by Mitsubishi. It was basically named TCL in 1990. In any case, now the framework has advanced into Mitsubishi's cutting-edge Active Skid and Traction Control (ASTC) framework [2].

There were numerous early cautioning frameworks that have been endeavored as ahead of schedule as the late 1950s. For instance, Cadillac, which built up a prototype vehicle named the Cadillac Cyclone and utilized the new innovation of radar to distinguish protests in the front of the auto with the radar sensors, which was mounted inside the "nose cones". It was resolved too expensive and the model was thusly dropped. In 1995, the primary current show, for really useful forward collision avoidance, was performed by a group of researchers and architects at Hughes Research Laboratories in Malibu, California. It was financed by Delco Electronics and was going by HRL physicist Ross D. Olney. This innovation was named for promoting purposes as "Admonish". The framework [3] was radar-based—an innovation that was promptly accessible at Hughes Electronics, however, not economically somewhere else. A little custom-manufactured radar-head was created particularly for this car application at 77 GHz.

So also there have been numerous different endeavors on numerous different parts of smart vehicles like utilizing ultrasonic sensors, and cruise control for advanced driver framework. Here in this venture, we have adjusted numerous methodologies endeavored before and have attempted to make a brilliant vehicle that is smart in the sense it is simple, however, does its work without quite a bit of client input. At the beginning of autos, paces of vehicles were low and there were generally a couple of vehicles out and about. The expansion in the number of vehicles and their speed has just outpaced the enhancements on streets and the other activity offices, bringing about hazardous street travel. The principal part of night driving is the headlight which gives better vision. On the other, the hand vision of the driver is influenced by the headlights of approaching vehicles. In addition, the action moves at a speedier rate during the evening, this prompts head-on impacts. Night driving is a wonder administered by various components. So to give ideal enlightenment at all separations when vehicles are in the region, a controller is planned to such an extent that speed, separation of a vehicle from another vehicle, [4] driver activity, climate condition, kind of street (terrain, bend, highway) while driving are considered to create the compelling yield power free of glare. Since the data sources are fluffy in nature, a fuzzy controller is outlined.

Obstacle detection framework is an exceptionally reasonable framework that can be utilized as a part of moving or stationary frameworks. It can even be utilized to help outwardly weakened individuals. It is additionally relevant to anything that moves, incorporating robot controllers alongside kept an eye on or unmanned vehicles for arriving, ocean, air, and space. Obstacle detection and risk recognition—these are synonymous terms with comparable results, however, are once in a while connected in various areas; for instance, obstacle detection is normally connected to ground vehicle route, through hazard detection is regularly connected to flying machine or rocket during the time spent arriving, as in "landing hazard detection." It is a framework issue that comprises of sensors that [5] investigate the world around, world models that speak to the sensor information in a pertinent shape, numerical models of the association between objects around and the vehicle, and algorithms to process the majority of this to induce obstruction avoidance feature to the vehicle.

What it truly does is, utilize sensors both at the front and back of the vehicle guards and afterward send radar and sonar signals outward. These outward moving signals when striking against an impediment, they return back to the vehicle guards, or all the more exactly to their source. This path traveled by the waves is then estimated and the separation between the vehicles and the obstacle is considered. Once the separation between the vehicle and the object is resolved, the suitable move is then made to keep up the security of the vehicle. As a rule, the vehicles, if the question is inside the range, convey an alert to the driver who at that point acts likewise. In any case, there are [6] additionally some advanced variants of this framework where the vehicle quickly applies a brake to stop the auto before crash. Light-dependent resistors (LDRs) or photoresistors are most of the time used as a piece of circuits where it is imperative to distinguish the proximity or the level of light. They can be depicted by a grouping of names from the light-dependent resistor (LDR), photoresistor, or even photocell, photocell, or photoconductor. But unique contraptions, for instance, photo diodes or photo transistor can in similarly be used, LDRs or photoresistors are a particularly beneficial equipment fragment to use. They give considerable change in security from changes in light level.

In perspective of their minimal effort, the simplicity of fabricating, and convenience LDRs have been utilized as a part of a wide range of uses. At one time, LDRs were utilized as a part of photographic light meters, and even now they are as yet utilized as a part of an assortment of uses where it is important to recognize light levels. LDRs are extremely helpful parts that can be utilized for an assortment of light-detecting applications [7]. As the LDR resistance ranges over such a wide range, they are especially valuable, and there are numerous LDR circuits accessible past any appeared here. With a specific end goal to use these parts, it is important to know something of how an LDR functions.

2.1 Automobile Manufacturers

i. **Volvo**:

2006: Volvo's "Impact Warning with Auto Brake", created in collaboration with Mobileye, was presented on the 2007 S80. This framework is fueled by a radar/camera sensor combination and gives a notice through a head up show that outwardly looks like brake lights. On the off chance that the driver does not respond, the framework pre-charges the brakes and expands the brake help affectability to boost driver braking execution. Later forms will naturally apply the brakes to limit person on foot impacts. In a few models of Volvos, the programmed stopping mechanism can be physically killed. The V40 likewise incorporated the primary walker airbag, when it was presented in 2012.

2013: Volvo presented the primary cyclist identification framework. All Volvo vehicles now come standard with a lidar laser sensor that screens the front of the roadway, and if a potential impact is recognized, the seat straps will withdraw to decrease overabundance slack. Volvo now incorporates this security gadget as a discretionary in FH arrangement trucks.

2015: "Intelli Safe" with auto brake at crossing point. The Volvo XC90 highlights programmed braking if the driver hands over front of an approaching auto. This is a typical situation at occupied city intersections and additionally on roadways, where as far as possible are higher. [src: wikipedia]

ii. **Audi:**

2010: "Pre sense" self-governing crisis slowing mechanism utilizes twin radar and monocular camera sensors and was presented in 2010 on the 2011 Audi A8. "Pre sense in addition to" works in four stages. The framework initially gives cautioning of an approaching mischance, initiating danger cautioning lights, shutting windows and sunroof, and pretensioning front safety belts. The notice is trailed by light braking to stand out enough to be noticed. The third stage starts self-governing incomplete braking at a rate of 3 m/s^2 (9.8 ft/s^2). The fourth stage builds braking to 5 m/s^2 (16.4 ft/s^2) trailed via programmed full braking power, generally a large portion of a moment before anticipated effect. "Pre sense raise", is intended to diminish the results of backside crashes. The sunroof and windows are shut and safety belts are set up for affect. The seats are advanced to secure the auto's inhabitants. 2015 presented the "shirking partner" framework that mediates in the guiding to enable the driver to maintain a strategic distance from a snag. On the off chance that a mischance happens the "turning collaborator" screens contradicting movement when turning left at low speeds. In basic circumstance, it brakes the auto. "Multicollision brake help" utilizes controlled braking moves amid the mischance to help the driver. The two frameworks were presented on the Second era Q7. [src: wikipedia]

iii. **Ford**:

Starting on the 2012 Ford Focus, Active City Stop was offered on the range topping Titanium show, under the discretionary Sports Executive Pack. The framework utilized windscreen mounted cameras, radars, and lidars to screen the street ahead. The framework doesn't give a notice, rather, it can keep a crash happening at speeds in the vicinity of 3.6 and 30 kph. This speed was later raised to 50kph, and was accessible on all models, the Trend, Sport, Titanium, ST, and RS (Limited Edition as it were.) [src: wikipedia]

iv. **Honda**:

2003: Honda presented a self-sufficient braking (Collision Mitigation Brake System CMBS, initially CMS) front impact evasion framework on the Inspire and later in Acura, utilizing a radar-based framework to screen the circumstance ahead and give brake help if the driver responds with deficient power on the brake pedal after a notice in the instrument bunch and a fixing of the seat belts. The Honda framework was the primary creation framework to give programmed braking. The 2003 Honda framework additionally joined an "E-Pretensioner", which worked in conjunction with the CMBS framework with electric engines on the safety belts. Whenever initiated, the CMBS has three cautioning stages. The principal cautioning stage incorporates capable of being heard and visual notices to brake. In the event that disregarded, the second stage would incorporate the E-Pretensioner's pulling on the shoulder bit of the safety belt a few times as an extra material cautioning to the driver to make a move. The third stage, in which the CMBS predicts that an impact is unavoidable, incorporates full safety belt slack takeup by the E-Pretensioner for more powerful safety belt insurance and programmed use of the brakes to reduce the seriousness of the anticipated crash. The E-Pretensioner would likewise work to diminish safety belt slack at whatever point the brakes are connected and the brake help framework is actuated.) [src: wikipedia]

3 Algorithms

INPUT: List of sensor implemented in the vehicle sense the information/data from the environment and transmits to the microcontroller.
OUTPUT: The microcontroller gives the perfect commands to the motor driver for actuation.

1: Initialize the PIN of Arduino UNO board for the list of sensors and D.C motor.
2: Command is given by a smartphone via Bluetooth to control the vehicle in the Forward, backward Left and Right directions. Here smart phone is used in place of driver to control the vehicle.
3: if command equal to "F" and obstacle detected in the way then,
   Target distance between the vehicle and obstacle will be calculated as given bellow:
   a. Ping-Time = Ping-Time/1,000,000*3600;
   b. Target-distance = speed of sound * Ping-Time;
   c. Target-distance = Target-distance/2;
   d. Target-distance = Target-distance * 63,360*2.54//Target-distance is converted into cm.
4: if Target-distance >40 cm then,
   Vehicle starts moving smoothly and message will display in the LCD display as "No obstacle in the Way".
5: if Target-distance will be >20 cm and <=40 cm then,
6: Speed of the vehicle reduced slowly and message will display in the LCD display as "obstacle detected in the way (e.g. 33 cm) ahead from the vehicle, Drive slowly".
7: if Target-distance will be equal to 20 cm.
8: Vehicle stops automatically with warning sound and message will display in the LCD display as "obstacle is too nearer to the vehicle".
9: if any objects with high Intensity detected with in/equal to 40 cm form the vehicle then,
10: Intensity of a Vehicle Headlight becomes dim and becomes in the original state after crossing the objects. (Object may be static or dynamic).
11: if command equals to "R" then,
12: Vehicle starts moving in the reverse direction.

13: if command equals to "TL" then,
14: Vehicle turns in the left direction.
14: if command equals to "N" then,
15: Vehicle STOP.
16: if LDR status equals to ON then, Head Light TURN ON,
17: Else, Head Light TURN OFF.

Note: Here distance is kept in cm according to the ability of components used in our project. For real-time implementation, the distance can be adjusted.

4 Proposed Implementation

4.1 Block Diagram

- The power supply is a rechargeable battery source which supplies power to the DC motor and different components of the vehicles like the microcontroller and sensors.
- The motor driver is an IC which receives power from the battery and signal from the Arduino, microcontroller, to move the four wheels of the vehicle.
- Arduino is a microcontroller device which along with its own processor and ROM supports various peripherals like the sensors, buzzers, and Bluetooth.
- Sensors represent the three different sensors used in this project. LDR sensor, IR sensor, and ultrasonic sensor for performing variously proposed functionalities of the system.
- Bluetooth module is a master/slave configured device used for controlling the vehicle with a remote handheld controlling device.
- LED is a small light which reflects the action taken by the IR sensor and LDR sensor.
- A buzzer is a sound device which gives a warning sound when certain criteria are met by the vehicle.
- LCD is a displaying device that shows the information provided by the various vehicular components (Fig. 1).

4.2 Adaptive Headlight

The central part of night driving is the headlight which gives better vision. On the other, the hand vision of the driver is influenced by the headlights of approaching vehicles. Also, the activity moves at a quicker rate around evening time, this prompts head-on impacts. Night driving is a marvel administered by various variables. So to give ideal light at all separations when vehicles are in the region, a controller

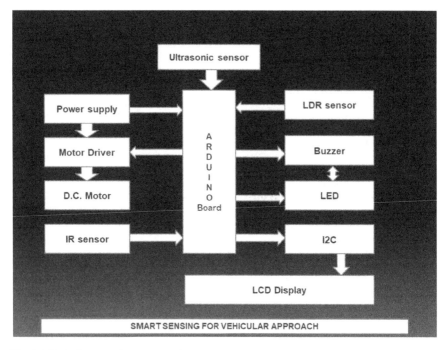

Fig. 1 Block diagram of the proposed work

is planned with the end goal that speed, separation of a vehicle [8] from another vehicle, driver activity, climate condition, kind of street (landscape, bend, highway) while driving are considered to create the successful yield force free of glare. Since the sources of info are fluffy in nature, a fluffy controller is composed.

Application:

When driving on twisting streets during the evening, amid dusk, or in other low-light conditions, adaptive headlights are very helpful. Or maybe they can be necessities on occasion. They can be utilized as a part of the accompanying circumstances and situations.

- A stray creature amidst the street, during the evening.
- Sudden passageway of a vehicle in your path while turning.
- Hill peaking on a restricted and bowed street, we can not affirm if a vehicle is coming or not.
- The headlights may turn the approaching rider bling amid turning.

4.3 Electronic Stability

Accessible in numerous new autos, this innovation enables drivers to keep up control of their vehicle amid extraordinary controlling moves by keeping the vehicle headed in the driver's planned course, notwithstanding when the vehicle nears or surpasses the breaking points of street footing.

Electronic stability control (ESC) causes drivers to keep away from crashes by diminishing the threat of slipping or losing control because of overdirecting. ESC ends up dynamic when a driver loses control of their auto [9]. It utilizes PC-controlled innovation to apply singular brakes and helps bring the auto securely back on track, without the threat of fish-following.

Australian research demonstrates that ESC decreases the danger of:

- Single auto collisions by 25%
- Single 4WD crashes by 51%
- Single auto collisions in which the driver was harmed by 28%
- Single 4WD crashes in which the driver was harmed by 66%*

4.4 Obstacle Detection

Obstacle detection framework is an exceptionally common sense framework that can be utilized as a part of moving or stationary frameworks. It can even be utilized to help outwardly disabled individuals. It is additionally material to anything that moves, incorporating robot controllers alongside kept an eye on or unmanned vehicles for arrive, ocean, air, and space. Hindrance recognition and peril discovery—these are synonymous terms with comparable results, however, are at times connected in various regions; for instance, impediment identification is generally connected to ground vehicle route, while danger location is regularly connected to flying machine or shuttle during the time spent arriving, as in "landing risk identification." It is a framework issue that comprises of sensors that examine the world around, world models that speak to the sensor information in an applicable shape, scientific models of the cooperation between objects around and the vehicle, and calculations to process the greater part of this to gather deterrent evasion highlight to the vehicle.

What it truly does is, utilize sensors both at the front and back of the vehicle guards and afterward send radar and sonar flags outward. These outward moving signs when strike against a deterrent, they return back to the vehicle guards, or all the more accurately to their source. This way went by the waves is then estimated and the separation between the vehicles and the question is resolved. Once the separation is resolved, suitable move is then made to keep up the security of the vehicle. Much of the time, the vehicles, if the protest is inside the range, convey an alert to the driver who at that point demonstrations in like manner. In any case, there are likewise some propelled variants of this framework where the vehicle promptly applies brake to stop the auto before crash.

4.4.1 Forward Collision Avoidance

Forward collision avoidance frameworks are a dynamic security include that cautions drivers in case of an unavoidable frontal crash. At the point when the FCW framework prepared vehicle comes excessively near another vehicle before it, a visual, capable of being heard, or potentially material flag jumps out at alarm the driver to the circumstance.

FCW frameworks are likewise alluded to as "Pre-safe Braking," "Collision Warning with Auto-Brake," "Pre-Crash Warning Systems," "Collision Mitigation Braking System," "Predictive Forward Collision Warning," and different names. The capacity and limits of these frameworks can fluctuate incredibly, notwithstanding a typical general objective to keep a forward impact. To discover the name of the particular framework in your vehicle, counsel your proprietor's manual or the dealership [9].

How it can help you avoid a crash
FCAT can be successful by:

- Warning the driver of a potential crash danger while going along an expressway or in rush hour. On the off chance that the driver does not react to the notices, the framework can lessen the vehicle's speed.
- Emergency halting if a walker strolls before it
- Preventing a stationary vehicle from driving forward into the back of another stationary vehicle.

Different innovations have been created for FCAT frameworks, varying in:

- Detection go
- Responsiveness.

4.5 Light-Dependent Resistor

Light-dependent resistors (LDRs) or photoresistors are frequently utilized as a part of circuits where it is important to distinguish the nearness or the level of light. They can be portrayed by an assortment of names from lightward resistor, LDR, photoresistor, or even photograph cell, photocell, or photoconductor. Albeit different gadgets, for example, photodiodes or photograph transistor can likewise be utilized, LDRs or photoresistors are an especially advantageous hardware segment to utilize. They give substantial change in protection from changes in light level. In perspective of their minimal effort, simplicity of make, and usability LDRs have been utilized as a part of a wide range of uses. At one time, LDRs were utilized as a part of photographic light meters, and even now they are as yet utilized as a part of an assortment of uses where it is important to distinguish light levels. LDRs are extremely helpful segments that can be utilized for an assortment of light-detecting applications [10]. As the LDR protection fluctuates over such a wide range, they are especially valuable, and there

Table 1 Reading of ultrasonic sensor

Distance	Actions
0–20	Vehicle will stop automatically
21–40	Speed will reduced by 40 rpm
>=41	No obstacle in the way

Table 2 Reading of LDR sensor

LDR status	LED status	Distance (cm)
300	ON	0–10
500–800	ON	25–35
>=800	ON	>35

are numerous LDR circuits accessible past any appeared here. Keeping in mind the end goal to use these parts, it is important to know something of how a LDR functions.

5 Results

The vehicle is successfully able to move in all the directions, and the integration of various components is quite successful too. We successfully connect the LCD display to the microcontroller to display the warning messages. And as a driver, we are able to control the vehicle with smart phone via Bluetooth. Along with the driver assistant, we have implemented smart sensing through sensors to provide smart features.

5.1 Test Cases

Description

- Table 1 explains the working of the ultrasonic sensor in which it determines the action with respect to the distance calculated or given. When the vehicle is between 0 and 20 cm then the vehicle will stop, between 20 and 40 cm the speed gets slowed down and greater than 40 cm the vehicle will be running normally.
- Table 2 explains the working of LDR sensor that determines the LDR status along with the LED status with respect to the distance of the light from the LDR. On 0–10 cm, the resistance is between 300Ω and the LED is ON, between 25 and 35 cm with nearly 500–800Ω the LED will glow, similarly for the distance greater than 35. On minimum distance between 0 and 10, the LED will be OFF.
- Table 3 determines the LED by the IR sensor. Obstacle on less than 40 cm, the LED will glow with less intensity but distance over 40 cm the LED will glow with high intensity depending on day and night.

Table 3 Reading of IR sensor

Distance (cm)	LED status
Less than equal to 40	Light intensity becomes LOW
Greater than 40	Light intensity becomes HIGH

Fig. 2 Detecting obstacle in fog (low visibility)

Fig. 3 Obstacle detected by the vehicle 22.75 cm ahead

5.2 Snap Shot

See Figs. 2, 3, 4, 5.

Fig. 4 Vehicle stopped because the distance between obstacle and vehicle is less than 20 cm

Fig. 5 Vehicle state after demo

6 Conclusion

Road accidents and crashes are now facing an all-time high number in modern India, and there are various reasons for this—both natural and situational. And to counter this increasing toll of lost lives, we have tried to present a smart-sensing vehicle with Arduino as its microcontroller and DC motor for actuation of the vehicle. We have considered some major reasons for road accidents like weather conditions, night driving, or vehicle's loss of control over its wheels. The project focuses on adaptive headlights which can alter the level of the brightness of the headlights as per timing of the day and light by sensing the amount of light it is receiving. The obstacle detection feature is probably the most helpful as it measures the distance between the obstacle and the vehicle and helps to take proper action.

The advancement in road transportation and its decreasing cost is enabling people to take upon the automobiles and thus the increased traffic is a major reason for the number of people dying on the roads for vehicle accidents. Thus, with this project, we have called upon engineers to take up action to save those lives. While this project gives a rudimentary prototype of where the vehicles have reached in regard to the safety of vehicles, further methods can be developed to help with this aim.

References

1. Kimura, Y., Naito, T., & Ninomiya, Y. (2010). Detection of LED light by image processing for the visible light communication system. *Transaction Society of Automation and Engineering Japan, 41*(4), 889–894.
2. Bachir, A., & Benslimane, A. (2003). A multicast protocol in networks inter-vehicle geocast. *Vehicular Technology Conference*.
3. Caballero-Gil, P., Caballero-Gil, C., Molina-Gil, J., & Fúster-Sabater, A. (2010). On privacy and integrity in vehicular ad hoc networks. In *Proceedings of the International Conference on Wireless Networks (ICWN'10)*, Las Vegas, USA, July 2010.
4. Caballero-Gil, P., Hernández-Goya, C., & Fúster- Sabater, A. (2009, July). Differentiated services to provide efficient node authentication in VANETs. In *Proceedings of the International Conference on Security and Management SAM-WorldComp2009*, pp. 184–187, Las Vegas, Nevada, USA.
5. Barghi, S., Benslimane, A., & Assi, C. (2009, June). A lifetime-based routing protocol for connecting VANET's to the internet. In *2009 IEEE International Symposium on a World of Wireless, Mobile and Multimedia Networks Workshops. Wow, Mom*.
6. Reddy, B. M., Anumandla, K. K., Tiwari, V. K. (2017, January 05–07). Optimization of smart vehicle ad hoc network (SVANET) communication for traffic related issues with a security. In *2017 International Conference on Computer Communication and Informatics (ICCCI -2017)*, Coimbatore, India.
7. Al Ridhawi, I., Aloqaily, M., Karmouch, A., & Agoulmine, N. (2009). A location-aware user tracking and prediction system. In *Global information infrastructure symposium, GIIS'09*.
8. Aloqaily, M., Kantarci, B., & Mouftah, H. T. Vehicular clouds: State of the art, challenges and future directions. In *2015 IEEE Conference on Applied Electrical Engineering and Computing Technologies (AEECT)*.
9. Aloqaily, M., Kantarci, B., & Mouftah, H. T. (2017). Fairness-aware game theoretic approach for service management in vehicular clouds. In *IEEE 86th Vehicular Technology Conference*.

10. Srinivasan, R., Sharmili, A., Saravanan, S., Dr., Jayaprakash, D. Smart Vehicles with Everything. In *2016 2nd International Conference on Contemporary Computing and Informatics (ic3i)*.

Android Malware Detection Techniques

Shreya Khemani, Darshil Jain and Gaurav Prasad

Abstract Importance of personal data has increased along with the evolution of technology. To steal and misuse this data, malicious programs and software are written to exploit the vulnerabilities of the current system. These programs are referred to as malware. Malware harasses the users until their intentions are fulfilled. Earlier malware was major threats to the personal computers. However, now there is a lateral shift in interest toward Android operating system, which has a large market share in smartphones. Day by day, malware is getting stronger and new type of malware is being written so that they are undetected by the present software. Security parameters must be changed to cope up with the changes happening around the world. In this paper, we discuss the different types of malware analysis techniques which are proposed till date to detect the malware in Android platform. Moreover, it also analyzes and concludes about the suitable techniques applicable to the different type of malware.

Keywords Android malware · Static analysis · Hybrid analysis · Detection techniques · Dynamic analysis

1 Introduction

Cybersecurity is a very challenging task which is becoming cumbersome day by day. We live in the era of smartphones and laptops, which has made our life easier in certain ways but also it has resulted in making all our information and data more vulnerable. Cybercrimes are increasing at a rapid rate because the vulnerability has increased largely over the time.

According to a survey conducted by We are Social [1], 66% of the total population use mobile phones. Moreover, Statista [2] reported, out of these 66% people, 47% own a smartphone, and out of these 47% people, 87% use Android OS in their smartphones. That means, two out of every seven people use Android platform in

S. Khemani (✉) · D. Jain · G. Prasad
Department of CCE, Manipal University Jaipur, Jaipur 303007, India
e-mail: Shreya.khemani16@gmail.com

© Springer Nature Singapore Pte Ltd. 2019
N. R. Shetty et al. (eds.), *Emerging Research in Computing, Information, Communication and Applications*, Advances in Intelligent Systems and Computing 906,
https://doi.org/10.1007/978-981-13-6001-5_36

Fig. 1 Total malware samples found in mobile sector

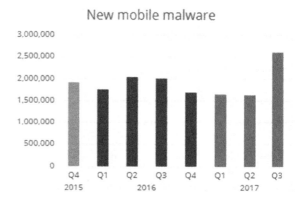

Fig. 2 New malware samples found quarterly

their smartphones. This is one of the reasons that attract malware writers to attack on this particular smartphone operating system.

Kaspersky Lab reported that nearly 40 million attacks by malicious mobile software in 2016 [3]. McAfee Labs reported that new malware rose by 60% from the previous quarter in 2017 on Android platform [4]. Figure 1 shows the total malware samples quarter by quarter since 2015. The total malware is increasing quarter by quarter and thus increasing the chances of increase in attacks.

Figure 2 depicts the amount of new malware found. New malware found in 2017 Q3 is very high, this means writers are finding new ways on how getting the malware undetected is increasing. This scenario urges us to increase the security of all platforms to keep it's users safe from these malware. Hence, this malware must be detected and removed for the betterment of users and proper functioning of Android OS.

In order to detect this malware, the methods proposed are static analysis, dynamic analysis, and hybrid analysis. The remainder of the paper is organized as follows: Sect. 2 discusses different types of malware detection techniques; Sect. 3 presents the concluding remarks.

2 Malware Analysis

2.1 Static Analysis

Static analysis tests and evaluates an application by examining its source code, without executing it. This analysis checks even those parts of the code, which may not get invoked during execution.

Permission-Based Approach

These different types of malware are harmful to the systems, and hence, permissions are inculcated in the Android system to ensure that that application gets access to limited components and data of the smartphone. A user has the privilege of choosing whether to install the application on the basis of the permissions requested by it. All the permissions that are required by an app are mentioned in the AndroidManifest.xml file. The manifest file contains inherent information about the application, the information that the system must have before it can run any of the application's code.

Seth et al. [5] proposed malware detection using Android permission. Apktool is used to extract the manifest file from an apk file. After extraction of permissions, a feature vector is created for each application based on 35 permissions. Once the feature vector is created, the accuracy of malware detection is evaluated by applying three supervised machine learning algorithm namely K-Nearest Neighbor (KNN), Decision Tree, and Support Vector Machine (SVM). Supervised machine learning learns a function that maps an input to an output, based on already present input–output pairs. KNN is a simple nonparametric algorithm that stores all obtainable cases and classifies new cases on the basis of similarities. A decision tree is a map of the viable outcomes of a series of interconnected choices. These can effectively deal with large, complicated datasets without imposing a complicated parametric structure (Fig. 3).

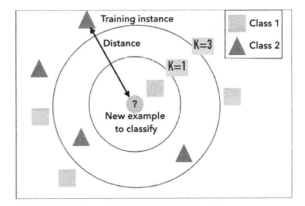

Fig. 3 kNN Algorithm

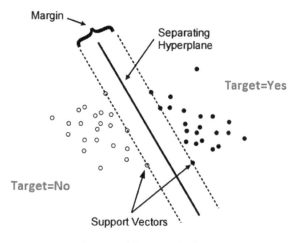

Fig. 4 Support vector machine

Support Vector Machines are based on the concept of hyperplanes that define decision boundaries. A hyperplane is one that divides the sets of objects having unrelated class memberships. In other words, the algorithm results into an optimal hyperplane which classifies new instances. For observation, a total of 90 applications were taken into account, out of which 60 applications were used as training data, and 30 applications were used as test data. Decision tree classifier was claimed to obtain the best results, with an accuracy of 93.33% (Fig. 4).

Xing et al. [6] proposed a two-layered permission-based detection. They analyzed xml file and extracted features such as requested permissions, requested permission pair, used permissions, used permission pair as their entities. They have extracted 20,548 benign apps and 1136 malicious apps as training set, and 800 benign apps and 400 malicious apps as test set. A dataset comprises of a training set and a test set. A training is built to formulate a model, while a test set is to validate the model built. Thereafter, to evaluate the performance of the proposed method, the following standard metrics such as true positive rate, false positive rate, precision, and accuracy are taken into account. True positive rate measures the proportion of positives that are correctly identified. False positive rate measures the proportion of negatives that are correctly identified. Precision is the fraction of relevant instances among the retrieved instances, while accuracy is the fraction of relevant instances that have been retrieved over the total amount of relevant instances. Both precision and accuracy are, therefore, based on the measure of relevance. Hereafter, to classify benign and malicious applications, they have used Weka which is a collection of machine learning algorithms for data mining tasks. It contains tools for data preprocessing, classification, regression, clustering, association rules, and visualization. At last, they have concluded that permission-based model can be used as the first step of malware detection, but it certainly requires further detection to correctly determine whether the app is malicious or not.

Shahrier et al. [7] proposed Latent Semantic Indexing (LSI). It is an analysis technique through which the concepts are produced by analyzing the relationship between the set of documents and the words that are contained in those documents. LSI works on a concept that the words whose meanings are close should occur in similar pieces of text. To implement this technique, a matrix is computed where rows are a set of permission keywords and columns are set of documents. The set of short listed aberrant applications (documents) are identified and common set of permissions are obtained from their respective categories. Further, singular-value decomposition (SVD) is applied to reduce the matrix and get relevant data as the outcome. Therefore, if permissions of an application significantly match with the common set of relevant data, then the app is marked as an anomalous application, which then requires dynamic analysis to detect its behavior. This approach has the potential to discover new anomalous applications and holds scope for future work. It works on knowledge-based analysis. It continuously detects attacks and can notify the user about them.

Signature-Based Approach

Faruki et al. [8] proposed AndroSimilar. The malware that is crafted using code obfuscation technique and repackaging are detected using AndroSimilar. A variable length signature is generated using this tool for the application which is under observation. The signature is matched with the AndroSimilar malware database to identify the app as malicious or benign on the basis of matched percentage. A total of 24,441 applications were taken for testing, out of which 15,993 apps were taken from Google Play Store, 5139 apps were taken from third-party App Store, and 3309 were malicious apps. As a result, 76% of the samples were detected correctly. To increase the efficiency, the database of signatures should be large enough so that there are sufficient entries to match the known malware signature. The limitation it holds is that this method cannot detect unknown malware.

2.2 Dynamic Analysis

It is the type of analysis where the application or the program is executed and it's behavior, interaction with the system is monitored. It is implemented in an isolated environment such as a virtual box, a sandbox so that it doesn't infect the actual machine. Static analysis cannot detect new or obfuscated malware since it examines the code and not the nature of it. However, dynamic analysis is very effective in detecting them.

In [9], Egele presented a comprehensive overview of different dynamic malware analysis, techniques, and tools for categorizing benign and malware apps. Dynamic analysis approach is potent against obfuscation techniques such as metamorphic and polymorphic still; it requires more resources.

Anomaly Detection

It is based on monitoring the behavior of the device and then classifying it as malicious or benign.

Iker et al. [10] proposed Crowdroid, a behavior-based malware detection system. Applications are inspected on the basis of system calls made. Strace [6] tool is used to collect these Linux kernel system calls. Also, crowdsourcing is used to obtain the non-personal behavior-related data of each application user use and then this data is passed onto a remote server. A system call vector with respect to user's interaction with the applications is created. Then, each dataset is clustered using the partial clustering algorithm. Moreover, k-means algorithm is used at the server end, where the value of $k = 2$, as an input parameter to determine the nature of the application as benign or malicious. As a result, Crowdroid is capable of distinguishing applications which have the same name and version. Crowdroid uses deep analysis. However, it requires the installation of Crowdroid client application to perform detection, and it also gives incorrect results if legitimate app invokes more system calls.

Portokalidos et al. [11] proposed Paranoid a platform where complete malware analysis can be performed with the help of mobile phone replicas in cloud. The exact mobile phone replicas need to be executed in a secure virtual environment which limits the number of applications running to not greater than 105 simultaneously. Paranoid is preferred for detecting continuous attacks like worm infection, DoS, whereas it reduces the battery life of the device by 30%.

Shabtai et al. [12] proposed Andromaly, a framework for anomaly detection. It constantly monitors features and patterns that indicate device status for instance battery, etc., while it is executing. It works on knowledge-based analysis. It continuously detects attacks and can notify the user about them. Andromaly also detects continuous attacks. Although, only four self-created malware was used for testing and depletes the battery very quickly.

Taint Analysis

It determines what type of data is sent where. Enck et al. [13] proposed TaintDroid, an information flow tracking system. It keeps track third-party applications that use sensitive data like location of device, camera, microphone data. Data is assigned a taint marking (or a label) indicating the data type. When the app is tainted, data leaves the system at taint sink. It informs the users if their data is compromised. It provides efficient tracking of sensitive information but does not perform control flow. It ends it tracking, the moment data leaves the device. When the tracked data contains configuration identifiers, it causes significant false positives. Taintdroid is an effective method for tracking private information. However, it does not track control flows and the information that leaves the device and returns as a network reply.

Emulated Analysis

Amos et al. [14] proposed STREAM, a System for Training and Evaluating Android Malware. It runs on a single server. It is an automation framework and simulates user input to trigger malicious activity. Also, it implements random fuzz testing. Android's monkey is used to generate random stream of events as an input to an application. However, it may not be that accurate as compared to a real user input. The information collected is about battery consumption, binder, memory, network, and permissions. Many machine learning algorithms were used to distinguish. However, Bayes net classifier was the one with suitable results.

2.3 Hybrid Analysis

It is the combination of both dynamic and static analyses. This analysis consists of merging both static features obtained analyzing the application and dynamic features obtained from executing the applications. The accuracy of the detection rate can be increased using this approach.

This analysis consists of merging both static features obtained analyzing the application and dynamic features obtained from executing the application. The accuracy of the detection rate can be increased using this approach.

Yang et al. [15] proposed a two-stage malware detection for Android platforms using hybrid analysis. Static analysis is used to extract permission features, triggering mechanism, software's package feature, and component feature. Moreover, dynamic analysis is being performed simultaneously to extract the behavior characteristics of the app. DroidBox [16] a sandbox tool is used for performing dynamic analysis. MonkeyRunner script is used to simulate user's clicks and system events. Features extracted from dynamic analysis will be based on dynamic loading behavior, services started, transmitting and retrieval of network data, reading and writing file, opening the network connection, closing the network connection, sending text messages, making calls, recording type of encryption algorithm, and encryption keys. The common permissions of a malicious software along with 22 dangerous rights in Android software permissions provided by Google in it's developer documentation are used in feature vector. From the component feature, only the service and broadcast receiver component, 10 triggering events which malware usually listens to are also used in feature vector. For dynamically extracted features, dynamic behavior monitoring uses 13 fields. In the two-stage classification, the first stage uses classifiers to determine the malicious nature of the software. Furthermore, on the second stage, random forest-based multi-classifier to determine the family which the malware belongs to. Drebin [3] is used for malware samples. Support Vector Machine, Random Forest, Naïve Bayes, Decision Tree J48, Logistic Regression are the machine learning algorithms used in stage one. The criteria for evaluating these algorithms will be on the basis of true positive rate (also known as recall rate), false positive rate, precision, F-Measure, area under curve. Eigenvectors of all samples containing the static and dynamic characteristics are stored in a local file. 256-dimensional eigenvector is used. Moreover, 10-fold cross-validation is used to divide the dataset. As a result, random forest has the best optimal performance with accuracy of 95.9% and recall rate of 95.1%. In the second stage, all the benign samples of the first stage are removed and use random 70% of malware for the second stage. Therefore, random forest was able to classify the families with the accuracy of 94.8% and recall rate 95%.

3 Conclusion

Smartphones have become an important part of our day-to-day life, which is why attackers are targeting this sector. With Android's open-source platform and significant market share, it's easy for attackers to exploit the vulnerabilities of this OS. Techniques are devised to create a more secure environment in order to prevent damages. This paper briefly discusses different techniques and it's categories for prevention and detection of malware.

References

1. *Digital in 2017: Global overview.* March 2018, URL: https://wearesocial.com/special-reports/digital-in-2017-globaloverview.
2. *Android-statistics and facts.* March 2018. URL: https://www.statista.com/topics/876/android/.
3. *Mobile malware evolution 2016.* Retrieved March 2018 from https://securelist.com/mobile-malware-evolution-2016/77681/.
4. *McAfee labs threat report December 2017.* Retrieved March 2018 from https://www.mcafee.com/us/resource/reports/rp-quartely-threats-dec-2017.pdf.
5. Seth, R., Kaushal, R. (2015). Permission based malware analysis and detection in android.
6. Liu, X., Liu, J. (2014, April). A two-layered Permission-based android malware detection scheme. In *Proceedings of 2nd IEEE International Conference on Mobile Cloud Computing, Services, and Engineering*, Oxford, UK (pp. 142–148).
7. Shahriar, H., Islam, M., Clincy, V. (2017). Android malware detection using permission analysis. In *Proceedings of Southeast Con, Charlotte, NC, USA*.
8. Faruki, P., Laxmi, V., Bharmal, A., Gaur, M. S., & Ganmoor, V. (2015). AndroSimilar: Robust signature for detecting variants of android malware. *Journal of Information Security and Applications, 22*, 66–80.
9. Egele, M., Scholte, T., Kirda, E., & Kruegel, C. (2012). A survey on automated dynamic analysis tools and techniques. *ACM Computing Surveys, 44*(2), 1–42.
10. Burguera, I., Zurutuza, U., & Nadjm-Tehrani, S. (2011). Crowdroid: Behaviour-based malware detection system for android. In *Proceedings of the 1st ACM Work. Security and Privacy in Smartphones and Mobile Devices-SPSM'11* (p. 15).
11. Portokalidis, G., Homburg, P., Anagnostakis, K., Bos, H. (2010).Paranoid android: Versatile protection For smartphones. In *Proceedings of the 26th Annual Computer Security Applications Conference, ASCAC'10* (pp. 347–356).
12. Shabtai, A., Kanonov, U., Elovici, Y., Glezer, C., & Weiss, Y. (2011). Andromaly: A behavioral malware detection framework for android devices. *Journal of Intelligent Information Systems*, 1–30.
13. Enck, W., Gilbert, P., Chun, B. G., Jung, J., McDaniel, P., & Sheth, A. N. (2010). Taintdroid: An information flow tracking system for real-time privacy monitoring on smartphones. In Osdi'10 (Vol. 49, pp. 1–6).
14. Amos, B., Turner, H., & White, J. (2013). *Applying machine learning classifiers to dynamic android malware detection at scale* (pp. 1666–1671).

15. McLaughlin, N., Del Rincon, J. M., Kang, B., Yerima, S., Miller, P., Sezer, S. ... Ahn, G. J. (2017). Deep android malware detection. In *Proceedings of CODASPY, Scottsdale, Arizona, USA.*
16. Demontis, A., Melis, M., Biggio, B., Maiorca, D., Arp, D., Rieck, K. ... Roli, F. (2017). Yes, Machine learning can be more secure! A case study on android malware D=detection. *IEEE Transactions on Dependable and Secure Computing.*
17. Idrees, F., Rajarajan, M., Conti, M., Chen, T. M., & Rahulamathavan, Y. (2017). PIndroid: A novel android malware detection system using ensemble learning methods. *Computer and Security, 68,* 36–46.
18. Wei, F., Li, Y., Roy, S., Ou, X., & Zhou, W. (2017). Deep ground truth analysis of current android malware. In *International Conference on Detection of Intrusions and Malware, and Vulnerability and Assessment (DIMVA)* (pp. 252–276).

A Comparative Study of Machine Learning Techniques for Emotion Recognition

Rhea Sharma, Harshit Rajvaidya, Preksha Pareek and Ankit Thakkar

Abstract Humans share emotions which they exhibit through facial expressions. Automatic human emotion recognition algorithm in images and videos aims at detection, extraction, and evaluation of these facial expressions. This paper provides a comparison between various multi-class prediction algorithms employed on the Cohn-Kanade dataset (Lucey in The extended Cohn-Kanade dataset (CK+): a complete dataset for action unit and emotion-specified expression, pp. 94–101, 2010 [1]). The different machine learning algorithms can be used to provide emotion recognition task. We have compared the performance of K-nearest neighbors, Support Vector Machine, and neural network.

Keywords Emotion recognition · Machine learning · Cross-validation · Performance analysis

1 Introduction

The process of recognizing human emotions from pictures and facial expressions can be performed almost instantaneously by humans but due to the generation of a large amount of data, automated tools are required to perform emotion recognition task. Recognizing emotions is crucial in many applications for which the system requires to gain a deeper understanding of a particular subject. The categorization of emotions is also an important aspect in the field of classification and is widely used in

---

R. Sharma · H. Rajvaidya · P. Pareek (✉) · A. Thakkar
Institute of Technology, Nirma University, Ahmedabad 382481, Gujarat, India
e-mail: 14bit015@nirmauni.ac.in

R. Sharma
e-mail: 14bit044@nirmauni.ac.in

P. Pareek
e-mail: preksha.pareek@nirmaunni.ac.in

A. Thakkar
e-mail: ankit.thakkar@nirmauni.ac.in

© Springer Nature Singapore Pte Ltd. 2019
N. R. Shetty et al. (eds.), *Emerging Research in Computing, Information, Communication and Applications*, Advances in Intelligent Systems and Computing 906,
https://doi.org/10.1007/978-981-13-6001-5_37

fields where customer relationship is paramount. Machine learning-based methods are widely used for classification. This paper presents an analysis of different machine learning approaches of multi-class classification using Cohn-Kanade Dataset (CK+) [1].

For an automatic emotion recognition task, preprocessing techniques generally falls under four main categories: knowledge-based, feature invariant, template matching, and appearance-based [2]. The knowledge-based method is multi-resolution rule-based method which provides the rules from the knowledge base which is prepared by the experts from the field [3]. Feature invariant method performs preprocessing using the grouping of edges using perceptual grouping using spatial feature points [4]. For template matching statistical local features such as Local Binary Pattern is used which performs texture analysis [5]. In the appearance-based method, invariant features (under different lighting and background conditions) for faces are identified [6].

In this paper, we have used an appearance-based feature which is robust against different lighting and background conditions. Euclidean distance between the left and right eye, the mouth and the angle between the eyes provide useful information about the appearance of a face [7]. Dimensionality reduction of these facial features is performed using Principal Component Analysis (PCA).

2 Methodology of Emotion Recognition

Emotion classification can be performed using intuitive learning such as Kekule-Archimedean learning. It derives from the Archimedean moment of discovery, known as the Eureka moment. For automated emotion classification, this task will be tedious and time-consuming.

Machine learning techniques are popular for multi-class classification [8]. The problem statement for emotion recognition requires solving multi-class classification problem which uses an algorithm that works well with both, supervised learning, and multi-class classification [8]. In this paper, list of methods that are reviewed includes K-Nearest Neighbor (K-NN) [9], Support Vector Machines (All-vs-All), Support Vector Machines (One-vs-Rest), neural networks (All-vs-All) and neural networks (One-vs-Rest) [10]. The working of these methods is as follows (Fig. 1).

In K-NN method, each sample from the dataset is treated as an n-dimensional point where n is the number of the features for each sample. This algorithm finds k closest points to a given point that represents the k classes into which the data is to be classified.

SVM is a widely used algorithm for multi-class classification which finds optimal hyperplane to separate the classes [11]. For experimentation with Cohn-Kanade dataset [1], we have used 39-dimensional vectors (number of input features) each corresponding to the Action Units (AU) along with their intensity. The dataset [1] has 7 class labels (Number of output classes available with the dataset). Thus, so

A Comparative Study of Machine Learning Techniques ... 461

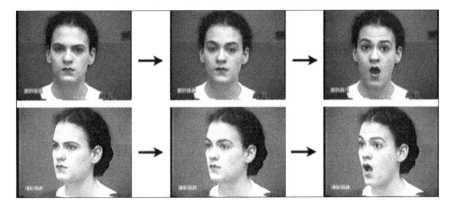

Fig. 1 Frontal and 30° views from the Cohn-Kanade dataset (taken from [1])

we cannot implement the generic SVM binary classifier. Hence, we compared the performance of the One-vs-Rest method and All-vs-All [12].

Neural networks are widely used in many industrial applications as they offer extremely efficient and highly accurate models to use in machine learning problems [8]. These networks are modeled analogously to the neurons and their connections through synapses in the human brain. A node in an artificial neural network is called perceptron and is similar to the neuron in biological terms. The output of each node is assigned a weight, which denotes the importance of a particular node to the overall output. Weights are initialized by assigned random values. The purpose of a neural network training algorithm is to adjust these weights in such a way so as to guarantee a prediction closest to the expected output [13].

Neural network constructed using AUs as nodes is shown in Fig. 2 (taken from [7]). Each node works by creating different combinations of the inputs by computing a nonlinear function of the sum of all its inputs. Many such nodes constitute a layer,

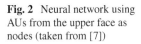

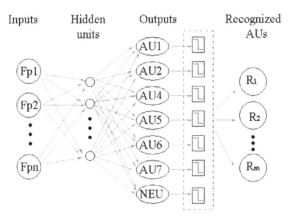

Fig. 2 Neural network using AUs from the upper face as nodes (taken from [7])

Table 1 Impact of learning rate, number of nodes in hidden layers, activation function, and optimizer on Cohn-Kanade dataset

Learning rate	Number of nodes in 3 hidden layer	Activation function	Optimizer	Accuracy (in %)
Adaptive	(42, 22, 22)	tanh	adam	98.6975058
Adaptive	(42, 22, 22)	tanh	sgd	98.6298673
Adaptive	(42, 22, 22)	ReLu	adam	98.6102597
Adaptive	(42, 22, 22)	ReLu	sgd	98.0153405
Adaptive	(40, 30, 20)	tanh	adam	98.5044672
Adaptive	(40, 30, 20)	tanh	sgd	98.1788945
Adaptive	(40, 30, 20)	ReLu	adam	98.6013195
Adaptive	(40, 30, 20)	ReLu	sgd	97.5406853
Constant	(42, 22, 22)	tanh	adam	98.4185531
Constant	(42, 22, 22)	tanh	sgd	98.3261710
Constant	(42, 22, 22)	ReLu	adam	98.9255603
Constant	(42, 22, 22)	ReLu	sgd	97.7478002
Constant	(40, 30, 20)	tanh	adam	98.6115917
Constant	(40, 30, 20)	tanh	sgd	98.19916806
Constant	(40, 30, 20)	ReLu	adam	98.68408955
Constant	(40, 30, 20)	ReLu	sgd	98.22798670

which sends its output to the next layer. The final output is computed and the error is sent back to the nodes. To accommodate this difference in output, the weights at each node may be changed. Choosing the number of layers and nodes is almost always an iterative hit and trial method. A higher number of nodes would lead to an overfitted model that will not generalize well while a smaller number of nodes may give a model with lower accuracy.

The effect of number of nodes in hidden layers, learning rate (adaptive and constant == 0.0001), and activation function is shown in Table 1.

ReLu is a linear activation function and it gives better accuracy with neural networks as compared to sigmoid functions.

Dataset used

The Cohn-Kanade Dataset ($CK+$) [1] contains a huge amount of subjects' faces going through a range of emotions. It uses a combination of features called AU to make the distinction between different emotions (Happiness, Sadness, Anger, Contempt, Disgust, Neutral, Surprise, and Fear) (see Fig. 1). A large number of actors are used for this purpose to obtain diversity in the image samples. An AU is a combination of Facial Action Coding (FAC) units [7] and is an extremely popular method of identifying facial features in the study of emotions [14].

Table 2 Accuracy of different classifiers

Classifier	Accuracy (in %)
K-nearest neighbors	45.36
All-vs-All SVM classifier	82.56
RBF Kernel SVM classifier for each emotion	93.49
Polynomial Kernel SVM classifier for each emotion	89.16
SVM with RBF as Kernel (with $k = 3$)	94.23
Neural network with linear activation function	98.92
Neural network with sigmoid activation function	98.41
Neural network (All-vs-All)	93.27

3 Experimental Setup and Result Discussions

One-vs-Rest OR All-vs-All: In One-vs-Rest, the classifier is built for each and the rest of the classes. However, this can lead to imbalanced classifiers [15]. In such cases to incorporate regularization, cross-validation can be used on the data. Stratification is performed to make the folds for an accurate representation of the distribution of all different emotions in the actual dataset.

We have conducted experimentation on Cohn-Kanade dataset [1]. The accuracy of different methods is summarized in Table 2. The SVM method in *nltk* [16] creates an All-vs-All classifier by default and using it was quick but gave an accuracy of 82.56%. Similarly, the multilayer perceptron also gave a reduced accuracy of 93.27% as compared to a similar One-vs-Rest classifier with an accuracy of 98.92%.

4 Conclusion

In this work, we have performed the emotion recognition task using machine learning techniques. We have analyzed the performance of ANN, SVM, and K-NN. The solution provided by a neural network with linear activation function provides the best accuracy on Cohn-Kanade database for the identified set of experiments. However, the performance of the classifier may vary subject to the other datasets.

References

1. Lucey, P., Cohn, J. F., Kanade, T., Saragih, J., Ambadar, Z., & Matthews, I. (2010). The extended Cohn-Kanade dataset (CK+): A complete dataset for action unit and emotion-specified expression. In *2010 IEEE Computer Society Conference on Computer Vision and Pattern Recognition Workshops (CVPRW)* (pp. 94–101). New York: IEEE.

2. Hemalatha, G., & Sumathi, C. (2014). A study of techniques for facial detection and expression classification. *International Journal of Computer Science and Engineering Survey, 5*(2), 27.
3. Yang, M.-H., Kriegman, D. J., & Ahuja, N. (2002). Detecting faces in images: A survey. *IEEE Transactions on Pattern Analysis and Machine Intelligence, 24*(1), 34–58.
4. Yow, K. C., & Cipolla, R. (1997). Feature-based human face detection. *Image and Vision Computing, 15*(9), 713–736.
5. Bartlett, M. S., Movellan, J. R., & Sejnowski, T. J. (2002). Face recognition by independent component analysis. *IEEE Transactions on Neural Networks/a Publication of the IEEE Neural Networks Council, 13*(6), 1450.
6. Turk, M. A., & Pentland, A. P. (1991). Face recognition using eigenfaces. In *IEEE Computer Society Conference on Computer Vision and Pattern Recognition, 1991. Proceedings CVPR'91* (pp. 586–591). New York: IEEE.
7. Tian, Y.-I., Kanade, T., & Cohn, J. F. (2001). Recognizing action units for facial expression analysis. *IEEE Transactions on Pattern Analysis and Machine Intelligence, 23*(2), 97–115.
8. Donalek, C. (2011). *Supervised and unsupervised learning.* In *Astronomy Colloquia* USA.
9. Cunningham, P., & Delany, S. J. (2007). k-nearest neighbour classifiers. *Multiple Classifier Systems, 34*, 1–17.
10. Ding, C. H., & Dubchak, I. (2001). Multi-class protein fold recognition using support vector machines and neural networks. *Bioinformatics, 17*(4), 349–358.
11. Durgesh, K. S., & Lekha, B. (2010). Data classification using support vector machine. *Journal of Theoretical and Applied Information Technology, 12*(1), 1–7.
12. Guo, B., Gunn, S. R., Damper, R. I., & Nelson, J. D. (2008). Customizing kernel functions for SVM-based hyperspectral image classification. *IEEE Transactions on Image Processing, 17*(4), 622–629.
13. Anand, R., Mehrotra, K., Mohan, C. K., & Ranka, S. (1995). Efficient classification for multi-class problems using modular neural networks. *IEEE Transactions on Neural Networks, 6*(1), 117–124.
14. Ekman, P., & Rosenberg, E. L. (1997). *What the face reveals: Basic and applied studies of spontaneous expression using the Facial Action Coding System (FACS).* Oxford: Oxford University Press.
15. Rifkin, R., & Klautau, A. (2004). In defense of one-vs-all classification. *Journal of machine learning research, 5*(Jan), 101–141.
16. Bird, S., & Loper, E. (2004). NLTK: The natural language toolkit. In *Proceedings of the ACL 2004 on Interactive poster and demonstration sessions* (p. 31). Association for Computational Linguistics.

IoT-Based Smart Parking System

G. Abhijith, H. A. Sanjay, Aditya Rajeev, Chidanandan, Rajath
and Mohan Murthy

Abstract Post liberalization, Indian cities are growing at an exponential growth rate. The rapid growth of the towns is giving birth to many socioeconomic problems. With the increase in the number of personal vehicles and shrinking parking spaces, the problem of parking vehicles at wrong parking spaces is steadily increasing which causes home and business establishment owners a lot of discontents, time wastage, and unnecessary chaos. By adopting latest technologies, the parking issue can be addressed more smartly. In this work, we have designed, developed, and tested an IoT-based smart parking solution. We have conducted rigorous testing of our solution in real life under various circumstances and observe that our approach provides a practical solution to the wrong parking issue.

Keywords IoT · Smart parking · Raspberry Pi

1 Introduction

In the last 25 years, Indian automobile industry has seen significant growth. An average of 25 lakhs cars sold in India every year [1]. At the same time, the population of the urban areas is increasing at an exponential growth rate. The increase in people per square kilometer and the number of vehicles has resulted in the scarcity of parking spaces. Due to the limited legal parking spaces, sometimes people tend to park their cars at others parking spaces (wrong parking space). This can happen knowingly or unknowingly, in either case, parking a vehicle at the wrong parking space creates a lot of nuisance and frustration to the actual owners of the parking spaces. This problem is getting severe day by day and is bound to become worse in the future. Due to the limited personnel of the traffic law department, they are unable to cope up with this increase in the wrong parking of the vehicles in front of homes, government offices, and business establishments. Since there is no sign of reduction in the growth rate of population and number of vehicles, the coming days will be worst.

G. Abhijith · H. A. Sanjay · A. Rajeev · Chidanandan · Rajath · M. Murthy (✉)
Department of ISE, Nitte Meenakshi Institute of Technology, Bengaluru, India
e-mail: motgharemm@rknec.edu

© Springer Nature Singapore Pte Ltd. 2019
N. R. Shetty et al. (eds.), *Emerging Research in Computing, Information, Communication and Applications*, Advances in Intelligent Systems and Computing 906,
https://doi.org/10.1007/978-981-13-6001-5_38

We have conducted a survey based on a quantitative questionnaire to get a better understanding of the problems faced by building owners and traffic authorities. Following are the survey questions:

1. Have you faced an issue of illegally parked vehicles near your house/establishment?
2. How often do you face this issue?
3. How does it affect your daily routine/business?
4. How easy is it to report these vehicles to the authorities?

88.9% of people participated in the survey confirmed that they had faced wrong parking issue. 25% of persons reported they are facing the issue every week, and 12.5% of persons reported they are facing the issue once in a month. Most of the responses to the third question talk about the time waste and unnecessary anxiety, and there were few responses which talk about the loss of business by a blocking vehicle, where people cannot see shops. 55.6% of the persons said it is difficult to report such incidents to the corresponding authorities, rest of them have not reported such incidents yet. The result of our study confirmed the issues resulted by wrong parking of a vehicle.

To address the wrong parking of four-wheelers, we proposed a smart parking system, which helps to manage the parking problems more efficiently. The system first sounds an alarm when any vehicle is parked in the wrong parking zone and clicks an image of the vehicle's number plate. We have used a Raspberry Pi camera to capture the image of the wrongly parked vehicle; this number plate image is sent to a server running Automatic License Plate Recognition (ALPR) software which processes the number plate details and sends these details to the registered client's mobile application. The client can choose to report this incident to the corresponding authorities or white list the vehicles if required. The mobile application also offers different options to control the various aspects of the outdoor unit. We have conducted various experiments by setting up our system in different real-life circumstances and observe that our system helps in minimizing the no-parking issues.

Rest of the paper is organized as follows. Section 2 sheds light on the existing literature. Section 3 explains the proposed system in detail. Section 4 briefs implementation. Section 5 gives details on the experiments conducted and result obtained followed by a conclusion.

2 Related Work

The currently available systems are either sensor based [2–4] or RFID based [5–7], and most of them are focused toward finding a vacant parking spot or determining the vacancy status of the parking lots. There are no many solutions which address the above-mentioned issues. The work [8] uses features such as color, edges, and size of the plate for character extraction. But most of the papers argue that one of the major reasons for inaccurate results is the character extraction phase [8–11]. Hence, there is a pressing need to invest an extended amount of time and effort on research activities on increasing the accuracy of the system by trying to improving the extraction phase [8, 10]. A major problem of using color features is that they are not powerful enough due to the vastly varying license plate colors [8, 12]. Some papers used edge detection to segment and partition the license plate to obtain significant information [8, 11, 12], but the use of edge detection is highly dependent on background noise and thus inaccuracies tend to creep in during plate extraction phase [12, 13]. Another solution tried to use a mix of morphological procedures and adjusted Hough changes approach for plate extraction [13], but it concentrates mainly on the characters and not the license plates fringes. This approach had some major drawbacks like requirement of high calculation time, huge memory space, and not being reasonable progressive application [13, 14].

3 Proposed System

The proposed system consists of three layers as shown in Fig. 1.

Each of these layers handles a specific set of tasks.

Outdoor Unit (Layer 1): The outdoor unit comprises the components that are placed in the physical location where we intend to place our system. The outdoor unit is instrumental in sensing the presence of vehicles in the no-parking zone; once any vehicle is sensed, a sound clip warning is played and an image of the vehicle's license plate is captured. These images are converted to gray scale and sent to the server along with the time details.

Server (Layer 3): The grayscale image is pre-processed to reduce background noise, and only the license plate region is selected. Now character segmentation process separates the different characters of the number plate and the region to

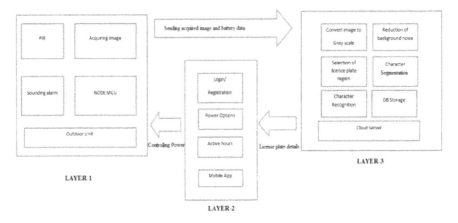

Fig. 1 Proposed system

which the vehicle is registered. The individual characters are recognized and stored in a local database before being sent to the client application.

Client Application (Layer 2): The registered building owner will get push notifications on his mobile device indicating a wrongly parked vehicle in the wrong parking zone. The client can choose to white list the license plate, report the incident to the authorities, or not do anything. The application can be used to set up the working hours of the device, and it will also indicate the battery level of the outdoor unit.

We have used the following hardware to build our system.

Raspberry Pi: The Raspberry Pi is a miniature-size single board computing device, which can be used to interface the various sensors and other devices to the Internet. We use the Raspberry Pi as it is small in size, independent, extensible (software support), and economical.

PIR Sensor: The passive infrared sensor is a commonly used sensor for motion detection. It works on the principle that all objects emit energy in the form of IR radiations, and the PIR sensor detects the changes in this IR radiation to detect motion.

Node MCU: Node MCU is an open-source, programmable, low cost, and Wi-Fi enabled Arduino-like hardware device. Lua scripts are used to code the programs on the ESP8266. The Node MCU is used to control the power of the Raspberry Pi.

Wi-Fi Module: The 802.11 Wi-Fi module is used to connect the Raspberry Pi with the Internet. Instead of the Wi-Fi module, we can also make use of an Ethernet cable. The Wi-Fi module is preferred in our project as it is a mobile.

The connections between the different components are as shown in Fig. 2.

We have used Ionic mobile app framework to develop the mobile app. Ionic is an open-source software development kit for hybrid mobile application development. Ionic is built using Apache Cordova and Angular JS.

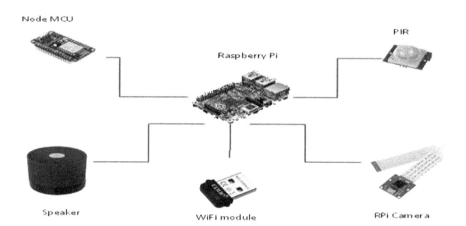

Fig. 2 Connections between the components

4 System Implementation

The pins of the PIR sensor are connected to the GPIO pins of the Raspberry Pi. The camera module is attached to the Raspberry Pi camera slot. When the PIR sensor detects movement, the camera module captures the image and transmits the image to the server. The power settings of the Pi are controlled by the mobile app through the Node MCU.

The steps that take place to obtain the license plate from the acquired image in the server are:

A. Image acquisition and conversion to gray scale

The outdoor unit has a PIR sensor that detects for any movement in its coverage area, when the sensor is triggered after a warning sound clip; an image of the vehicle is clicked. This captured image is resized and converted to gray scale to increase the efficiency of ALPR and to reduce the amount of data being transmitted to the server. Algorithm for grayscale conversion is given below.

Algorithm 1.1 Algorithm for Grayscale Conversion

```
Algorithm Grayscale_conversion(r,g,b,n):

i=0
red = pixel.red
green = pixel.green
blue = pixel.blue
gray = (red * 0.4 + green * 0.6 + blue * 0.10)
for i=0 to m do
    pixel.red = gray
    pixel.blue = gray
    pixel.blue = gray
end for
```

B. Reduction of background noise and selection of license plate region

The grayscale image consists of a lot of noise that might interfere and reduce the efficiency of the ALPR algorithm; thus, these background noises have to be removed. Once the noise is removed, the image is scanned to identify and select only the region of the image corresponding to the license plate of the vehicle. We are referring the official template of the license plate [15] to identify the license plate and contents in it. The rest of the area is discarded, and selected region is enhanced and based on the dimensions of the selected region, the state to which the license plate is registered is found out.

C. Character segmentation and character recognition

Now the license plate region is split into the individual characters and fed to the recognition algorithm. The correct separation of characters in the segmentation process is essential to get the most accurate results from the character recognition algorithm; hence, each of the separated characters is flanked by dark pixels. Pattern matching is used to identify each of these individual segmented characters. Initially, each of the segmented characters are skeletonized by using edge detection algorithms and then matched with a predefined input set of characters to obtain the correct character. Once the license plate details are extracted, this data is sent to the client's mobile application where the license plate number is matched with a white list database maintained on the mobile application; if there is a match, then the image is discarded else a warning message is sent to the client via push notifications.

5 Setup and Results

We have designed and implemented an experimental setup of the system at the college campus using Raspberry Pi, Node MCU, and PIR sensors.

Table 1 discusses the performance and efficiency of our system for a total of 90 different license plates from different states of the country. From the tabulated data, we can observe that each step has excellent success rate. Higher the success rate, greater the accuracy.

Table 2 demonstrates the efficiency of our system during daytime and nighttime; from the tabulated data, we can understand that our system gives more accurate results at daytime when there is more natural light and lesser reflection on the license plate.

Table 3 represents the effect of reflection on the number plates. The light reflections and illuminations on the license plate are treated as white pixels and are not detected

Table 1 Accuracy of the proposed system

Step #	No. of inputs	No. of successful outputs	Success rate (%)
Reduction of noise	90	86	95.55
Selection of license plate region	86	82	95.34
Character segmentation	82	81	98.78
Character recognition	81	76	93.82

Table 2 Accuracy of different steps based on time of the day

Image type	No. of images	Noise reduction	Region selection	Segmentation	Recognition
Daytime	64	64	63	63	61
Nighttime	26	22	19	18	15

Table 3 Effect of reflection on license plate

Image type	No. of images with reflection	Noise reduction	Region selection	Segmentation	Recognition
Daytime	12	12	11	11	11
Nighttime	20	16	14	14	12

by the character recognition algorithm which causes slight reduction in the accuracy during nighttime.

From the experiments and results, we can conclude that our system detects the vehicles and recognizes the license plates in different circumstances with a decent accuracy.

6 Conclusion and Future Work

Increased vehicles and reduced parking spaces in urban areas are resulting in unwanted wrong parking incidents. Our proposed system aims to reduce such incidents by alerting the registered user about wrong parking in front of their property. The system also gives a warning sound to the impeaching driver beforehand to avoid rising a traffic violation ticket. The experimental tests prove that our system consistently gives success ratios of above 95% in the four steps after image acquisition, i.e., noise reduction, selection of license plate region, character segmentation, and character recognition. The system identifies the license plates of the vehicles (four-wheelers) in various circumstances with a decent accuracy.

In future, we will be supporting the two-wheeler and three-wheeler plates. We are also planning to enhance our system so that it can be placed in public no-parking zones, and wrongful parking is directly notified to the traffic enforcement authorities.

References

1. https://www.financialexpress.com/economy/indians-buy-25-lakh-cars-per-year-but-only-24-lakh-people-earn-over-rs-10-lakh/487838/.
2. Moon, J.-H., & Ha, T. K. (2013). A car parking monitoring system using wireless sensor networks. *International Journal of Electrical Robotics, Electronics and Communications Engineering, 7*(10).
3. Wang, H., & He, W. (2011). A reservation-based smart parking system. In *The First International Workshop on Cyber-Physical Networking System* (pp. 701–706). New York: IEEE.
4. Yang, J., Portilla, J., & Riesgo, T. (2012). Smart parking service based on wireless sensor networks. In *IECON 2012-38th Annual Conference on IEEE Industrial Electronics Society*. New York: IEEE.
5. Pala, Z., & Inanc, N. (2007). Smart parking applications using RFID technology. In *RFID Eurasia, 2007 1st Annual*. New York: IEEE.

6. Rahman, M. S., Park, Y., & Kim, K.-D. (2009). Relative location estimation of vehicles in parking management system. In *11th International Conference on Advanced Communication Technology, 2009. ICACT 2009* (Vol. 1). New York: IEEE.
7. Shang, H., Lin, W., & Huang, H. (2007). Empirical study of parking problem on university campus. *Journal of Transportation Systems Engineering and Information Technology, 7*(2), 135–140.
8. Ibrahim, N., Kasmuri, E., Jalil, N., Norasikin, M., & Salam, S. (2013). License plate recognition (LPR): A review with experiments for Malaysia case study. *The International Journal of Soft Computing and Software Engineering, 3*(3), 83–93.
9. Sulaiman, N., Mohammad Jalani, S., Mustafa, M., & Hawari, K. (2013). Development of automatic vehicle plate detection system. In *2013 IEEE 3rd International Conference on System Engineering and Technology*, Shah Alam (pp. 130–135).
10. Khalifa, O., Khan, S., Islam, R., & Suleiman, A. (2007). Malaysian vehicle license plate recognition. *The International Arab Journal of Information Technology, 4*(4), 359–364.
11. Ng, H., Tay, Y., Liang, K., Mokayed, H., & Hon, H. Detection and recognition of Malaysian special license plate based on SIFT features.
12. Ng, S., & Choong, F. (2013). Automatic car-plate detection and recognition system. In *EURECA, 2013* (pp. 113–114).
13. Ganapathy, V., & Lui, W. A Malaysian vehicle license plate localization and recognition system. *Systemic Cybernetic and Informatics, 6*(1), 13–20.
14. Soon, C., Lin, K., Jeng, C., & Suandi, S. (2012). Car number plate detection and recognition system. *Australian Journal of Basic and Applied Sciences, 6*(3), 49–59.
15. http://www.htp.gov.in/Annexure-I.pdf.

Smart Agricultural Monitoring System Using Internet of Things

H. V. Asha, K. Kavya, S. Keerthana, G. Kruthika and R. Pavithra

Abstract India is one of the largest agricultural countries with a population of 1.3 billion. Farming in India is labor intensive and absolute. 70% of India's residents are dependent on farming, and one-third of nations' funds come from agriculture. Even after decades of cultivation practice, it is lagging behind in maximizing the yield thereby hampering the progress of the nation. In order to overcome this, there is a need for promoting cultivation practice for high yield of crops. With the availability of IT and internet, Internet of Things is proliferating at an unprecedented rate. The perception of agricultural IoT (Internet of things) utilizes networking equipment in farming construction. The hardware part of this project includes processors with data processing capability and sensors which are used to measure various parameters like temperature, humidity, and water level. In this paper, the sensor node is designed to monitor the environmental conditions that are vital for the proper growth of crops. The collected data received are analyzed for proper monitoring and improving the yield of the crop. The result depicts the data being stored and retrieved on Agri Cloud (https://en.wikipedia.org/wiki/Internet_of_things [1]).

Keywords IoT · Sensors · Agri Cloud · Smart agriculture

1 Introduction

The IoT [2] is one of the ever-growing technologies contributing to the smarter world. IoT is a giant network of people and things which includes many bodily strategy, vehicles, domicile appliance, software, actuators, and sensors. The network-established connectivity enables these interconnected objects to connect and exchange data [3]. Today pioneering and ground-breaking appeal are being developed in IoT—smart cities, smart farming, automation home, connected cars [4], efficient industries, etc. Nevertheless, the impact of IoT in agriculture is spell bound. The Internet of Things

---

H. V. Asha (✉) · K. Kavya · S. Keerthana · G. Kruthika · R. Pavithra
Nitte Meenakshi Institute of Technology, Bangalore, India
e-mail: Asha.hv@nmit.ac.in

© Springer Nature Singapore Pte Ltd. 2019
N. R. Shetty et al. (eds.), *Emerging Research in Computing, Information, Communication and Applications*, Advances in Intelligent Systems and Computing 906,
https://doi.org/10.1007/978-981-13-6001-5_39

is remodeling the agribusiness generation by engaging ranchers and producers to manage the difficulties they confront each day [5].

Indian economy is principally in view of farming, and the climatic conditions are isotropic. Farming has been a high-chance, work concentrated, low-remunerate industry. Ranchers are probably going to be affected by startling natural changes, financial downturns, and numerous other hazard factors [6]. One of the solutions to this problem is smart farming by modernizing the traditional methods of farming. The Internet of Things has opened up to a great degree beneficial approaches to develop soil with the utilization of modest, simple to introduce sensors and a plenitude of quick information they offer. IoT-based smart farming is offering high-precision crop control, useful data collection, and automated farming techniques [2]. In brilliant cultivating, a framework is worked for observing the product field with the assistance of sensors to measure various parameters such as light, humidity, temperature, soil dampness, and so on and mechanizing the water system framework. With this framework the farmer is now able to screen the field conditions from anyplace and whenever [7]. IoT based cultivating is very productive when compared with regular approach. According to recent survey [6], India hosts population of 1.3 billion among which around 190.7 million people stay hungry on a daily basis. Every year the crop yields are lower when compared to countries like USA, Europe, and China. India is continuously facing challenges like starvation of financial resources, continued neglect by the government [8], weather patterns, and water availability. The solution for all these agriculture-related difficulties is to augment the conventional practices with technology and implement automation into agriculture. IoT in smart farming helps in collection of data and monitoring, and the gathered data can be further used to improvise the next cycles.

In this paper, we propose the use of sensors and microcontroller to monitor the environmental conditions that are vital for the growth of crops with an aim to maximize farming yield while maintaining quality. The reminder of this paper is organized as follows. In Sect. 2, the literature work related to the proposed system is summarized. Section 3 gives the details of the system implementation that constitutes different sensors, monitoring System and, Agri Cloud. Experimental scenarios and results are explained in Sect. 4. Finally, in Sect. 5, the conclusions and future extensions are discussed.

2 Literature Survey

The existing and conventional methods of agriculture use manual methods for checking the parameters. For example, soil moisture content is evaluated by looking at soil physical properties like soil color, texture, density, and structure. Manual evaluations are completely dependent on the farmer's knowledge and experience and this increases the probability of false predictions [4]. It focuses on automating the agricultural practices by deploying sensors like temperature sensor, moisture sensor, and

PIR sensor (it is for detecting the movement of people). The information gathered from these sensors are associated with microcontroller, which additionally checks the got information with the limit esteems. If the data exceeds the threshold value, a buzzer is switched ON and alarm message is sent to the farmer. Web page is developed to generate the given set of values and the farmer gets the full description of the information collected [9]. Present survey on Smart Drip Irrigation System is using Raspberry Pi and Arduino. The monitoring system [2] is an embedded Linux board used to collect information from sensor node continuously, store it in the data warehouse, and provide the Web interface to the user. With the help of the Web interface and database, the user can now easily understand the collected data, monitor, and control the system which minimizes the human intervention and manual labor. The scenario of decreasing water tables and appropriate use of water to adapt up to the utilization of temperature and dampness sensor at reasonable areas for observing of yields is actualized in [10]. Gondchavar and Kawitkar [7] has proposed IoT-based smart agriculture that includes smart GPS-based remote control robot to perform tasks like Weeding, spraying, moisture sensor, bird, and animal scaring, etc. It also incorporates shrewd water system with keen control in view of constant field information and stockroom administration for temperature upkeep, mugginess support and burglary location utilizing interfacing sensors, Wi-Fi modules, cameras and actuators with small scale controller, and monitoring framework. A smart administration of agrarian nursery in light of Internet of Things is introduced in [3]. A brilliant water system framework utilizing soil temperature and dampness sensor is proposed in [10].

Most of the works done in literature survey on agriculture and IoT have concentrated on one parameter say either irrigation, temperature, or soil moisture and very few have considered all aspects of farming say from weeding to yielding. And, technology used is very costly and convoluted for regular agriculturists to utilize and get it. Hence, there is a need and requirement to develop a cost effective and efficient system to make farming automate which would increase the efficiency of the farm.

3 Implementation

Our project work is motivated by our farmers who work day and night in their farm lands. They have been using several irrigation techniques which are performed through the manual control in which the ranchers flood the land at consistent interims by turning the water pump ON/OFF when required. They do not have any proper method or system for monitoring and controlling the temperature, humidity, soil moisture. and water level which are vital for proper growth of any crop. Hence, the current project proposed a novel smart agriculture model based on IoT which assists farmers to get real-time data such as temperature, soil moisture, water content, and light availability. This ensures efficient environment monitoring, enabling them to do smart farming and increases their overall yield and quality of products. The proposed

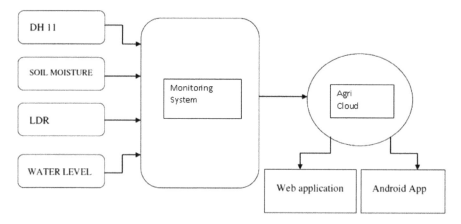

Fig. 1 Block diagram

system mainly consists for three components: sensors, monitoring system, and Agri Cloud. The block diagram of the same is shown in Fig. 1.

I. **Sensors: Various Sensors are deployed to obtain the data.**

- DHT11: The temperature and the humidity of the environment in which the crops are grown is monitored using DHT11 [2] sensor. This sensor is connected to gpio pin 2 of Raspberry Pi 3. We obtain new data from it once in every 2 s.
- Water level detector [11]: A float switch is a gadget used to identify the level of fluid inside a tank which is given a power supply of 5 V. The switch is utilized to control the water level.
- Soil moisture sensor [12]: This sensor is supplied with 3 V power supply from monitoring system. Data can be obtained by inserting this rugged sensor into the soil to be tested.
- LDR [13]: A photo resistor (or light-subordinate resistor, LDR, or photograph conductive cell) is used to detect the presence and absence of sunlight. It is provided with 5 V power by connecting it to Monitoring System.

II. **Monitoring System**

We have implemented the monitoring system by using the Raspberry Pi 3 [14] microcontroller which is a powerful single board computer. After successful data acquisition using sensors, it is sent from credit card-sized computer monitoring system to Agri Cloud after local processing.

III. **Agri Cloud**

It is an IoT application and has API to store and recover information from things utilizing the HTTP convention over the Internet or by means of a local area network. Application is implemented using ThingSpeak [15] open source which empowers the formation of sensor logging applications, area following applications, and an

Smart Agricultural Monitoring System ... 477

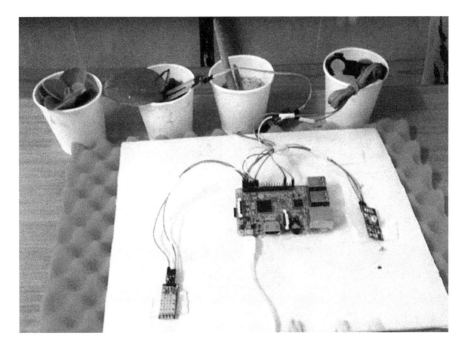

Fig. 2 Connecting the different sensors to monitoring system

informal community of things with notices. This allows us to aggregate, observe, and examine live data streams in the cloud. The farmer can visualize and monitor the environmental parameters by creating an account. Figure 2 shows the connection between various sensors and microcontroller.

The system as shown in Fig. 2 is integrated with monitoring system which uses Raspbian operating system. This free operating system is based on Debian which is Unix-like computer operating system.

Since the objective is to monitor environmental parameters wirelessly, we have made use of Virtual Network Computing (VNC). VNC is a sort of remote-controlled programming that makes it possible to control another PC over a framework affiliation. Keystrokes and mouse clicks are transmitted beginning with one PC then onto the following. VNC server is installed on remote computer, i.e., monitoring system and the VNC client is installed on the client computer which is used by the farmer (PC or laptop). Android IP scanner is used to obtain the IP address of monitoring system microcontroller. Both server and customer must be coordinated with TCP/IP and have open ports permitting movement from the IP locations of gadgets. This guarantees the network among customer and server.

The circuit is designed by connecting various sensors to microcontroller using GPIO.BCM (Broadcom SOC channel) mode of pin numbering. The interfacing of sensors is achieved using python programming language where LDR, water level,

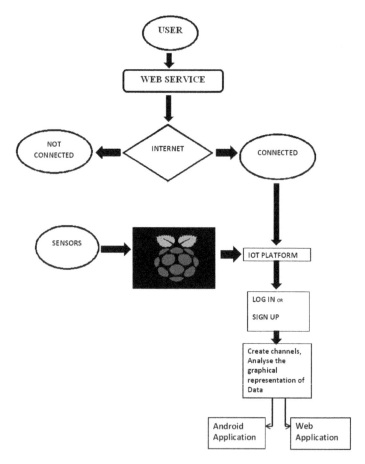

Fig. 3 Flowchart

and soil moisture sensors are set as inputs. The output of these sensors is analog values which are converted to digital values using Analog-to-digital converter. Parameters like temperature and humidity are obtained as analog value from DH11 sensor. Different functions are used to read data from different functions. In case of DH11 sensor, we have written a function which converts the Celsius value to Fahrenheit value. In the main function, the data are sent to Agri Cloud using Wi-Fi Module which is in-built in Raspberry Pi 3. This cloud is identified using base URL which locates the resources on World Wide Web. Each module/function is tested separately by writing suitable test cases. Agri Cloud is used to visualize and analyze live data streams in the cloud. Farmer or the concerned user can check the data acquired from the fields in any device which supports IoT-based application. Using login credentials, Read API and Write API keys one can log into their account on any device. Each account holder can have various channels identified by channel ID, each for one field. In this way, user can have the secure access to their account. Upon successful login, he can analyze the data obtained with respect to the range, time, and date. The work flow process involved while retrieving the sensed data on Web application is depicted in Fig. 3. The obtained value of each parameter is sent from microcontroller to Web application through Wi-Fi module which is in-built in Raspberry Pi [14]. Once Internet connection is established, farmers can create an account or login to the existing one in order to monitor the environmental conditions. The user can create channels for each of his field area and can access it securely. Secure access is due to the authentication required for Read API and Write API keys. Each channel displays the graphical representation of the values obtained over a period of time. X-axis represents the time stamp and data can be obtained for a particular time. Y-axis represents the range of value and hence user can identify the abnormal environmental conditions by setting an allowed range for each sensor values. This platform can be accessed either on Web or android application using log-in credentials.

4 Results

All the various sensors are connected to microcontroller using GPIO. The experimental results are viewed using Agri Cloud. Figure 4 is the snapshot of the cloud stored data. This data is updated once in 15 s in the database. It provides the graphical representation of the sensor data. The data read from the various sensors assist the condition of the farm. Read data is sent to monitoring system which is a mini computer and the same will be updated to cloud simultaneously through Wi-Fi. Data collected from LDR are plotted as graph using Agri Cloud as shown in Fig. 5. The concerned person can read the data on any IoT platform. LDR sensor is used where there is a need to sense the presence and absence of light is necessary. Soil moisture sensor determines the water content in the soil. Presence of water content in the soil is determined by the value 1 from the sensor else value 0. Similarly, the results obtained by water level detector are of digital value. Value 0 corresponding to low water level and value 1 to high water level. Humidity data read using DH11 is as shown in Fig. 6.

In the same way, data from all the sensors are read periodically, graphs are plot, and suitable action will be taken. The variations in the graph can be obtained by setting the scale value of X-axis which is a time stamp. In Y-axis, a particular range of values can be set, out of which the user can identify the abnormal conditions.

	A	B	C	D	E	F	G
1	created_a	entry_id	field1	field2	field3	field4	
2	2018-03-2:	34	0	1	0	36	
3	2018-03-2:	35	0	1	0	36	
4	2018-03-2:	36	0	1	0	36	
5	2018-03-2:	37	0	1	0	36	
6	2018-03-2:	38	0	1	0	36	
7	2018-03-2:	39	0	1	0	36	
8	2018-03-2:	40	0	1	0	35	
9	2018-03-2:	41	0	1	0	36	
10	2018-03-2:	42	0	1	0	36	
11	2018-03-2:	43	0	1	0	36	
12	2018-03-2:	44	0	1	0	36	
13	2018-03-2:	45	0	0	0	30	
14	2018-03-2:	46	0	0	0	31	
15	2018-03-2:	47	0	1	0	31	
16	2018-03-2:	48	0	1	0	30	
17	2018-03-2:	49	0	1	0	31	
18	2018-03-2:	50	0	0	0	31	
19	2018-03-2:	51	0	0	0	31	
20	2018-03-2:	52	0	0	0	31	
21	2018-03-2:	53	0	0	0	31	
22	2018-03-2:	54	0	0	0	31	
23	2018-03-2:	55	0	0	0	31	

Fig. 4 Cloud stored data

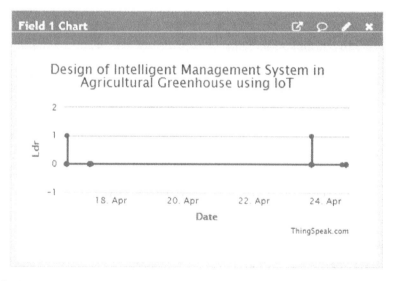

Fig. 5 Live data of LDR

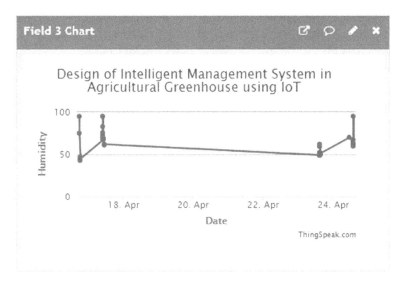

Fig. 6 Live humidity data of DH11

5 Conclusion

An attempt has been made to develop a smart agriculture system that will improve the growth of agricultural yield. IoT-based smart agriculture is very helpful for farmers who are required to monitor and analyze the environmental conditions without physically present in farm field. We can monitor the greenhouse from anywhere using internet connection in our smart phones. The sensors are properly interfaced with the monitoring system and the data relating to the temperature, humidity, water level, soil moisture, and LDR are being represented graphically on an IoT platform. The given set of corresponding values changes every 15 s and is updated hereafter. The analog signals of sensors are converted into the digital values using microcontroller and the data from sensors give the characteristic of environmental factors which is helpful for farmers. After successful data acquisition, data is displayed in the graphical form wherein the farmers can analyze the environmental conditions.

Usage of such a framework in the field can enhance the yield of the harvests and general creation. In this way, the IoT agrarian applications are making it feasible for ranchers to gather significant information and break down. Expansive landowners and little ranchers must comprehend the capability of IoT showcase for horticulture by introducing keen advancements to build intensity and maintainability in their creations. The interest for developing populace can be effectively met if the farmers and in addition little ranchers execute agrarian IoT arrangements in a fruitful way. One of the limitations of this project is that Internet connectivity is required to user and which is costly for farmers and this can be overcome by extending the system to send notification via SMS directly to the farmer on his mobile using GSM module instead of using mobile app. Weather data from the meteorological department can be used along with the sensed data to predict more information about the future which can help farmer plan accordingly to that and improve his livelihood. In this project, we have used open source cloud. Instead we can develop our own cloud to enhance the GUI.

References

1. Available [online] at https://en.wikipedia.org/wiki/Internet_of_things.
2. Available [online] at https://www.adafruit.com/product/386.
3. Li, Z., et al. (2017). Design of an intelligent management system for architectural greenhouses based on the internet of things. In *IEEE International Conference on Computational Science and Engineering (CSE) and IEEE International Conference on Embedded and Ubiquitous Computing (EUC)*.
4. Suma, N., Samson, S. R., Saranya, S., Shanmugapriya, G., & Subhashri, R. (2017, February). IOT based smart agriculture monitoring system. *International Journal on Recent and Innovation Trends in Computing and Communication, 5*(2).
5. Internet of things: Science fiction or business fact? (PDF). Harvard Business Review. November 2014. Retrieved 23 October 2016.
6. Available [online] at https://www.indiafoodbanking.org/hunger.
7. Gondchawar, N., & Kawitkar, R. S. (2016, June). IOT based smart agriculture. *International Journal of Advanced Research in Computer and Communication Engineering, 5*(6).
8. About FAO. FAO. Retrieved 18 February 2014.
9. Silvister, S., Rai, R., Yadav, M., Bopshetty, S., & Sagar, P. (2017). A survey on remote monitoring and controlling of drip irrigation methodologies. *International Journal of Advanced Research in Computer Engineering & Technology, 6*(1).
10. Nandurkar, S. R., Thool, V. R., & Thool, R. C. (2014). Design and development of precision agricultural system using wireless sensor network. *IEEE International conference on Automation, Control, Energy systems (ACES)*.
11. Available [online] at http://www.nskelectronics.com/water_float_switch.html.
12. Available [online] at https://en.wikipedia.org/wiki/Soil_moisture_sensor.
13. Available [online] at https://en.wikipedia.org/wiki/Photoresistor.
14. Available [online] at https://www.raspberrypi.org/products/raspberry-pi-3-model-b.
15. Available [online] at https://thingspeak.com/.

Detecting Healthiness of Leaves Using Texture Features

Srishti Shetty, Zulaikha Lateef, Sparsha Pole, Vidyadevi G. Biradar, S. Brunda and H. A. Sanjay

Abstract Agriculture is the dominant sector of our economy and contributes in various ways but the yield in the productivity leads to a significant reduction in the farmer's income. Monitoring crop health is important to increase the quality and quantity of the yield. But this requires manually monitoring the crops and also expertise in the field. Hence, automatic disease detection using image texture features is used for ease and to detect the disease at an early stage. The proposed methodology for the project is to design and implement the algorithm on two sets of databases: firstly, a locally generated leaf database which contains images of leaves and secondly, a standard database which is a common test database. The basic steps for crop disease detection include image acquisition, image preprocessing, image segmentation, feature extraction, and classification using image processing techniques. The acquired leaf images are preprocessed by removing undesired distortion and noise, and then, the processed image is further subjected to K-means-based segmentation. The segmented image is further analyzed using Haar wavelet transform and GLCM based on its texture by extracting feature vector. SVM is used for classification of image. Thus, the presence of diseases in leaf is identified along with all the features values of the leaf. It also calculates the accuracy rate of the prediction made by the system.

Keywords Image processing · Gray-level co-occurrence matrix · Support vector machine-nearest neighbor · Haar wavelet · K-means

---

S. Shetty · Z. Lateef · S. Pole · V. G. Biradar · S. Brunda (✉) · H. A. Sanjay
Department of ISE, NMIT, Bangalore 560064, India
e-mail: brunda.s@nmit.ac.in

1 Introduction

The main aim of a development project is to change the current situation into a better one over time. Though industry has been playing an important role in Indian economic scenario, the contribution of agriculture still cannot be denied [1]. Agriculture is one of the biggest shares of economy in India, in terms of generating employment as well as provision of food for the ever-increasing population. The recognition and classification of crop diseases are of major economical and technical importance in agricultural industry. Maize popularly known as "corn" is one of the most versatile and major cash crops having wider adaptability under varied climatic conditions. Maize is one of the major cereal crops and is the third most contributing major crop in India after rice and wheat. Estimated losses due to major diseases of maize in India are about 13% of which fungal diseases cause major yield losses [2]. As per survey, bacterial blight, anthracnose, leaf spot, and rust are major commonly occurring diseases affecting maize crops in northern parts of Karnataka. They affect the photosynthesis with adverse reduction in the yield to an extent of 2891%. These diseases affect the quality as well as quantity of the agricultural products and lead economic losses. The leaves are the first one to be affected in case of any disease, generally before or after blooming which effects the growth plants. Our main objective is to concentrate on maize leaf disease detection based on the texture of the leaf.

2 Related Work

See Table 1.

3 Methodology

Image processing methods involve versatility, repeatability, and the preservation of original data precision. The various image processing techniques are:

1. Image acquisition
2. Image preprocessing
3. Image segmentation
4. Feature extraction
5. Image classification

   (i) **Image Acquisition**
   The images of the maize leaves are captured using a digital camera in JPEG format. The size of each image is 1500 × 1500. The samples are collected from a field located in the University of Agricultural Sciences, Dharwad. Three

Table 1 Literature review

Author name	Preprocessing	Feature extraction	Result	Name of the data set used
Dheeb Al Bashish, Malik Braik, Sulieman Bani-Ahmad [10], 2010		CCM SDGM Neural networks	93%	Early scorch Cottony mold, ashen mold, late scorch, tiny whiteness
Sanjay B. Dhaygude, Nitin P. Kumbhar [8], 2013		CCM	Satisfied	Self
Kiran R. Gavhale, Ujwalla Gawande, Kamal O. Hajari [3], 2014	DCT $L*a*b$ YCbCr CES	GLCM	96%— SVMRBF 95%— SVMPOLY	Created 300
Aakanksha Rastogi, Ritika Arora, Shanu Sharma [4], 2015			Satisfied	Hydrangea leaf Maple leaf
Sachin D. Khirade, A. B. Patil [9], 2015	Histogram equalization	CCM	Satisfied	Wheat
Shivaputra S. Panchal, Rutuja Sonar [6], 2016	Image enhancement	GLCM	Satisfied	Pomegranate leaf, image database consortium
Anand R., Veni S., Aravinth J. [5], 2016	Histogram equalization, $L*a*b$	CCM SDGM	Satisfied	
R. Meena Prakash, G. P. Saraswathy, G. Ramalakshmi, K. H. Mangaleswari, T. Kaviya [2], 2017	$L*a*b$	GLCM	0.9–1.0	Self-created database 60 citrus leaves using camera
Amruta Ambatkar, Ashwini Bhandekar, Avanti Tawale, Chetna Vairagade, Ketaki Kotamkar [7], 2017	Histogram equalization	CCM GLCM SDGM	Satisfied	Self
Ahmad Nor Ikhwan Masazhar, Mahanijah Md Kamal [1], 2017	$L*a*b$	GLCM	Chimaera 97% Anthracnose 95%	Self

hundred images of each category: healthy, northern blight, southern blight, and rust, are collected by keeping constant distance of 1 foot. One hundred and fifty images of each category are used for training, and remaining 150 images are used for testing. Thus, we have 600 images for training and 600 images for testing.

(ii) **Image preprocessing**
 1. The images are of varying contrast. Preprocessing is necessary in order to correct nonuniform illumination. First, RGB image is converted into $L*a*b$ [1–3] color space and the luminance component (L) is extracted. Then, contrast limited adaptive histogram equalization (CLAHE) is applied to L component and concatenated with 'a' and 'b' components, and finally, the image is converted from $L*a*b*$ color space to RGB space (Fig. 1).

(iii) **Image segmentation**
Image segmentation is the way toward partitioning a propelled image into various bits (sets of pixels, generally called super-pixels). The goal of division is to contemplate the picture in detail and also change the features of an image into something that is more vital and less complex to analyze. Image segmentation is consistently used to discover objects and edges (lines, curves, etc.) in images. More definitively, image segmentation is the path toward consigning a label to every pixel in an image to such a degree, to the point that pixels with a comparative feature share certain characteristics. The yield of this division is a course of action of portions that cover the entire image. Each pixel in a fragmented zone is similar with respect to some basic element, for instance, color, texture, intensity, and contrast. Neighboring locales are diverse relating to similar characteristics.

(iv) **Feature Extraction using Haar wavelet and GLCM**
Texture analysis plays an important role in image analysis, such as disease detection, medical imaging, machine vision, and content indexing of image databases. Texture analysis can be done by wavelet transform. The Haar transform has been used as a necessary tool in the wavelet transform for feature extraction. The method is based on the application of Haar wavelet on RGB image of the preprocessed image to obtain horizontal, vertical, and diagonal coefficients. Then, gray-level co-occurrence matrix [1, 3, 6, 7] (GLCM) is constructed into two directions (0° and 90°) for each of the coefficients. GLCM matrix produces four statistical features like contrast, correlation, energy, and homogeneity.

(v) **Classification using SVM classifiers**
The concept of support vector machine (SVM) was presented by Vapnik and collaborators. It gets predominance as it offers alluring highlights and competent equipment to handle the issue of order. The SVM depends on measurable learning hypothesis. SVM's better speculation execution depends on the standard of structural risk minimization (SRM). The idea of SRM is to expand the edge of class partition. The SVM was characterized for two-class issue and it searched for ideal hyperplane, which amplified the separation, the edge, between the closest cases of the two classes, named SVM. At exhibit, SVM is

Detecting Healthiness of Leaves Using Texture Features

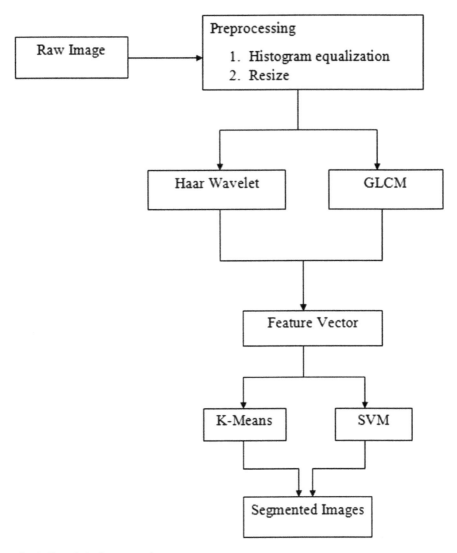

Fig. 1 Steps in leaf segmentation

mainstream grouping instrument utilized for design acknowledgment and other characterization purposes. The standard SVM classifier takes the arrangement of information and predicts to characterize them in one of the main two unmistakable classes. For multiclass order issue, we decay multiclass issue into various paired class issues, and we plan reasonable joined numerous parallel SVM classifiers.

4 Data set

For the purpose of this project, focus is laid mainly on maize leaves. The samples are gathered from a field situated in University of Agricultural Sciences, Dharwad. A total of 1200 deals were gathered with 600 images utilized for preparing and 600 for testing. In any case, non-dominant part of this maize is exhausted clearly by individuals. A bit of the maize production is used for corn ethanol, animal support, and other maize substances, for instance, corn starch and corn syrup. Some of the normal ailments found in maize leaves are northern blight, southern blight, rust, anthracnose, etc. These sicknesses are essentially found in Southern Indian districts.

5 Software Requirements

- The system will identify whether the leaf image is healthy or diseased.
- The system will detect different types of diseases that commonly occur in maize.
- The system is able to detect the percentage of the disease-affected area.
- The system is able to suggest the right treatment plan for the identified disease in maize leaf.

6 Nonfunctional Requirements

1. **Performance**: Fast and accurate detection of disease. (3–5 s).
2. **Usability**: The system will be useful for farmers and agricultural analysts.
3. **Reliability**: The system will be able to detect the three commonly occurring diseases in maize leaves and suggest the remedy.

    a. **SVM**: 93–95% reliable with training data set of 400 images.
    b. **KNN**: 86–90% reliable with training data set of 400 images.

4. **Availability**: The proposed system is readily available on any platform and provides on-demand service.
5. **Simplicity**: The proposed system is user friendly.

7 Algorithms Used

1. **Histogram equalization**

Histogram shows the frequency of pixel intensity values. It is used for analysis of image and thresholding. Histogram equalization is used for contrast adjustment.

2. **K-Means Clustering**

The K-means clustering is utilized for division and characterization of leaves in view of an arrangement of features, into K number of groups. The classification is done by minimizing clusters.

The algorithm for K-means clustering:

- Choose the center of K cluster, randomly or based on some logic.
- Relocate each pixel in the leaf image to the cluster that minimizes the distance between the pixels and the cluster center.
- Recompute the cluster centers by averaging all of the pixels in the cluster.
- Repeat steps 2 and 3 until effective clustering is achieved.

3. **Haar Wavelet**

The Haar wavelet is used for texture analysis. In this method, the preprocessed image is used to obtain the horizontal, vertical, and diagonal coefficients. The obtained coefficients are used for texture analysis.

4. **Gray-Level Co-occurrence Matrix (GLCM)**

It is a statistical method of examining texture that considers the spatial relationship of pixels. GLCM is used to calculate how often a pixel with gray-level value 'i' occurs either horizontally, vertically, or diagonally to an adjacent pixel with gray-level value 'j'.

Statistics derived from the GLCM include contrast, correlation, energy, homogeneity, etc., of the image.

5. **Support Vector Machine (SVM)**

Linear support vector machine is used for classification of leaf diseases. SVM is a binary classifier which uses a hyperplane called the decision boundary between two classes. It maximizes the margin around the separating hyperplane.

1. Mean: It measures the mean value of all pixels in the relationships that contributed to the GLCM. It can be measured using the following formula:
   where

   P_{ij} Element i, j of the image.
   N Number of gray levels

$$\mu = \sum_{i,j=0}^{N-1} i(P_{ij}) \qquad (1)$$

2. Variance: The variance of the intensities of all the reference pixels is calculated using the following formula:

$$\sigma^2 = \sum_{i,j=0}^{N-1} P_{ij}(i-\mu)^2 \qquad (2)$$

3. Contrast: Contrast measures the quantity of local change in an image. It reflects the sensitivity of the textures in relation to the changes in the intensity. It returns the measure of intensity contrast between a pixel and its neighborhood. Contrast is 0 for a constant image. It can be measured using the following formula:

$$\sum_{i,j=0}^{N-1} P_{ij}(i-j)^2 \qquad (3)$$

4. Energy: Energy also means uniformity. It returns the sum of squared elements in the GLCM. The more homogenous the image is, the larger the value. When energy equals to 1, the image is believed to be a constant image. It can be measured using the following formula:

$$\sum_{i,j=0}^{N-1} (P_{ij})^2 \qquad (4)$$

5. Correlation: This feature is used to measure how correlated a pixel is to its neighborhood. Correlation is 1 or −1 for a perfectly positive or negative correlated image. It can be measured using the following formula:

$$\sum_{i,j=0}^{N-1} P_{ij} \frac{(i-\mu)(j-\mu)}{\sigma^2} \qquad (5)$$

6. Homogeneity: Homogeneity measures the similarity of pixels. A diagonal gray-level co-occurrence matrix gives homogeneity of 1. It becomes large if local textures only have minimum changes. It can be measured using the following formula:

$$\sum_{i,j=0}^{N-1} \frac{P_{ij}}{1+(i-j)^2} \qquad (6)$$

8 Result

Sample 1 (Fig. 2).

Detecting Healthiness of Leaves Using Texture Features 491

Sample 2 (Fig. 3).
Sample 3 (Fig. 4).
Sample 4 (Fig. 5).
Sample 5 (Fig. 6).
Sample 6 (Fig. 7).
Sample 7 (Fig. 8).
Sample 8 (Fig. 9 and Table 2).

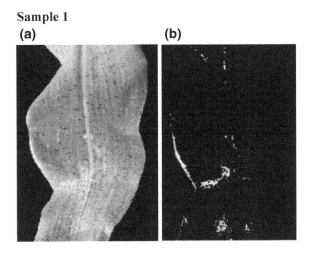

Fig. 2 a Input leaf and b Segmented leaf

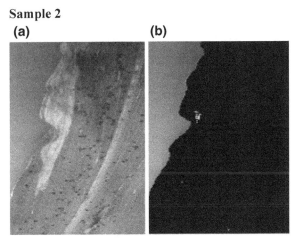

Fig. 3 a Input leaf and b Segmented leaf

Fig. 4 a Input leaf and **b** Segmented leaf

Sample 3
(a)   (b)

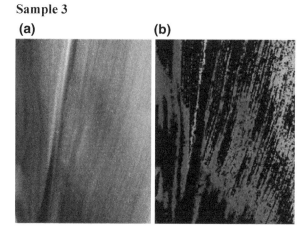

Fig. 5 a Input leaf and **b** Segmented leaf

Sample 4
(a)   (b)

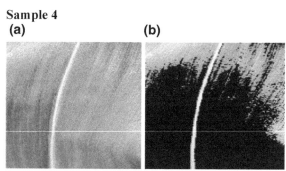

Fig. 6 a Input leaf and **b** Segmented leaf

Sample 5
(a)   (b)

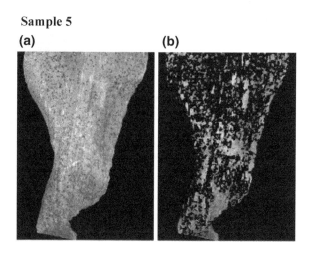

Detecting Healthiness of Leaves Using Texture Features 493

Fig. 7 **a** Input leaf and **b** Segmented leaf

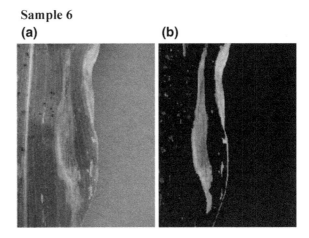

Fig. 8 **a** Input leaf and **b** Segmented leaf

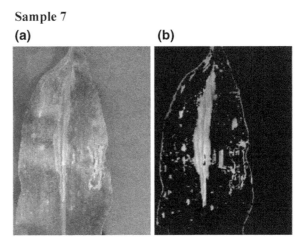

Fig. 9 **a** Input leaf and **b** Segmented leaf

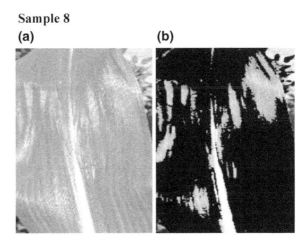

Table 2 GLCM values of different samples

	Mean	S.D	Entropy	Variance	Contrast	Correlation	Energy	Homogeneity
Sample 1	5.3614	26.3431	1.313	674.395	0.371148	0.953555	0.153258	0.867639
Sample 2	23.8789	49.098	1.89308	1136.33	0.256279	0.871701	0.206531	0.882174
Sample 3	31.4184	41.643	4.00298	1416.95	0.223564	0.884036	0.210835	0.898197
Sample 4	49.5676	74.519	2.97791	5088.52	0.402556	0.828097	0.128442	0.820587
Sample 5	29.887	57.3747	2.40529	2835.69	0.347677	0.960773	0.246748	0.879001
Sample 6	15.0888	41.6127	1.29395	1111.79	0.182057	0.908318	0.218444	0.914992
Sample 7	18.5059	48.7076	1.41064	1767.55	0.252881	0.825716	0.222579	0.886234
Sample 8	42.8826	79.4477	2.42895	5831.5	0.321507	0.80492	0.246217	0.858014

9 Conclusion

This work deals with crop yield detection based on the healthiness of leaves using image texture features. The texture features are Harlick features in addition to Haar wavelet features. The classification algorithm is a combinational one, SVM and K-means classification. The experiment is carried out on maize leaves database which is collected from Dharwad University. The results are found satisfactory. However, the accuracy may be improved by novel algorithms for texture features.

References

1. Masazhar, A. N. I., & Kamal, M. M. (2017). Digital image processing technique for palm oil leaf disease detection using multiclass SVM classifier. In *Proceedings of the 4th IEEE International Conference on Smart Instrumentation, Measurement and Applications*.
2. Meena Prakash, R., Saraswathy, G. P., Ramalakshmi, G., Mangaleswari, K. H., & Kaviya, T. (2017). Detection of leaf diseases and classification using digital image processing. In *International Conference on Innovations in Information, Embedded and Communication Systems (ICIIECS)*.
3. Gavhale, K. R., Gawande, U., & Hajari, K. O. (2014). Unhealthy region of citrus leaf detection using image processing techniques. In *International Conference for Convergence of Technology*.
4. Rastogi, A., Arora, R., & Sharma, S. (2015). Leaf disease detection and grading using computer vision technology and fuzzy logic. In *2nd International Conference on Signal Processing and Integrated Networks*.
5. Anand, R., Veni, S., & Aravinth, J. (2016). An application of image processing techniques for detection of diseases on Brinjal leaves using K-means clustering method. In *Fifth International Conference on Recent Trends in Information Technology*.
6. Panchal, S. S., & Sonar, R. (2016). Pomegranate leaf disease detection using support vector machine. *International Journal of Engineering and Computer Science, 5*(6), 16815–16818.
7. Ambatkar, A., Bhandekar, A., Tawale, A., Vairagade, C., & Kotamkar, K. (2017). Leaf disease detection using image processing. In *International Conference on Recent Trends in Engineering Science and Technology (IJRITCC)* (Vol. 5, No. 1, pp. 333–336).
8. Dhaygude, S. B., & Kumbhar, N. P. (2013, January). Agriculture plant leaf disease detection using image processing. *International Journal of Advanced Research in Electrical, Electronics and Instrumentation Engineering, 2*(1), 599–602.

9. Khirade, S. D., & Patil, A. B. (2015). Plant disease detection using image processing. In *International Conference on Computing Communication Control and Automation* (pp. 768–771). New York: IEEE.
10. Al Bashish, D., Braik, M., & Bani-Ahmad, S. (2010). A framework for detection and classification of plant leaf and stem diseases. In *International Conference on Signal and Image Processing* (pp. 113–118). New York: IEEE.

A Survey on Different Network Intrusion Detection Systems and CounterMeasure

Divya Rajput and Ankit Thakkar

Abstract Recent studies have pulled tons of research in the domain of cloud security and various intrusion detection systems (IDSs). This is because of advancement in the different types of attacks on computer systems. Distributed denial of service (DDoS) attack is one of them wherein the attackers can compromise the cloud system by exploiting vulnerabilities. Initially, during the multi-step exploration, vulnerability with low frequency along with the virtual machine which is identified and compromised are included in DDoS attacks. In this context, various IDSs have been surveyed with different countermeasure techniques including some effective techniques to minimize the malicious activities within end systems or networks. The main aim of IDSs is to detect different attacks within networks and end systems or to be precise against any information systems which are very difficult to maintain in a secure state for a long duration. Some studies have shown that the use of host-based systems and the network-based systems help to improve the attack detection. This paper focuses on the study of various well-known IDS and various techniques to minimize malicious activities within the system.

Keywords Graph model · Zombie detection · Network security · Cloud computing · Intrusion detection systems · DDoS attacks · Cloud security · Survey

---

D. Rajput (✉) · A. Thakkar
Institute of Technology, Nirma University, Ahmedabad 382481, Gujarat, India
e-mail: 16mcei18@nirmauni.ac.in

A. Thakkar
e-mail: ankit.thakkar@nirmauni.ac.in

© Springer Nature Singapore Pte Ltd. 2019
N. R. Shetty et al. (eds.), *Emerging Research in Computing, Information, Communication and Applications*, Advances in Intelligent Systems and Computing 906,
https://doi.org/10.1007/978-981-13-6001-5_41

1 Introduction

Cloud Security Alliance (CSA) has performed studies that briefly describes all security issues such as loopholes, exploitation, and malware where attack and vulnerability are on the top consideration of security threat wherein the attackers within the cloud exploit vulnerabilities and utilize the resource available on cloud systems to deploy attacks. In the traditional approach, the system admin has the overall control over machines including vulnerabilities detection and patching which is performed in a centralized fashion. In some cases, fixing known gaps in cloud server may not work efficiently due to the clients rule introduced in managed Virtual Machines (VMs) [1] and as a result, it causes damage to service level agreement (SLA). Moreover, a defenseless programming is introduced by the cloud clients that adds security features to clouds. The main task is to set up an effective vulnerability-free/Attack-free environment so that security breaches in the cloud clients would be minimized.

Different models have shown a better progress in the current IDSs/ intrusion prevention systems (IPS) arrangements by utilizing programmable virtual network administration approach that enables the framework to form a dynamic re-configurable IDS [2]. A network attack graph approach for any attack location and countermeasure action proposes most powerful countermeasures. It also enhances the versatility of VM and attacks exploration discovery without affecting existing cloud admin. This is because mirroring-based approach does not intercept with user traffic, compare to proxy-based approach. The survey presented in this paper describes various network intrusion detection systems (NIDS) approach to counter Zombie attacks. For more accuracy and improvement, host-based intrusion detection systems (HIDS) is also covered to complete the whole spectrum of cloud security.

2 Related Work

The survey focuses on describing few exceptionally related research zones to network intrusion detection and countermeasure selection in the virtual private system in cloud environment including Zombie detection and counteractive action, attack graph development and security investigation, programming characterized systems for attack countermeasures [3]. To avert DoS, system administrators need high accuracy in DDoS detection. There are various sorts of intrusion detection systems, they are categorized as network-based detection systems and host-based detection system. Wireless sensor networks use the intrusion detection systems for high security [4]. It is been observed that wireless sensor networks require high security due to their nature and operation [5].

Identifying malicious activities within any process has been investigated well in the literature. A recent work focuses on identifying compromised hosts that are acting as spam zombies [6]. Their approach addresses the shortcoming of payload-inspection-based IDS and standard traffic-based IDS by proposing traffic-based IDS build on randomized data partitioned learning model (RDPLM). This model depends on features set, and techniques of feature selection are simplified sub-spacing and multiple randomized meta-learning methods.

Using attack graphs to organize vulnerabilities makes an intrusion detection system more efficient [7]. The nodes in an attack graph represent a possible exploit. System's most vulnerable components are been identified by the shortest path in an attack graph. Attack graphs play an important role in vulnerability analysis in large networks.

Sheyner et al. [8] proposed a method of building attack graph that utilizes NuSMV based on binary decision diagram (BDD) to generate multi-stage, multi-host attack graphs wherein scalability is a major issue. Ou et al. [9] addressed the shortcoming of NuSMV-based method and proposed another attack graph generation method using MulVAL—a logic programming-based network security analysis engine. To give the security evaluation and secure connection highlights, HIDS [10] and firewall are generally used to filter and identify malicious activities within the system.

The various attacks against the networks, computer systems, and other distributed application have become more diverse and sophisticated [11]. Security is the major problem and the detection of attack is also difficult in the cloud environment. Thus, detection and mitigation of attacks are a priority in order to avoid disastrous outcomes. Dealing with such complex new attacks becomes difficult due to which several solutions were proposed for IDS/IPS and web application firewalls (WAF). The paper discussed several existing countermeasures on network intrusion and their limitations. Zonouz et al. [12] presented attack response trees (ARTs) that can be used to demonstrate how a system may be attacked. An ART is an enhanced attack tree that includes the attack consequences. This method models a stochastic game against the attacker and determines lower level attack consequences and their counteractive measures by applying attack response trees (ART) on security events at the system level within host computers. However, ARTs suffers from the ill effects of the state-space blast issue. ARTs [12] demonstrate an ideal countermeasure against a single target only.

Roy et al. [13] provide an improved method called attack countermeasure tree (ACT) that allows probabilistic analysis of an attack on a node in the attack tree. Specifically, some examination and improvement were managed as many individuals trust it is difficult to acquire probability calculations for the attack, discovery and their mitigation. However, the ideal security countermeasure set can be chosen from the pool of protection systems utilizing a non-state-space approach which is considerably less costly than the state-space approach as presented by Zonouz et al. Greedy procedures and certain identification methods (Divide and Conquer) are utilized to register the ideal countermeasure set for different target work under various requirements. Although their countermeasure process is robust, they are computationally heavy and unsuitable for deployment in a large number of virtual hosts where the resources are dedicated to the consumers [12]. A survey on IDS and countermeasure is presented in Table 1.

Table 1 Survey on IDS and countermeasures

Article/Paper	Problem	Solution	Model Used	IDS	Accuracy
NICE [14]	Distributed DOS on Cloud System	NICE (network intrusion detection and countermeasure selection) mechanism	Attack Graph Model	NIDS	High (98.476%)
Non-intrusive process-based monitoring system to mitigate and prevent VM vulnerability explorations [1]	Targeted weak nodes of vulnerable virtual machines	A hybrid intrusion detection framework	VMM-based model	NIDS	High
An Improved Feature Selection Algorithm Based on MAHALANOBIS Distance for Network Intrusion Detection [2]	Unidentified attacks due to complex MAHA- network structures	MAHALANOBIS [2] Distance feature Method	KDD- CUP 1999 data-sets using SVM classifier and KNN classifier	NIDS	High
Detecting Anomalous Network Traffic in Organizational Private Networks [15]	Malicious traffic and Anomalies	–	–	NIDS/ HIDS	Medium (80%)
Out-VM Monitoring for Malicious Network Packet Detection in Cloud [16]	Attacks on Virtual Machine (VM) within cloud environment	Malicious Network Packet Detection (MNPD) [1]	VM monitoring [17] /MNDP	NIDS	Medium
An advanced method for detection of botnet traffic using Intrusion Detection System [4]	Payload identification with in a packets	Sub-spacing and multiple randomized meta-learning techniques [18]	Randomized data partitioned learning model	PIDS/ TIDS	High (99.984%)
NIDSV: Network-based Intrusion Detection and Countermeasure Exception in Virtual Environment using AODV protocol [19]	Attack detection in different cloud services like PaaS, SaaS, and IaaS	NIDSV, which is designed based on Alert Correlation Graph and Countermeasures	Alert Correlation Graph model	NIDS/ HIDS	High (93.768%)

(continued)

Table 1 (continued)

Article/Paper	Problem	Solution	Model Used	IDS	Accuracy
A Centralized HIDS Framework For Private Cloud [20]	Traditional host-based IDS for cloud computing consumes a large number of system resources	A centralized host-based IDS framework to reduce the use of the resources	log stash tool	HIDS	High (92.765%)
Unified, Multi-level Intrusion Detection in Private Cloud Infrastructures [3]	Security flaws in cloud environment due to traditional IDSs (collaborative [21]) can either notice too late or never at all a potentially-costly intrusion	Traditional approaches to safeguards and detection mechanisms that leverage knowledge of typical/correct private cloud operations	Multi-level IDS architecture	HIDS	High
Security risk analysis of enterprise networks using probabilistic attack graphs [22]	Security flaws in Enterprise systems	Methodology based on probabilistic attack graph for strengthen the security	Probabilistic attack graph model	HIDS/NIDS	Medium
Modeling and analysis of attacks and counter defense mechanisms for cyber physical systems. [23]	Anamalous conduct and guard for the digital physical systems	An investigative model to catch the progression of physical systems	Digital physical framework	HIDS	Medium
Developing and evaluating a Hands-on lab for teaching local area network vulnerabilities [24]	Common vulnerability in ARP	Hands-on lab to enable understudies to figure out how ARP satirizing assault functions	ARP Spoofing Tools and Defense	–	High
Distributed denial of service attacks in software-defined networking with cloud computing [25]	Attacks on Software Defined Networks	A bundle conglomeration strategy which goes for making attack marks and utilize them to pervent attacks on SDNs	A conglomeration strategy	NIDS	Medium

3 Review of the Models Used for the IDS

3.1 Attack Graph Model [14]

The concept of Attack Graph Model has been used in several research work. As per network security perspective, a *graph* is the natural choice to show the detailed view of vulnerability within a given system which was firstly introduced by Dacier to show the detailed view of vulnerabilities within a network system. The concept was proposed by Philips and Swiler in 1998. The node within the privileged graph represents the privilege set of users and vulnerabilities are been represented by edges. For a given path of an attack tree, an initial stage has been represented by a leaf node through which an attacker can reach to root (final state) of the tree by exploring different possible vulnerability. Attack Graph model is the consolidated representation of a given attack tree graph merged with different attack paths containing common nodes. In real-time scenario, edges represents the change of state not only caused by an attacker to perform the attack but also weighted on the basis of attackers efforts required to be successful, and the nodes represent all the possible system state during execution of an attack.

3.2 Virtual Machine Manager (VMM)-Based Model [26]

The malware instances try to compromise the services provided by the VM, i.e., OS kernel due to which VMM Model came into the picture to avoid such a scenario to take place. VM security can be compromised by malware that may result in corrupted VM. The VMM model is a code-based observation of a VMM, i.e., smaller the VMM more stable the OS. Further, it provides a limited interface to all the untrusted VMs in an abstracting underlying physical resource form. As compared to other research efforts on VM, this approach has proved to be the most consistent ones.

3.3 Alert Correlation Graph Model [27]

Alert Correlation Graph Model is based on the concept of correlating the alerts raised during the attacks. This model has three different categories:

– **Probabilistic Alert Correlation**—In this category, alerts are been correlated based on alert attributes similarities(i.e., alerts with the same source and destination IP address). It is an effective approach but it fails to discover the full casual relationships between similar alerts.
– **LAMBDA and the data mining approach**—It is an attack scenario specified approach which can be human specified or can be a fully trained dataset. This

method is limited to attack scenarios only. Variation is also been done by specifying which kind of attack can be followed from a given attack.
- **JIGSAW approach**—Its preconditions and consequences of an attack approach wherein it correlates the preconditions of some future alerts that will satisfy the consequences of the past alerts. This approach can potentially explore all the causal relationship between alerts and it is not restricted to the known attacks.

3.4 Probabilistic Attack Graph Model [28]

This model has been widely preferred during large-scale attack graphs. It describes the approximate probability of Common Vulnerability Scoring System Security (CVSSS) [29] and attack graph. Initially, there will an approximate estimation of maximum and a minimum probability of every node in an attack graph, where the probability is either 0 or 1. Afterward, this model will be used for calculation of maximum and a minimum probability of every node.

3.5 Network Intrusion Detection and Countermeasure Selection (NICE) Model [14]

An agent NICE-A: NICE-A is an agent within a network-based intrusion detection system which is installed in the cloud server. The main task is to scan the network traffic passing through the Linux bridge as well as the different VMs from the physical cloud server. Snort [30] is a type of NICE agent which used to capture the packets at the initial phase of implementation. The main task of NICE-A is to sniff all the ports in each virtual bridge in open switch.

VM profiling: VM profiling gives information about all the vulnerable and non-vulnerable VM ports. After sniffing all the ports by NICE-A, VM profiler will scan all the ports by running port scanning mechanism. The detailed information of any open ports comes and the history will tell how vulnerable they are and also the VM.

- **Attack Graph Generator**—During generation of attack graph, the information of detected vulnerability is been added to the corresponding VMs database entry.
- **An agent NICE-A**—The VM profile database will record all the alerts involving the VM.
- **The Network controller**—The five tuples-based traffic pattern is generated which involves the VMs. The five tuples contain the source MAC address, destination MAC address, source IP address, destination IP address, protocol.

Attack Analyzer: The main function of attack analyzer is to handle the analysis operation and alert correlation. The two major function of attack analyzer is constructing Attack Correlation Graph (ACG) and providing the solution to the network controller for virtual network reconfiguration by providing threat information and appropriate countermeasures. Based on the severity of the results, the selected countermeasure process is been applied to the network controller.

Network Controller: Due to the internal cloud discovery modules which use the protocol like Domain Name System (DNS), Dynamic Host Configuration Protocol (DHCP), Link Layer Discovery Protocol (LLDP) [31, 32] and flow initiations, the ability to discover the network connectivity information by network controller from Open vSwitch (OVS) and Open Flow Switch (OFS) is easy. The network connectivity information like each switch's current data paths and the associated flow information such as MAC and TCP/IP header. The changes in topology and the network flow will be sent to the controller automatically and after that, it is been delivered to the attack analyzer for attack graph reconstruction. Network controller also assists the attack analyzer module. Open Flow protocol states that the first packet received by the controller, it holds it and check the complying traffic policies in the flow table.

In NICE, there is a mutual communication between the network controller and the attack analyzer about the setting up of filtering rules for flow access on the corresponding OFS [10] and OVS. During the admission of traffic flow, the upcoming packets are not handled by the network controller instead of that they have been monitored by agent NICE-A. Applying countermeasure from attack analyzer has also been done by a network controller.

4 Conclusion

In general scenarios usually, the virtualization is what the cloud infrastructure relies on. Without having much knowledge of guest OS configurations and security, the cloud providers run the VMs due to which it may corrupt the entire cloud environment. In order to prevent such scenarios related to security of VMs, an efficient and effective approach that uses various intrusion detection systems are surveyed and discussed in the paper. To provide a solution, the cloud provider offers security-as-a-service based on VM introspection which promises both effective protection and efficient centralization. In this paper, the overview of NICE has been presented which uses attack graph model for detection and prevention of attacks in the cloud environment. The countermeasure discussed provides the enhancement and accuracy of detecting any malicious attacks and provide attack-free cloud environment.

References

1. Chung, C.-J., Cui, J. S., Khatkar, P., & Huang, D. (2013). Non-intrusive process-based monitoring system to mitigate and prevent VM vulnerability explorations. In *2013 9th International Conference Conference on Collaborative Computing: Networking, Applications and Worksharing (Collaboratecom)* (pp. 21–30). IEEE.

2. Yongli, Z., Yungui, Z., Weiming, T., & Hongzhi, C. (2013). An improved feature selection algorithm based on MAHALANOBIS distance for network intrusion detection. In *2013 International Conference on Sensor Network Security Technology and Privacy Communication System (SNS & PCS)* (pp. 69–73). IEEE.
3. Humphrey, M., Emerson, R., & Beekwilder, N. (2016). Unified, multi-level intrusion detection in private cloud infrastructures. In *IEEE International Conference on Smart Cloud (SmartCloud)* (pp. 11–15). IEEE.
4. Koli, M. S., & Chavan, M. K. (2017). An advanced method for detection of botnet traffic using intrusion detection system. In *2017 International Conference on Inventive Communication and Computational Technologies (ICICCT)* (pp. 481–485). IEEE.
5. Alrajeh, N. A., Khan, S., & Shams, B. (2013). Intrusion detection systems in wireless sensor networks: A review. *International Journal of Distributed Sensor Networks, 9*(5), 167575.
6. Wong, K., Dillabaugh, C., Seddigh, N., & Nandy, B. (2017). Enhancing Suricata intrusion detection system for cyber security in SCADA networks. In *2017 IEEE 30th Canadian Conference on Electrical and Computer Engineering (CCECE)* (pp. 1–5). IEEE.
7. Roschke, S., Cheng, F., & Meinel, C. (2011). A new alert correlation algorithm based on attack graph. In *Computational Intelligence in Security for Information Systems* (pp. 58–67). Berlin: Springer.
8. Sheyner, O., Haines, J., Jha, S., Lippmann, R., & Wing, J. M. (2002). Automated generation and analysis of attack graphs. In *2002 IEEE Symposium on Security and privacy, 2002. Proceedings* (pp. 273–284). IEEE.
9. Ou, X., Boyer, W. F., & McQueen, M. A. (2006). A scalable approach to attack graph generation. In *Proceedings of the 13th ACM Conference on Computer and Communications Security* (pp. 336–345). ACM.
10. Souissi, S. (2015). Toward a novel rule-based attack description and response language. In *2015 11th International Conference on Information Assurance and Security (IAS)* (pp. 44–49). IEEE.
11. Abduvaliyev, A., Pathan, A.-S. K., Zhou, J., Roman, R., & Wong, W.-C. (2013). On the vital areas of intrusion detection systems in wireless sensor networks. *IEEE Communications Surveys & Tutorials, 15*(3), 1223–1237.
12. Zonouz, S. A., Khurana, H., Sanders, W. H., & Yardley, T. M. (2014). RRE: A game-theoretic intrusion response and recovery engine. *IEEE Transactions on Parallel and Distributed Systems, 25*(2), 395–406.
13. Roy, A., Kim, D. S., & Trivedi, K. S. (2010). Cyber security analysis using attack countermeasure trees. In *Proceedings of the Sixth Annual Workshop on Cyber Security and Information Intelligence Research* (p. 28). ACM.
14. Chung, C.-J., Khatkar, P., Xing, T., Lee, J., & Huang, D. (2013). Nice: Network intrusion detection and countermeasure selection in virtual network systems. *IEEE Transactions on Dependable and Secure Computing, 10*(4), 198–211.
15. Vaarandi, R. (2013). Detecting anomalous network traffic in organizational private networks. In *2013 IEEE International Multi-Disciplinary Conference on Cognitive Methods in Situation Awareness and Decision Support (CogSIMA)* (pp. 285–292). IEEE.
16. Mishra, P., Pilli, E. S., Varadharajan, V., & Tupakula, U. (2017). Out-VM monitoring for malicious network packet detection in cloud. In *Asia Security and Privacy (ISEASP), 2017 ISEA* (pp. 1–10). IEEE.
17. Payne, B. D., Martim, D. P. A., & Lee, W. (2007). Secure and flexible monitoring of virtual machines. In *Computer Security Applications Conference, 2007. ACSAC 2007. Twenty-Third Annual* (pp. 385–397). IEEE.
18. Stefanova, Z., & Ramachandran, K. (2017). Network attribute selection, classification and accuracy (NASCA) procedure for intrusion detection systems. In *2017 IEEE International Symposium on Technologies for Homeland Security (HST)* (pp. 1–7). IEEE.
19. Ingle, L., & Pakle, G. K. (2016). NIDSV: Network based intrusion detection and countermeasure excerption in virtual environment using AODV protocol. In *International Conference on Inventive Computation Technologies (ICICT)* (Vol. 3, pp. 1–6). IEEE.

20. Wang, Z., & Zhu, Y. (2017). A centralized HIDS framework for private cloud. In *2017 18th IEEE/ACIS International Conference on Software Engineering, Artificial Intelligence, Networking and Parallel/Distributed Computing (SNPD)* (pp. 115–120). IEEE.
21. Jin, R., He, X., & Dai, H. (2017). On the tradeoff between privacy and utility in collaborative intrusion detection systems—a game theoretical approach. In *Proceedings of the Hot Topics in Science of Security: Symposium and Bootcamp* (pp. 45–51). ACM.
22. Singhal, A., & Ou, X. (2017). Security risk analysis of enterprise networks using probabilistic attack graphs. In *Network Security Metrics* (pp. 53–73). Berlin: Springer.
23. Mitchell, R., & Chen, R. (2016). Modeling and analysis of attacks and counter defense mechanisms for cyber physical systems. *IEEE Transactions on Reliability*, 65(1), 350–358.
24. Xu, J., Yuan, X., Yu, A., Kim, J. H., Kim, T., & Zhang, J. (2016). Developing and evaluating a hands-on lab for teaching local area network vulnerabilities. In *Frontiers in Education Conference (FIE), 2016 IEEE* (pp. 1–4). IEEE.
25. Yan, Q., & Yu, F. R. (2015). Distributed denial of service attacks in software-defined networking with cloud computing. *IEEE Communications Magazine*, 53(4), 52–59.
26. Jiang, X., Wang, X., & Xu, D. (2007). Stealthy malware detection through VMM-based out-of-the-box semantic view reconstruction. In *Proceedings of the 14th ACM Conference on Computer and communications Security* (pp. 128–138). ACM.
27. Ning, P., Cui, Y., & Reeves, D. S. (2002). Constructing attack scenarios through correlation of intrusion alerts. In *Proceedings of the 9th ACM Conference on Computer and Communications Security* (pp. 245–254). ACM.
28. Yun, Y., Xi-shan, X., & Zhi-chang, Q. (2011). A probabilistic computing approach of attack graph-based nodes in large-scale network. *Procedia Environmental Sciences*, 10, 3–8.
29. Hong, J. B., & Kim, D. S. (2016). Assessing the effectiveness of moving target defenses using security models. *IEEE Transactions on Dependable and Secure Computing*, 13(2), 163–177.
30. Roy, A., Kim, D. S., Trivedi, K. S. (2012). Scalable optimal countermeasure selection using implicit enumeration on attack countermeasure trees. In *2012 42nd Annual IEEE/IFIP International Conference on Dependable Systems and Networks (DSN)* (pp. 1–12). IEEE.
31. Padhy, R. P., Patra, M. R., & Satapathy, S. C. (2011). Cloud computing: Security issues and research challenges. *International Journal of Computer Science and Information Technology & Security (IJCSITS)*, 1(2), 136–146.
32. Ateniese, G., & Mangard, S. (2001). A new approach to DNS security (DNSSEC). In *Proceedings of the 8th ACM conference on Computer and Communications Security* (pp. 86–95). ACM.

Compressed Sensing for Image Compression: Survey of Algorithms

S. K. Gunasheela and H. S. Prasantha

Abstract Compressed sensing (CS) is an image acquisition method, where only few random measurements are taken instead of taking all the necessary samples as suggested by Nyquist sampling theorem. It is one of the most active research areas in the past decade. In this age of digital revolution, where we are dealing with humongous amount of digital data, exploring the concepts of compressed sensing and its applications in the field of image processing is very much relevant and necessary. The paper discusses the basic concepts of compressed sensing and advantages of incorporating CS-based algorithms in image compression. The paper also discusses the drawbacks of CS, and conclusion has been made regarding when the CS-based algorithms are effective and appropriate in image compression applications. As an example, reconstruction of an image acquired in compressed sensing way using l_1 minimization, total variation-based augmented Lagrangian method and Bregman method is presented.

Keywords Compressed sensing · Image compression · Nyquist sampling theorem

1 Introduction

Nyquist sampling theorem [1] states that it is possible to reconstruct the signal with less ambiguity from its samples if the rate of sampling is greater than or equal to twice the highest frequency content in the signal. The drawback of Nyquist sampling theorem is in some applications it is not possible to acquire large volume of input samples as required by Nyquist sampling theorem. For example, in MRI [2] imaging, input samples need to be taken in a quick span of time as it is very inconvenient for the patient and acquiring all the samples is not economically feasible. In some applications like satellite image processing, acquiring all the input samples will lead to high computational and storage cost. In contrast to Nyquist sampling theorem proposed by Shannon, compressed sensing [3] states that it is possible to reconstruct the signal

S. K. Gunasheela (✉) · H. S. Prasantha
Department of ECE, Nitte Meenakshi Institute of Technology, Bengaluru 560064, India
e-mail: gunasheela.ks.2012@gmail.com

with considerable accuracy by collecting very few samples when the signal is sparse in some basis, e.g. Fourier basis, DCT basis. There are many applications of compressed sensing in image processing. The main applications are image compression [4], single-pixel camera [5], MRI, astronomy, sensing networks, etc.

Image compression can be broadly classified into lossless and lossy compression. In most of the applications, lossy compression is acceptable since it reduces the storage overhead by reproducing the image of considerable quality. Conventional lossy compression algorithms are based on transform coding techniques. In general, the conventional compression algorithms encode and decode the image in the following fashion. First, the analog signal is captured by the digital camera, which converts it into a digital signal. According to sampling theorem, the analog signal is sampled to obtain sufficiently large number of input samples. To perform compression, first the digitized input signal samples are transformed into a sparse domain, e.g. DFT, DCT and wavelet domain. Sparse domain means that the energy in the digitized input samples is concentrated in very few coefficients; remaining coefficients are either zero or have negligible value. Therefore, for encoding process, only the location and value of significant coefficients are considered. Significant coefficients are quantized and then entropy coded. In the reconstruction process, entropy coding is reversed and inverse transform is applied to get back the original image with considerably good image quality. This is the basic principle of conventional lossy compression algorithms. Now arises the question, anyway we are discarding the insignificant coefficients in the transform domain and coding only significant coefficients, so why do we need to sense all the samples in the first place to discard it later? Can we just only sense significant coefficients where the locations of significant coefficients are unknown when the image is not acquired yet? The answer to the above questions is yes. This is where compressed sensing-based algorithms come into picture. The aim of this paper is to provide some insights into how compressed sensing works and why it is sensible to use compressed sensing-based algorithms in image compression applications.

2 Related Work

Compressed sensing is first introduced in the literature as an abstract mathematical idea [6–8], where the authors prove that it is possible to reconstruct the signal with less number of samples than as is required by the sampling theorem when the original signal is sparse is some basis. Based on the reconstruction strategy used to recover the original image, compressed sensing recovery algorithms can be classified into four different kinds. They are greedy algorithms, l_1 minimization algorithms, Total variation minimization algorithms and Bregman distance minimization algorithms.

Greedy algorithms are based on the principle that every iteration finds the best local optima in the immediate neighbourhood, and it is expected to find the global optima at the end of the iterations. It works in certain applications, but it is not guaranteed that the algorithm finds the global optima in all applications [9]. Examples of greedy algorithms are matching pursuit (MP [10]) and orthogonal matching pursuit (OMP

[11]). Matching pursuit and its variants like orthogonal matching pursuit work on the principle that when the signal is represented as linear combinations of atoms in a redundant dictionary, some selected atoms match the structure of the original signal. The advantages of using matching pursuit algorithms are they are easy to implement and their convergence rate is also very high. The drawbacks are there is no guarantee in theory regarding its capability to attain true sparse representation. These drawbacks led to the development of l_1 minimization-based algorithms.

l_1 minimization was first used to reconstruct impulse train [12], later comes the several papers [13–17] regarding how and why under certain conditions minimization works. Based on those early results, a newly compressed sensing framework has been proposed which provides the theoretical guarantee for minimization. In [18–20], it is proved that l_0 minimization is equivalent to minimization under certain conditions. This new compressed sensing framework leads to improvement in performance. The restricted isometry property (RIP) introduced in [21] provides the theoretical guarantee for the recovery. It is proved that if the measurement matrix satisfies RIP to a certain extent then it guarantees the sparse recovery of the signal. The problem with RIP is that it is very difficult to verify practically. But it is very likely the matrix satisfies the RIP when it is random. Later, a detailed research is proposed which investigates the reconstruction without RIP is performed in [22]. The formulation of basis pursuit which seeks the solution by l_1 minimization is an example of l_1 minimization algorithms, and the advantage of basis pursuit [23, 24] is it works when greedy algorithms fail under certain circumstances. Many applications of l_1 minimization in compressed sensing framework have been discussed in the literature [25–29]. The drawback of l_1 minimization techniques is they do not preserve the edge information very accurately leading to blurring of edges and sharp corners. This led to the development of reconstruction techniques based on total variation minimization.

Total variation (TV) minimization [30] was first introduced in the literature for image denoising by Rudin, Osher and Fatemi, which is famously known as ROF model for image denoising. From then on, this technique is popularly used for image restoration purposes. Detailed discussion on TV minimization can be found in [31–33]. Even though TV minimization preserves the sharp edges and prevents blurring, the TV functions are not linear and not differentiable, which makes them mathematically more complex compared to l_1 minimization. In [34–36], different variations in TV minimization have been developed for image restoration and image deblurring. In [37], a robust TVAL3 algorithm is presented to reconstruct the images acquired by the single-pixel camera.

Bregman distance was first used by the mathematician L. Bregman in 1967 [38]. In [39], the application of Bregman iteration for l_1 regularization and compressed sensing-based applications is discussed. Bregman iteration is also used for denoising, and a fast variant of Bregman iteration called linearized Bregman algorithm is reported in [40]. In [41], split Bregman algorithm has been proposed which can be used in image denoising, MRI applications and compresses sensing applications. Split Bregman method has many advantages with regard to the computational complexity. Due to its parallelizing nature, it can be efficiently implemented to have faster computation. The error-forgetting and error-cancellation properties of Bregman iteration are presented in [42].

3 Compressed Sensing Overview

This section presents the brief overview of compressed sensing. First of all, let us see what is compressed sensing. As the name suggests, it captures the compressed form of signal. To illustrate this with example, consider an image of size 1 MB. Conventional image compression algorithms like JPEG transform the image into DCT domain; then consider only significant coefficients. The location and value of significant coefficients are encoded. Now, one can store the image in just few kilobytes. The principle of compressed sensing is, rather than collecting million samples, collecting just few samples, i.e. in CS, number of samples taken is proportional to the compressed size of an image rather than its original size. The beauty of compressed sensing is that one is still able to reconstruct the original image almost exactly. To understand the concept, let us revisit the classical problem in linear algebra:

$$Ax = b \tag{1}$$

where A is an $m \times n$ matrix. It is assumed that the matrix A has full rank. That is, the columns of matrix A are linearly independent. x is the n-dimensional vector, which is unknown. In general, x can be any real or complex data like an image, sound wave, etc., based on application. b is the m-dimensional measured vector. In compressed sensing, one measures b not x, i.e. few linear combination of signal $x(m \ll n)$. The aim is to reconstruct x from b.

In real-world applications, it is not possible to have exact measurements. There are so many perturbations involved like instrument error, round off error, sensing or measurement error, transmission error, etc. The actual reconstruction problem is:

$$b = Ax + N \tag{2}$$

N refers to perturbations, which vary with different applications. To simplify the discussion, consider the noise-free ideal case of Eq. (1). Since very few measurements are considered in CS ($m \ll n$), the number of unknowns is greater than the number of equations. Clearly, it is an underdetermined system of linear equations. From results of linear algebra, an underdetermined system of equations either has no solution (when the solution is not spanned by the columns of A) or has infinitely many solution. It is assumed that matrix A has full rank, i.e. columns of A spans the entire space R^n, therefore the problem has infinitely many solutions. So, now the problem is, how can one select just one solution out of many possible solutions available. The answer to the question is different algorithms use different strategies to do this. It depends on application and how exactly one wants to recover the data. In general, this process is called regularization.

Let $R(x)$ be the regularization function. The problem in (1) can be now rewritten as:

$$\min_x R(x) \text{ such that: } b = Ax \qquad (3)$$

The choice of $R(x)$ varies with application. If one considers $R(x) = x_0$, the problem in (1) can be written as:

$$\min_x \|x\|_0 \text{ such that: } b = Ax \qquad (4)$$

Even though the above equation seems to represent the ideal sparsest solution possible for a given underdetermined system of equations, there is no theoretical proof to guarantee the uniqueness of the solution. Moreover, this is combinatorial problem and it is NP hard [43]. As a result, it is not practical to use l_0 norm as a regularization function.

The classical solution to Eq. (1) is to choose $R(x)$ as the squared l_2 norm. When $R(x) = x_2^2$, by using the concept of Lagrange multipliers, Eq. (1) can be written as:

$$L(x) = \|x\|_2^2 + \lambda(Ax - b) \qquad (5)$$

λ denotes the Lagrange multipliers. Solution of Eq. (5) is popularly known as pseudo-inverse solution or least squares solution, which is given by:

$$x' = A^T \left(AA^T\right)^{-1} b = A^{\text{psuedo-inverse}} b \qquad (6)$$

The choice of l_2 norm for regularization has been extensively used in the research community. It is great in some applications as well, and it is mathematically not complex. But, for even better reconstruction results, it is necessary to explore different choices for the regularization term. This is an important and significant topic of research in the past few years. Among many available choices for regularization term, l_1 is very much popular because of its tendency to make solution sparse.

4 Compressed Sensing Compression and Reconstruction

Consider an input image X, which has m rows and n columns. The image is transformed into a single vector $x = \text{vex}(X)$, where X has mn components. Consider an $mn \times mn$ measurement matrix M. The measurement matrix M should satisfy the restricted isometry property [21] and other necessary conditions. Common examples are matrices with Gaussian or Bernoulli i.i.d. entries and orthogonal DFT matrix. Depending on the number of measurements N chosen ($N \ll mn$), N rows are selected at random from M. Now we have the matrix K (compressed sensing matrix) where the rows in K are randomly selected N rows from M. First the original signal is multiplied with the sparsifying matrix M'. The resulting signal Y is multiplied with the matrix K. The resulting signal is R. Figures 1, 2, 3 and 4 represent compressed sensing methodology using toy diagram. Figure 1 shows the vectorization of input image matrix. Figure 2 shows the selection of N random rows in the measurement matrix M. Figure 3 shows the multiplication of image vector with the measurement matrix. Figure 4 shows the multiplication of Y with the sensing matrix.

Now the reconstruction is performed using optimization techniques. Starting point Y' can be calculated as $Y' = K^{inverse} R$. With this Y', K and R, it is possible to reconstruct the signal X by iterative optimization algorithms. Now the reconstruction problem is solving a system of underdetermined system of equations, $X = KY'$. An underdetermined system of equations can have infinitely many solutions. In order to choose a solution from a set of infinitely possible solutions, the condition of sparse solution is imposed. That is, the solution which has less number of nonzero coefficients is considered. Sparse solution constraint is employed by minimizing the

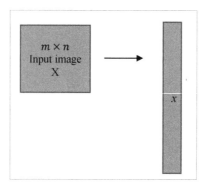

Fig. 1 Vectorization of image matrix X

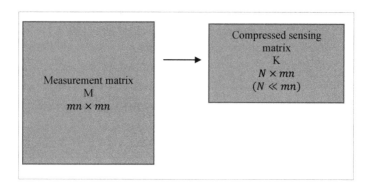

Fig. 2 Generation of compressed sensing matrix K

Fig. 3 Image sparsification

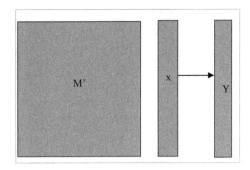

Fig. 4 Sparse reconstruction using compressed sensing matrix

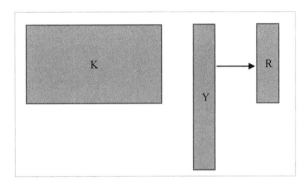

l_0 norm. In [18], authors have proved that in many cases, l_1 norm is equivalent to l_0 norm. Therefore, one can use l_1 norm instead of l_0 as it is easy to implement.

5 Results

This section describes the results of image reconstruction using three solvers, one is based on l_1 minimization, TV minimization and one more based on Bregman distance minimization. The data sets used are 256 × 256 Barbara image, 256 × 256 phantom image and 512 × 512 Lena image. These images are reconstructed using YALL1 [44] solver, TVAL3 [37] and Bregman solver [41], respectively. Figure 5 shows the phantom image reconstructed with only 0.25 sampling rate. Figure 6 shows Lena image reconstructed using Bregman solver with only 0.65 sampling rate. Figure 7 shows Barbara image reconstructed using YALL1 solver with only 0.6 sampling rate. Sampling rate corresponds to the number of samples used out of all the samples. Experiments are performed in MATLAB 2017b version, Mac operating system with 1.6 GHz Intel Core i5 processor and 8 GB RAM.

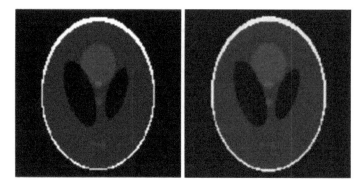

Fig. 5 Reconstruction using TVAL3 solver, original image(left), reconstructed image (right)

Fig. 6 Reconstruction using Bregman solver, original image (left), reconstructed image (right)

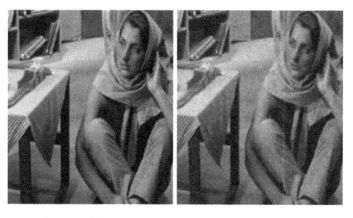

Fig. 7 Reconstruction using YALL1 solver, original image (left), reconstructed image (right)

6 Conclusion

The paper briefly describes the basic concepts and different methods for reconstruction of images acquired in compressed sensing way. Compressed sensing is a very effective way for image compression since it uses considerably less number of input samples to reconstruct the image. It is an excellent alternative for conventional image compression algorithms in case of medical imaging and satellite image compression. The main drawback of compressed sensing algorithms is reconstruction quality and computational time during reconstruction. The reconstruction quality can be enhanced by choosing appropriate reconstruction technique for a particular application. Computational time can be improved by using graphic processing units (GPUs) and parallel architectures. Improving the computational time during reconstruction by manipulations in algorithms construction or by using new optimization strategies is the scope of research in the area of compressed sensing.

References

1. Marks, R. J., II. (1991). *Introduction to Shannon sampling and interpolation theory*. Berlin: Springer.
2. Lustig, M., Donoho, D., & Pauly, J. (2007). Sparse MRI: The application of compressed sensing for rapid MR imaging. *Magnetic Resonance in Medicine, 58*(6), 1182–1195.
3. Donoho, D. L. (2006). Compressed sensing. *IEEE Transactions on Information Theory, 52*(4), 1289–1306.
4. Martin, G., Bioucas-Dias, J. M., & Plaza, A. (2015). HYCA: A new technique for hyperspectral compressed sensing. *IEEE Transactions on Geoscience and Remote Sensing, 53*, 2819–2831.
5. Takhar, D., Laska, J. N., Wakin, M. B., Duarte, M. F., Baron, D., Sarvotham, S., Kelly, K. F., & Baraniuk, R. G. (2006, January). A new compressed imaging camera architecture using optical-domain compression. *Computational Imaging IV, 6065*, 43–52.
6. Candès, E., Romberg, J., & Tao, T. (2006). Robust uncertainty principles: Exact signal reconstruction from highly incomplete frequency information. *IEEE Transactions on Information Theory, 52*(2), 489–509.
7. Donoho, D. (2006). Compressed sensing. *IEEE Transactions on Information Theory, 52*(4), 1289–1306.
8. Candès, E., & Tao, T. (2006). Near optimal signal recovery from random projections: Universal encoding strategies? *IEEE Transactions on Information Theory, 52*(12), 5406–5425.
9. Black, P. E. (2005, February). "Greedy algorithm" in Dictionary of Algorithms and Data Structures [online], U.S. National Institute of Standards and Technology.
10. Mallat, S. G., & Zhang, Z. (1993). Matching pursuits with time-frequency dictionaries. *IEEE Transactions on Signal Processing, 41*(12), 3397–3415.
11. Pati, Y. C., Rezaiifar, R., & Krishnaprasad, P. S. (1993). Orthogonal matching pursuit: Recursive function approximation with applications to wavelet decomposition. In *Twenty-Seventh Asilomar Conference on Signals, Systems and Computers*.
12. Santosa, F., & Symes, W. W. (1986). Linear inversion of band-limited reflection seismograms. *SIAM Journal on Scientific and Statistical Computing, 7*(4), 1307–1330.
13. Donoho, D. L., & Stark, P. B. (1989). Uncertainty principles and signal recovery. *SIAM Journal on Applied Mathematics, 49*, 906–931.
14. Donoho, D. L., & Logan, B. F. (1992). Signal recovery and the large sieve. *SIAM Journal on Applied Mathematics, 52*, 577–591.

15. Donoho, D. L., & Huo, X. (2001). Uncertainty principles and ideal atomic decomposition. *IEEE Transactions on Information Theory, 47*, 2845–2862.
16. Donoho, D. L., & Elad, M. (2003). Optimally sparse representation in general (nonorthogonal) dictionaries via l1 minimization. *Proceedings of the National Academy of Sciences of the United States of America, 100*, 2197–2202.
17. Elad, M., & Bruckstein, A. M. (2002). A generalized uncertainty principle and sparse representation in pairs of RN bases. *IEEE Transactions on Information Theory, 48*, 2558–2567.
18. Candes, E., & Tao, T. (2006). Near optimal signal recovery from random projections: Universal encoding strategies. *IEEE Transactions on Information Theory, 52*(12), 5406–5425.
19. Candes, E., Romberg, J., & Tao, T. (2006). Robust uncertainty principles: Exact signal reconstruction from highly incomplete frequency information. *IEEE Transactions on Information Theory, 52*(2), 489–509.
20. Donoho, D. (2006). Compressed sensing. *IEEE Transactions on Information Theory, 52*(4), 1289–1306.
21. Candes, E., & Tao, T. (2005). Decoding by linear programming. *IEEE Transactions on Information Theory, 51*(12), 4203–4215.
22. Zhang, Y. (2008, July). *On theory of compressed sensing via $\ell 1$-minimization: Simple derivations and extensions*. CAAM Technical Report TR08-11, Department of Computational and Applied Mathematics, Rice University.
23. Chen, S. S. (1995). *Basis pursuit*, Ph.D. thesis, Stanford University, Department of Statistics.
24. Chen, S. S., Donoho, D. L., & Saunders, M. A. (2001). Atomic decomposition by basis pursuit. *SIAM Review, 43*(1), 129–159.
25. Rudelson, M., & Vershynin, R. (2005). Geometric approach to error-correcting codes and reconstruction of signals. *International Mathematics Research Notices, 64*, 4019–4041.
26. Donoho, D., & Tanner, J. (2005). Neighborliness of randomly-projected simplices in high dimensions. *Proceedings of the National Academy of Sciences, 102*(27), 9452–9457.
27. Candles, E. J., Wakin, M. B., & Boyd, S. (2008). Enhancing sparsity by reweighted $\ell 1$ minimization. *Journal of Fourier Analysis and Applications, 14*(5), 877–905.
28. Tropp, J. (2006). Just relax: Convex programming methods for identifying sparse signals. *IEEE Transactions on Information Theory, 51*, 1030–1051.
29. Tsaig, Y., & Donoho, D. (2005). Extensions of compressed sensing. *Signal Processing, 86*(3), 533–548.
30. Rudin, L., Osher, S., & Fatemi, E. (1992). Nonlinear total variation based noise removal algorithms. *Physica D, 60*, 259–268.
31. Chambolle, A. (2004). An algorithm for total variation minimization and applications. *Journal of Mathematical Imaging and Vision, 20*, 89–97.
32. Chan, T. F., Esedoglu, S., Park, F., & Yip, A. (2004). *Recent developments in total variation image restoration*. CAM Report 05-01, Department of Mathematics, UCLA.
33. Geman, D., & Reynolds, G. (1992). Constrained restoration and the recovery of discontinuities. *IEEE Transactions on Pattern Analysis and Machine Intelligence, 14*(3), 367–383.
34. Geman, D., & Yang, C. (1995). Nonlinear image recovery with half-quadratic regularization. *IEEE Transactions on Image Processing, 4*(7), 932–946.
35. Yang, J., Yin, W., Zhang, Y., & Wang, Y. *A fast algorithm for edge-preserving variational multichannel image restoration*. Technical Report 08-09, CAAM, Rice University, Submitted to SIIMS.
36. Yang, J., Zhang, Y., & Yin, W. *An efficient TVL1 algorithm for deblurring of multichannel images corrupted by impulsive noise*. TR08-12, CAAM, Rice University, Submitted to SISC.
37. Li, C. (2009). *An efficient algorithm for total variation regularization with applications to the single pixel camera and compressed sensing*.
38. Bregman, L. (1967). The relaxation method of finding the common points of convex sets and its application to the solution of problems in convex optimization. *USSR Computational Mathematics and Mathematical Physics, 7*, 200–217.
39. Yin, W., Osher, S., Goldfarb, D., & Darbon, J. (2008). Bregman iterative algorithms for L1-minimization with applications to compressed sensing. *SIAM Journal on Imaging Sciences, 1*, 142–168.

40. Cai, J. F., Osher, S., & Shen, Z. *Linearized Bregman iterations for compressed sensing*. UCLA CAM Report, 08-06.
41. Goldstein, T., & Osher, S. *The Split Bregman method for L1 regularized problems*. UCLA CAM Report, 08-29.
42. Yin, W., & Osher, S. (2012). Error forgetting of Bregman iteration. *Journal of Scientific Computing, 54*(2), 684–698.
43. Natarajan, B. K. (1995). Sparse approximate solutions to linear systems. *SIAM Journal on Computing, 24*(2), 227–234.
44. Zhang, Y. (2009, May). *User's guide for YALL1: Your algorithms for L1 optimization*. Technical Report TR09-17, Department of Computational and Applied Mathematics, Rice University.

Intrusion Detection System Using Random Forest on the NSL-KDD Dataset

Prashil Negandhi, Yash Trivedi and Ramchandra Mangrulkar

Abstract In the modern world of interconnected systems, network security is gaining importance and attracting a lot of new research and study. Intrusion detection systems (IDSs) form an integral part of network security. To enhance the security of a network, machine learning algorithms can be applied to detect and prevent network attacks. Taking advantage of the robust NSL-KDD dataset, we have employed the supervised learning algorithm random forests to train a model to detect various networking attacks. To further increase the classification accuracy of our model, we have employed the use of famous data mining technique of feature selection. Smart feature selection using Gini importance has been employed to reduce the number of features. Experimental results have shown that our model not only runs faster but also performs with a higher accuracy.

Keywords NSL-KDD · Machine learning · Random forest · Classification · Computer networks · Cybersecurity

1 Introduction

Due to the development of widespread highly connected systems, the Internet is slowly changing how people live, study, and work. Every year, even more networks are established for social, business, and government purposes. However, with the benefits of such a system come some drawbacks. With such huge-scale development, securing these networks is getting harder day by day. This is evident from the fact that with each passing year there is an increasing number of hacking and intrusion incidents. Thus, to prevent these breaches, research in cybersecurity is quickly gaining importance.

P. Negandhi · Y. Trivedi (✉) · R. Mangrulkar
Department of Computer Engineering, Dwarkadas J. Sanghvi College of Engineering, Mumbai 400056, India
e-mail: yashtrivedi618@gmail.com

© Springer Nature Singapore Pte Ltd. 2019
N. R. Shetty et al. (eds.), *Emerging Research in Computing, Information, Communication and Applications*, Advances in Intelligent Systems and Computing 906,
https://doi.org/10.1007/978-981-13-6001-5_43

The concept of intrusion can be succinctly explained as the breaching of a computer network's security and forcing it to enter into an insecure state. Once this happens, the attackers can then exploit the resources of the network such as confidential data and use it for malicious purposes. To secure against such events, organizations usually use a firewall, which acts as an efficient first line of defense that protects private networks. Firewalls work by monitoring the incoming and outgoing packets in a computer network and based on a predefined set of rules allow or block specific connections. However, there are myriad ways to bypass firewalls, a common one being passing malicious packets via ports that are usually left open and unguarded by the network (e.g. SMTP, HTTP). This is where the importance of intrusion detection systems gains prominence as an efficient second line of defense.

An intrusion detection system (IDS) can be a software or a device application that manages the security of a computer network by monitoring the network traffic for malicious activities or policy violations. It works by gathering data from various devices in the network and analyzing this information to identify potential security breaches. There are two main types of IDS [1]:

(a) Signature-based IDS (SBIDS): They work by searching network traffic for specific patterns and signatures (e.g., byte sequences) or known harmful sequences frequently used by malware. This list of signatures is stored within the system and compared with the ongoing traffic to detect intrusions. Although such systems can easily catch known attacks, they find it impossible to detect novel attacks for which no historic pattern is available.
(b) Anomaly-based IDS (ABIDS): These systems are quickly gaining prominence due to their ability to detect hitherto unknown attacks. They commonly work by leveraging machine learning algorithms to train a model that classifies network traffic as either legitimate or malicious. This results in a more robust IDS that can perform better in uncertain environments.

In this research paper, we have developed an intrusion detection system based on the principles of anomaly-based IDS. We have leveraged the machine learning algorithm of random forest classification due to the robustness it offers on large datasets [2]. Random forest models tend to adapt to sparsity; that is, their convergence rate depends mostly on the strong features present in the dataset. In spite of this, the importance of preprocessing and prior feature selection cannot be ignored. Thus, to further improve the accuracy of our model, smart feature selection using Gini importance has been deployed. To train the model, we have used an improved version of the famous KDD dataset [3], called the NSL-KDD [4] dataset. Testing of the proposed model has yielded much higher accuracy than existing systems.

2 Related Work

With the recent advances in machine learning, especially deep learning, their application in novel domains has intensified. Ranging from geosciences to computer networks, machine learning algorithms are being applied to increase the performance of existing systems by leaps and bounds. The famous deep learning model of recurrent neural networks has been proposed by the researchers of [5] to classify network attacks. They have trained their model on NSL-KDD to achieve both binary as well as multiclass classification of networking attacks. Although a high amount of accuracy was achieved by the researchers, training the model required a lot of time (more than 10,000s for most combinations). This would be a huge disadvantage considering the dynamic and ever-changing nature of networking. Each day, potentially millions of new data points could be captured. Thus, integrating this new data to the existing model would require retraining the whole model and would be time-consuming as well as tedious.

An approach to use incremental learning to solve the intrusion detection problem has been proposed in [6]. The researchers have proposed the use of equality constrained-optimization-based ELM (C-ELM) to the problem of network intrusion detection (NID). However, as explained by them, an effective C-ELM model needs to be constructed by trial and error which is very time-consuming. Moreover, the model could produce high accuracy on a dataset, but sometimes the constraints of a custom networking system might necessitate a reconstruction of the model. This might not be feasible for the organization. The use of artificial neural networks (ANNs) on various networking datasets has been studied by the researchers in [7]. After finding that NSL-KDD dataset was one of the best datasets, they applied preprocessing techniques and feature reductions before training their model. The model was trained for both intrusion detection and attack classification with accuracies of 81.2 and 79.9% which would not be up to the mark for securing networks that require a high level of security like government and military networks.

Due to the huge amount of features in NSL-KDD dataset, smart feature selection for further analysis has also attracted a lot of attention from researchers. In [8], it was found that correlation-based feature selection (CFS) combined with information gain (IG) leads to the highest accuracy for classification of networking attacks. However, the features were reduced to such an extent that some relevant features (especially host-based traffic features) were left out. Although the time spent to train the model decreased slightly, the accuracy of the model also decreased. An improved version of the model used in [8] was proposed by the researchers in [9]. Adding an extra selection parameter of gain ratio (GR) was done to select more features and boost the accuracy of the model further.

3 Dataset Description

The famous KDD dataset [3] had many errors as explained by Tavallaee et al. in [10]. The NSL-KDD dataset was thus proposed to solve these issues. It has the below-mentioned advantages over the previously popular KDD dataset:

1. It has no redundant records present in the training set; thus, the classifiers applied on it will not tend to be biased toward records that are more frequent.
2. It does not have duplicate records present in the test sets also. Thus, the classifier performance is not skewed in favor of methods that have better detection rates on the more frequent records.
3. There is an inversely proportional relation between the number of selected records of each difficulty level group and the percentage of records in the original KDD dataset. This leads to a wider range of classification rates for ML algorithms. Thus, researchers can get an accurate assessment of the different learning techniques.
4. The size of the train and test sets is within reason, which results in researchers being able to perform experiments on the whole set without the necessity of selecting a small portion randomly. Consequently, results from applying the same algorithm would vary less as a result.

Each record of the NSL-KDD dataset has 41 attributes followed by a label assigned to them to give its class, associated attack or normal. The characteristics of all the attributes, that is, the attribute name, attribute description, and sample data present, are listed in Tables 1, 2, 3, and 4. The 42nd attribute in each record contains data about whether the record corresponds to an attack type or is a legitimate connection

Table 1 Basic features of each network connection vector

Attribute number	Attribute name	Attribute description	Sample data
1	Duration	Length of the connection's time duration	0
2	Protocol_type	Protocol used by the connection	udp
3	Service	Service used by the destination network	ftp_data
4	Flag	Status of the current connection: error or normal	SF
5	Src_bytes	Number of bytes of data transferred from source to destination in one connection	146
6	Dst_bytes	Number of bytes of data transferred from destination to source in one connection	0
7	Land	This variable takes the value 1 if source and destination IP addresses in addition to port numbers are equal. Value is 0 otherwise	0
8	Wrong_fragment	Total number of wrong fragments present in this connection	0
9	Urgent	Total number of urgent packets in this connection. Urgent bit is only activated for urgent packets	0

Table 2 Content-related features of each network connection vector

Attribute number	Attribute name	Attribute description	Sample data
10	Hot	Total number of 'hot' indicators in the content such as creating programs, entering a system directory and executing programs	0
11	Num_failed_logins	Number of failed login attempts	0
12	Logged_in	Indicates login status. It is 1 if successfully logged in, 0 if not	0
13	Num_compromised	Number of compromised conditions in the connection	0
14	Root_shell	If root shell is acquired, it is 1, 0 otherwise	0
15	Su_attempted	If 'su root' command has been tried or used, it is 1, 0 otherwise	0
16	Num_root	Number of root accesses or number of operations executed as a root in the connection	0
17	Num_file_creations	Number of file creation operations performed in the connection	0
18	Num_shells	Number of shell prompts	0
19	Num_access_files	Number of operations performed on files with access control	0
20	Num_outbound_cmds	Number of outbound commands given in a session which is ftp	0
21	Is_hot_login	If the login falls under the 'hot' list that is root or admin, it is 1, 0 otherwise	0
22	Is_guest_login	If the login falls under 'guest' login, it is 1, 0 otherwise	0

(normal). Table 5 contains the grouping of all the different attacks found in the dataset into four attack classes, namely DoSs, Probe, R2L, and U2R.

The NSL-KDD dataset is arguably one of the few open-source datasets which has a very comprehensive collection of labeled intrusion events. It provides very intriguing characteristics on the distribution of networking events and the dependencies between different attributes. These features have made it as an appropriate benchmark for intrusion detection research.

Table 3 Time-related traffic features of each network connection vector

Attribute number	Attribute name	Attribute description	Sample data
23	Count	Number of connections to a destination host which is the same as current connection, for the past two seconds	2
24	Srv_count	Number of connections to a service (port number) which is the same as current connection for the past two seconds	2
25	Serror_rate	The percentage of connections triggering the Flag (4) s0, s1, s2, or s3, for the connections accumulated in Count (23)	0
26	Srv_serror_rate	The percentage of connections triggering the Flag (4) s0, s1, s2 or s3, for the connections accumulated in Srv_count (24)	0
27	Rerror_rate	The percentage of connections triggering the Flag (4) REJ, for the connections accumulated in Count (23)	0
28	Srv_rerror_rate	The percentage of connections triggering the Flag (4) REJ, for the connections accumulated in Srv_count (24)	0
29	Same_srv_rate	The percentage of connections to the same service, for the connections accumulated in Count (23)	1
30	Diff_srv_rate	The percentage of connections to different services, for the connections accumulated in Count (23)	0
31	Srv_diff_host_rate	The percentage of connections to different destination machines, for the connections accumulated in Srv_count (24)	0

Table 4 Host-based traffic features in a network connection vector

Attribute number	Attribute name	Attribute description	Sample data
32	Dst_host_count	Number of connections that have the same destination host IP address	150
33	Dst_host_srv_count	Number of connections that have the same port number	25

(continued)

Table 4 (continued)

Attribute number	Attribute name	Attribute description	Sample data
34	Dst_host_same_srv_rate	The percentage of connections to the same service, for the connections accumulated in Dst_host_count (32)	0.17
35	Dst_host_diff_srv_rate	The percentage of connections to different services, for the connections accumulated in Dst_host_count (32)	0.03
36	Dst_host_same_src_port_rate	The percentage of connections to the same source port, for the connections accumulated in Dst_host_srv_count (33)	0.17
37	Dst_host_srv_diff_host_rate	The percentage of connections to different destination machines, for the connections accumulated in Dst_host_srv_count (33)	0
38	Dst_host_serror_rate	The percentage of connections triggering the Flag (4) s0, s1, s2 or s3, for the connections accumulated in Dst_host_count (32)	0
39	Dst_host_srv_s error_rate	The percentage of connections triggering the Flag (4) s0, s1, s2 or s3, for the connections accumulated in Dst_host_srv_count (33)	0
40	Dst_host_rerror_rate	The percentage of connections triggering the Flag (4) REJ, for the connections accumulated in Dst_host_count (32)	0.05
41	Dst_host_srv_r error_rate	The percentage of connections triggering the Flag (4) REJ, for the connections accumulated in Dst_host_srv_count (33)	1.00

Table 5 Attack class with attack types

Attack class	Attack type
DoS	Neptune, Pod, Teardrop, Smurf, Back, Land
Probe	Ipsweep, Satan, Portsweep, Nmap, Mscan
R2L	Ftp_write, Imap, Guess_Passwd, Warezclient, Spy, Phf, Multihop, Warezmaster
U2R	Loadmodule, Buffer_overflow, Rootkit, Sqlattack, Perl

4 Proposed Intrusion Detection System

The proposed intrusion detection system uses the NSL-KDD dataset to train a model to classify all the 22 attacks present in the dataset. The supervised machine learning algorithm, random forest, is used. The trained model then undergoes testing, and its performance is measured. The steps in the proposed model are as shown in Fig. 1.

A. **Overview of the Proposed System**

Step (1)—Split the full NSL-KDD dataset into two parts: training data (75%) and testing data (25%).
Step (2)—Perform preprocessing of the data.
Step (3)—Select appropriate features from all the attributes of the dataset.
Step (4)—Apply random forest classification on the resultant data to obtain a trained model.
Step (5)—Evaluate the performance of the model by giving it testing data as input.

B. **Preprocessing and Feature Selection**

The entire NSL-KDD dataset is randomly split into two parts: training data (75%) and testing data (25%). The training data are labeled, while the testing data are unlabeled. It was found that three attributes of the dataset [Protocol_type (2), Service (3) and Flag (4)] have string values. Since all classifiers need attributes in numeric format, they must first be converted into equivalent numeric format. A python script that factorizes these columns in the dataset is implemented. It ensures that each unique entry is replaced by a different integer in the dataset. Moreover, the mapping between the unique entries and the replacing integers is also returned so that we can use this later on to ensure that the testing data are also consistent with this format.

Feature selection is an important process that selects a subset of relevant features from all the features after applying some evaluation criteria. Although RF has an implicit feature selection, applying feature selection to high-dimensional data boosts the performance of random forest classification algorithm [11]. Feature selection is done by 'Gini importance' [12] for our model by following the iterative approach. Firstly, each features worth in the dataset was calculated by using Gini importance criteria as shown in Fig. 2. Secondly, using this feature importance, noise and other

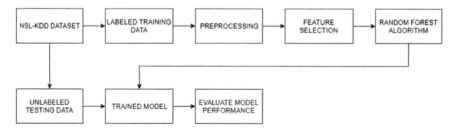

Fig. 1 Steps in proposed system

Intrusion Detection System Using Random Forest ...

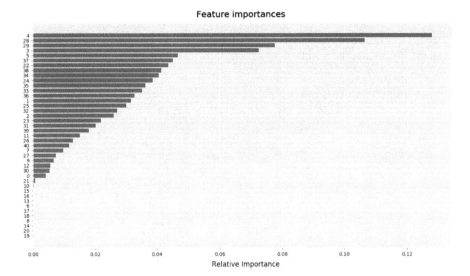

Fig. 2 Relative importance of all attributes

low importance features were eliminated. Thirdly, the final subset of features is evaluated to find if it is better than the whole. This process is done iteratively by keeping on eliminating features with the lowest importance. The accuracy of each subset was evaluated and formulated in Table 6, and the best subset was the final subset of features.

Gini Importance:

This is defined as a feature significance score which gives us a relative ranking of all the attributes. It is actually a consequence of the training associated with the RF classifier. An optimal split, at a node 't' inside the binary trees T of the RF, is achieved using the Gini impurity $i(t)$. Gini impurity is actually a computationally efficient approximation for the entropy of the system. It measures how well a prospective split would separate the data belonging to two separate classes at that node.

Now, with $p_k = \frac{n_k}{n}$ indicating the n_k sample points from class $k = \{0, 1\}$ from a total of n samples at a node 't', Gini impurity, $i(t)$, is computed as follows:

$$i(t) = 1 - p_1^2 - p_0^2$$

Its decrease Δi that results from the splitting and sending of samples to the two sub-nodes t_l (left) and t_r (right) with the sampling fractions $p_l = \frac{nl}{n}$ and $p_r = \frac{nr}{n}$ by a threshold th_Θ on an attribute Θ is defined as:

$$\Delta i(t) = i(t) - p_l i(t_l) - p_r i(t_r)$$

Table 6 Evaluation of different feature combinations

Serial No.	Selected features	Accuracy (%)
1	[4,38,29,34,37,28,25,35,3,2,22,1,32,5,36,23,33,40,31,39,7,9,24,0,12,30,11,26,27,21]	99.822
2	[4,38,29,34,37,28,25,35,3,2,22,1,32,5,36,23,33,40,31,39,7,9,24,0,12,30,11,26,27]	99.833
3	[4,38,29,34,37,28,25,35,3,2,22,1,32,5,36,23,33,40,31,39,7,9,24,0,12,30,11,26]	99.845
4	[4,38,29,34,37,28,25,35,3,2,22,1,32,5,36,23,33,40,31,39,7,9,24,0,12,30]	99.855
5	[4,38,29,34,37,28,25,35,3,2,22,1,32,5,36,23,33,40,31,39,7,9,24,0,12]	99.880
6	[4,38,29,34,37,28,25,35,3,2,22,1,32,5,36,23,33,40,31,39,7,9,24,0]	99.843
7	[4,38,29,34,37,28,25,35,3,2,22,1,32,5,36,23,33,40,31,39,7,9,24]	99.828
8	[4,38,29,34,37,28,25,35,3,2,22,1,32,5,36,23,33,40,31,39,7,9]	99.8303
9	[4,38,29,34,37,28,25,35,3,2,22,1,32,5,36,23,33,40,31,39,7]	99.8270
10	[4,38,29,34,37,28,25,35,3,2,22,1,32,5,36,23,33,40,31,39]	99.823

At the node, a comprehensive search is done over all of the variables Θ available there (surprisingly RF has a property that restricts this search to a randomly generated subset of the available features [12]), and over all the possible thresholds th$_\Theta$, the pair $\{\Theta, \text{th}_\Theta\}$ which leads to a maximal Δi is found out. The decrease in $i(t)$ resulting from this optimal split $\Delta i_\Theta(t, T)$ is calculated and aggregated for all nodes t in all the trees T present in forest, independently for all attributes Θ:

$$I_G(\Theta) = \sum_T \sum_t \Delta i_\Theta(t, T)$$

This is the quantity that gives us the Gini importance I_G which finally is an indicator of how often a specific attribute Θ was chosen for a split and how big is its cumulative discriminative value under the classification problem currently being studied.

As can be observed from Table 6, the combination at serial no. 5 generates the highest accuracy and thus the features selected for making the model are the attributes: [4,38,29,34,37,28,25,35,3,2,22,1,32,5,36,23,33,40,31,39,7,9,24,0,12].

C. Random Forest Classifier

A random forest classifier is an extremely flexible, easy to implement, supervised machine learning algorithm that produces exceptional results even without parameter tuning. It is simplicity, and popularity is further enhanced by the fact that it can be used for both classification and regression tasks, which form the majority of current machine learning tasks. Random forests are an ensemble learning method that builds multiple decision trees and merges them together such that the model results in a more accurate and stable prediction of the data's class. Random forests overcome the overfitting disadvantage that decision trees have. A diagram explaining the working of RF is shown in Fig. 3.

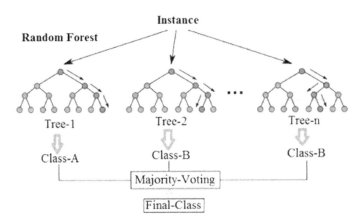

Fig. 3 Simplified explanation of random forest classification

Table 7 Evaluation result of proposed system

Parameters	Basic RF model	Proposed model
Accuracy (%)	99.752	99.880
Inaccuracy (%)	0.248	0.120

After selecting the features in the previous step, the training data are given to a script that runs a RF classification algorithm on the training data. Since the number of features is high, a higher number of decision trees are needed. The classifier was run with 1000 trees due to this.

D. Evaluating Model Performance

After applying the random forest classification, we get a trained model that is then used for the proposed intrusion detection system. The part of the dataset that was initially set aside for testing purposes is now used to evaluate the performance of the dataset. The testing data are factorized such that they are consistent with the training data and the relevant features used to train the model are only used by the model to calculate the accuracy of the intrusion detection system (IDS). The accuracy of the system is tabulated in Table 7.

5 Conclusion and Future Work

Intrusion detection systems (IDSs) play a crucial role in the realm of network security. Thus, in this paper, we have proposed a robust and highly accurate IDS that uses smart feature selection based on Gini importance. This boosts the accuracy of the system to much higher levels than not just a naive RF model but also other existing systems. Future work will be based on modeling a hybrid IDS with binary classification (intrusion detection) as a first step followed by multiple classification (attack-type detection) which happens simultaneously with alerting the network analyst to take a closer look at suspicious connections in a network.

References

1. Gyanchandani, M., Rana, J. L., & Yadav, R. N. (2012). Taxonomy of anomaly based intrusion detection system: A review. *International Journal of Scientific and Research Publications*, *2*(12), 1–13.
2. Biau, G. (2012). Analysis of a random forests model. *Journal of Machine Learning Research*, *13*, 1063–1095.
3. http://archive.ics.uci.edu/ml/datasets/kdd+cup+1999+data.
4. http://nsl.cs.unb.ca/NSL-KDD/.
5. Yin, C., Zhu, Y., Fei, J., & He, X. A deep learning approach for intrusion detection using recurrent neural networks. In *IEEE Access* (Vol. 5).

6. C.-H. Lee, Su, Y.-Y., Lin, Y.-C., & Lee, S.-J. (2017). Machine learning based network intrusion detection. In *2017 2nd IEEE International Conference on Computational Intelligence and Applications*.
7. Ingre, B., & Yadav, A. *Performance analysis of NSL-KDD dataset using ANN*. In SPACES-2015, Department of ECE, K L University.
8. Wahba, Y., ElSalamouny, E., & ElTawee, G. (2015, May). Improving the performance of multi-class intrusion detection systems using feature reduction. *IJCSI International Journal of Computer Science Issues, 12*(3).
9. Chae, H., Jo, B., Choi, S.-H., & Park, T. Feature selection for intrusion detection using NSL-KDD. *Recent Advances in Computer Science*.
10. Tavallaee, M., et al. (2009). A detailed analysis of the KDD CUP 99 data set. In *Proceedings of the Second IEEE Symposium on Computational Intelligence for Security and Defence Applications 2009*.
11. Breiman, L. (2004). Consistency for a simple model of random forests in *Technical Report 670*. Technical report, Department of Statistics, University of California, Berkeley, USA.
12. Breiman, L. (2001). Random forests. *Journal of Machine Learning, 45*, 5–32. https://doi.org/10.1023/A:1010933404324.

Dual-Mode Wide Band Microstrip Bandpass Filter with Tunable Bandwidth and Controlled Center Frequency for C-Band Applications

Shobha I. Hugar, Vaishali Mungurwadi and J. S. Baligar

Abstract This paper presents a unique approach for designing dual-mode wide band BPF with tunable bandwidth and controlled center frequency for C-band (4–8 GHz) applications. The proposed filter is designed using radial stub-loaded dual-mode $\lambda_g/2$ resonator to get wide passband. The dual-mode behavior of the resonator, i.e., odd- and even-mode resonance frequencies are realized by inserting a radial stub at the center of the resonator and further the size of filter is reduced by folding the resonator. A modified feed structure which embraces the two arms of the resonator is used to obtain two transmission zeros in upper stop band. By keeping all calculated dimensions of filter fixed and by varying only radial angle θ (in degrees) of radial line stub, FBW is tuned while controlling center frequency. From simulation results, it is observed that the designed filter has very good passband characteristics, a wide 3-dB passband from 4.4 to 7.8 GHz with center frequency at 6 GHz, fractional bandwidth of 56.6%, return loss S11 more than 13 dB, and transmission loss S21 better than 0.3 dB, respectively.

Keywords SIR—Stepped impedance resonator · BPF—Bandpass filter · FBW—Fractional bandwidth

1 Introduction

Modern wireless communication technology demands compact, light weight, sharp-frequency-selective, planar RF/microwave filters with wide stop band. Dual-mode bandpass filters provide good features like excellent passband performance, good frequency selectivity, and small size compared to conventional bandpass filters. A sym-

S. I. Hugar (✉)
Department of ECE, Sapthagiri College of Engineering, Bangalore 560057, India
e-mail: Shobha_hugar@yahoo.co.in

V. Mungurwadi
Department of ECE, Sarvajanik College of Engineering, Surat 395011, Gujarat, India

J. S. Baligar
Department of ECE, Dr. AIT, Bangalore 560056, India

metrical microstrip structure with perturbation element at symmetrical plane forms a generic dual-mode microstrip resonator. A half wave length microstrip line loaded centrally with open circuit/short circuit stub or grounded via forms a dual mode resonators [1]. In literature, ring resonator [2], square-ring resonator [3], multi-arc resonators [4], open-loop stub-loaded resonators [5], stepped impedance resonator [6], meander loop resonator [7], etc. have been used to design dual-mode resonators.

T-shaped stub-loaded microstrip resonator has been proposed in [8] to configure dual-mode broadband bandpass filter (BPF) with multiple controllable transmission zeros. In this work, five controllable transmission zeros have been achieved by using two capacitive and inductive S-L couplings. The proposed filter was designed for center frequency 5.8 GHz, insertion and return losses of 1.75 and 28.18 dB, respectively, and 3-dB fraction bandwidth of 8.9% at center frequency are reported.

In [6], symmetrical T-shaped stub-loaded stepped impedance resonator has been used to design dual-mode high-frequency-selective BPF. The designed filter has fractional bandwidth of 10.1%, centered at 5.23 GHz. The filter has wide upper stop band in range of 5.9–12.9 GHz with harmonic suppression level of more than 20 dB.

A pair of symmetrical T-type tap-connected open-ended stubs have been used as load in a half wavelength stepped impedance resonator (SIR) [9] to design miniature dual-mode microstrip bandpass filter at the central frequency of 15.5 GHz. Symmetrical T-type stubs have been used to obtain good in-band and out of band performances. The in-band return loss and insertion losses of greater than 12 and 2 dB, respectively, are reported. The filter has wide stop band from 18 to 40 GHz with attenuation level of 20 dB.

Authors Hong-Shu Lu, Qian Li, Jing-Jian Huang, Xiao-Fa Zhang, and Nai-Chang Yuan have proposed a meander loop the resonator with Minkowski-like pre-fractal geometry with first-order iteration embedded at the center to design dual-mode bandpass filter to achieve improved BPF performance in [7].

In [10] by employing an open-circuited stub which can be switched by a pin diode, a switchable notched band microstrip bandpass filter has been developed. A notched band at 5.8 GHz has been achieved by switching an open-circuited stub via pin diode. The designed filter has notched band centered at 5.75 GHz with 5-dB rejection and FBW of 3.2%. The designed filter has 3-dB pass band width of 3.4 GHz centered at 4.9 GHz and insertion loss of 0.5 dB in passband.

In this work, a novel approach for designing dual-mode wide band pass filter based on radial line stub-loaded $\lambda_g/2$ resonator for C-band applications is proposed. In literature, many authors have used uniform impedance open-stub or short-stub to design dual-mode resonators [6, 8–10]. But in this work, radial line stub is used since they found to work better than low-impedance rectangular stubs [11]. The advantages of using radial line stub compared to conventional stubs are that it provides spurious-free broad passband. At high frequency, its size becomes relatively small compared to conventional stubs [11]. By introducing modified stepped impedance feed lines which embraces two horizontal arms of resonator and by adjusting gap 'g' between feed lines two transmission zeros are realized in upper stop band.

The paper is organized as follows, in Sect. 2, designing of proposed dual-mode resonator is discussed. Section 3 illustrates dual-mode characteristics of the resonator.

Section 4 validates designing of dual-mode wide bandpass filter from proposed resonator. ADS momentum results are discussed in Sect. 5. Section 6 is conclusion.

2 Dual-Mode Resonator

Figure 1 shows layout of proposed dual-mode resonator composed of radial line stub at center of $\lambda_g/2$ resonator where λ_g guided wavelength at center frequency. The proposed filter is fabricated on RT/Duroid 6010 substrate with dielectric constant 10.8 and thickness of 1.27 mm. The dimensions of proposed filter are listed in Table 1.

3 Dual-Mode Characteristics of Proposed Resonator

In literature, it is reported that even- and odd-mode resonance frequencies of dual-mode resonator are controlled by varying either width or length of the open-stub [6, 12]. But in this work, even- and odd-mode resonance frequencies of the resonator are studied by varying radial angle θ (in degrees) of stub keeping its length and width fixed. It is observed that by varying radial angle θ odd-mode resonance frequency remains uninterrupted, whereas even-mode resonance frequency varies. Further it is also noticed that narrow radial angle results into larger bandwidth. To observe odd- and even-mode resonances, full-wave EM simulation is carried out using ADS.

As shown in Fig. 2, two coupling capacitors of value 0.2 pf are used for loose coupling. By varying radial angle θ, it is observed that odd-mode frequency remains

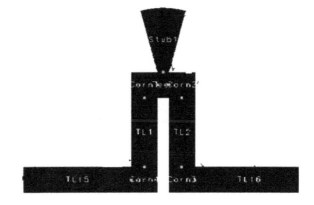

Fig. 1 Proposed dual-mode resonator

Table 1 Dimensions of proposed resonator in mm

W	L1	L2	L3	Θ in degrees	W3
0.67	2.47	2.75	1.67	10, 20, 30	0.3

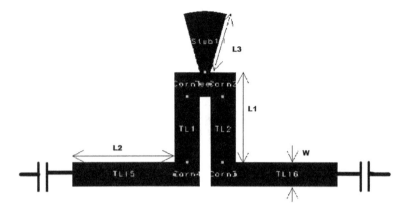

Fig. 2 Proposed dual-mode resonator with coupling capacitors for loose coupling

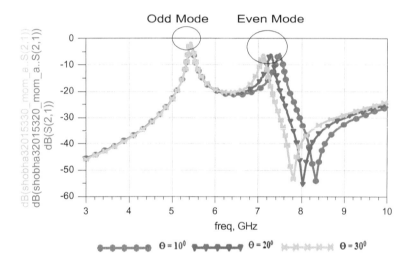

Fig. 3 Variation of even-mode frequency with radial angle θ (Θ in degrees)

unchanged while even-mode frequency changes with radial angle θ in degrees (Fig. 3).

4 Dual-Mode BPF Design

To validate design concept, two-pole wideband bandpass filter is designed from proposed dual-mode resonator. 50 ohms transmission line is used at input and output. To achieve two transmission zeros in upper stop band, a modified feed structure is

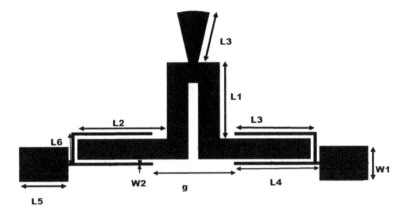

Fig. 4 Dual-mode wide bandpass filter

Table 2 Dimensions of proposed BPF in mm

L4	L5	L6	W1	W2	Θ in degrees	W3	L7
2.6	1.5	2.51	1.12	0.1	30	0.3	0.95

used which embraces two arms of resonator as shown in Fig. 4. The dimensions of proposed filter are listed in Table 2.

5 Result and Discussion

Figure 5 demonstrates frequency response of designed filter. From ADS momentum results, it is observed that filter has wide 3-dB passband from 4.4 to 7.8 GHz centered at 6 GHz. A transmission zero at 10 GHz with attenuation level of 34 dB is found at upper edge of passband. The transmission loss S21 in passband is better than 0.3 dB and return loss more than 13 dB. Two poles at 5.3 and 7 GHz are obtained. The proposed filter reports 56.6% fractional bandwidth at center frequency.

Figure 6 demonstrates tuning of 3-dB bandwidth by varying radial angle θ. Further it is observed that by varying θ, 3-dB fractional bandwidth can be tuned without interrupting the center frequency and shift in location of transmission zeros in upper stop band.

The filter response is observed for three different values of θ, and results are tabulated in Table 3.

This work exploits the fact that narrow radial angle results into wider bandwidth.

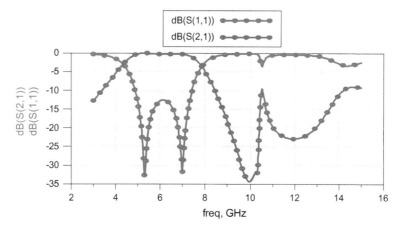

Fig. 5 S_{21} and S_{22} parameters of proposed filter

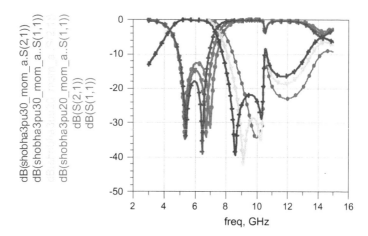

Fig. 6 Demonstrates tuning of 3-dB bandwidth by varying radial angle θ in degrees

Table 3 Tuning FBW by varying θ

Θ in degrees	FBW in %	Center frequency f_c in GHz	Location of transmission zeros (GHz)
10°	56.6	6	10
20°	51.6	6	9.125, 10.25
30°	48	6	8.594, 10.29

6 Conclusion

A dual-mode wide bandpass filter is designed using RT/Duroid6010 substrate with dielectric constant 10.8 and thickness 1.27 mm. The designed filter has spurious-free wide passband from 4.4 to 7.8 GHz with center frequency at 6 GHz, fractional bandwidth of 56.6%, return loss S11 more than 13 dB, and transmission loss S21 better than 0.3 dB, respectively. Bandwidth tuning is achieved by varying radial angle θ, while without changing center frequency. From the results, it is observed that bandwidth can be tuned by varying radial angle θ, without disturbing center frequency and wider bandwidth can be achieved by narrow radial angle. Designed filter is suitable for C-band (4–8 GHz) applications.

Acknowledgements The authors would like to thank Mr. Sumit Shatabdi Bharat Electronics limited Bangalore, for his support in this work.

References

1. Lee, K. C., Su, H. T., Haldar, M. K. (2015, June). A review of centrally loaded multimode Microstrip Resonators for band pass filter design. *International Journal of Electronics and Communications*. https://doi.org/10.1016/j.aeue.2015.06.006. Journal ISSN: 1434-8411.
2. Tan, B.-T., Yu, J.-J., Chew, S.-T., Leong, M.-S., & Ooi, B.-L. (2005, January). A miniaturized dual-mode ring bandpass filter with a new perturbation. *IEEE Microwave and Wireless Components Letters, 53*(1), 343–345.
3. Huang, X.-D., & Cheng, C.-H. (2006, January). A novel coplanar-waveguide bandpass filter using a dual-mode square-ring resonator. *IEEE Microwave and Wireless Components Letters, 16*(1), 13–15.
4. Kang, W., Hong, W., & Zhou, J.-Y. (2008, March). Performance improvement and size reduction of Microstrip dual-mode bandpass filter. *Electronics Letters, 44*(6), 421–422.
5. Hong, J.-S., Shaman, H., & Chun, Y.-H. (2007, August). Dual-mode microstrip open-loop resonators and filters. *IEEE Transactions on Microwave Theory and Techniques, 55*(8), 1764–1770.
6. Hu, C., Xiong, X., Wu, Y., & Liao, C. (2013). Design dual-mode bandpass filters based on symmetrical T-shaped stub-loaded stepped impedance resonators with high frequencies selectivity. *Progress in Electromagnetics Research B Journal, 55*, 325–345.
7. Lu, H.-S., Li, Q., Huang, J.-J., Zhang, X.-F., & Yuan, N.-C. (2016, August). Dual-mode dual-band microstrip bandpass filter with high selection performance. In *Progress in Electromagnetic Research Symposium (PIERS)*, Shanghai, China.
8. Yao, Z., Wang, C., & Kim, N. Y. (2014). A dual-mode bandpass filter with multiple controllable transmission-zeros using t-shaped stub-loaded resonators. *The Scientific World Journal, 6*, Article ID 572360.
9. Liu, X., & Hu, H. (2011). A miniature dual-mode microstrip bandpass filter at 15.5 GHz. In *Cross Strait Quad-Regional Radio Science and Wireless Technology Conference*.
10. Rabbi, K., & Budimir, D. (2013). Miniaturized dual-mode microstrip bandpass filter with a reconfigurable notched band for UWB applications. In *IEEE International Conference 2013*. https://doi.org/10.1109/MWSYM.2013.6697586.

11. Giannini, F., Sorrentino, R., & Vrba, J. (1984). Planar circuit analysis of microstrip radial stub. *IEEE Transactions on Microwave Theory and Techniques, 32*(12), 1652–1655.
12. Hua, C., Chen, C., Miao, C., & Wu, W. (2011). Microstrip bandpass filters using dual mode resonators with internal coupled lines. *Progress in Electromagnetics Research C Journal, 21,* 99–111.

ADHYAYAN—An Innovative Interest Finder and Career Guidance Application

Akshay Talke, Virendra Patil, Sanyam Raj, Rohit Kr. Singh, Ameya Jawalgekar and Anand Bhosale

Abstract ADHYAYAN is an innovative mobile application which determines a user's interest in a particular domain and nurtures them effectively so that they can pursue career in the field which they are interested in. The system takes into account social media posts, results of a test and application activity to find out the interest of users in different fields and then assists, guides and evaluates them continuously to improve their skills in these fields. ADHYAYAN is a three-tier system which consists of a front-end, middle layer, and back-end. Front-end is an Android application which provides personalized GUI for each user. Middle layer is Firebase, while back-end is a server hosted on 'Google Cloud Platform'. An algorithm has been developed for ADHYAYAN which calculates the ratio of user's interest in different domains and eventually feeds are generated in the same ratio on user's profile. To cater the increasing need of skilled employees in different fields and promote interest-based learning, ADHYAYAN has been proposed to overcome various limitations and drawbacks of existing solutions.

Keywords Unemployability · Social media · Continuous evaluation · Test · Feeds · Profile · Skills · Career · Short term profile · Long term profile · Personalized · Real-time

---

A. Talke (✉) · V. Patil · S. Raj · R. K. Singh · A. Jawalgekar · A. Bhosale
International Institute of Information Technology (IIIT), Pune, India
e-mail: akshaytalke@gmail.com

V. Patil
e-mail: viren21096@gmail.com

S. Raj
e-mail: sanyamraj22@gmail.com

R. K. Singh
e-mail: rohit.12.ks@gmail.com

A. Jawalgekar
e-mail: jawalgekar007@gmail.com

© Springer Nature Singapore Pte Ltd. 2019
N. R. Shetty et al. (eds.), *Emerging Research in Computing, Information, Communication and Applications*, Advances in Intelligent Systems and Computing 906,
https://doi.org/10.1007/978-981-13-6001-5_45

1 Introduction

Education system is the most important aspect to look upon for a country's future development. As today's children are tomorrow's driving force for the economy, it is very important that these students are trained and mentored in the field of their liking so that they are interested and passionate about their work, to gain maximum efficiency. In today's scenario, most of the people are dissatisfied with their jobs because the nature of their job is not in line with their passion, reducing the overall happiness index of the country. Today, students are not able to do critical analysis of anything and are not able to look at things with their own perspective [1]. They take decisions influenced by their elders or peers and end up taking a career path which is not of their interest. This leads to unemployability, college dropouts, suicides, etc., thus affecting the youth of the country in an adverse manner. Taking into account all of the problems mentioned and the current existing methodologies related to such issues, we have come up with a system that tries to address these issues to a greater extent. Our system collects data from various popular social media platforms and results from a test which is based on 'Gardner's theory of Multiple Intelligences' [2]. It helps users identify their area of interest, nurtures them in that field and also helps them to build a career in it by providing various career options.

2 Existing Methodologies

There are some methodologies that are taking care of this issue, but all of them have some limitations.

To determine student's interest and motivate them, some privileged schools appoint psychologists or mentors. These facilities incur extra charges in schools which is not affordable to major sector of the society. Also, psychologists or mentors make decisions based on personal opinions which may be biased [3].

Different online Web sites and mobile applications elaborate career options available but for this, the users must know their interest as a prerequisite.

Various start-ups or companies visit schools and take aptitude tests, based on which the student's interest is determined. This one-time assessment does not provide continuous evaluation and is also expensive and time-consuming.

To overcome these limitations of the present solutions, we have proposed a system which is an android application available to everyone at a very low cost, is personalized for each user and is automated. Moreover, it is a continuous evaluation process to give more precise interest prediction.

3 Background and Related Work

There are systems proposed to address these issues. A system was proposed which classifies students based on their school levels, personal and professional tendencies and then builds a profile using semantic classification in ontology [4]. But this system is restricted to only one domain, i.e., Academics. Another system was proposed in which the system collects Facebook post data using 'Facebook Graph API' and translate the data from Thai language to English language using the 'Google Translate API', then it uses 'Microsoft Cognitive Services API' to find the useful keywords which can be mapped to the words in their self-created static dictionary. Based on the mapping, they find field similar to user's interest [5]. This system is limiting itself to determination of interest only, no further action is taken. Also, this system is only doing text mapping to static dictionary without text analysis and only considers textual posts.

Whereas our system is not limiting itself just to academics but expands into various other domains which can be categorized into the general umbrellas namely Arts, Sports and Academics. Our system takes into account textual as well as non-textual posts (images, videos, etc.). It also does continuous evaluation and provides feeds to motivate students and enhance their knowledge about different career options that they can take up.

4 Implementation

Our system consists of three layers namely front layer, middle layer, and end layer.

4.1 Front Layer

The user registers through Email ID and Password. User has to then provide Facebook username, Instagram username and Twitter handle if they have an account on the mentioned platforms. Then the user proceeds to the questionnaire section designed on the basis of 'Gardner's Theory of Multiple Intelligences' which is a set of defined questions that need to be answered by the user. On the basis of the above-collected information, a personalized profile is generated for the user in our mobile application. Some of the features present in our UI is mentioned below:

- The main profile feed page which would show the user the feeds, i.e. the suggestions by the system. These suggestions could be some information about the field of interest, trending topics, career options, success stories in their field of interest to motivate them, places, and events they could visit.

- There would also be a timeline to track the list of events happening around the user and for the same, a calendar would be maintained to suggest future events in the area of user's interest.
- A Google custom search bar to search for any query.

4.2 Middle Layer

Firebase serves as the middle layer for our system. Firebase provides us with the user's authentication services with its features like refreshing access tokens on its own for 'Facebook Login' and other authentication services without the need of maintaining a server for the same. 'Firebase Cloud Functions' is used to pass the data collected from the front-end to our back-end system and call the back-end server API hosted on 'Google Cloud Platform (GCP)'. We have a virtual machine created on GCP which hosts our back-end server. GCP and Firebase complement each other fairly well as both of them belong to Google.

4.3 End Layer

Our back-end server uses Facebook User's 'Access Token' to call 'Facebook Graph API' and deduce relevant information from posts, feeds, pages liked, video, statuses, etc., of the user. The back-end server gets the Instagram username and passes it to the Instagram Crawler (a Python Script). The acquired 'Twitter handle' is passed to an R script to get the tweets on the user's timeline using 'Twitter API' [6]. The activities of the user on our application are evaluated continuously and passed on to the API hosted on back-end. This data helps in providing more and more relevant feeds. The back-end also evaluates the result of the test based on 'Gardener's Theory of Multiple Intelligences.' It collects all the images from all the above sources and passes to 'Google Cloud Vision API' which gives five labels to each image [7]. The labels are then passed to 'DatumBox's Topic Classification API' which helps us in categorizing the images into either of academics, arts, or sports. Keywords extracted from the textual data like tweets, Facebook posts, etc., are passed to 'DatumBox's Keyword Extraction API' which helps us in categorizing again [8]. The total number of entities in each category are counted and fed to our Algorithm henceforth, referred as 'Profile & Domain Relevance (P.D.R)' Algorithm. P.D.R generates output as a ratio which tells us the proportion of a user's interest in the three mentioned domains which helps us in providing feeds to the user in the calculated ratio. All the data from social media and our application is gathered, updated and fed to P.D.R in real time (Fig. 1).

Feed Generation The feeds are categorized into suggestions and motivation. Suggestions can be categorized as trending topics, tips, and events or competitions being held at or around the user's location. In motivation, career opportunities for different domains are presented to users with the success rates and success stories in that

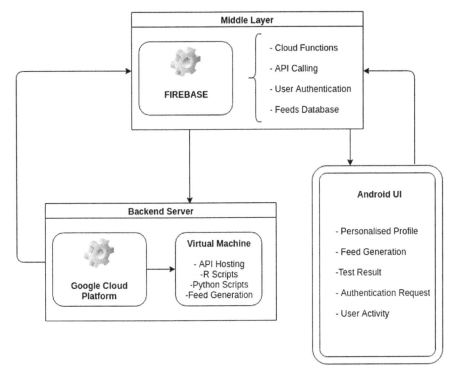

Fig. 1 System Architecture of ADHYAYAN

particular domain. All this information is presented in the form of quizzes, fruitful games, textual and non-textual feeds.

The feeds are generated by the following methods.

- meetup API: It gives nearby events happening around the location of the user. The events are labeled into 24 categories by the Web site itself [9].
- news API: It provides latest news and feeds from various websites based on technology, sports, and entertainment [10].
- Real-time match statuses using API's from different sports Web sites.
- And some manually entered static feeds in the form of textual and non-textual information(based on our research on specified domains).

5 Profile and Domain Relevance (P.D.R) Algorithm

5.1 Aim

To determine the Ratio of Interests in Domains (i.e. Interest (D_1): Interest (D_2):...: Interest (D_n))

5.2 Variables Used and Explanation

1. "m" : It denotes the number of sources from which data is collected. As of now, the system has five sources(i.e. Facebook, Instagram, Twitter, Test [2] and Activities performed on our APP)
2. "n" : It denotes the number of domains in which user interest is classified. As of now, the system has three domains (i.e., Academics, Arts and Sports)
3. "S_i" : Denotes the ith Source
4. "D_j" : Denotes the jth Domain
5. "W" : Weightage Matrix

$$W = \begin{bmatrix} W_{S_1 D_1} & W_{S_1 D_2} & .. & W_{S_1 D_n} \\ W_{S_2 D_1} & W_{S_2 D_2} & .. & W_{S_2 D_n} \\ \vdots & \vdots & & \vdots \\ W_{S_m D_1} & W_{S_m D_2} & .. & W_{S_m D_n} \end{bmatrix} \quad (1)$$

   (a) '$W_{S_i D_j}$' in Eq. (1) denotes the weightage given to data from Source 'i' in Domain 'j'.
   (b) The purpose behind using W matrix and giving weightage to data is that people don't post on social media with the same interest level as in their real life. For example, an intelligent student interested in maths Olympiads may participate in 10 Olympiads, but posts only 1 Olympiad post on social media. This 1 post has to be given a high weightage in order to reduce the gap between number of post on social media and number of real-life activities done.
   (c) 'W' matrix is universal and would be used for every user.
   (d) We get values of '$W_{S_i D_j}$' by studying Posts from 100 profiles from Facebook, Twitter, Instagram each and then classifying them into domains. The values of 'W' matrix are presented in Table 1.

6. "P_s/P_l" : Profile Matrix.

$$P = \begin{bmatrix} P_{S_1 D_1} & P_{S_1 D_2} & .. & P_{S_1 D_n} \\ P_{S_2 D_1} & P_{S_2 D_2} & .. & P_{S_2 D_n} \\ \vdots & \vdots & & \vdots \\ P_{S_m D_1} & P_{S_m D_2} & .. & P_{S_m D_n} \end{bmatrix} \quad (2)$$

Table 1 Table representing values of 'W' Matrix for 4 sources and 3 domains

Source/Domain	Academics	Arts	Sports
Facebook	22.62	12.24	15.43
Instagram	90.95	7.75	12.66
Twitter	6.80	11.36	15.87
Test	2	1	2

(a) $P_{S_i D_j}$ in Eq. (2) denotes the number of Posts of a User from Source 'i' in Domain 'j'.
(b) P_s : Profile Matrix(Short Term)
    i. P_s is active for 3 days. A snapshot of P_s is taken on every third day. After which new P_s is initialized. In all, five snaps are taken in span of 15 days.
    ii. After 15 days a new cycle starts.
    iii. Use of P_s : It helps in dealing with sudden spike of number of topics discussed in a particular domain. Such spike in social media are caused due to a popular event in real world. For example, recent death of Avicii(a Swedish musician, DJ) [11] caused number of people posting posts related to him (Note: system classifies such posts into arts). But, in reality these people may not be highly interested in music. People tend to talk/post/share on the same topic on which their peers are talking /posting /sharing, resulting in sudden spike of posts related to a specific domain.
(c) P_l : Profile Matrix (Long Term)
    i. P_l Is active infinitely. It is updated every 15 Days, by aggregating the values of all 5 Snapshots of P_s and then aggregating this value (aggregated value of 5 P_s) with previous value of P_l.
(d) Unlike W matrix, P is user specific.
(e) Larger weightage is given to P_l than P_s while determining the ratio of final feeds.

7. "P.W" (dot product of matrices) : Calculates Profile domain relevance

    (a) P_s.W : Short-term Profile domain relevance
    (b) P_l.W : Long-term Profile domain relevance

8. "TD_i" : Total Interest in i^{th} Domain

$$TD_i = \sum_{a=1}^{m}((W_{S_a D_i}).(P_{S_a D_i})) \qquad (3)$$

Note: Formula remains same for Short Term (P_s) and Long Term (P_l).

9. "P(TD_i)": Percentage of Total Interest in i^{th} Domain

$$P(TD_i) = \frac{\sum_{a=1}^{m}((W_{S_a D_i}).(P_{S_a D_i}))}{\sum_{b=1}^{n}(\sum_{a=1}^{m}((W_{S_a D_b}).(P_{S_a D_b})))} \qquad (4)$$

Note: Formula remains same for Short Term (P_s) and Long Term (P_l)

5.3 Algorithmic Steps

START

1. Read data for i^{th} Source(such that i = 1:m)
2. Generate P_s
3. IF(Cycle of 15 Days Completes)

    (a) Generate P_l
    (b) Start new cycle for P_s

4. Compute $(P_s).(W)$ and $(P_l).(W)$ (Refer Eqs. (1) and (2))
5. Compute $P(TD_i)$ [for all i=1:n] (Refer Eq. (4))
6. Round off each $P(TD_i)$ to multiples of 5.
7. Generate feeds for i^{th} domain(i = 1:n) in percentage determined by corresponding $P(TD_i)$

END

5.4 Capping Rules

1. Capping rule ensures that users are always exposed to feeds from all the domains at all the times. This gives them opportunity to explore new domains.
2. Upper cap: no $P(TD_i)$ can exceed 90%
3. Lower cap : no $P(TD_i)$ can fall below 5%.

6 Working of Algorithm using a Pseudo User

- A pseudo user profile was created on Instagram, Twitter, and Facebook. The ratio of academics, arts, and sports posts was kept 1:1:8.
- This profile has major interest in sports and minor interest in arts and academics.
- Data was collected from these sources and then categorized into different domains with help of different machine-learned modelś API(s) mentioned previously.
- Table 2 is generated at the end of categorizing images from posts.
- "Image ID" denotes the id of image which is unique to every image.
- "$Lable_1...Label_5$" are the labels generated for a particular image using 'Google Cloud Vision API'. These labels are classified into Academics, Arts, Sports and Others labeled as 1, 2, 3 and 4 respectively (i.e. 1-Academics, 2-Arts, 3-Sports and 4-Others), using 'Datumbox Topic classification API'.
- "Category" stands for final category in which the image is classified based on the label classification.
- Table 3 is generated after categorizing textual data from posts.

Table 2 Table representing image data classification into specific domain

Image ID	Label$_1$	Label$_2$	Label$_3$	Label$_4$	Label$_5$	Category
1041...7799	Player (3)	Football player (3)	Sport venue (3)	Team sport (3)	Soccer player (3)	3
1041...2122	Sports (4)	Team sport (3)	Fun (3)	Football player (3)	Player (3)	3
1041...8823	Barechested (2)	Male (2)	Eyebrow (2)	Muscle (4)	Leisure (4)	2
1041...2174	Face (4)	Hair (4)	Sport venue (4)	Blond (2)	Human hair color (4)	4
1041...2179	Eyewear (4)	Vision care (4)	Human hair color (4)	Eyebrow (4)	Glasses (2)	4

Table 3 Table representing textual data classification into specific domain

S. No.	Textual data	Text domain
1	Former Arsenal goalkeeper David Seaman, who was a key cog in the 1998 and 2002 double-winning sides, called on fans https://t.co/q4OlEAiXOx	3
2	Virat Kohli guides India to fighting total against New Zealand in first ODI	3
3	Apple offers free battery replacement for this MacBook model	1
4	The International Monetary Fund expects India's role in the Indo-Pacific region's development to continue to expand https://t.co/ahpNumtAqb	4
5	An officer of the Central Reserve Police Force (CRPF) was killed in a gunfight with the Maoists in a dense forest i https://t.co/p2okanYa00	4

- Number of Posts in each domain, from each source, is calculated and P matrix is updated as given in Table 4.
- According to the Step 4 of P.D.R Algorithm, we get following Values:
  - $P(TD_1)$: Percentage Interest in Academics is 22.45%
  - $P(TD_2)$: Percentage Interest in Arts is 10.32%
  - $P(TD_3)$: Percentage Interest in Sports is 67.39% (Table 5).
- Rounding Off Values to Nearest Multiple of 5:
  - $P(TD_1)$: Percentage Interest in Academics is 20%
  - $P(TD_2)$: Percentage Interest in Arts is 10%
  - $P(TD_3)$: Percentage Interest in Sports is 70%.
- Output : The generated ratio of Interests in Domains is:

$$Academics : Arts : Sports = 2 : 1 : 7$$

- Initial Feed can be generated in the same ratio as mentioned above.
- Once the feeds are displayed on APP, the user activities are tracked and updated on server, which act as a fifth source in the PDR.

Table 4 Table representing values of 'P_s' Matrix

Source/Domain	Academics	Arts	Sports
Facebook	3	5	15
Instagram	1	0	11
Twitter	1	1	8
Test	7	10	20

Table 5 Table representing values of Matrix ('P_s'.'W')

Source/Domain	Academics	Arts	Sports
Facebook	67.86	61.20	231.45
Instagram	90.95	0	139.26
Twitter	6.80	11.36	126.96
Test	14	10	40

7 Conclusion

An article from 'The Hindu' states 'Unemployability is a bigger crisis than unemployment' [12]. Statistics show that in all 57% of Indian youth has suffered from some degree of unemployability. Such huge crisis could have been avoided if individuals would have followed their passion and developed skill set in the field they wanted to. Our system finds student's interest area and makes constructive measures. As of now, we have taken into account only three general domains namely arts, sports, and academics as when we generalize the specific career options, we get these three as generic domains but in future, we aim at expanding these general domains into more specific ones.

Smartphones have become an integral part of a student's life and lot of time is invested in various uses of smartphone. So, we are trying to leverage this energy to make them do fruitful and productive activities.

References

1. Problems in Indian Education System. http://surejob.in/10-fundamental-problems-with-education-system-in-india.html. Last accessed on April 29, 2018.
2. Gardener's Theory of Multiple Intelligences. https://tinyurl.com/y98n7gyf. Last accessed on April 27, 2018.
3. Tiffany Iskander, E., Gore, P., Bergerson, A. A., & Furse, C. (2013). Gender disparity in engineering: Results and analysis from school counsellors survey and national vignette. In *2013 IEEE Antennas and Propagation Society International Symposium (APSURSI)*.
4. Alimam, M. A., Seghiouer, H., El Yusufi, Y. (2014). Building profiles based on ontology for career recommendation in E-leaming context. In *International Conference on Multimedia Computing and Systems (ICMCS)*.

5. Pisalayon, N., Sae-Lim, J., Rojanasit, N., & Chiravirakul, P. (2017). Identifying personal skills and possible fields of study based on personal interests on social media content. In *6th ICT International Student Project Conference (ICT-ISPC)*.
6. Twitter API. https://developer.twitter.com/en/docs. Last accessed on April 29, 2018.
7. Google Cloud Vision API. https://cloud.google.com/vision/docs/. Last accessed on April 29, 2018.
8. DatumBox API. http://www.datumbox.com/machine-learning-api/. Last accessed on April 29, 2018.
9. Meetup API. https://tinyurl.com/yaaqf7r4. Last accessed on April 27, 2018.
10. News API. https://newsapi.org/. Last accessed on April 29, 2018.
11. Avicii. https://en.wikipedia.org/wiki/Avicii. Last accessed on April 29, 2018.
12. Unemployability is a bigger crisis than unemployment, says Kalam. https://tinyurl.com/y9q65yc8. Last accessed on April 29, 2018.

Implementation of Cure Clustering Algorithm for Video Summarization and Healthcare Applications in Big Data

Jharna Majumdar, Sumant Udandakar and B. G. Mamatha Bai

Abstract The Data Mining Techniques provide useful ways to generate desired patterns from the large data and establish relations between them to solve problems using data analysis. This paper focuses on a data mining algorithm called CURE, and its applications on Health Care and Video data. Big Data consists of large volume, ever growing Datasets with multiple sources. Big Data in Health Care is an emerging area which helps healthcare organizations for their analytics and reporting needs. Data Mining Techniques, predictive analytics, and prescriptive analytics are some of the methods to analyze the healthcare data and derive useful information for several applications. On the other hand, Video Processing is an emerging area of research which gives rise to variety of applications like object tracking, shot detection, Video Summarization, etc. This paper discusses the application of CURE clustering algorithm on Video Processing for generating Video Summary and application of the same algorithm on Big Data Health Care Dataset for deriving disease related information.

Keywords Big Data Analytics · Video Processing · Data Mining Techniques · Health Care · CURE · Video Summarization

1 Introduction

Big Data Analytics is an emerging technology and an area of research with its vast application areas. With technology advancement, the era of data (text, numerical, and video) is growing abundantly and handling of such data is a major challenge in real-time scenario. This paper discusses the two applications of Health Care and

---

J. Majumdar · S. Udandakar · B. G. Mamatha Bai (✉)
Department of MTech CSE, Nitte Meenakshi Institute of Technology, Bangalore, India
e-mail: mamathamane@gmail.com

J. Majumdar
e-mail: jharna.majumdar@gmail.com

S. Udandakar
e-mail: sumantudandakar@gmail.com

Video Summarization on big data using Data Mining Techniques. The purpose of this paper is to deal with enormously produced data and predict the type of disease the patient is suffering with respect to Health Care, whereas in Video Summarization, the entire summary of the video for a given event is analyzed which helps the people to know the gist of the event in their busy schedule thereby updating the happenings around them.

1.1 Big Data in Health Care

Big data healthcare data pool consists of data belonging to a variety of sources. Healthcare data can be structured or unstructured. It has to be formatted and restructured properly for proper analysis at later stages. The data obtained thus is used in analytics to derive the information and relations to utilize the obtained relation in the application in order to get the desired result. The volume of information that one needs to negotiate has detonated to incomprehensible levels before decade, and in the meantime, the cost of information stockpiling has methodically decreased. Privately owned businesses and research foundations catch terabytes of information about their user's connections, business, online networking, and furthermore. The test of this time is to comprehend this ocean of information. This is where enormous information examination comes into picture.

The big data is described by five Vs [1]:

- Volume—The volume refers to the tremendous measure of the information which is generated each second that are bigger than the standard social database.
- Variety—The variety deals with the kind of attributes of the Dataset like numerical, continuous, categorical, ratio, and ordinal.
- Velocity—The velocity deals with the algorithm analysis for the computation of different types of attributes of process data.
- Veracity—The data accuracy of analysis depends on the veracity of the source data.
- Value—The value is the most important aspect of big data. The data is useless unless we convert it into value.

1.2 Video Summarization

In the recent years, there is an enormous growth in the ability of individuals to create, capture video, gradually leading to large personal, and also corporate video archives. In corporate field, there is a huge scope to Video Summarization. For example, a company may desire to summarize the CCTV video of the buildings so that only main events are included in the summary. An online education course material distributor may want to generate brief summaries of videos that concentrate on the most interesting snippets of the course.

A summary of a video must satisfy the following three principles:

- The video summary must consist of main events from the video. For example, a summary of a soccer game must show goals, as well as any other notable events such as the elimination of a player from the game.
- The summary should show some degrees of *continuity*.
- The summary should not contain *repetition* which can be difficult. Like, in soccer videos if the same goal to be replayed many times.

These three features, named the CPR (continuity, priority, and no repetition) is the basic goal of all Video Summarization methods.

2 Literature Survey

Raghupathi and Raghupathi [2] explains in detail the field of big data in Health Care and its advantages, and describes the examples found in the literature, different challenges, and comes up with conclusions. Big Data Analytics techniques can contribute to Health Care in many areas like

- Evidence-based medicine: clinical data and genomic data are analyzed to check matching treatments with outcomes and faster development of the more accurately targeted vaccines.
- Aiding diagnosis: predictive analytics are used in aiding diagnosis. With a series of tests performed, a set of symptoms, results, and observations could be provided for potential diagnosis and identifying diseases.

Using big data in Health Care and using analytics to identify and manage high-risk and high-cost patients are discussed in [3]. The patients with high risk are predicted using previous patients' data with same conditions as the current patient.

A proposed system which can help in diagnosing diseases like asthma, high/low blood pressure, diabetes which are most prevalent and costly chronic conditions which cannot be cured easily is presented in [4]. But if investigated early, they can prevent the disease to not cross threshold. In this system, implemented k-means algorithm and multi-rank walk algorithm used, on tweet streams to provide useful tweets, apply filtering technique with regard to topics. Then whenever user searches information on particular diseases that can help the user to get an overview of diseases quickly. In paper [5], the automated healthcare management system is proposed for a healthcare organization. It contains different modules like patient's module, doctor's module, and pharmacist's module which work together to manage the patient's data.

The various challenges and hurdles for big data healthcare sector are explained in detail in [6]. The patient's privacy, availability of accurate data, and security of data are few of the challenges. The CURE algorithm [7] stands for **C**lustering **U**sing **Re**presentatives, considered as an efficient algorithm for clustering of large databases. The algorithm starts by initial-clustering a sample of the whole data, uses representative points obtained from the sample to assign the remaining points in the Dataset. The clusters of the sample are produced by an agglomerative clustering maintaining the representative points in each cluster, also the nearest neighbor of every cluster.

The Video Summarization using k-means algorithm is presented in [8]. The paper explains a competent algorithm using k-means clustering and RGB histograms for summarizing video. It concentrates mainly on low-quality videos, e.g., videos from YouTube.

The proposed method of the system discussed is as follows:

1. Divide the input file into different number of segments for every k seconds such as frame 0 (f_0) ... frame n (f_n).
2. Consider first frame of every segment as representatives assigning values x_0 ... x_n.
3. Generate the histograms for x_0 ... x_n and call them as y_0 ... y_n.
4. The histograms are grouped into k groups with Euclidean distance being the error function.
5. Iterate the process for k groups and randomly pick a segment.
6. Add the picked segment to the list until the desired no. of frames are selected.
7. Group all the frames in the list to generate the final video summary.

The Local Binary Patterns (LBP) [9] proposes a method for feature extraction. The LBP is nonparametric descriptor which efficiently summarizes the local structures of an image. The LBP operator uses decimal numbers to name the pixels of an image, known as *Local Binary Patterns* or *LBP codes*, that encodes the local neighbors surrounding each pixel.

3 Proposed Methodology

The proposed system works for two applications for CURE Algorithm: One for Video Summarization and another for big data in Health Care. For Video Summarization, the input is the video shots corresponding to an event and the extracted frames are used to extract the LBP features. Then the clustering technique is applied on the features to get the clusters and extract the key frames and then generate the video summary.

For big data, the input Dataset is normalized and then the clustering techniques are applied to the Dataset to generate the clusters. Then the clusters are analyzed by plotting the graph which predicts the type of Thyroid disease such as hypothyroid, normal, and hyperthyroid conditions (Fig. 1).

Implementation of Cure Clustering Algorithm for Video ...

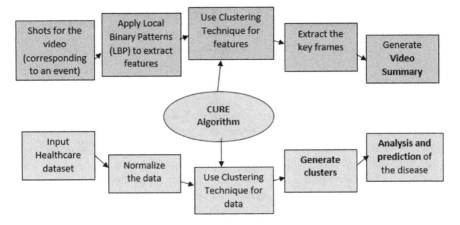

Fig. 1 Flow diagram of the proposed system

3.1 CURE Algorithm

The **C**lustering **U**sing **RE**presentatives (CURE) algorithm uses a hierarchical clustering approach. In CURE, random sample of the input Dataset is generated and the data is divided into different partitions. A constant number of sufficient distant points within a cluster are chosen and elected as representatives of the cluster. The clusters with the nearest pair of representatives are merged at each step of the algorithm.

CURE (Total no of points, n)

Input: Set of points, S
Output: n clusters

(a) For each cluster p, p.mean contains the mean and a set of c representative points wherein, m.closest holds the clusters nearest to m.
(b) The set of input points are included in a n-d tree "T."
(c) Considering every input point to be a separate cluster, calculate m.closest value for every m, where m and m.closest distances are stored in ascending ordering.
(d) It is further followed by inserting each cluster into the heap Q.
(e) While the heap size $Q > n$
(f) Eliminate top point of hcap (denoted by "m") and combine it with its nearest cluster m.closest "l" and then compute new representative points for the combined cluster e.
(g) Eliminate m and l from tree T and heap Q.
(h) For each cluster y in Q, update x.closest, displace x and insert it into the heap Q.
(i) Repeat until no more updates.

4 Applications

This section describes the unique applications of the CURE algorithm on two different domains: big data in Health Care and Video Summarization. The data mining technique is applied on the medical data to get the disease prediction and on video to get the meaning of the video through its summary. The inference obtained from this work helps people to identify the early signs of the disease in particular by diagnosing the symptoms, wherein, the Video Summarization gives a meaningful summary of the corresponding video input so that the highlights reach the concerned people. This work in particular focuses on the mindset and lifestyle of the society by giving them what they need at any instant of time so that they can make their further decisions. This also helps the healthcare organizations to prescribe the right medication in the early stages of the disease or when a symptom is encountered.

4.1 Application for Big Data Health Care

The proposed system is to analyze and predict the presence of Thyroid disease using early symptoms using CURE Algorithm. To analyze and predict the type of Thyroid Disease using Data Mining Techniques which helps doctors to diagnose and suggest the treatment. It also helps the patients to know about their health condition at an earlier stage. The proposed system is the implementation of the CURE algorithm which is applied to the healthcare Dataset that is Thyroid Dataset.

4.1.1 Design of Implementation

The modules of the proposed system include: The random sampling, drawing kd-tree from the sample, and choosing representatives and then final clusters. These all modules are implemented in the proposed system which gives final clusters which correspond to the outcome of the symptoms. The kd-tree data structure is very essential and is helpful to calculate the nearest neighbor of each point. The modules of the proposed system are the important routines mentioned in the algorithm. The output of the algorithm gives the different clusters of the given Dataset and the corresponding elements which correspond to a particular type of Thyroid disease. The clinical data on which the algorithm is applied consist of patient's records which correspond to different tests conducted and their values from them. The Dataset is taken from UCI machine learning repository [10]. Totally 21 attributes are there in the Dataset out of which 4 attributes are key attributes for deciding the disease condition as shown in Table 1. There are three types of conditions: hyperthyroid, hypothyroid, and normal.

The modules of the Clustering algorithm are:

Implementation of Cure Clustering Algorithm for Video ... 559

Table 1 Attributes of Thyroid Dataset

S. No.	Attributes	Possible values/types	S. No.	Attributes	Possible values/types
1	Age	Continuous	12	Lithium	False, true
2	Sex	Male, female	13	Goiter	False, true
3	On thyroxine	False, true	14	Tumor	False, true
4	Query on thyroxine	False, true	15	Hypopituitary	False, true
5	On anti-thyroid medication	False, true	16	Psych	False, true
6	Sick	False, true	17	Thyroxine stimulating hormone (TSH)	Continuous
7	Pregnant	False, true	18	Total tri-iodothyronine test (T3)	Continuous
8	Thyroid surgery	False, true	19	Thyroxine test (TT4)	Continuous
9	I131 treatment	False, true	20	T4U test	Continuous
10	Query hypothyroid	False, true	21	Free thyroxine index (FTI)	Continuous
11	Query hyperthyroid	False, true			

- **Random sampling**: Random sampling favors big Datasets. Usually, the random sample is saved in primary memory and includes a trade-off between the accuracy and efficiency.
- **Initial Partitioning**: Initial step is to partition the sample into p no. of partitions. The no. of clusters is user defined in the first pass. For second pass, only the representative points are chosen and used. Thus, partitioning the input reduces the execution time drastically.
- **Labeling data**: This step uses only representative points for k no. of clusters, where all the residual data points are assigned to clusters. Here, a set of randomly chosen representative points are selected for every cluster and the data point is passed on to the cluster containing representative point nearest to it.

4.2 Application for Video Summarization

The proposed system is to summarize a video by using LBP feature extraction and CURE clustering method to obtain key frames corresponding to different shots and then combining the key frames to generate video summary. The video summary thus obtained contains brief highlights about the input video. This application is useful where videos are large and the user wants to see only the important parts of the video. The summary enables the user to only see the highlights without missing important events.

4.2.1 Design of Implementation

The input to the system is the set of video shot frames of the input video. The video contains large no. of frames and shots. A single shot has similar frames. The task is to extract the feature vectors from each frame by using a relevant method (LBP-local binary patterns in this case) and then group similar frames in the corresponding clusters using a data mining technique called CURE. The key frame from each cluster which represents the whole cluster is then extracted using centroid of every cluster. The frame closest to the centroid is chosen as the key frame. The LBP feature vectors are obtained for each pixel in each frame. The histogram of LBP features of a single frame represents the feature vector of that frame. Equation to compute LBP code is given by:

$$\text{LBP}_{P,R}(x_c, y_c) = \sum_{P=0}^{P-1} s(g_p - g_c) 2^P \qquad (1)$$

The feature vectors are then computed for all the frames of the video. Then the file containing the feature vectors of all the frames is then passed as input to the clustering algorithm. The clustering algorithm groups the similar frames of a shot as one cluster. The clustering algorithm CURE computes the nearest neighbors of the frames and clusters them. The farthest frames are chosen as representatives, and then the key frames are chosen for each cluster. All the key frames are combined to get the video summary.

5 Experimental Results

This section shows the results obtained from the two applications on medical and video data. The output screenshots of the two applications in their respective domain are given below.

5.1 CURE Algorithm on Thyroid Data

Figure 2 shows the final clusters obtained from the Thyroid Dataset. The no. of points in the Dataset depends on the random sample taken by the algorithm. The

Implementation of Cure Clustering Algorithm for Video ... 561

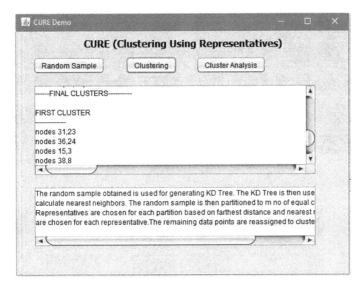

Fig. 2 Clusters obtained for Thyroid Dataset

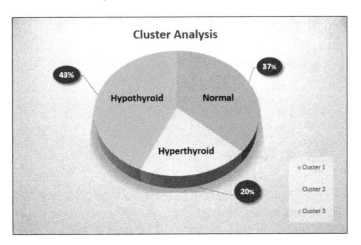

Fig. 3 Clusters analysis

final clusters are obtained by recalculating the distance between the representatives and remaining points. The smallest distance is merged into new cluster.

Figure 3 shows the analysis part of the clustering output. The analysis shows three categories of patients suffering from Thyroid disease. 43% patients suffer from hypothyroid, 37% are normal, and 20% suffer from hypothyroid.

Fig. 4 Summarised output of entertainment video

Fig. 5 Summarised output of nature video

5.2 Video Summarization

The video samples namely entertainment and nature are taken from the YouTube website. The duration of each video is 60 seconds. The entertainment video contains 1850 frames and the nature video contains 1087 frames in total.

Figures 4 and 5 show the video summary of a sample video, each frame represents a shot and is the key frame representing that shot. The key frames of all the shots are combined to generate the video summary. The video summary thus obtained highlights the main parts of the video. If there are any redundant frames they are eliminated in cluster-merging phase.

6 Analysis and Conclusion

The proposed system focuses on the application of CURE algorithm on two different domains, namely, big data in Health Care and Video Summarization, respectively.

The CURE Algorithm is efficient for large databases and is efficient to outliers. The application on big data helps in getting useful information from the Health Care, i.e., patient data and analyze it for future purposes. The application greatly helps healthcare organizations to analyze patient's data. On the other hand, Video Summarization deals with highlighting main parts of the video by choosing key frames for each shot. This helps the user to quickly view the main events in the video without browsing through the whole video. This is especially helpful when the video duration is high.

Thus, we conclude that the application of CURE algorithm on video and big data Health Care can produce two different applications which may aid the healthcare organizations and also in Video Summarization. This work can be further extended by considering symptoms of various diseases to predict the presence of the disease. Also, for Video Summarization, the longer duration videos can be summarized to get the highlights of the video.

Acknowledgements The authors express their sincere gratitude to Prof. N. R. Shetty, Advisor and Dr. H. C. Nagaraj, Principal, Nitte Meenakshi Institute of Technology for giving constant encouragement and support to carry out research at NMIT.
The authors extend their thanks to Vision Group on Science and Technology (VGST), Government of Karnataka, to acknowledge our research and providing financial support to set up the infrastructure required to carry out the research.

References

1. Ishwarappa Anuradha, J. (2015). A brief introduction on big data 5V's characteristics and Hadoop technology. In *International Conference on Intelligent Computing, Communication & Convergence (ICCC-2015)*. Elsevier. https://doi.org/10.1016/j.procs.2015.04.188.
2. Raghupathi, W., & Raghupathi, V. (2014). Big data analytics in healthcare: promise and potential. *Health Information Science and Systems Journal*.
3. Bates, D., Saria, S., & Escobar, G. (2014). Big data in health care: Using analytics to identify and manage high-risk and high-cost patients. *Article in Health Affairs*.
4. Archana Bakare, M., & Argiddi, R. V. (2016). Prediction of diseases using big data analysis. *The International Journal of Innovative Research in Computer and Communication Engineering publication, 4*(4).
5. Ravindra, Ch., Rajesh, G., & Annapurna, G. (2016). Automated health care management system using big data technology. *Journal of Network communications and Emerging Technologies, 6*(4).
6. Patel, S., & Patel, A. (2016). Big data in health care sector challenges, opportunities and technological advancements. *International Journal of Information Sciences and Techniques, 6*(1/2).
7. Guha, S., (Stanford), & Rastogi, R., (Bell labs). (1995). CURE: An efficient algorithm for large databases. In *International Conference on Knowledge Discovery in Databases and Data Mining* (*KDD-95*), Montreal, Canada, August 1995.
8. Chheng, T. *Video Summarization Using Clustering*. Department of Computer Science, University of California, Irvine.

9. Huang, D., Shan, C., Ardebilian, M., Wang, Y., & Chen, L. (2011). Local binary patterns and its application to facial image analysis: A survey. *IEEE Transactions on Systems, Man, and Cybernetics, Part C (Applications and Reviews) 41*(6).
10. http://archive.ics.uci.edu/ml/machine-learning-databases/thyroid-disease.

Redundancy Management of On-board Computer in Nanosatellites

Shubham, Vikash Kumar, Vishal Pandey, K. Arun Kumar and S. Sandya

Abstract Satellite bus has subsystems like attitude determination and control system (ADCS); electrical power system (EPS); and communication, command and data handling (C&DH) for operations at different phases of the mission. These subsystems' requirements are processed and controlled by the on-board computer (OBC) subsystem in the satellite. On-board computer (OBC) subsystem plays a vital role in the functioning of the satellite system; a small malfunction in this system might result in the entire mission failure. For such critical subsystem where on-board manual intervention to repair or replace a failed component is difficult, it is very much essential to have a redundant mechanism. Using redundancy concepts to improve the reliability of systems or subsystems is a well-known principle. This paper describes the architecture of OBC and the redundancy configuration in the nanosatellite. Further, the redundancy management between master and redundant subsystem is explained. This is achieved by first detecting the failure by using an external watchdog timer that monitors master OBC unit along with the redundant. The isolation of the fault signals from the failed unit is controlled by the power control switch.

Keywords OBC (on-board computer) · ARM (advanced RISC machine) · MCU (master control unit) · RCC (reset and clock control) · USART (universal synchronous/asynchronous receiver/transmitter) · GPIO (general-purpose input/output) · CAN (controller area network) · C&DH (command and data

---

Shubham · V. Kumar · V. Pandey · K. Arun Kumar (✉)
Centre for Small Satellite, Nitte Meenakshi Institute of Technology, Bengaluru 560064, India
e-mail: arunatstudsat2@gmail.com

Shubham
e-mail: shubhamkumarsk44@gmail.com

V. Kumar
e-mail: kumar.vikashroy100@gmail.com

V. Pandey
e-mail: vshl765@gmail.com

S. Sandya
Department of ECE, Nitte Meenakshi Institute of Technology, Bengaluru 560064, India
e-mail: sandya.prasad@nmit.ac.in

© Springer Nature Singapore Pte Ltd. 2019
N. R. Shetty et al. (eds.), *Emerging Research in Computing, Information, Communication and Applications*, Advances in Intelligent Systems and Computing 906,
https://doi.org/10.1007/978-981-13-6001-5_47

handling) · ADCS (attitude determination control system) · IPC (inter-process communication)

1 Introduction

OBC is a special-purpose on-board computer intended to screen and control the constant activities of a satellite through sensors, actuators and other components. It takes control of the satellite proximately after its launch and endures to remain in command until the end of the mission. The core component of the OBC subsystem is a microcontroller unit. This paper deals with the design of the failure detection system of an OBC for nanosatellite using an external watchdog timer. The work includes redundancy design development for OBC architecture as well as the development of the software for the microcontroller. The architecture detects the error and provides isolation from the failed system. This approach would ensures the system fault tolerant and fail-safe. To implement this mechanism, an external watchdog timer is used. The main advantage of using watchdog timer is that it can be used to automatically detect programming and equipment glitches and produces the timeout flag. Inheritance of a watchdog timer depends on counters that check down from beginning an incentive to zero, if the counter ever achieves zero preceding the product restart it, the product is ventured to glitch and the processor reset flag is affirmed.

Different redundancy methods like hot standby, warm standby and cold standby can be used in satellites. Here, a warm standby method has been used. The redundancy architecture of OBC will provide a one-to-one redundancy for the OBC subsystem in the satellites. The external watchdog timer is refreshed at regular intervals by the master OBC. If the OBC fails to do so, the watchdog generates a timeout. Furthermore, it sends an interrupt to the backup OBC, which will send a message through the primary CAN and wait for an acknowledgement. If acknowledgement is received from the master OBC, the main microcontroller is reset. Otherwise, the same procedure is repeated via the secondary CAN, assuming that the primary CAN has failed. If still the acknowledgement is not received, the reconfiguration of redundant OBC as master happens and the operations prevail.

2 Architecture of OBC

The command and data handling subsystem controls and handles information between various subsystems of satellites. The OBC has programming introduced that deals with the projects written to deal with different assignments. OBC considered for nanosatellite is 32-bit ARM Cortex M4-based STM32F407VGT6 microcontroller which has built-in digital signal processing (DSP) and floating point unit (FPU) features. The system operates at 168 MHz and is strong enough to execute 210 Dhrystone million instructions per second (DMIPS). The advantages of different

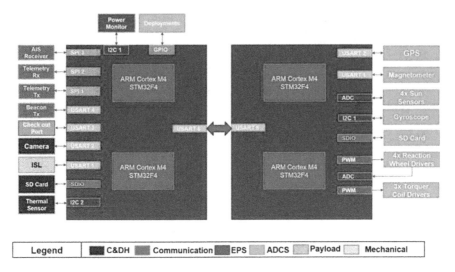

Fig. 1 Architecture of OBC

low power modes like stop, standby, sleep with real-time clock enable the system to switch between these power modes (Fig. 1).

3 Redundant Architecture of OBC

Redundancy is the replication of critical components of a system with the purpose of increasing consistency or availability of the system, usually in the case of a catastrophe or degradation. The needs for redundancy are:

- Unpredictable hardware faults in OBC cause a malfunction in the satellite system and would further lead to mission failure.
- Failures in the hardware could be due to environmental anomalies or any other physical fluctuations in system, subsystem and interfacing modules.
- As OBC is one of the most crucial systems in the satellite, care must be taken while designing the system.

Figure 2 shows the OBC architecture of nanosatellite where the master will be in run mode and the redundant system will be in sleep mode. Communication between the two is achieved through CAN bus, and IPC between C&DH and ADCS subsystem is achieved through UART interface.

The on-board real-time clock (RTC) in the redundant system allows it to auto-wake up from standby mode. Upon periodic wakeup, OBC will check the activeness of the master OBC. If the master OBC is active, the redundant system will continue to stay on standby; else, it will come into run mode and coup the satellite OBC

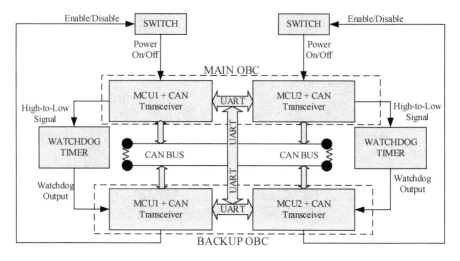

Fig. 2 Architecture of OBC redundancy

operations. Interface failure between master–redundant OBC systems and between C&DH and ADCS subsystem is evaded by offering alternative communication line.

3.1 Management of Master and Redundant OBC

The development of the OBC architecture has various factors related to it. The selection of the microcontroller for the OBC is based on the power consumption, software and performance parameters. The STM32F407VG ARM Cortex M4 board has the ability to switch between various power modes which is suitable for a nanosatellite mission sequence. Active mode consumes power as less as 100 mA with all the peripherals in run mode. In sleep mode, the power usage is only 2–3 μA which is best suited for the redundancy management purpose. Since two identical STM32 microcontroller boards are used, the software overheads can be minimized. Also, the same algorithm is used in both the microcontroller boards with minimum amount of changes for the main and redundant OBC. The performance of the OBC is also fast with a minimum delay of 2–3 s for the transfer of control from master to backup. The combination of watchdog timer and CAN approach is used to provide the best redundancy for this particular application.

3.1.1 Working of the Redundant Architecture

Initially, the main OBC will be configured to operate in run mode and continue the mission operations, while the backup OBC will be configured with minimum configurations and switched into deep-sleep mode until an on-board real-time clock (RTC) or external watchdog timer generates an interrupt. Failure of the two MCUs in main OBC is checked at two levels in order to avoid the false failure report at the single attempt. The first level is through external watchdog timer, which will be continuously monitoring the microprocessor activity and generates a signal if MCU operates incorrectly. The output signal from the watchdog timer is in turn fed to the MCUs in the backup OBC as shown in Fig. 2. At the second level, backup MCUs will broadcast message to main MCUs via primary CAN and confirm the activeness of main MCUs through the reception of acknowledgement message from the main microcontroller. If the acknowledgement is not received, same procedure is repeated via the secondary CAN assuming the primary CAN has failed. A reset signal will be sent to the MCUs in main OBC to reset the controllers when they fail to clear at first- and second-level checks. The reset will be done for several number of attempts in order recover the master OBC from the failure. In the redundancy architecture of OBC shown in Fig. 2, there are probabilities that either one of the main MCU or the entire OBC system would fail. Hence, the backup system will have the auto-reconfigurability feature at subsystem level and system level to replace the failed MCU or the main OBC. In order to isolate the erroneous output signals from failed MCU in main and avoid unnecessary consequences in the mission cycle, the MCUs in backup OBC will send the disable signal to main and cut off the power distribution to the main board through a switch.

4 Results

See Figs. 3 and 4.

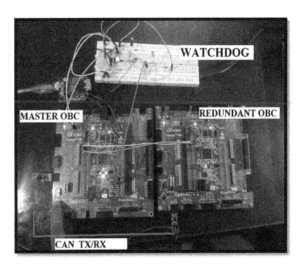

Fig. 3 Experimental set-up

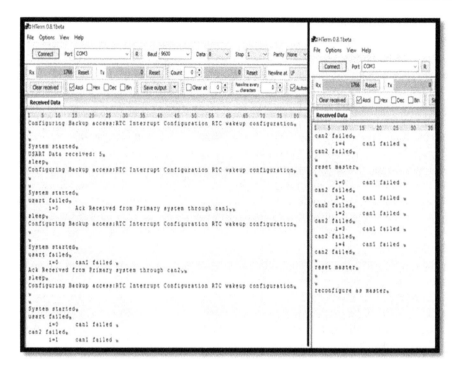

Fig. 4 Output in H-term terminal window

5 Conclusion

This paper explains the redundancy concept of the on-board computer in a nanosatellite. Redundancy management of OBC in nanosatellite is designed with a watchdog timer to improve the reliability of OBC by providing redundancy at the system-level, subsystem-level and communication-level interface. The watchdog timer reduces the power consumption as the redundant OBC does not need to wake up regularly to monitor the health of the master OBC. Thus, the communication between the OBCs has a minimum delay (5–10 ms) after the master OBC starts malfunctioning. Hence, the redundancy in the nanosatellite makes it robust and reliable for mission-critical tasks.

A Fault Tolerant Architecture for Software Defined Network

Bini Y. Baby, B. Karunakara Rai, N. Karthik, Akshith Chandra, R. Dheeraj and S. RaviShankar

Abstract Software defined network is a single-point control architecture where the control plane and data plane are disaggregated. It has a centralized controller, switches and hosts. Here, the OVS switches which act as the data-forwarding plane are connected to the controller having forwarding details. In the above-said architecture, if the controller goes down due to bottleneck problems that arise because of packet injection or any other attacks, then the network experiences performance drawbacks. Hence in this paper, we propose a fault tolerant approach for software defined networks. Once the controller fails, OVS switches are made as controllers by using a switch type by name "UserspaceSwitch in namespace" which separates switches' namespace from the controller. We have used Mininet an emulator for simulation of software defined networks with POX, a remote controller. The proposed fault tolerant approach for software defined networks is also simulated using the mentioned software. The hosts are made as client and server, and the data traffic is generated between them using UDP which gives different parameters as output for analysing the performance. Further, graphs are plotted considering three-parameter delay, packet loss and throughput for both SDN with centralized control and fault tolerant approach to analyse the performance.

Keywords Software defined networking · Network bottleneck · SDN controller · DHT-switch

1 Introduction

Similar to Google, Microsoft is also trying to use SDN with the help of an open-source tool called OpenDaylight for providing a better Skype end-user experience in branch offices. Gap Inc is one such company which uses software defined network

---

B. Y. Baby (✉) · N. Karthik · A. Chandra · R. Dheeraj · S. RaviShankar
Department of ISE, NMIT, Bengaluru, Karnataka, India
e-mail: biniybabe@gmail.com

B. Karunakara Rai
Department of ECE, NMIT, Bengaluru, Karnataka, India

© Springer Nature Singapore Pte Ltd. 2019
N. R. Shetty et al. (eds.), *Emerging Research in Computing, Information, Communication and Applications*, Advances in Intelligent Systems and Computing 906,
https://doi.org/10.1007/978-981-13-6001-5_48

approach for processing the credit card transactions more efficiently and securely. FOX Broadcasting Company uses SDN, as SDN has better control over the network packets. FOX uses it to broadcast the compressed videos to different stations around the country efficiently. The Former CEO John Chambers of Cisco declared victory about company's battle with SDN due to the increase in popularity of their Application-Centric Infrastructure (ACI), which is described as a "business-relevant software defined networking policy model". Many manufacturers are entering into the market to compete against Cisco. HP as of late presented a new line of open switches and expanded its organization with VMware, which demonstrates that it is considering the SDN amusement quite important.

Software defined network, a centralized network architecture, helps the network administrator to programmatically control, change and manage the behaviour of the network with the help of interfaces. In software defined network, both the data plane (network devices) and the control plane (controller) are decoupled. The data plane is used to control the forwarding of the network traffic, whereas the control plane is logical entity which receives the information from the applications. In SDN, we use different programmable devices such as switches and routers. SDN contains a centralized controller that manages the flow of the network. Since SDN has a centralized controller, it provides various features like dynamic programmability, easy network management and control and complexity reduction. Though SDN is centralized, it encounters various challenges such as link failure, attacks, bottleneck problem and many more that are to be taken care off.

The SDN architecture as shown in Fig. 1 consists of three layers, i.e. infrastructure layer, control layer and application layer. The infrastructure layer comprises network devices such as switches and routers for network traffic forwarding. The devices in the infrastructure layer are controlled by the controller of SDN. Control layer is where the logic of the network resides, which controls the entire network infrastructure. There are many controllers in the market such as POX, OpenDaylight and floodlight. It is one such area where currently most of the network vendors are working on to come up with their own product. The control layer or the controller lies in between the application layer and the infrastructure layer; therefore to connect to the upper and lower layer, the control layer exposes two interfaces called as northbound interface and southbound interface. Northbound interface is used to connect the application layer and the control layer. There are no any standard northbound interfaces yet; hence, REST-based API is used. Southbound interface is used for the communication between the control layer and the infrastructure layer. OpenFlow was the first and most widely used southbound interface, and then came many other interfaces such as OSPF, BGP and MPLS. Application layer is where different applications are developed for network monitoring, network troubleshooting, network policies and security. SDN applications are developed based on the different network information available such as network state, network stats, network topology and more. Brocade is one such vendor which provides applications such as brocade flow optimizer, brocade virtual router and brocade network advisor. Also, there are other applications like TechM server load balancer, HPE Network Visualizer and Aricent SDN Load balancer.

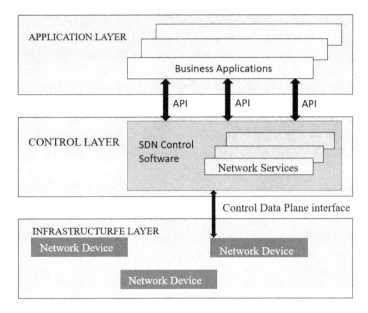

Fig. 1 SDN architecture

Due to the SDN-based framework adaptability, they accompany with performance penalty. Regardless of SDN's growth and popularity, there are only few studies regarding the performance of SDN models. Performance of any network is affected by considering different parameters. Due to the functionalities provided by the SDN, it enhances the, in general, efficiency of the network. But the performance mainly depends on the SDN model and implementation rather than on the functionalities and complexity of it. Therefore, programmable generic forwarding element (ProGFE), an adaptable and reconfigurable stage which allows us to deploy different network services, protocols and architectures, was introduced. This ProGFE platform is configured through an XML-based API. ProGFE is fundamentally a superset of OpenFlow regarding capacities and gives a wealthier usefulness than OpenFlow. It is a completely different technology compared to OpenFlow which is based on the IETF standard which targets on separating the data and control plane. When performance analysis was done between SDN with OpenFlow and ProGFE, it was observed that the throughput in ProGFE was higher compared to OpenFlow. The throughput and latency of SDN are affected due to the complexity in workload. The jitter in OpenFlow is higher compared to ProGFE and hence results in decreased throughput, even though both OpenFlow and ProGFE are having similar latency.

Advantages

SDN provides a centralized controller which distributes the security and policy information throughout the network. It is an important challenge to create a controller which is highly secure. SDN operating costs can be reduced by centralizing and automating the network administration issues. This results in the better server utilization and better control of virtualization. We can also reduce the overall infrastructure cost by using the available commercial off-the-shelf (COTS) devices. In SDN, centralized controller has the capability to see into the resource requirements of the applications and also the available resources at lower levels. Due to the network visibility feature of SDN, the hardware which is overprovisioned is made to work precisely, thus reducing the capital expenditures of the network. In a traditional architecture, the instructions and policies were embedded into the network elements like routers and switches. But in an SDN approach, these instructions and policies are embedded into the centralized controller, which reduces the complexity of network elements.

Disadvantages

Since SDN contains a centralized controller, it is prone to single-point failure which results in network degradation. Software defined networking involves programming of network through software. But software usually has security issues that are to be taken care; even SDN suffers from these security issues. SDN requires virtualization which is costly and takes more time for deployment. If SDNs are implemented in data centres, when there is huge data traffic between the nodes. SDN should handle this traffic efficiently, but due to huge data the controller may suffer from bottleneck problem, which is one of the important challenges of SDN. In SDN networking, tasks can be characterized as data plan or control plane. With continuous activities, standard outlines that are given occasional cuts of a centre could present colossal latencies on movement.

As software defined networks are prone to a single-point failure, in our proposed work an approach to the software defined architecture is given where the switches in the network which are situated in data plane of software defined network architecture are converted to controllers on the occurrence of bottleneck problem at the centralized controller. This conversion is done by using a switch type by name "UserSpaceSwitch in namespace" which separates the switches (data plane) from the root directory of the controller and creates its own namespace which can be programmed as the programmer wants it to work as. We used this type of the switch and made the controller run in the switches by opening its namespace. All these simulations are carried out using available open-source software called Mininet which has an inbuilt editor MiniEdit where the topologies are built. The controller we used here is the one which came built in with the Mininet software called POX which is used as the remote controller that can be made to run both as the centralized controller and inside the switches namespace. After the topologies are built for SDN and our proposed work, we then used another open-source software by name gnuplot which when given the parameters of x- and y-axes plots the graph for us. We considered three performance parameters, bandwidth, delay jitter and packet loss for which we drew the values by running centralized controller as well as distributed controller and then plotted graphs using gnuplot. Once the graphs for both types of networks were plotted, we then analysed by comparing the performance of both types of networks. We found that SDN performance was better but there are performance parameters which we are yet to consider and analyse the performance. Basically, this approach to the SDN architecture was given in order to make the architecture fault tolerant where even if the controller goes down the switches in data plane take over the network until the centralized controller is ready and away from bottleneck problems.

2 Related Work

Motivation for SDN was because of the coupling of control plane and data plane in traditional network which resulted in inefficient network performance; hence, SDN was introduced where both the control plane and the data plane are decoupled. SDN architecture mainly comprises three layers: application layer, control layer and an infrastructure layer [1]. To overcome a DoS attack for SDN, a distributed firewall with intrusion prevention system (IPS) for SDN has been proposed [2]. Since SDN is used in many vulnerable areas, both the controller and the communication devices can be attacked. For mitigating these attacks in software defined networks, a policy-based security application is proposed that has the capability to protect and control the SDN domain behaviour. The application is developed using an ONOS SDN controller to test these attack scenarios [3]. Moving target défense (MTD) presents a continually altering condition with a particular true objective to defer or hinder attacks on a system [4]. Software defined network uses programmable devices like switches and routers. As of in the traditional network, even SDN is prone to bugs that are inevitable and ruin the software quality; hence, many SDN debugging tools are provided. The capabilities of the tools are to check stability of stream tables, back hints of system movement, checking switch sending rules, recording and replaying of all or chose activity and identification of network configuration changes [5, 6].

Software defined network is a novel networking architecture that disaggregates network control and forwarding functions from the data plane; therefore, many novel attacks are introduced, namely packet injection attack in SDN. A packet injection attack is where an unauthorized user injects malicious packets continuously into the network which results in the blockage of certain network services. As a result, a prototype called packet checker is implemented using a floodlight controller to defend against these packet injection attacks [7]. Path restoration mechanism is where a bypass path is used to provide solution against link failure [8]. Different technologies are provided which allows user to control and remotely access network devices that provides real-time information from intelligent devices [9].

In software defined network (SDN), the switches are statically doled out to controllers, which cause stack unevenness among controllers. In this firm, we consider a spread of SDN controlling switches running OpenFlow tradition. We propose an add-on algorithm realized on the controllers. Reproductions exhibited that such algorithms upgrade execution and unflinching quality, and it is an absolute necessary in a distributed SDN. Failure handling and load balancing is to be the key features of such algorithm. This algorithm works as a disturbed SDN, in which an immense system is managed by various SDN controllers. The dynamic switch remapping algorithm empowers controllers to impart among themselves to choose an origin controller.

The algorithm can be effortlessly sent as an additional component for controllers, and it chips away the controller's inscription given in the controller design record, permitting the presence of different controllers which will not utilize the algorithm in the network that tends to be commendatory. As a subsequent work, more estimations likewise, features will be installed to update load balancing, picking switches S_k that should be remapped. Moreover, a troubleshooting and taking a short at the versatility safety measures of the estimation will be done to make it perfect for tremendous courses of action on data centres [10].

Software defined networking (SDN) builds up an incorporated control plane to deal with the entire system, yet the customary unified control plane experiences the issues of unwavering quality and adaptability. Although a few techniques explain the issues, they do not balance load among controllers. It puts forward a load-balancing component in the light of switches gathered for various controllers. The component not just stabilizes the load amid the controllers, yet it additionally tackles the load swaying and enhances time effectiveness. Simulation results have shown that the proposed component addresses the issue of stacks swaying among controllers and achieving the time proficiency. From the above research to unravel the difficulty of load swaying amid controllers and enhance network balancing, it sets forward a switch selection algorithm and a target controller selection formula, which guarantee that execution stays stable even under exceedingly unique movement conditions. In expansion, to enhance the time proficiency, we additionally plan a Target controller's selection algorithm for switch gathering. Reproductions demonstrate that our approach can explain the load-balancing issue and enhance network balancing in a period productive way [11].

Software defined networking (SDN) is as of now viewed as a standout among the most encouraging ideal models of future networks, in spite of, the fact that the accessibility, versatility, security challenges are not fully addressed. Analysis established on floodlight demonstrates that the system could adjust the load of every controller dynamically and decrease the season of load balancing. From the above analysis, we can conclude that the dissimilar load conveyance is an inescapable matter in sending arrangements of numerous controllers. To resolve this issue, we put forward an instrument in view of load-informing procedure to adjust the load amid controllers and diminish the time of adjusting. Later on, it is proceeded with respect to streamlining of our proposed load-balancing component, with the emphasis on upgrading load-informing component and balance decision component. In expansion, we mean to actualize our load-balancing mechanism amid numerously varied SDN controllers [12].

Software defined networking (SDN), empowered by OpenFlow, speaks to a change perspective from conventional system to the future Internet. Recreate or circulated controllers are put forward to mark the problem of accessibility and versatility that brought together controller experiences. Be that as it may, it does not have an adaptable instrument to adjust stack among circulated controllers. To mark this problem, this paper tells dynamic and adaptive algorithm for controller load balancing (DALB), a dynamic furthermore, adaptable algorithm for controller load balancing completely in view of conveyed engineering, with no brought together segment. This algorithm runs as a module of SDN controller. On the other hand, it receives a flexible load accumulation edge to decrease the elevated of trading messages for load accumulation, and then again it can make approach and race territory keeping in mind the end goal to diminish the choice postponement occurred by network transmission. It fabricates model framework on floodlight to show the outline and test the execution of algorithm. From the above analysis, we set forward a load-balancing technique mentioned DALB for SDN controller in the light of dispersed choice. We portray each segment of DALB, test the execution to our appropriated SDN controllers and test the DALB algorithm work. Later on, it centres to consideration on the testing algorithm in depth. In addition, if we pick diverse estimations of collection threshold (CT) and ρ, distinctive execution of the algorithm will show. It is essential for us to investigate on the most proficient method to choose the estimation of CT and ρ keeping in mind the end goal to get the ideal execution [13].

Software defined networking (SDN) is a system design that has got ample consideration lately. It speaks to the eventual fate of system industry. As the Internet keeps on surpassing desires of quick advancement, a solitary unified controller can be stretched out to circulated different controller designs. Be that as it may, the circulated different controller designs are confronting increasingly genuine trials in the parts of versatility, dependability, security. Sequentially to overcome these difficulties, we put forward a helpful stage for load balancing and security on SDN disseminated controllers, called smart cooperative platform for load balancing and security (SCPLBS). The cooperative stage is based on the control plane. A safe correspondence component in the light of message validation code is embraced between the helpful stage and the controllers. Cooperative stage utilizes an information accumulation calculation adjusting to information variance to gather the controllers' status and load data. Community stage takes system to accomplish the circulated controller stack adjusting and disappointment recuperation. The adequacy of the proposed system has been tested. Trial comes about to demonstrate that this plan can skilfully accomplish the heap adjusting and disappointment recuperation of the circulated controllers based on the protected correspondence between the helpful stage and the controllers. This paper proposes an agreeable stage for stack adjusting and security on SDN conveyed controllers, named SCPLBS. Floodlight controller has been utilized to build up the agreeable stage in the light of restlet structure. SCPLBS carries message confirmation module, information gathering module, load balancing and disappointment recuperation module. From examination and investigation, it demonstrates that the stage can understand the load balancing and disappointment recuperation of the circulated controller in the light of the protected correspondence. Later on, we will enhance the security encryption instrument in this plan to give more successful put stock in administration. Besides, we will accomplish more inside and out examination on the problems about security and versatility of multi-space disseminated controller [14].

Network-as-a-service (NaaS) is a cloud-based service model which provides necessary network connectivity, contingency and management of network services. We propose an SDN-based way to deal with the help of NaaS model. It actualizes a proof of idea (PoC) on a physical test bed and approves on test performance assessment. From the above analysis, we conclude that we have offered an SDN-based architecture for the network-as-a-service (NaaS) cloud-based model to be copied and have instrumented a proof of concept putting into effect upon a physical network test bed coupled with provisioning and business managers of basic network power to make connection services over it [15].

Together, software defined network (SDN) and big data have drawn in extraordinary attention from both industry and scholarly community. At any cost, on the other hand, the tremendous highlights of SDN can incredibly encourage big data procurement, transmission, stockpiling and preparing. Counting enormous information in data centre networks, information conveyance, joint enhancement, logical Big Data models and planning problems have necessitated the need of SDN. It demonstrates SDN will deal with the system adequately to enhance the execution of big data applications. Also, we demonstrate that big data can profit SDN too, in addition to activity building, cross-layer configuration, vanquishing security assaults and SDN-based intra- and inter-data centre networks. Recent trends have proved that Big Data applications show better performance in SDN environment. Initially, we establish some tremendous highlights of SDN that can profit enormous information applications, incorporating big data handling in cloud data centre, information conveyance, joint enhancement, logical big data structures and scheduling issues. It demonstrates that big data profits different parts of software defined network, together with activity building, cross-layer configuration, crushing security assaults, SDN-based intra- and inter-data centre. To whole up, the joint plan of big data and SDN can be a great answer for large information organizing. Step-by-step instructions to completely utilize SDN's favourable circumstances to improvise the execution of enormous information applications and to handle big data to improve SDN work and all the more adequately are dire issues that should be tended to [16].

Software defined network (SDN) demonstrated advantages, and there stays huge hesitance in embracing it. Among the issues that hamper SDNs appropriation, two emerge: unwavering quality and adaptation to non-critical failure (fault tolerance). At the centre of these issues is a course of action of predetermination sharing associations: among the controllers and SDN apps, the crash of the previous actuates a crash of the last mentioned, in this way influencing accessibility, later, between the network and SDN app, wherein a byzantine disappointment example: dark gaps, system circles, incites a disappointment in system, and in this way influencing system accessibility. The main goal is that accessibility is of most extreme concern—and just to security. It shows how these reflections can be utilized to enhance the dependability of an SDN domain, in this manner disposing of one of the obstructions to SDNs appropriation. Recent days, Packaging and implementing the applications in controller code as a solitary solid procedure has become the regular practice; failure of SDN app cuts down the whole controller. Moreover, an SDN app failure brings about a conflicting system, as the controller cannot move back system modification undergone by SDN app. We propose an arrangement of deliberations for improving controller accessibility: AppVisor (fault isolation) and Netlog (network transactions). We show the adequacy of our deliberations by building a framework, LegoSDN, that retrofits a current controller stage to help these reflections with no progressions to either of the controller or SDN apps. The framework enables SDN administrators to promptly send new SDN apps in their systems [17].

Software defined networking is getting to be increasingly predominant in data centre network for its programmability that empowers brought together system design and administration. Since switches are statically allocated to controllers, intensive activity between controllers brings unbalance in the load. As an outcome, a few controllers are not completely used, while switches associated with overburden controllers may encounter long reaction times. Considering Dynamic controller assignment with a specific end goal to constrain the ordinary response of control plane will enhance the overall performance of the system. We figure this issue as a stable planning issue with trades and propose a continuously two-stage algorithm that directs key thoughts from both organizing theory and coalition amusements to enlighten it capably. Theoretical examination shows that our count meets a nearby perfect Nash stable course of action inside a few accentuations. Wide re-enactments give the idea that our approach diminishes response time by around 86% and achieves better load changing among controllers appeared differently in relation to static errand. From the above examination we reason that the Dynamic Controller Assignment (DCA) approach will overcome the constrains in controller response time. The DCA issue is understood in two organized ways. Beginning a stable organizing is beneficially made among switches and controllers, which guarantees the response time in most critical situation. It fills in as a commitment to the second coalitional beguilement stage to also diminish the response time. The two-stage algorithm which achieves close perfect load altering among controllers lessens the controller response time by around 22 and 86% took a gander from an optimistic standpoint in class DCP-GK and the fundamental static task, independently [18].

3 Proposed Work

3.1 Workflow

Figure 2 defines the proposed workflow for the fault tolerant approach:

a. Decision to analyse the performance

The network performance analysis is a critical task where the different parameters of the network configuration are to be analysed with different workloads, and helps in knowing the quality of the network. SDN has centralized network infrastructure and suffers a single-point failure. When the controller goes down, the network goes down or crashes, so the performance of the network gradually comes down. So in our work, the performance analysis of SDN during the network functioning is done considering the parameters such as throughput, delay and packet loss. The proposed work here, carries out performance analysis of traditional SDN for QOS parameters and the same procedure is repeated for the SDN with distributed controller

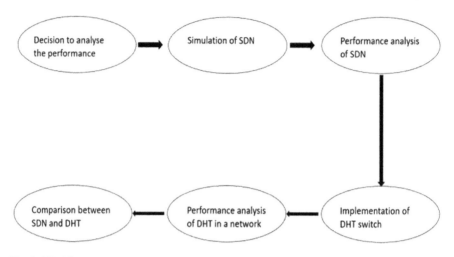

Fig. 2 Workflow

architecture. Conclusions are drawn based on the comparision of the results obtained in these procedures. Both the results on performance analysis and comparison are done between the results.

b. Simulation of SDN

The simulation of SDN is done using Mininet with POX controller. The network consists of elements such as controller, switches and hosts. The network is built using a custom topology which consists of two OpenvSwitch and four hosts and POX controller. The network topology is built using python programming language. The traffic in the network can be generated by the communication of the hosts. The hosts will transfer data to another host. The controller specifies the flow rules to the switches based on which the switches forward the data to the destination. The simulation of SDN in the Mininet environment on the given topology is done, and the traffic is generated in the network to do the performance analysis.

c. Performance analysis of SDN

The performance analysis of SDN network in Mininet environment using POX controller is done on the example topology which consists of one controller, two switches and four hosts. For analysing the performance of the network, we consider some of the parameters such as throughput, packet loss, packet delay. We create a client and a server application in the host so that we establish a communication between the hosts and observe the performance of the network. During the simulation fixed size of data-packets are transmitted from client to server. The packet loss, utilization of bandwidth and the delay experienced by the packets is estimated and evaluated for the transmission. The result is stored in a log file.

d. Implementation of DHT

The DHT is implemented using separate namespace in the OpenvSwitch. The SDN network when controller goes down the user space in namespace switch is used to change OVS to DHT. The user space in namespace switch has a separate network namespace which separates the OVS switch from the controller and creates its own namespace. The switch then acts as both controller and switch, and network is maintained without suffering any network issues. This approach can serve the SDN even when controller fails.

e. Performance analysis of DHT in a network

The performance analysis of DHT in network using Mininet environment using user space in namespace switch is done on the example topology which consists of one controller, two switches and four hosts. For analysing the performance of the network, we consider some of the parameters such as throughput, packet loss, packet delay. We create a client and a server application in the host so that we establish a communication between the hosts and observe the performance of the network. During the simulation fixed size of data-packets are transmitted from client to server, the packet loss during this transmission, effective utilization of bandwidth and the delay experienced by the packets during transmission is evaluated. The result is stored in a log file.

f. Comparison between SDN and DHT

The results of the performance analysis of SDN and DHT are stored in two log files, the values observed during the performance analysis are taken, and graphs are drawn considering the parameters such as throughput, delay and packet loss against time. The graphs drawn are observed to have a clear comparison between the SDN and the DHT; the proposed approach is evaluated to see whether the objectives have been achieved.

3.2 Implementation

The network topology considered for the simulation is shown in Fig. 3; the network consists of a controller, two switches and four hosts. The POX controller is used to configure and specify the flow rules. C0 is the controller in the network. OVS switch is used to forward the packets in the network; S1 and S2 are the two OVS switches used for the simulation. The four hosts H1, H2, H3 and H4 are used. Hosts are end users which generates the traffic in the network. Data packets are sent from one host to another host to communicate with each other.

The host can generate traffic by doing the ping test to another host. During the ping from one host to another host, if the controller is not functioning, then the path to the destination host will be unreachable. The controller when functioning the host sends a packet to another host, the packet is first forwarded to switch, and it checks the flow

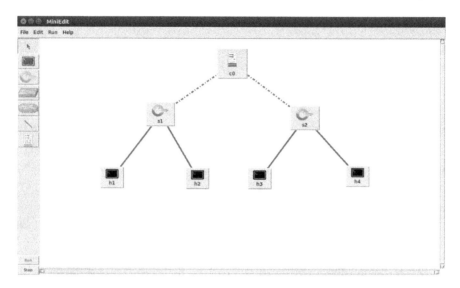

Fig. 3 Network topology for simulation

rules stored in the table. If it finds a path to the destination, then it forwards the packet; if it does not find a path, then it sends a packet-in message to the controller asking for the path to the destination. The controller then specifies a flow rule and sends the path to the destination. If the flow rules are not available, then the first packet to reach destination takes a greater time than the following packets to the destination. In our network topology, we have implemented client–server applications in hosts H1 and H2. The client application is hosted in H1 and a server application in host H2, respectively. A UDP connection is used to transfer the data packets from client to server. During the simulation, the host H1 sends packet to the host H2 for duration of ten seconds. Iperf command is used to set up the UDP connection and send packets of fixed size to the server from client. The network analysis is done considering the parameters such as throughput, delay and packet loss; for every one second, the performance of the network for the parameters is analysed and the results are stored in log file. Next, the network with DHT is analysed with same parameters by separating the switch from controller and implementing DHT in the switch. The results are also stored in the log file. The results are then compared to check the performance of both the networks.

4 Simulation Results

We have analysed the performance of both SDN and DHT-switches by plotting a graph, considering the parameters such as packet loss, throughput and delay. First,

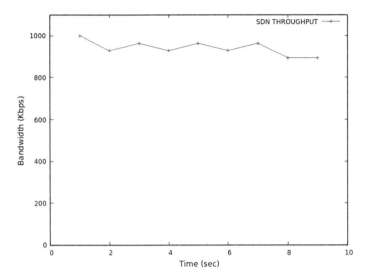

Fig. 4 Throughput with SDN

we will analyse the performance of SDN. The below three graphs will explain about the performance of SDN.

Figure 4 explains the throughput of the network with SDN where the graph is plotted by taking time (s) on X-axis and bandwidth (Kbps) utilization on the Y-axis. According to Fig. 4, the bandwidth utilization for interval of ten seconds is considered and at every second an average of 0.9 Mbps of bandwidth is utilized.

Figure 5 explains the packet loss in the network with SDN where the graph is been plotted with time (s) on X-axis and number of packets on Y-axis. During the simulation, UDP connection between the hosts was established and around 90 packets of data were sent each second with an interval of ten seconds. The packet lost during this transmission due to network link was an average of 10 packets lost for every second. So in the network with SDN, the packet loss was around 10% overall.

Figure 6 explains the packet delay in the network with SDN where the graph is plotted with time (s) on X-axis and time (ms) to deliver packet to destination on Y-axis. As we observe from the graph, the line tells delay experienced for every second by the packets to reach the destination. In SDN network, the delay is less ranging between 0 and 0.1 ms. So, we can observe almost straight line without much variation giving a constant delay in the network.

Figure 7 explains the throughput of the network with DHT-switch when the controller has gone down due to some bottleneck issues, where the graph is been plotted by taking time (s) on X-axis and bandwidth (Kbps) utilization on the Y-axis. According to Fig. 4, the bandwidth utilization for interval of ten seconds is considered and at every second an average of 0.9 Mbps of bandwidth is utilized. The throughput of the network with DHT-switch is almost the same as seen in SDN network. There is not much of variation when compared to the throughput of both of the networks.

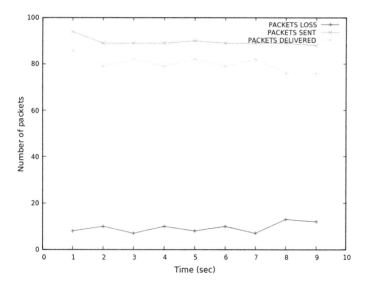

Fig. 5 Packet loss with SDN

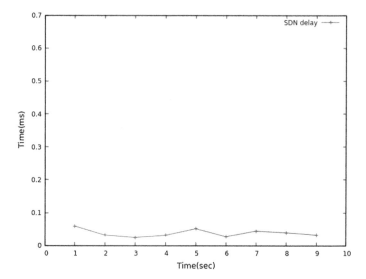

Fig. 6 Packet delay with SDN

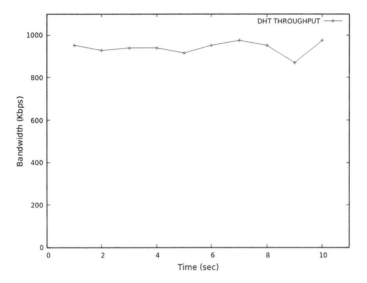

Fig. 7 Throughput with DHT-switch

Figure 8 explains the packet loss in the network with DHT-switch when the controller has gone down facing bottleneck issues, where the graph is plotted with time (s) on X-axis and number of packets on Y-axis. In this network also during the simulation, we established a UDP connection between the hosts and transfer of data packets was done. Even here, around 90 packets of data were sent every second over an interval of ten seconds. The packet loss in this network was found to be around 12–15 packets every second resulting in an average of 13% packet loss in the network. Compared with SDN network, the DHT-switch has a greater packet loss but still there was difference of overall 3% more packet loss.

Figure 9 explains the packet delay in the network with DHT-switch during controller facing bottleneck issues, where the graph is been plotted with time (s) on X-axis and time (ms) to deliver packet to destination on Y-axis. Graph here tells us about the time delay experienced by the packets to get transmitted to the destination in the DHT network. The delay is varied between 0 and 0.7 ms which is more compared to the SDN network. The packet delay for every second is not constant in DHT network producing varying delay to the packets.

Here, we can see from the above graphs that SDN network has a better performance compared to the DHT-switch, times when controller is down and facing bottleneck issues the DHT-switch can be used to maintain the network in the absence of the controller and avoid the network from crashing or going down. Even though the same performance as like in SDN cannot be rendered, we can give a performance that is similar or almost near to that of an SDN using the DHT-switch in the absence of the controller. It is also important to note down that both are producing the same throughput in the network but varying in other parameters such as packet loss and packet delay.

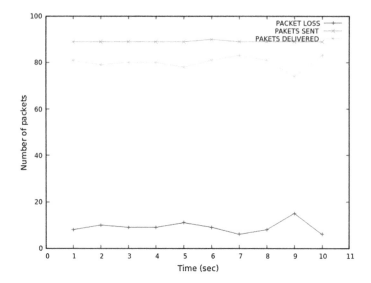

Fig. 8 Packet loss with DHT-switch

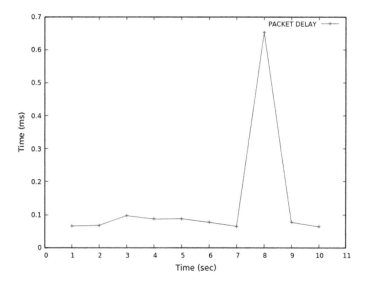

Fig. 9 Packet delay with DHT-switch

5 Conclusion

In contrary to the traditional architecture for networks where control plane and data plane were together, software defined network was developed having control plane and data plane decoupled where the control is centralized and data plane is distributed among the switches. As the controller is centralized in software defined network, it is prone to a single-point failure. If the controller goes down, then network experiences drawbacks and hence a decentralized approach is given to software defined networks wherein the switches in the network are made as controller so that even if the controller experiences bottleneck problems the switches made as controller can manage network.

References

1. Malik, A., Aziz, B., Ke, C.-H., Liu, H., & Adda, M. (2017). Virtual Topology partitioning towards an efficient failure recovery of software-defined networks. In *International Conference Machine Learning and Cybernetics (ICMLC)* (pp. 646–651). China.
2. Rengaraju, P., Ramanan V. R., & Lung, C.-H. (2007). Detection and prevention of DoS attacks in software-defined cloud networks. In *IEEE Conference on Dependable and Secure Computing, 7–10 August 2017* (pp. 217–223).
3. Karmakar, K.K., Varadharajany, V., & Tupakula, U. (2017). Mitigating attacks in software-defined network (SDN). In *Fourth International Conference Software Defined Systems (SDS)* (pp. 112–117). IEEE.
4. Kampanakis, P., Perros, H., & Beyene, T. (2014). SDN-based solutions for moving target defense network protection. In *IEEE 15th International SymposiumWorld of Wireless, Mobile and Multimedia Networks (WoWMoM)* (pp. 1–6).
5. Nde, G. N., & Khondoker, R. (2016). SDN testing and debugging tools: a survey. In *5th International Conference Informatics, Electronics and Vision (ICIEV)* (pp. 631–636). IEEE
6. Kreutz, D., Ramos, F. M., Verissimo, P. E., Rothenberg, C. E., Azodolmolky, S., & Uhlig, S. (2015). Software-defined networking: A comprehensive survey. *Proceedings of the IEEE, 103*(1), 14–76.
7. Deng, S., Gaol, X., & Lu, Z. (2018, March) Packet injection attack and its defence in software-defined networks. *IEEE Transactions on Information Forensics and Security, 13*(3), 695–705.
8. Hegde, S., Koolagudi, S. G., & Bhattacharya, S. (2017). Path Restoration in source routed software-defined networks. In *Ninth International Conference Ubiquitous and Future Networks (ICUFN)* (pp. 720–725). IEEE.
9. Bera, S., Misra, S., & Vasilakos, A. V. (2017, December). Software-defined networking for internet of things: A survey. *IEEE Internet of Things Journal, 4*(6), 1994–2008.
10. Ammar, H. A., Nasser, Y., & Kayssi, A. (2017) Dynamic SDN controllers-switches mapping for load balancing and controller failure handling. In *International Symposium on Wireless Communication Systems (ISWCS)* (pp. 216–221). IEEE.
11. Zhou, Y., Wang, Y., Yu, J., Ba, J., & Zhang, S. (2017). Load balancing for multiple controllers in SDN based on switches group. In *19th Asia-Pacific Network Operations and Management Symposium (APNOMS)* (pp. 227–231). IEEE.
12. Yu, J., Wang, Y., Pei, K., Zhang, S., & Li, J. (2016). A load balancing mechanism for multiple sdn controllers based on load informing strategy. In *18th Asia-Pacific Network Operations and Management Symposium (APNOMS)* (pp. 1–4), IEEE.

13. Zhou, Y., Zhu, M., Xiao, L., Ruan, L., Duan, W., Li, D., Liu, R., Zhu, M. (2014). A load balancing strategy for sdn controller based on distributed decision. In *IEEE 13th International Conference on Trust, Security and Privacy in Computing and Communications* (pp. 851–856).
14. Zhong, H., Sheng, J., Cui, J., & Xu, Y. (2017). SCPLBS:A smart cooperative platform for load balancing and security on sdn distributed controllers. In *3rd IEEE International Conference on Cybernetics (CYBCONF)* (pp. 1–6).
15. Manthena, M. P. V., van Adrichem, N. L. M. van den Broek, C., & Kuipers, F. (2015). An SDN-based architecture for network-as-a-service. *1st IEEE Conference on Network Softwarization (NetSoft)* (pp. 1–5).
16. Cui, L., Yu, F. R., & Yan, Q. (2016). When big data meets software-defined networking: sdn for big data and big data for SDN. *IEEE Network*, *30*(1), 58–65.
17. Chandrasekaran, B., & Benson, T. (2014). Tolerating SDN application failures with legoSDN. In *13th ACM Workshop on Hot Topics in Networks* (pp. 1–7).
18. Wang, T., Liu, F., Guo, J., & Xu, H. (2016). Dynamic SDN controller assignment in data center networks: stable matching with transfers. In *35th Annual IEEE International Conference on Computer Communication, IEEE INFOCOM 2016* (pp. 1–9).

Optimal Thresholding in Direct Binary Search Visual Cryptography for Enhanced Bank Locker System

Sandhya Anne Thomas and Saylee Gharge

Abstract Visual cryptography (VC) is one of the strongest cryptographic method present. The main advantage of this system is that the decryption doesnot need any specific requirements for decoding other than human eyes. Using halftoning techniques binary images are obtained for grayscale and color images, this technique is applied in Halftone VC. In this paper, direct binary search (DBS) is implemented and initial images are modified for better quality of recovered images. The concept is proposed for bank locker systems. Comparison has been made using parameters like PSNR, Correlation, UQI and SSIM.

Keywords Visual cryptography · Halftone visual cryptography · Direct binary search · Color images · Bank lockers · Security

1 Introduction

Security is an undeniable problem, with the rapid need use and development of modern technologies. Security for bank lockers is important and must be customer friendly. Creating and sending a key is an issue because of that visual cryptography (VC) is the best method to send information hidden. In visual cryptography [1] the visual information is covered in such a manner that the decryption can be done using human eyes which is the biggest advantage of this method. The person need no previous knowledge about the cryptography or computers. Visual cryptography operates on binary image. Images are available in monochrome, grayscale and color format. Conversion of a grayscale and a color image to binary image is done using

---

S. A. Thomas (✉) · S. Gharge
Department of Electronics and Communication Engineering,
VESIT, Chembur, Mumbai, Maharashtra, India
e-mail: sandhyastanley@gmail.com

S. Gharge
e-mail: saylee.gharge@ves.ac.in

© Springer Nature Singapore Pte Ltd. 2019
N. R. Shetty et al. (eds.), *Emerging Research in Computing, Information, Communication and Applications*, Advances in Intelligent Systems and Computing 906,
https://doi.org/10.1007/978-981-13-6001-5_49

halftoning techniques. A commonly used halftone techniques for color image is Direct Binary Search (DBS) which is implemented and improved in this paper.

The VC principle is that, in a (s, t) VC scheme a secret image is divided into t shares which is to be divided among t participates. Each share shows a random noise arrangement of black and white and doesn't disclose any information of the secret image by itself. In this scheme, s is the minimum qualifying participants required to decode the secret image. Fewer participants that are s-1 will be considered has false participant and decryption will fail and cannot occur even if they have high computational knowledge. Qualified share will be stacked/superimposed to decrypt the secret image.

Several methods have been developed by researches for VC in literature [2]. To analyze the contrast of the recovered image for a (k, n)-threshold VCS, Blundo [3] proposed an optimal contrast. Ateniese [4] developed a general scheme. Blundo [5] implemented images in grayscale instead of the traditional VC method. A gray level image is converted into a halftone image to generate a halftone share [6] which is then used in VC. While a secret information is transferred, it is important to have a cover image which can avoid intrusion. Extended visual cryptography developed by Ateniese [7] got shares which carries secret information and cover images know as meaningful shares that enhances the security. To get better visual images good halftone shares must be generated using halftone methods in VC [8]. Error diffusion [9] is a halftone method which is a neighborhood process, this technique is applied by Inkoo [10] in Color Extended Visual Cryptography. Another commonly used halftone method is Direct Binary Search (DBS) [11].

This paper also proposes the implemented VC for bank locker with enhanced security. Presently, there are four types of lockers. The traditional and most commonly used is the key-based locker systems which involves human operated locks. For the operation of this type of locker a human operated key is required. This is comparatively simple to either access or duplicate. Access to the locker is easy since it has no other verification or authentication involved. The second locker which is available is known as Digital Locker Systems (DLS). In a DLS, a low-cost number system controls the lock to the locker instead of a key. The system consist of a small screen mounted on the system and the system is attached to an embedded controller which will manage the operations of locker using a verified password set by the user. The third type of locker is called GSM-based locker systems. Here, every locker is granted with a digital system which is linked to an electronic brain that has data collection information of the users. A disordered number that is exclusive to the user is generated using their personal details. To operate the locker they have to see and send this disordered number through their registered cellular number to the admin computer. GSM technology is the technology used for transferring messages. The last type of locker which is available is called the RFID Based Locker System. This system consists of a microcontroller to which the RFID reader is jointed. The reader reads the ID number from the passive tags, if the information is correct then the message is sent to the authenticated cellular number for the actual password for locker operations. The password is entered using a keyboard which is verified and matched by the microcontroller if the provided password and typed password match the locker is operational.

2 Methodology

Encryption: In this paper, a better recovered and secure image is implemented. The implementation is done by using two different inputs to the Direct binary search algorithm. VC works only on binary images, therefore, it is necessary that it works on all types on images. Halftoning helps in converting a color or a grayscale image into a binary image that is required for VC. A DBS algorithm is a search-based algorithm and is also known as iterative algorithm. In a DBS system, an initial halftone image is given has it input. In this paper, two such inputs for the algorithm are taken. The first input is taken using conventional halftoning like screening, error diffusion and the second by taking an optimal thresholding method.

Encryption for a (2, 2) VC scheme is done using extended VC [10]. In order to convert a grayscale image to the best possible binary image, it first takes some initial (random) binary image and it then iteratively tries to modify this initial binary image by toggling any random pixel of it. Then, it checks in every iteration whether the binary image has any improved or not in term of any metric. This process is repeated a number of times which is equal to the number of iterations given. Finally, the required binary image is got at the end of the last iteration. Here, the conventional method taken is error diffusion [10]. However, it is observed that the final binary output image depends on the initial binary image taken. Therefore, an initial binary image is proposed such that it gives a better output after the DBS procedure. In order to do so, a Otsu [12, 13] thresholding technique is used. It is applied on the original the grayscale image. It returns a binary image that is more suitable for the initial binary image that is required at the starting of the DBS algorithm (Fig. 1).

A secret image of size $m \times m$ is taken and converted to its RGB plane and given to the Otsu's thresholding. This will give an initial binary image which is then fed to the DBS algorithm where the toggling of the pixel takes place for n iterations. During

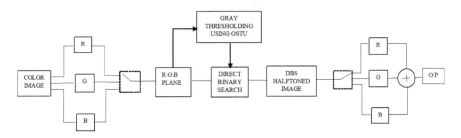

Fig. 1 Direct binary search using Otsu

each iteration, the parameters are calculated and analyzed. The DBS halftone image is taken based on this analysis. Share image of size $n \times n$ is taken for producing meaningful share which is twice the size of the secret image. This also needs to be halftoned using the previously explained optimized DBS method. The halftoned secret image and the halftoned shares and given to VC. Now the Encrypted share 1 and share 2 are generated which carries both secret information and meaningful images. These shares are ready to be given as qualified shares to the participants.

Decryption: The decryption remains same for all VC schemes. The qualified participants on stacking/superimposing will recover the original secret image. The stacking/superimposing is executed using the $\otimes$ the 'OR' Boolean function.

The parameter metrics used for analyzing the encrypted shares with the original halftoned share are PSNR, Correlation, UQI [14] and SSIM [15]. By considering more parametrics measures better understanding of the quality of the desired image can be achieved. Therefore, correlation, UQI and SSIM are taken into consideration.

3 Results and Discussion

In this section, simulation results of 2-out of-2 Halftone VC is constructed using DBS with error diffusion and DBS with Otsu. The secret images of size *128 × 128* to be encoded are *10 numbers (1 2 3 4 5 6 7 8 9 0)*, 10 symbols (α β λ η σ π μ Σ Δ Ω), *8 lettered word (PHOTONIC)*. Two images with text *'Mumbai university'* and *'Engineer'* are taken as meaningful images having a size of *256 × 256*.

Table 1 shows results of Encrypted share 1 using DBS with error diffusion and DBS with Otsu. Secret image taken here is 10 symbols α β λ η σ π μ Σ Δ Ω for different font sizes 12, 14, 16, 18 and 20.

Table 2 shows results of Encrypted share 1 using DBS with error diffusion and DBS with Otsu for 10 numbers *1 2 3 4 5 6 7 8 9 0* secret image. The secret image is taken in three different font types arial black, calibri and verdana having different font sizes 12, 14, 16, 18 and 20.

Table 3 shows the results of Encrypted share 1 using DBS with error diffusion and DBS with Otsu for a eight lettered word, the word taken has secret here is *PHOTONIC*

Table 1 Encryption for share 1 using DBS with error diffusion and DBS with Otsu for 10 symbols

Type	Size	DBS with error diffusion share I				DBS with Otsu share I			
		PSNR	Correlation	UQI	SSIM	PSNR	Correlation	UQI	SSIM
Symbols	12	51.595	0.2990	0.1235	0.0817	**52.137**	**0.3712**	**0.2012**	**0.0942**
	14	51.616	0.3006	0.1276	0.0867	52.131	0.3714	0.1999	0.0933
	16	51.601	0.2997	0.1246	0.0825	52.125	0.3705	0.1990	0.0928
	18	51.597	0.2988	0.1241	0.0827	52.13	0.3712	0.1997	0.0922
	20	51.604	0.2993	0.1253	0.0827	52.125	0.3704	0.1992	0.0927

Table 2 Encryption for share 1 for DBS with error diffusion and DBS with Otsu for 10 numbers

Type	Size	DBS with error diffusion share I				DBS with Otsu share I			
		PSNR	Correlation	UQI	SSIM	PSNR	Correlation	UQI	SSIM
Arial black	12	51.608	0.2999	0.1260	0.0837	52.125	0.3705	0.1990	0.0917
	14	51.619	0.3003	0.1283	0.0859	51.574	0.3043	0.1156	0.0794
	16	51.601	0.2993	0.1246	0.0827	52.13	0.3709	0.1999	0.0929
	18	51.612	0.2999	0.1269	0.0849	52.124	0.3710	0.1988	0.0918
	20	51.605	0.2998	0.1253	0.0828	52.103	0.3697	0.1950	0.0876
Calibri	12	51.595	0.2990	0.1235	0.0817	**52.137**	**0.3712**	**0.2012**	**0.0942**
	14	51.616	0.3006	0.1276	0.0867	52.131	0.3714	0.1999	0.0933
	16	51.616	0.3001	0.1276	0.0851	52.124	0.3708	0.1989	0.0925
	18	51.593	0.2988	0.1229	0.0804	52.126	0.3709	0.1991	0.0921
	20	51.593	0.2991	0.1230	0.0804	52.122	0.3702	0.1987	0.0916
Verdana	12	51.401	0.3075	0.0724	0.0464	51.297	0.2919	0.0554	0.0371
	14	51.374	0.3061	0.0666	0.0411	51.32	0.2931	0.0603	0.0412
	16	51.383	0.3066	0.0686	0.0427	51.297	0.2919	0.0553	0.0366
	18	51.359	0.3052	0.0632	0.0366	51.3	0.2920	0.0559	0.0372
	20	51.38	0.3064	0.0679	0.0417	51.277	0.2908	0.0510	0.0325

secret image. The secret image is taken in three different font types arial black, calibri and verdana having different font sizes 12, 14, 16, 18 and 20.

From the above three tables, it can be observed that for symbols the highest value is at font size 12 which has a PSNR of 52.137, correlation of 0.3712, UQI of 0.2012 and SSIM of 0.0942. For 10 numbers font type calibri with a size 12 gave high values for PSNR of 52.137, correlation of 0.3712, UQI of 0.2012 and SSIM of 0.0942. In the eight lettered word, a PSNR of 52.138 correlation of 0.3718, UQI of 0.2010 and SSIM of 0.0936 is the highest values in Table 3 for arial black font type having a font size of 12. Same results have been obtained for share 2 also.

Table 4 shows the final selection to form any secret message related to bank locker system.

3.1 Proposed Model for Bank Lockers

Figure 2 shows a proposed GUI model. The user will request the bank for locker operation. With this request the bank will generate the shares using the improved DBS algorithm on their systems. The model above is taken for two shares but the number of shares can be varied. More the number of shares higher is the security. A share is randomly generated and send to the user (to a registered device) and another share to the locker operator. When the user comes to the locker room the shares can

Table 3 Encryption for share 1 using DBS with error diffusion and DBS with Otsu for 8 letters

Type	Size	DBS with error diffusion share I				DBS with Otsu share I			
		PSNR	Correlation	UQI	SSIM	PSNR	Correlation	UQI	SSIM
Arial black	12	51.606	0.3000	0.1255	0.0828	**52.138**	**0.3718**	**0.2010**	**0.0936**
	14	51.612	0.3005	0.1267	0.0841	52.13	0.3709	0.1999	0.0930
	16	51.617	0.3005	0.1278	0.0857	52.126	0.3711	0.1989	0.0922
	18	51.592	0.2992	0.1227	0.0809	52.114	0.3705	0.1967	0.0890
	20	51.61	0.2999	0.1264	0.0839	52.126	0.3710	0.1992	0.0922
Calibri	12	51.6	0.2993	0.1245	0.0823	52.127	0.3708	0.1994	0.0930
	14	51.608	0.2994	0.1261	0.0841	52.126	0.3706	0.1994	0.0917
	16	51.594	0.2991	0.1232	0.0813	52.123	0.3706	0.1986	0.0908
	18	51.608	0.2999	0.1262	0.0841	52.121	0.3706	0.1982	0.0918
	20	51.6	0.2990	0.1245	0.0830	52.125	0.3704	0.1992	0.0925
Verdana	12	51.616	0.3001	0.1276	0.0851	52.124	0.3708	0.1989	0.0925
	14	51.593	0.2988	0.1229	0.0804	52.126	0.3709	0.1991	0.0921
	16	51.593	0.2991	0.1230	0.0804	52.122	0.3702	0.1987	0.0916
	18	51.608	0.2999	0.1260	0.0837	52.125	0.3705	0.1990	0.0917
	20	51.619	0.3003	0.1283	0.0859	52.11	0.3699	0.1963	0.0900

Table 4 Template for any bank locker system

Type	Font type	Font size
Symbol	–	12
10 numbers	Calibri	12
8 Letter word	Arial black	12

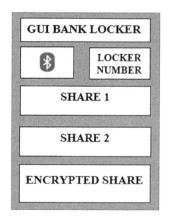

Fig. 2 GUI model for bank locker system

be transferred from the Bluetooth operated device to the GUI, which will read and decrypt the shares. The share will be displayed if it is qualified share otherwise the alarm will turn on. Based on the above tabulation a template for the proposed GUI is made and implemented.

Figure 3 shows illustrative results using DBS with error diffusion. A template for secret image is considered as shown in Table 4 made by taking the best values from Tables 1, 2, 3 and 4. That is for symbols font size 12 is taken, for numbers calibri with size 12 for 10 letters verdana of font size 18 is taken Figure (a) is the Secret image which is to be hidden, Figures (b) and (c) are Share images. Figure 3d–f and halftoned image using improved DBS. Figure (g) is Encrypted Share 1 and Figure (h) is Encrypted Share 2. Figure (i) is the decoded secret image.

Figure 4 shows illustrative results using DBS with Otsu. A template for secret image is considered as shown in Table 4. Figure 4a is the secret image which is to be hidden, Figures 4b and c are Share images. Figure 4d–f and halftoned image using

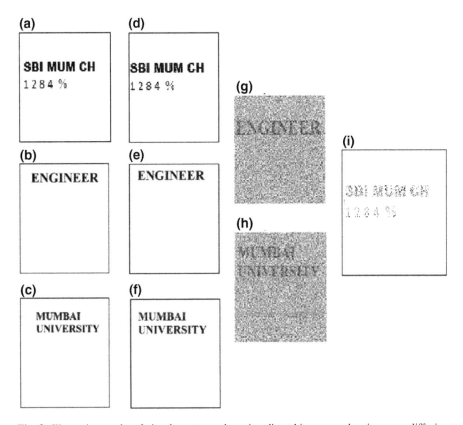

Fig. 3 Illustrative results of visual cryptography using direct binary search using error diffusion. **a** Original secret image, **b** and **c** Original cover share image respectively, **d** Halftoned secret image **e** Halftoned share1, **f** Halftoned share 2, **g** Encrypted share 1, **h** Encrypted share 2, **i** Decoded secret image

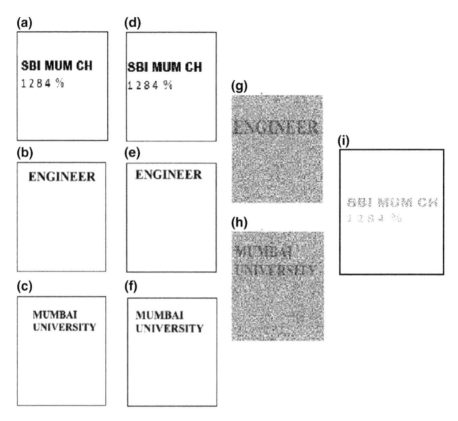

Fig. 4 Illustrative results of visual cryptography using Direct binary search using Otsu. **a** Original secret image, **b** and **c** Original cover share image respectively, **d** Halftoned secret image **e** Halftoned share 1, **f** Halftoned share 2, **g** Encrypted share 1, **h** Encrypted share 2, **i** Decoded secret image

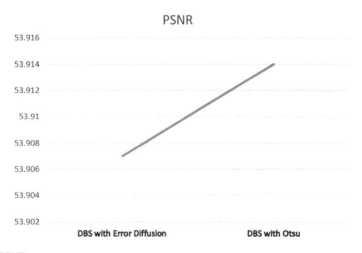

Fig. 5 PSNR

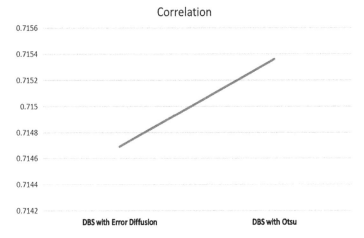

Fig. 6 Correlation

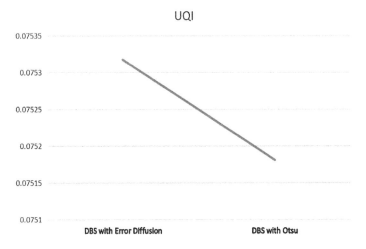

Fig. 7 UQI

improved DBS. Figure 4g is Encrypted Share 1 and Figure 4h is Encrypted Share 2. Figure 4i is the decoded secret image.

Graphical representation of PSNR, correlation, UQI and SSIM is shown in Figs. 5, 6, 7 and 8 for the secret image using DBS with error diffusion and DBS using Otsu

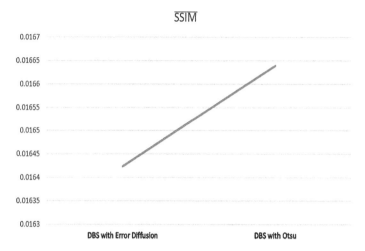

Fig. 8 SSIM

4 Conclusion

Visual cryptography allows the transfer of images with high security in bank locker system. Images in color and text have been encoded and decoded using DBS with error diffusion and DBS with Otsu. For secret image 'SBI MUM CH 1284%' both these methods have been implemented. DBS with Otsu shows better decoded images as compared to DBS with error diffusion. Otsu thresholding method is less complex than error diffusion thus DBS with Otsu is less complex and gives better results. According to performance measure like PSNR, Correlation, UQI and SSIM DBS with Otsu is better. There is a 50% increase found in SSIM. It has been observed that UQI decreases by a small value in case of DBS with Otsu but all the other parameter values show increase and are better than DBS using error diffusion. This method enhances the security which can be used in bank locker system. Even with highest authority power it is not possible to get any information from the shares which carries the secret. Also unlike the traditional method the access or duplication is not possible. GSM technology not required. Personal details are not required for generation of shares unlike in RFID tags. There by using the GUI model, it is possible to have better security.

References

1. Naor, M., & Shamir, A. (1994). Visual cryptography. In *Proceeding EUROCRYPT* (pp. 1–12).
2. Thomas, S. A., & Gharge, S. (2017). Review on various visual cryptography schemes. In *2017 International Conference on Current Trends in Computer, Electrical, Electronics and Communication (CTCEEC)* (pp. 1164–1167). Mysore.

3. Blundo, C., DArco, P., De Santis, A., & Stinson, D. R. (2003). Contrast optimal threshold visual cryptography schemes. *SIAM Journal on Discrete Mathematics, 16*, 224–261.
4. Ateniese, G., Blundo, C., Santis, A. D., & Stinson, D. R. (1996). Visual cryptography for general access structures. *Information and Computation, 129*, 86–106.
5. Blundo, C., Santis, A. D., & Naor, M. (2000). Visual cryptography for grey level images. *Information Processing Letters, 75*, 255–259.
6. Hou, Y. C. (2003). Visual cryptography for color images. *Pattern Recognition, 36*, 1619–1629.
7. Ateniese, G., Blundo, C., Santis, A., & Stinson, D. R. (2001). Extended capabilities for visual cryptography. *Theoretical Computer Science, 250*, 143–161.
8. Zhou, Z., Arce, G. R., & Crescenzo, G. D. (2006). Halftone visual cryptography. *IEEE Transactions on Image Processing, 18*, 2441–2453.
9. Thomas, S. A., & Gharge, S. (2018). Halftone visual cryptography for grayscale images using error diffusion and direct binary search, in *2018 2nd International Conference on Trends in Electronics and Informatics (ICOEI)* (pp. 1091–1096). Tirunelveli, India.
10. Kang, I., Arce, G. R., & Lee, H.-K. (2011). Color extended visual cryptography using error diffusion. *IEEE Transactions on Image Processing, 20*.
11. Wang, Z. M., Arce, G. R., & Di Crescenzo, G. (2006). Halftone visual cryptography via direct binary search. *EUSIPCO*.
12. Otsu, N. (1979). A threshold selection method from gray-level histograms. *IEEE Transactions on Systems, Man, and Cybernetics, 9*(1), 62–66.
13. Thomas, S. A., & Gharge, S. (2018). Enhanced security for military grid reference system using visual cryptography, in *2018 9th International Conference on Computing, Communication and Networking Technologies (ICCCNT)* (pp. 1–7). Bangalore.
14. Wang, Z., & Bovik, A. C. (2002). A universal image quality index. *IEEE Signal Processing Letters, 9*.
15. Wang, Z., Bovik, A. C., Sheikh, H. R., & Simoncelli, E. P. (2004). Image quality assessment: from error visibility to structural similarity. *IEEE Transactions on Image Processing, 13*.

Comparison Between the DDFS Implementation Using the Look-up Table Method and the CORDIC Method

Anish K. Navalgund, V. Akshara, Ravali Jadhav, Shashank Shankar and S. Sandya

Abstract An efficient communication system requires synchronization between the transmitter and the receiver, which is achieved by generating the same local carrier frequency. Direct Digital Frequency Synthesizer (DDFS) is one of the methods to generate various frequencies, centered around a reference frequency. This paper presents the comparison between the DDFS implementation using the look-up table (LUT) method and the CORDIC, a multiplier-less algorithm. The implementation has been carried out in Simulink and various parameters have been analyzed.

Keywords DDFS · LUT · CORDIC

1 Introduction

A communication system is used to transmit a signal or information from the transmitter to the receiver, through a channel. For a reliable transmission, the frequency of operation at both the ends is expected to be identical. A frequency synthesizer is used to generate various frequencies which are centered on a reference frequency, to establish the desired synchronization between the transmitter and the receiver by keeping the transmitter signal frequency as the reference for the receiver. Direct Digital Frequency Synthesizer is a type of frequency synthesizer that generates arbitrary waveforms from a single, fixed frequency reference clock, making it an ideal synthesizer for digital communication domain.

The DDFS can be implemented using the LUT method which is a table of predefined values corresponding to the sine function that aids in mapping the phase values obtained from the digitally controlled oscillator (DCO) to the respective amplitudes. An alternate method of implementation uses the CORDIC, a multiplier-less algorithm, which reduces the amount of hardware consumed when implemented on a field-programmable gate array as opposed to the conventional LUT method.

A. K. Navalgund (✉) · V. Akshara · R. Jadhav · S. Shankar · S. Sandya
Electronics and Communication, Nitte Meenakshi Institute of Technology, Bengaluru, Karnataka, India
e-mail: anishnaval96@gmail.com

© Springer Nature Singapore Pte Ltd. 2019
N. R. Shetty et al. (eds.), *Emerging Research in Computing, Information, Communication and Applications*, Advances in Intelligent Systems and Computing 906,
https://doi.org/10.1007/978-981-13-6001-5_50

2 CORDIC Algorithm

CORDIC, which stands for "**Co**ordinate **R**otation **Di**gital **C**omputer" was first coined by Jack E. Volder in the year 1959. It is used to compute various mathematical functions like trigonometric, logarithmic, and linear operations using iterative add and shift method superseding multiplication. Being a hardware-efficient algorithm, it uses micro-rotations of arc-tan function to provide an output. This algorithm works in two modes, namely the rotation and vector mode.

In vector mode, the initial vector is rotated to align it along X-axis and it calculates the angle by which the vector has rotated. In rotation mode, the angle is initialized and the coordinates of the vector which has rotated by that angle are the output. The flow of CORDIC algorithm is shown in Fig. 1.

Generalized CORDIC equations are:

$$x_{i+1} = x_i - m \cdot y_i \cdot d_i \cdot 2^{-i} \tag{1}$$

$$y_{i+1} = y_i + x_i \cdot d_i \cdot 2^{-i} \tag{2}$$

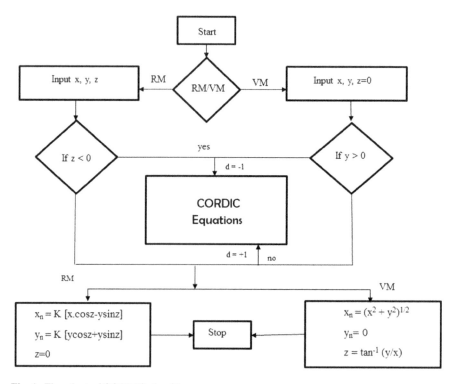

Fig. 1 Flowchart of CORDIC algorithm

$$z_{i+1} = z_i - d_i \cdot e_i \qquad (3)$$

where

$m = 1$ implies circular coordinate system
$m = 0$ implies linear system
$m = -1$ implies hyperbolic system

$$e_i = \tan^{-1}(2^{-i}) \quad \text{for } m = 1$$
$$e_i = (2^{-i}) \quad \text{for } m = 0$$
$$e_i = \tanh^{-1}(2^{-i}) \quad \text{for } m = -1$$

where,
e_i is the elementary angle of rotation, for iteration i in the selected coordinate system.

Finally, these equations are multiplied with CORDIC system processing gain A_n, as given by Eq. 4, which is the product of iterative CORDIC gains K_i. This product approaches 0.6073 as the number of iterations goes to infinity [1, 2].

$$A_n = \prod_n \sqrt{1 + 2^{-2i}} \qquad (4)$$

3 Methodology

3.1 DDFS Using LUT

The essential blocks of the DDFS implemented using LUT has the phase accumulator, a look-up table containing a range of predefined values of amplitudes corresponding to a $-\pi$ to $+\pi$ range of a sinusoidal waveform stored in a read-only memory device and, if required, it also has the digital to analog converter (DAC) followed by a filter [3].

The input to the phase accumulator is a frequency control word, of N-bit word length, which is a constant and decides the resolution of the output frequency. This accumulator, which is a combination of an adder and a register, stores up the phase information corresponding to the N-bit input frequency, thus limiting the maximum stored value to $2^N - 1$. The output of the accumulator is a continuous ramp containing the phase information in the form of steps which is fed to the ROM. The ROM compares every phase step coming as the input with the indices of the look-up table to give analogous sine amplitude. The output from the ROM is the digitized sine wave, which can be smoothened by using an analog filter [3].

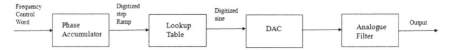

Fig. 2 DDFS using LUT method

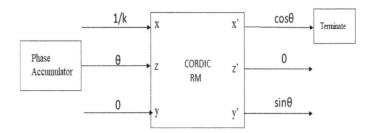

Fig. 3 DDFS using CORDIC algorithm

3.2 DDFS Using CORDIC

CORDIC can be utilized in applications that involve trigonometric functions as discussed in the previous section. In DDFS (which works in rotation mode), the look-up table is replaced with the CORDIC block that takes three inputs x, y, and z where x, y correspond to the axes of the Cartesian coordinate system and z to the angle in the range of $-\pi$ to $+\pi$ which is acquired from the phase accumulator, for which a corresponding sine and cosine value can be obtained. The values of x and y are 1/k or 1 and 0, respectively. As also discussed, CORDIC aids in reducing the hardware utilized by performing repeated shifts and additions, and hence, if implemented on the FPGA will considerably reduce the hardware utilization of the DDFS [4] (Figs. 2 and 3).

4 Implementation

Every DDFS implementation has the sampling frequency, output frequency, the word length, and frequency control word which to be considered as the design parameters. The output frequency follows the Nyquist criterion and hence can be evaluated as half the sampling frequency. From equation:

$$F_o = (F_s * k) / 2^N \quad (5)$$

where

F_o Output frequency of the DDFS in Hertz
F_s Sampling frequency in Hertz

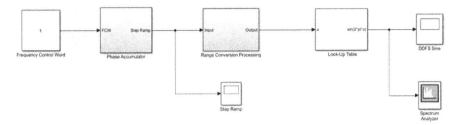

Fig. 4 Simulink model of DDFS using LUT

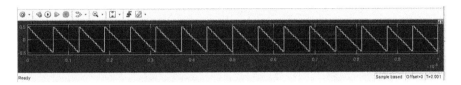

Fig. 5 Ramp output of phase accumulator

k Frequency control word
2^N Number of bits

4.1 DDFS Using LUT

Design considerations here are,

$$F_s = 16 \text{ KHz};$$
$$k = 1$$
$$N = 4$$
$$2^N = 16$$

The input to the phase accumulator is a frequency control word of value 1, of 4-bit word length, which is a constant and decides the resolution of the output frequency. The Simulink model designed with these assumptions is shown in Fig. 4. Here, the phase accumulator gives a range of phase steps from 0 to 2^4-1. The range conversion processing block converts this range to a stepped ramp of range -0.5 to $+0.5$ as shown in Fig. 5. These values are fed to the lookup-table block for mapping phase to amplitudes. The design results in an output frequency of 1 Hz as shown in Fig. 6.

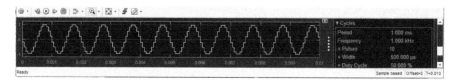

Fig. 6 Results of DDFS using LUT

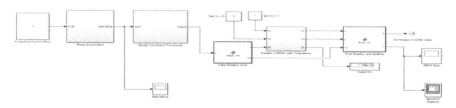

Fig. 7 Simulink model of DDFS using CORDIC

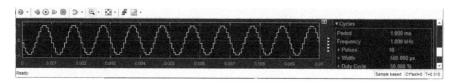

Fig. 8 Results of DDFS using CORDIC

4.2 DDFS Using CORDIC

Design considerations here are,

$$F_s = 16 \text{ KHz};$$
$$k = 1$$
$$N = 4$$
$$2^N = 16$$

The input to the phase accumulator is a frequency control word of value 1, of 4-bit word length, which is a constant and decides the resolution of the output frequency. Here, the phase accumulator produces a range of phase steps from 0 to 2^4-1. The range conversion processing block converts this range to a range between -0.5 and $+0.5$ and is the same as the ramp generated by the LUT as is shown in Fig. 5.

The Simulink model and results for this is shown in Figs. 7 and 8 as, respectively.

Table 1 DDFS using CORDIC

k	Fs (Khz)	Fout (Khz)	Period (μs)	THD (dBc)	SNR (dBc)	SINAD (dBc)	SFDR
1	128	8	125	−79.88	77.96	75.8	77.96
1	256	16	62.5	−79.88	77.96	75.8	77.96
1	512	32	31.25	−79.88	77.96	75.8	77.96

Table 2 DDFS using LUT

k	Fs (Khz)	Fout (Khz)	Period (μs)	THD (dBc)	SNR (dBc)	SINAD (dBc)	SFDR
1	128	8	125	−91.93	105.19	91.73	93.52
1	256	16	62.5	−91.93	105.19	91.73	93.52
1	512	32	31.25	−91.93	105.19	91.73	93.52

5 Observations

The parameters that have been used for the comparison between the two methods of implementation have been displayed in Tables 1 and 2. These values were obtained from the spectrum analyzer in Simulink.

6 Results and Conclusions

Direct Digital Frequency Synthesizer has been successfully implemented using LUT method and CORDIC method. The results from the spectrum analyzer have been analyzed, and it can be concluded that CORDIC method of implementation gives an identical output to the lookup-table with a slightly increased distortion with lesser consumption of hardware and thus can be used as an alternative for LUT.

References

1. Volder, J. E. (1959). The CORDIC trigonometric computing technique. *IRE Trans. Electron. Comput. 3*, 330–334.
2. Andraka, R. (1998). A survey of CORDIC algorithms for FPGA based computers. In *Proceedings of the 1998 ACM/SIGDA sixth international symposium on Field programmable gate arrays*. ACM.
3. Sharma, S., Kulkarni, S., & Lakshminarasimhan, P. (2009). Implementation and application of CORDIC algorithm in satellite communication. In *15th National Conference on Communication, January 2009*. Guwahati, India.
4. Khan, S. A. (2011). *Digital design of signal processing systems: a practical approach*. Wiley (2011).

Adding Intelligence to a Car

Komal Suresh, Svati S. Murthy, Usha Nanthini, Shilpa Mondal and P. Raji

Abstract The automobile business has been globalized from its initial days. Carmakers and innovation firms are investigating every possibility in their joint endeavors to upgrade the execution of keen automobile stages. Mischances are expanding everywhere; pace and different advancements are being utilized to diminish it. Utilizing generally straightforward programming and changes in accordance with existing equipment, we can accomplish an exceedingly secure automobile. This venture builds up a framework that the majority of its activities are controlled by smart programming inside the ARM LPC 2148. It expects to plan and build up a framework which can be controlled from the outside world utilizing Bluetooth and furthermore guarantees the driver well-being by utilizing a contrasting option to air bags with the assistance of rack and pinion framework. At the point when the automobile is being utilized by any unapproved individual, a message containing the automobile area with the assistance of GPS and GSM will achieve the proprietor quickly. The proprietor derives about the security rupture and tries to control and stop the automobile with Bluetooth. Adding to this, when the temperature of the motor is raised past a specific point of confinement, the automobile is made to stop naturally until the point that the temperature is under control.

K. Suresh · S. S. Murthy · U. Nanthini · S. Mondal · P. Raji (✉)
Department of ECE, Nitte Meenakshi Institute of Technology,
Karnataka, India
e-mail: raji.parappil@gmail.com

K. Suresh
e-mail: komalsuresh9@gmail.com

S. S. Murthy
e-mail: svati.s.murthy@gmail.com

U. Nanthini
e-mail: ushananthinisy@gmail.com

S. Mondal
e-mail: shilpasonu47@gmail.com

© Springer Nature Singapore Pte Ltd. 2019
N. R. Shetty et al. (eds.), *Emerging Research in Computing, Information, Communication and Applications*, Advances in Intelligent Systems and Computing 906,
https://doi.org/10.1007/978-981-13-6001-5_51

Keywords ARM LPC 2148 · Bluetooth HC-05 · Rack and pinion system · Global positioning system (GPS) · Global system of mobile communication (GSM) · Temperature sensor LM35

1 Introduction

From telephones to automobiles to spans, inserted advances are progressively influencing the things we need to utilize more quickly and consistently. As the progression of vehicular innovation enhances, so does the need of giving more secure and more effective vehicles for transportation. Vehicular innovation is utilized not just fill in as an effect to our each life, however to give a support of a more secure condition.

As the automobile innovation progresses, the innovation to take it likewise propels. Conventional automobile alerts are not excessively full of feeling any longer since individuals are so used to hearing them go off accidently that the overall population does not in any case investigate at an automobile with a caution actuating any longer. However, following frameworks, latent immobilizers and individual caution pagers offer high-tech choices or add ones to the customary alert that make the automobile significantly harder to take and less demanding to recoup. Mishaps are avoided today by different innovations. Air bags be worthwhile if there should arise an occurrence of impacts to shield the driver from wounds yet these frameworks are restricted to extravagance show automobiles.

This system uses ARM LPC 2148 which can be mounted in any hidden section of the car. After the installation of this system, the owner will be able to track and stop his vehicle in case of any unauthorized use. There are two states—active and inactive. If the car is unlocked using the registered mobile Bluetooth, which is paired with Bluetooth HC-05, the system remains inactive. In the second case, if the system is unlocked by any other unauthorized manner, the system goes into active state and sends the message to the owner via GSM module. The owner receives the location of his vehicle. He can then control and stop the car by disconnecting the Bluetooth in his phone. The engine is slowed down with the help of a relay interlocked with the DC motor and the wheels of the car. In addition to this, we know that modern vehicles incorporate a vast array of technologies to reduce the likelihood of injuries and fatalities in an event of a crash. To increase safety measures and as a substitute for the air bags, we have implemented a rack and pinion mechanism using IR sensors.

2 Related Work

In [1], Ramani and Valarmathy "Vehicle Following and Bolting Framework In view of GSM and GPS" portrayed when the burglary distinguished, the mindful individuals send SMS to the smaller scale controller, at that point issue the control signs to stop the motor engine. This plan will persistently watch a moving vehicle and report the status of the vehicle on request.

In [2], the equipment and programming of the GPS and GSM organized were produced. The proposed GPS-/GSM-based framework has the two sections; first is a portable unit, and another is controlling station. The versatile unit and control stations are working effectively with the framework forms, interfaces, associations, information transmission, and gathering of information. These outcomes are good with GPS advancements.

In vehicle, the following framework [3] is a gadget, which is introduced in a vehicle to empower the proprietor or an outsider to track the vehicle. In this paper, the outline chips away at the premise of GPS and GSM innovation. This framework depends on inserted framework. Following and situating of any vehicle are distinguished by utilizing worldwide situating framework (GPS) and worldwide framework for versatile correspondence (GSM). The status of the moving vehicle is transferred habitually on request.

In [4], Chen and Chiang portrayed to track the burglary vehicle by utilizing GPS and GSM innovation. This framework puts into the resting mode after the vehicle gets took care of by the proprietor or approved people through the reset catch over it.

In [5], Thin Zar Thein Hlaing depicted the component of rack and pinion directing framework. This explanatory examination is mostly in light of Lewis push formula. It is centered around twisting and contact worries of the pinion apparatus and rack bowing pressure utilizing diagnostic and limited component investigation.

3 Proposed Method

If the car is accessed by Bluetooth, then it is considered to be in the safe mode. If it is not accessed through the registered Bluetooth, then a message through GSM is sent to the owner which contains the latitude and the longitude of the car. The owner uses Arduino Bluetooth controller app from his phone to stop the car. In this project, we have also taken temperature of the engine into consideration. If the temperature exceeds the threshold value of thermistor 103, the car stops automatically. We even take the driver's safety into account by using an IR sensor. If the IR sensor is set high, the relay acts as a switch and activates the DC motor. This in turn moves the seat back by the rack and pinion.

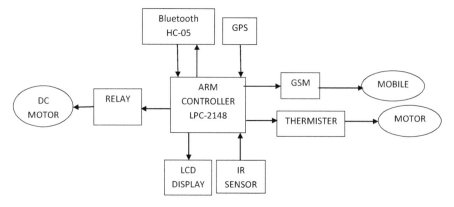

Fig. 1 Block diagram of the system

4 Design

The block diagram of the designed system is shown in Fig. 1; 5 V supply is given to every one of the parts of the framework. At the point when the automobile is utilized by some other unapproved individuals, the proprietor can stop the motor utilizing Bluetooth and he can likewise get the area of the vehicle at the same time. Notwithstanding this, we are utilizing a rack and pinion framework to move the seat in reverse in the event of a crash.

5 Hardware Components

5.1 Arm LPC 2148

LPC 2148 is a generally utilized IC from ARM-7 family. It has 8 t 40 KB of on-chip static slam and 32–512 KB of on-chip streak memory. Its 128 piece-wide interface/quickening agent empowers rapid 60 MHz activity. It deals with 3.3 V power; however, the fundamental peripherals like LCD, motor driver, and so on take a shot at 5 V. One or two 10-bit A/D converters give an aggregate of 6/14 simple contributions inside.

5.2 Bluetooth HC-05

HC-05 module is a Bluetooth SPP (serial port convention) module which is utilized for straightforward remote serial association. The design utilized as a part of HC-05 Bluetooth module is an ace or a slave setup. This is a completely qualified Bluetooth V2.0 +EDR (upgraded way rate) 3 Mbps tweak with finish 2.4 GHz radio handset and baseband. HC-05 can be designed by AT Summons. The slave modules cannot start an association with another Bluetooth gadget, yet it can acknowledge associations.

5.3 GSM—SIM900

GSM module is an ultra-minimized and solid remote module which works at recurrence of 900 MHz. It is a breakout board and least arrangement of SIM900 quad-band GSM module. It speaks with the controller by means of AT summons. It has free serial port associating, and it can be associated with equipment/programming serial port control. It utilizes supercapacitor control supply for RTC. The inner module is overseen by AMR926EJ-S processor which controls telephone correspondence, information correspondence, and the correspondence with the circuit interfaced with the phone itself.

5.4 Global Positioning System (GPS)

The worldwide situating framework is a satellite-based route framework. It causes the client to decide their two-dimensional position. It has three sections, space fragment, client portion, and control section. In the space portion, it comprises 24 satellites, each in its own circle. The client fragment comprises a collector, which is held in the automobile. The control section comprises ground station and ensures that the satellite is working legitimately. The radio signs are been transmitted by the GPS satellites which empower the GPS beneficiary in your automobile to gauge the satellite area. It additionally finds the separation between the satellite and the vehicle.

5.5 103 Thermistor

These negative temperature coefficient thermistors are resistors with a negative temperature coefficient and can be utilized as present restricting devices and resistive temperature sensors. They can be utilized for temperature estimation, temperature control, temperature remuneration, control supply fan control, and PCB temperature checking.

5.6 IR Sensor

IR sensors can be worked for aloof or dynamic conditions. Infrared finders are essentially aloof infrared sensors. Latent infrared sensors are typically used to distinguish vitality radiated by deterrent in the field of view. Infrared source and infrared identifiers are two components of dynamic infrared sensors. Infrared source comprises drove or infrared laser, and an infrared finder comprises photodiode or phototransistor. The vitality produced by infrared source I reflected by a protest and falls on infrared locator.

5.7 Relay

Relays are straightforward switches which are worked both electrically and mechanically. The exchanging system is done with the assistance of the electromagnet. Single shaft single throw (SPST) has an aggregate of four terminals. Out of these two terminals can be associated or disconnected. The other two are required for the loop.

6 Flowchart

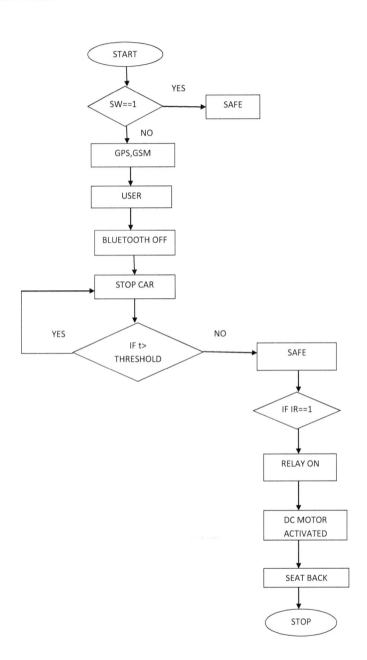

Fig. 2 Message sent to the owner of the car in case of unauthorized user

7 Result

The hardware was tested and gave accurate results on testing as shown with the help of figures below. All the stages from receiving the message till the complete prototype of the car are shown. With the help of model, we are able to deliver a system where the owner can control his car from outside with the help of Bluetooth and also ensure the driver safety with the help of rack and pinion system. In addition to this, we also control the engine based on the temperature hike (Fig. 2).

8 Conclusion

Remote control is a standout among the most fundamental requirements for every single living being. Be that as it may, tragically because of a lot of information and correspondence overheads, innovation is not completely used. In this paper, we have made utilization of Arduino Bluetooth application for controlling the auto. This proposed strategy is extremely modest and effectively accessible. For this application to work, the android versatile client needs to introduce the application on his/her portable. At that point, the client needs to turn on the Bluetooth in the versatile for availability. The client can utilize different summons to turn on and off the motor of the vehicle which are sent from the android portable. The vehicle has a recipient unit which gets the charge and offers it to the microcontroller circuit to control the engines. Android Bluetooth empowers telephones and Bluetooth modules by means of HC-05 and imparts among the Bluetooth gadget. Additionally, controlling the auto based on temperature and execution of rack and pinion framework is finished effectively.

9 Future Scope

The undertaking can be actualized on an extensive scale utilizing Wi-fi technology and Raspberry Pi. As of now, the model is being created. In the future, it should be possible on ongoing cars.

References

1. Ramani, R., & Valarmathy, S. (2013). Vehicle tracking and locking system based on GSM and GPS. Published Online August 2013 in MECS.
2. Asaad, M. J., & Talib, I. (2012). Experimentally Evaluation of GPS/GSM Based System Design. *Journal of Electronic Systems, 2*,(2), 1–8.
3. Maurya, K., Singh, M., & Jain, N. (2012). Real time vehicle tracking system using GSM and GPS technology—an anti-theft tracking system. *International Journal of Electronics and Computer Science Engineering, 1*,(3), 1103–1107. K. Elissa, "Title of paper if known," unpublished.
4. Chen, H. Chiang "Real time vehicle ceasing and tracking using GSM and GPS technology" 2(1), [1].
5. Hlaing, T. Z. T., Win, H. H., & Thein, M. (2017). Design and analysis of steering gear and intermediate shaft for manual rack and pinion steering system. *International Journal of Scientific and Research Publications, 7*(12), 861–881. ISSN 2250-3153.

Thermal Care and Saline Level Monitoring System for Neonatal Using IoT

Huma Kousar Sangreskop

Abstract A newborn baby usually has a problem to adapt the change in temperature, be it a full-term healthy baby or a preterm baby or low-birth-weight babies. Neonatal usually has little body fat, and they are too immature in handling and regulating the body temperature. A temperature ranging between 37.5° and 36.5° is considered to be normal body temperature; according to the WHO, any newborn whose temperature is below the normal range and drops below 32° is considered as a risk leading to hypothermia condition in the newborn. In such conditions, babies are kept in incubators so that the babies can regulate their body temperature and get adjusted to the environment. Hypothermia in neonates is associated with increased mortality rate. Thermal management of babies is a vital and critical part of neonatal care. And frequent check on the level of saline status is a must when given to any neonatal which cannot be neglected or show inattentiveness which may lead to life-risking condition. With the advancement of technology and IOT in the boom, this paper provides the health monitoring system of neonatal care using the IoT, a system which can monitor and maintain the necessary temperature of the neonates and monitoring of the saline bottle from a distant place.

Keywords Neonates · Incubators · Saline · Internet of things (IoT) · IR Sensors · Arduino microcontroller

1 Introduction

The small infants require adequate warm environment as the newborn may have problem in maintaining their own body heat. In the early 1900s, the essentiality of infant's thermal control was identified. Due to less insulation, the neonatal thermal control was limited as compared to that of adult. Hypothermia is one of the major factors to the increasing rate of morbidity and mortality among neonates. More than

---

H. K. Sangreskop (✉)
Department of Computer Science, College of Sericulture UAS(B), Bangalore, Karnataka, India
e-mail: Huma.destiny@gmail.com

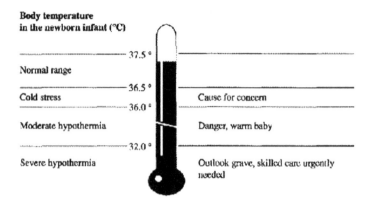

Fig. 1 Hypothermia in newborn infant

67% of high risk infants who were born outside of the hospital and according to a study conducted by WHO showed around 80% of infants born were hypothermic soon after birth. In developed countries, awareness of the problem has resulted in improved care of newborn especially of preterm and low-birth-weight infants.

When the body temperature drops below 36.5 °C, the condition is called as hypothermia. Similarly when the body temperature is between 32 and 35 °C, it is considered as moderate hypothermia, and when the body temperature goes below 32°, it is considered as server hypothermia (Fig. 1).

Due to the increasing demand in the IoT, several technologies have been introduced for automation. Using the recently introduced network connectivity solutions such as Ethernet and wireless LAN, data process can be modified with software programs. Most of the works for IoT connection may include the usage of Arduino with the help of Ethernet boards for expansion. As we require the Internet connectivity to generate alert system and for remote monitoring, Arduino alone cannot provide the solution due to its limitation of not executing multiple programs at once; therefore, a raspberry pi is the best suitable option available, and this in turn makes the massive growth for IoT.

2 Existing Approach

In the current healthcare system, there are complete dependencies on the professional nurses for managing and monitoring the neonatal especially when the newborn is kept in the incubators or in the case when the newborn is under observation for some treatment and is receiving saline. Presently, it is the duty of the nurses to maintain the appropriate temperature of the incubators and roller clamp is used for manually controlling the saline infusion and rate. If roller clamp rolls in one way, it compresses the intravenous tube more tightly which makes the tube more thin and allows saline fluid to flow through slower rate, and if it is rolled in other direction, it loosens or releases the saline tubing which makes the tube less thin and allows the saline fluid to flow through at a faster rate. In this era, there is no such monitoring system which will reduce the dependency of the neonatal neonates on the nurses and nurses frequently visiting the neonatal bed every time to check saline level status or to check the temperature of the incubator. Hence, there is a need for development of thermal care and saline level monitoring system for neonatal using IOT.

3 Proposed Methodology

The proposed methodology is going to use the smart health system with real-time data and pervasive computing. Wireless sensor network assisting in various healthcare solutions by measuring physiological parameters. With the massive growth of Internet of things and connections of things to the Internet with standard protocols, suitable architectural changes facilitate unobtrusive health monitoring for all day and any place. Modernized neonate centered monitoring system is the need of today's health care. Neonatal care is a very sensitive issue considering the utmost care requires in this phase of life although baby is normal or at risks. Every parent wants their just born fragile and tender baby should get non-disruptive care. Emerging of new technologies like bio-sensing devices and Arduino microcontrollers, it is possible to have complete neonatal care. The system requirements are as follows.

3.1 LM35 Temperature Sensor

The main object in temperature system is the reading of temperature value from LM35 temperature sensor. The primary use of LM35 temperature sensor is that it is the simplest of all the temperature sensors and it has an integrated circuit that gives an output as voltage that is proportional to the temperature in the degree Celsius.

3.2 IR Sensor

An infrared sensor [IR sensor] is an electronic device that emits in order to sense some aspects of the surroundings. IR sensor will be positioned at the critical level of the saline on the saline bottle to sense the critical level of saline as well as saline completion status.

3.3 Arduino Microcontroller

Arduino is an open-source microcontroller kit for building digital devices and interactive objects that can sense and control objects in the physical world. Arduino microcontroller will be used as processing and programming unit for sending instructions to the DC motor and buzzer.

3.4 DC Motor

DC motor is a rotary electrical machine that converts direct current electrical energy into mechanical energy. DC motor will function according to the commands given by the microcontroller and causes movement in the spring.

3.5 Buzzer

Buzzer is an audio signaling device. Buzzer will alert the nurses, caretakers, and doctors when saline reaches critical level and for replacement of saline bottle.

3.6 Power Supply Unit

Power supply unit converts main AC to low-voltage regulated DC power for the internal components of the computer. It will supply power to the rest of the components of the proposed system.

3.7 Clamp

Clamp will be attached to the spring. With stretching of the spring, the clamp will move in forward direction and pinch the intravenous tube and stop the reverse flow of blood into the saline bottle.

3.8 Spring

Spring is an elastic object that stores mechanical energy. When a spring is trenched from its resting position, the clamp attached to the spring will move toward the intravenous tube for stopping the reverse flow of blood into saline bottle. When a spring is compressed, it will return to its rest position.

3.9 Display

The display is used to show the current temperature of the incubator and helped in monitoring the body temperature of the neonates.

4 System Working

The system architecture is consisting of the following things as shown in the diagram (Fig. 2).

The proposed system functions as explained below: After saline gets consumed by the neonatal, the IR sensors sense when the saline reaches the critical level (Fig. 3).

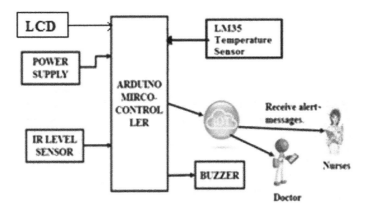

Fig. 2 System architecture

Fig. 3 Position of IR sensor

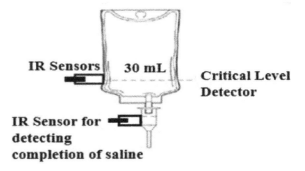

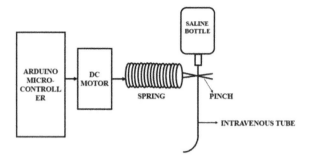

Fig. 4 Mechanism for stop reverse flow of blood

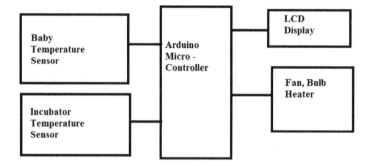

Fig. 5 Monitoring temperature of baby and incubator

This sensed output is sent to the microcontroller and buzzer starts ringing for alerting the nurses and doctors in the hospitals. An alert message is sent to the concerned nurses and doctors associated with the neonate through the use of the Internet. If the nurse attends the neonate, then she should stop the buzzer and reset the whole system (Fig. 4).

If the nurse fails to attend the neonate within the set time limit, the reverse flow of the blood into the saline bottle is stopped.

For this, a spring-DC motor arrangement will be made. The clamp will be attached to spring; along with the compression and stretching of spring, the clamp will also move in forward and backward directions. Again the IR sensor, at the neck of the saline bottle will sense that the saline is totally consumed and buzzer will again start ringing louder to notify the nurse that the saline is totally consumed, and there is a requirement for replacement of saline bottle. The instructions for Arduino will be sent to DC motor, and as per functioning of DC motor, the spring will be stretched and the clamp will move in forward direction and pinch the intravenous tube and stop the reverse flow of the blood in the saline bottle (Fig. 5).

On the other hand, the LM35 temperature sensor will constantly monitor the temperature of the neonate and the incubator temperature. A predetermined temperature would be set which will take appropriate action as directed by the Arduino microcontroller. Suppose if the body temperature of the baby reduces below the normal range, then the light bulbs will be set on which will increase the temperature of the incubator and it will increase the body temperature of the baby. And if the temperature of the baby is very high above the normal range, then the microcontroller will on the fan and regulate the temperature of the incubator and set it back to the specified and suitable temperature for the neonate.

5 Conclusion

With the more increasing usage of technology and growth of IoT, this system "thermal care and saline level monitoring system for neonatal using IOT" the manual effort on the part of nurses is saved. As this proposed system is automated, it requires very less human intervention. The special requirement of newborn to keep babies away from the germs which would be carried by frequent touch of humans/nurses can also help in the better development of the babies. Also, it will be advantageous at night as there will be no such requirement for the nurses to visit patient's bed every time to check the level of saline in the bottle since an alert notification will be sent to the nurses, doctors, and caretakers when saline reaches the critical level. It will save the life of the patients. This will reduce the stress in continual monitoring by the doctor or nurse at an affordable cost.

Home Security System Using GSM

P. Mahalakshmi, Raunak Singhania, Debabrata Shil and A. Sharmila

Abstract In areas where robbery and theft are a major issue, home security becomes a matter of prime importance to the residents of that area. Everyone in the locality is forced to take security measures to prevent their precious belongings from being stolen. It is therefore invincible that a technological solution has to be formulated to ensure the safety of the house. Hence, a security device has been designed to send an alert message to the owner of the house and to the security forces nearby in an attempt to void the theft taking place. The system is designed by interfacing sensor modules with a microcontroller to detect the motion in the house and a GSM module to send alert message to the owner of the house when the house is locked. This system uses low-cost sensors for motion detection and proves to be affordable. The installation of the system is easy and also the sensors and modules require very less space and consume low power when installed.

Keywords GSM · PIR · Arduino · SMS · IDE · Ultrasonic

1 Introduction

From most recent few years, home security has become a crucial necessity of family units to keep home safe from interlopers to get burglarized. So the analysts and organizations try to actualize the calculations and make some gradates that can keep your home safe from the burglars [1]. What's more there is a need to automate the home with the goal that client can exploit the mechanical headway in a manner that a man leaving his house unattended, does not need to consider his home security over and over again. It is subsequently this motivation behind this creation to give a gadget, which is able to give quick notice to the proprietor and administrations like police headquarters right away when the unwanted occasion may happen. This reason for existing is refined by means of utilization of a device which sends at least one SMS (Short Message Service) via GSM (Global System for Mobile) module

---

P. Mahalakshmi (✉) · R. Singhania · D. Shil · A. Sharmila
School of Electrical Engineering, Vellore Institute of Technology, Vellore 632014, India
e-mail: pmahalakshmi@vit.ac.in

© Springer Nature Singapore Pte Ltd. 2019
N. R. Shetty et al. (eds.), *Emerging Research in Computing, Information, Communication and Applications*, Advances in Intelligent Systems and Computing 906,
https://doi.org/10.1007/978-981-13-6001-5_53

to the proprietor and security administrations at the time of happening of theft [2]. This framework is minimal effort since it doesn't contain sensors which may be unaffordable and it is additionally simple to actualize as this system if installed will take very less space for establishment.

This work is mainly focused on the situation when the user is not present at the home. The GSM-based model will update the user about the current scenario of his/her house. In case of intrusion, the PIR or the ultrasonic sensors will detect the motion of the intruders, giving out an alarm and sending an alert text message to the user at the same time using the GSM.

2 Methodology

For home security, this paper proposes a novel approach using a microcontroller, sensors for motion detection and alert via GSM module and alarm via buzzer. Figure 1 shows the interfacing of the required components to be connected together to provide a security check at the house.

In order to practically layout the proposed model, the following components will be required:

- Microcontroller preferably Arduino.
- GSM for communication.
- PIR sensor.
- Ultrasonic sensor.
- Power supply.
- Buzzer as an alarm.

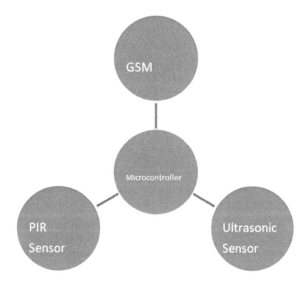

Fig. 1 Proposed layout for the home security system

Fig. 2 Arduino Uno microcontroller

Fig. 3 GSM module to be used

The following is the detailed description of components:

Arduino

The microcontroller Arduino Uno hardware is shown in Fig. 2. It has ATmega328P IC chip which is considered to be the brain of Arduino. It has a total of 14 digital input/output pins, a 16 MHz crystal oscillator for the clock cycle, a USB connector port, a power jack port and a reset button [3–5]. It has all the things needed for the functioning of the microcontroller. It can be connected to a PC with a USB wire or can be powered with an AC-to-DC adaptor [1]. One can play with Arduino UNO and test its limits and it will respond quite correctly, further, one can change the hardware for little amount and start the work again. Arduino can be used to create complex codes to interface various other modules, switches or sensors, and thereby figuring out to control various other outputs like buzzer, display, etc.

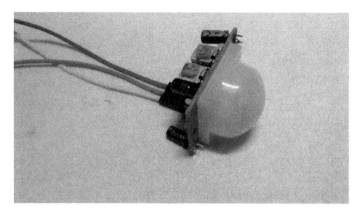

Fig. 4 View of PIR sensor

GSM

The Global System for Mobile (GSM) is an advanced innovation used for transmitting voice and information services. GSM is the most broadly used standard in media communications. GSM as shown in Fig. 3 is a GSM 800H, which is a quad-band GPS/GPRS module and works in frequency range of 850–1800 MHz [5, 6]. It requires a power supply of 3.4–4.4 V and does a current utilization of 0.7 mA in sleep mode of operation. It has a SIM card interface and can serially communicate with other devices using serial port. It works on "AT Commands" which it receives from the microcontroller. The GSM module can be used to send messages to users using AT Commands. This functioning ability of GSM to interact with a microcontroller and to send alert message to the user based on the microcontroller response makes it viable to be used in this type of security system [1].

PIR Sensor

Figure 4 is shown a passive infrared (PIR) sensor which has a variable resistance to align separation and deferral of timing. This sensor unit is definitely not hard to use rather it has a sensible cost. This sensor requires an operating current range of 100–150 μA and input voltage range of 3–5 V to be able to work. It can be calibrated to detect motion from 0.1 to 6 m. It also has the ability to work at a temperature range varying from −200 to 700 °C [1]. It works on a wavelength range of 7–14 μm. PIR sensors permit you to detect movement, additionally used to distinguish in the event that a person has stimulated in or out of the house. The PIR sensor has two regions from which the transmission and receiving of the electromagnetic waves take place [7]. When the PIR sensor is fitted in a region of space, and there is no motion of any body, a similar measure of IR from both the sensor openings is received, but when a bodily motion like that of a human comes in the sensor range, it automatically blocks first opening of the sensor, which tends to generate a positive differential change in current between the two openings of the sensor. Similarly, when the warm body tries to move away from the detecting range, the inversion takes place, making the sensor generate a negative differential change. These changes in beat are calculated and an

Fig. 5 Ultrasonic sensor as seen from top

output is given out by the sensor whether a motion has taken place or not. The PIR sensors are small, easily controlled and do not break easily. They are being used in various home automation and security devices.

Ultrasonic Sensor

This ultrasonic sensor provides a range measurement of 2–400 cm. It has a transmitter, receiver and a control unit. The module will be sending 8, 40 kHz signals [8]. The pulse in time of the signal is counted. This time is used to calculate the actual distance of the object from the module.

The following is the pin diagram of the ultrasonic sensor as shown in Fig. 5:

Power: +5 V
Trigger Input
Echo Output
Ground.

This module will measure the real-time distance of the objects. So if any intruder cuts its path, then the distance measured for a particular instance may drop below a set threshold, this would alert the microcontroller that an intruder has come inside the house. It will be fixed near the door or window, so any unwanted motion can easily be detected.

Power Supply

Figure 6 shows a 12 V, 1 A DC adaptor required to power up all the components.

Arduino IDE

The Arduino Integrated Development Environment is shown in Fig. 7. It has a very easy to work platform since execution of code is very fast. The programs can be easily written in C/C++, Java. The Arduino IDE assigns Arduino hardware pins to work as input and output pins accordingly as needed by the user thereby making interfacing the microcontroller with other modules a lot easier. The following two subparts of the program are the building blocks of the Arduino program [9]: void setup (): used for initialization and void loop (): used for implementing a task.

Fig. 6 Power supply unit

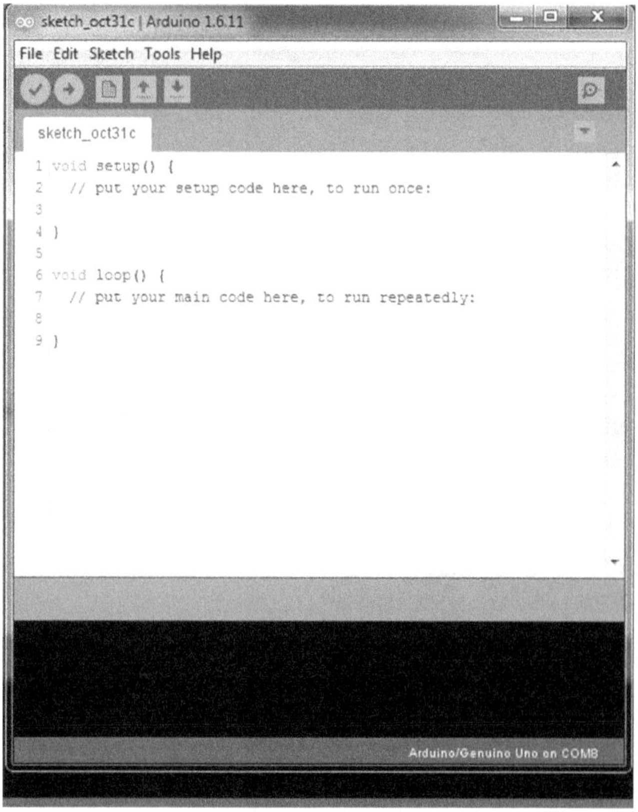

Fig. 7 Arduino IDE software

3 Implementation

See Fig. 8.

4 Results and Discussion

Once an intruder enters the room, either the PIR sensor fitted on the ceiling of the room or the ultrasonic sensor near the door and window, detects a motion and sends the data to Arduino. The Arduino processes the data and correspondingly sends an alert SMS to the owner of the house. The message reads as, "THE THIEVE IS IN THE HOUSE". Also a buzzer alarm goes high parallel to the SMS. Figure 9 shows the SMS received by the owner of the house.

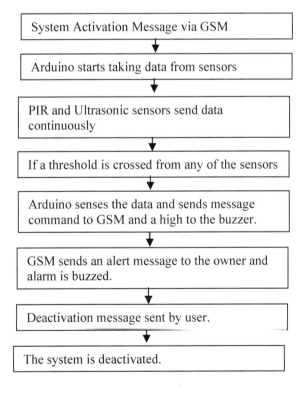

Fig. 8 Implementation of proposed home security system

Fig. 9 Alert SMS is sent to the registered user

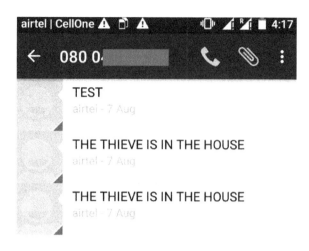

5 Conclusion

The proposed home security system is affordable and offers a standard security system to the people who have to go out of the house, no one being there to guard the house. It can easily be installed in homes and used by the people at any time since most of the people have access to mobile phones these days. Further, the efforts would be made in future to make the system cheaper and also to ensure that the power backup is provided to the system remotely which may last for long time to ensure security in case when there is no electricity at home.

References

1. Parab, A. S. (2015). Implementation of home security system using GSM module and the microcontroller. *IJCSIT, 6*(3), 2950–2953.
2. Budijono, S. (2014). Design and implementation of modular home security system with short messaging system. In *EJP Web of Conferences* (Vol. 68, p. 00025).
3. https://www.arduino.cc/en/Main/ArduinoBoardUno.
4. Chattoraj, S. (2015). Smart home automation based on different sensors and Arduino as master controller. *IJSRP, 5*(10), 1–4.
5. Patil, P. B. (2016). Home automation system using Android and Arduino board. *IJIRSET, 5*(4).
6. https://cdn-shop.adafruit.com/datasheets/sim800h_hardware_design_v1.00.pdf.
7. Salman, A. K. Proximity motion security system based on alert with multi-zone multi-responsible persons. *IJSR*. ISSN 2319-7064.
8. http://www.micropik.com/PDF/HCSR04.pdf.
9. Annapurna, L. B. Smart security system using Arduino and wireless communication. *IJEIR, 4*(2), ISSN 2277-5668.

Automatic Toll Tax Collection Using GSM

P. Mahalakshmi, Viraj Pradip Puntambekar, Aayushi Jain and Raunak Singhania

Abstract This paper proposes a very novel approach to implement the automatic toll tax collection system on the toll plazas using radio frequency identification (RFID) and global system for mobile (GSM). Nowadays, the cities and highways are bursting with traffic, and very often long queues of vehicles can be seen at various toll plazas so that they can pay the toll and then able to use the road or highway. So a system is proposed wherein the toll tax could be paid via cashless transactions and people wouldn't have to wait for a long time for the cash payment of the toll tax. This would save people's money and time simultaneously. It would also eliminate errors in cash transactions and further ease the job of the toll plaza companies. It would definitely bring down any of the corruptions occurring at the toll plazas. Finally, it would make the existing toll tax collection more efficient and ease our lives a bit more.

Keywords Arduino · RFID · GSM · Tags · Motor driver

1 Introduction

In our everyday life, we pay a certain measure of assessment through toll square to the administration. The toll entryways are for the most part found on national thruways and extensions and so on and we pay remaining over a line as money, despite the fact that the portability of vehicles gets hindered by this conventional technique which takes longer travel time and more utilization of fuel. Furthermore, the contamination level gets expanded in that locale [1].

Let us assume that the conventional toll collection system is quite efficient and the time taken by one vehicle to stop and pay toll tax is 1 min. Now let us suppose that 1000 vehicles will go through the toll booth every day. Then, the time consumed by a vehicle with an approximate stop of 1 min in a month is: $1 \times 30 = 30$ min and the total time taken in a year $= 30 \times 12$ min $= 360$ min. On a normal day each vehicle going through the toll court needs to sit tight for 6.0 h in the engine on state

P. Mahalakshmi (✉) · V. P. Puntambekar · A. Jain · R. Singhania
School of Electrical Engineering, Vellore Institute of Technology, Vellore 632014, India
e-mail: pmahalakshmi@vit.ac.in

© Springer Nature Singapore Pte Ltd. 2019
N. R. Shetty et al. (eds.), *Emerging Research in Computing, Information, Communication and Applications*, Advances in Intelligent Systems and Computing 906,
https://doi.org/10.1007/978-981-13-6001-5_54

condition. Additionally, every year 365,000 vehicles would be simply stopping for 6.0 h in engine on mode, subsequently expanding pollution and squandering fuel, time, and money. This is the situation when the framework is viewed as exceptionally productive; however, imagine a scenario in which every vehicle needs to hold up, say 3–4 min! Also, this is a figure in which we have considered only one toll plaza. Now considering the thousands of toll plazas, the above figure would be too big and the wastage of fuel, time, and money would also be very large [2].

Therefore, a new method is urgently required to reform this problem. The automation of the toll collection system can prove to be one of the easiest methods to provide a solution to the above-stated problem [3]. This framework does not require any manual operation of toll obstructions and gathering of toll sums; it is totally computerized toll accumulation framework. The vehicle proprietors are enlisted with their vehicle's appropriate data and their record is made, where they can renew their records with the required sum. At the point when the vehicle goes through the toll entryway, the data are shared between RFID tag and RFID reader and the sum is deducted from the proprietor's account and the balance amount message is sent to the user. This strategy lessens the travel time and decreases the fuel utilization [1].

2 Methodology

In order to implement the proposed system, we need the following hardware components and software tools for the design of our proposed system.

2.1 Arduino

Figure 1 shows Arduino Uno which is a microcontroller board having ATmega328P IC chip, 14 digital input/output pins, a 16 MHz crystal oscillator for clock, a USB port, a power jack, an ICSP header, and a reset catch [4, 5]. The microcontroller can essentially be associated with a PC with a USB port, and to power it an AC-to-DC connector or battery is required. You can play with your UNO without much thinking as you can buy a new one for a couple of dollars and start over again. Arduino can be used to create new projects with innovative ideas involved and also taking into consideration various switches or sensors, and thereby controlling an assortment of lights, engines, and other physical outputs. Arduino has many technical benefits which include open source platform and a very clear and simple programming environment.

Automatic Toll Tax Collection Using GSM

Fig. 1 Arduino Uno hardware module

Fig. 2 GSM module used

2.2 GSM

The GSM stands for global system for mobile. It is an advanced product innovated which is used for sending and receiving voice and text messages. GSM is widely acknowledged in the field of mass communication globally. As shown in Fig. 2, GSM module partitions every 200 kHz channel into eight 25 kHz time-openings. GSM works on the versatile correspondence groups including 900 and 1800 MHz in most parts of the world [6, 7].

GSM makes use of short band time division multiple access (TDMA). GSM fully supports voice and text messages and also gives facility for roaming which is the ability to use your unique GSM number registered on a network in another GSM network. GSM digitizes and packs information and then sends it down through proper channels [1].

Fig. 3 RFID reader

Benefits of using GSM module:

- The quality of speech gets improved very efficiently and effectively.
- The availability of the spectrum increases.
- The compatibility becomes more with mobiles.

2.3 RFID Reader

The RFID-RC522 reader is shown in Fig. 3. It has an operating frequency of 125 kHz, a detection range of 10 cm, and a wide operating temperature range and is cost effective also [8].

2.4 RFID Tags

The radio frequency identification system consists of three components [9]:

- An antenna
- A transceiver
- A transponder.

These RFID tags are classified as active tags and passive tags.

Fig. 4 RFID tags with its unique identity number fixed in vehicles

Passive RFID labels do not have their own particular power supply; the tag sends the reaction by the little electrical current actuated in the reception apparatus by means of radio recurrence filter. The response of passive RFID tag is only a UID number as shown in Fig. 4.

Active RFID labels have a power source and have longer ranges and bigger recollections than the passive labels; furthermore, they also have the ability to store the additional data sent by the transceiver. It may be noted that the technical differences between the tag types are not major concerns for collecting the data.

Additionally, the tags can be grouped on the basis of classes from Class 0 to Class 5. The classes have been controlled by electronic product code (EPC) Global Standard [9–11].

2.5 Motor

A DC motor is required as shown in Fig. 5 for rotating the gates, causing them to open and close whenever required [7].

Fig. 5 DC-powered motor

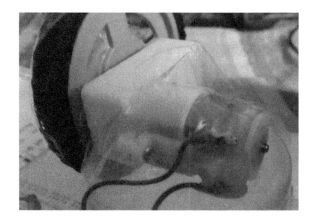

Fig. 6 L293D motor driver

2.6 Motor Driver

The L293D IC motor driver allows the DC motor to rotate in either direction as shown in Fig. 6. It is a set of 16 pins IC which has the ability to rotate the two DC motors simultaneously [6].

2.7 Power Supply

A power supply of 12 V, 1 A is expected to control the modules.

2.8 Arduino IDE Software

The Arduino Integrated Development Environment offers a straightforward and clear stage to execute codes and do nearly anything. The Arduino projects are composed in C/C++, Java. It has numerous inbuilt libraries. It allocates Arduino pins as input or output pins and makes collaboration with other modules and sensors a considerable measure easier. It has the accompanying two code scraps which are important to be composed dependably:

void setup (): used for setting up the initial conditions.
void loop (): used to run task infinitely.

3 Implementation

Figure 7 shows the working model of the proposed system, and Fig. 8 gives a description of the steps involved in the process of automatic toll tax collection.

4 Results

As the registered vehicle enters the toll plaza, the RFID reader reads the tag and the microcontroller deducts the balance and signals the traffic light to be green and also signals the motor driver to open the gate for the vehicle to pass on. It also sends a command to the GSM module to send message to the user as "Hey MR. XYZ. Your vehicle is: ABC. SIR, Your Remaining Balance Amount is: Rs. ----. HAVE A NICE DAY THANK YOU" as shown in Fig. 9.

Fig. 7 Working model of the proposed system

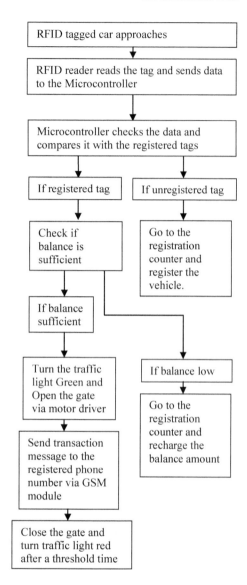

Fig. 8 Steps involved in the implementation of proposed system

5 Conclusion

The proposed system has been installed in a working model. Large-scale implementation of the proposed model can be done, and hence the traffic congestion can be avoided. By adopting this system, the manual effort could be reduced drastically. The emergency vehicles like ambulance, police cars, fire vans, etc., may be exempted from paying taxes. In this fast moving world, we need to update our technology for saving our precious time and money.

Fig. 9 Alert message being sent to the owner of the vehicle

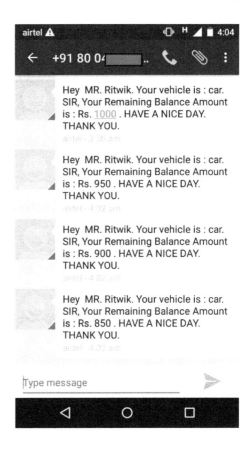

References

1. Andurkar, G. K., & Ramteke, V. R. (2015). Smart highway electronic toll collection system. *International Journal of Innovative Research in Computer and Communication Engineering, 3*(5).
2. Salunke, P., Malle, P., Datir, K., & Dukale, J. (2013). Automated toll collection system using RFID. *IOSR Journal of Computer Engineering (IOSR-JCE), 9*(2), 61–66. e-ISSN 2278-0661, ISSN 2278-8727.
3. Win, A. M., MyatNwe, C., & Latt, K. Z. (2014). RFID based automated toll plaza system. *International Journal of Scientific and Research Publications, 4*(6), 1. ISSN 2250-3153.
4. https://www.arduino.cc/en/Main/ArduinoBoardUno.
5. Chattoraj, S. (2015). Smart home automation based on different sensors and Arduino as master controller. *IJSRP, 5*(10), 1–4.
6. Narayanan, S. K., Thushara, C., Sandhya. C., Saranya, N., & Sreepriya, P. V. (2015). Automatic toll gate system using RFID and GSM technology. *IJSET, 3*(5).
7. Aphale, T., Chaudhari, R., & Bansod, J. (2015). Automated toll plaza using RFID and GSM. *International Journal on Recent and Innovation Trends in Computing and Communication, 2*(9). ISSN: 2321-8169
8. https://www.nxp.com/documents/data_sheet/MFRC522.pdf.

9. Giri, P., & Jain, P. (2013). Automated toll collection system using RFID technology. *International Journal of Advance Research in Science and Engineering, IJARSE, 2*(9).
10. Kamarulazizi, K., & Dr. ISMAIL, W. Electronic toll collection system using passive RFID technology. *Journal of Theoretical and Applied Information Technology*. © 2005–2010 JATIT & LLS.
11. Annapurna, L. B. (2015). Smart security system using Arduino and wireless communication. *IJEIR, 4*(2). ISSN 2277-5668.

Facial Expression Recognition by Considering Nonuniform Local Binary Patterns

K. Srinivasa Reddy, E. Sunil Reddy and N. Baswanth

Abstract Recognizing a face with an expression has paying attention due to its well-known applications in a broad range of fields like data-driven animation, human–machine interaction, robotics, and driver fatigue detection. People can vary significantly their facial expression; hence, facial expression recognition is not an easy problem. This paper presents a significant contribution for facial expression recognition by deriving a new set of stable transitions of local binary pattern by selecting the significant nonuniform local binary patterns. The proposed patterns are stable, because of the transitions from two or more consecutive ones to two or more consecutive zeros. For better recognition rate, the new set of patterns are combined with uniform patterns of local binary pattern. A distance function is used on proposed texture features for effective facial expression recognition. Preprocessing method is also used to get rid of the effects of illumination changes in facial expression by preserving the significant appearance details that are needed for facial expression recognition. The investigational analysis was done on the popular JAFFE facial expression database and has shown good performance.

Keywords Face · Expression · Pattern · Local binary pattern · Illumination · Preprocessing · Distance function and stable transition

1 Introduction

The face of human conveys a bunch of information concerning individuality and emotional situation. Facial expression is an instant and effectual part of message among humans which carries crucial information about the emotional, mental, and physical state. It is a desirable feature of next-generation computers, which can recognize facial expressions and responds accordingly and enables better human–machine interactions, driver state monitoring, medical and aiding autistic children. In modern years, automatic facial expression recognition has paying much attention [1–3].

K. Srinivasa Reddy (✉) · E. Sunil Reddy · N. Baswanth
Information Technology, Institute of Aeronautical Engineering, Hyderabad, Telangna, India
e-mail: kondasreenu@iare.ac.in

Much progress has been made in this area [4–8]; the expression recognition of a face with a high correctness remains hard due to the complexity and unpredictability of facial expressions. However, the intrinsic changeability of facial expression images is caused by different factors like variations in occlusions, pose, illumination, and alignment, which makes expression recognition a challenging task. Major categories of the facial expressions are annoyance (anger), disgust, terror (fear), happy, sad, surprise, and unbiased (neutral). A very fine variation in muscular movements causes different expressions, and hence, it is a critical task to do the local feature extraction to represent expressions. Two main types of approaches to extract facial expression features [9] are geometric feature-based methods [1, 2] and the appearance-based methods [3, 7]. Methods based on former one extract geometric information from the facial expression images; in later case, features are either extracted from the entire facial expression or specific regions in facial expression images. For more effectiveness, appearance-based approach was chosen. Appearance features evaluated from Gabor wavelets to be more effective than geometric features [6] even though computationally expensive.

Local Binary Pattern (LBP) [3, 4, 8] approach is empirically studied in this paper for person-independent facial expression recognition. The LBP features were proposed at first for texture analysis [10, 11] and are efficiently used in face detection and recognition [10]. Simplicity and tolerance against illumination changes are the most important properties of LBP. The experiment analysis was done with and without preprocessing technique [12, 13] to perform facial expression recognition using LBP features. By using LBP features, discriminative facial information in a compact representation can be retained in faster way with a single scan through the image. This paper proposes a facial expression recognition method by applying LBP operator and its variants on preprocessed images to extract features. This method is trained and tested on popular Japanese Female Facial Expression (JAFFE) database [14].

The remainder of this paper is structured as follows. We present a brief review of local binary pattern in the next section. Section 3 presents the proposed methodology, and Sects. 4 and 5 present the results, discussions, and conclusions.

2 Local Binary Pattern

The Local Binary Pattern (LBP) operator [11] is originally proposed for texture analysis. LBP is a gray-scale and rotation invariant texture primitive, easy, computationally competent, robust, and derives local attributes efficiently. Hence, LBP is extensively used in different domains like security and authentication for face recognition [10, 15] and facial expression recognition [3, 4, 7]. LBP operator can be represented as LBP(P,R) with different values of (P, R) where P represents the number of neighborhood pixels and R for radius. With $P = 8$ and $R = 1$, it is represented as LBP(8,1), i.e., eight-neighboring pixels on a 3 * 3 neighborhood which ranges from 0 to 255. Equation 1 describes the working principle of LBP operator.

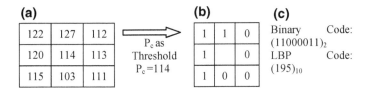

Fig. 1 Operation of LBP operator

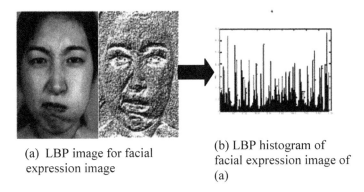

(a) LBP image for facial expression image

(b) LBP histogram of facial expression image of (a)

Fig. 2 LBP histogram. **a** LBP image for facial expression image. **b** LBP histogram of facial expression image of (**a**)

$$\text{LBP}(8, 1) = \sum_{n=0}^{7} 2^n S(P_c - P_n) \quad (1)$$

where 'n' is over the eight neighbors (0–7) of the central pixel C, P_c and P_n are the gray-level intensities at c, and n and $S(u)$ will be 1 if $u \geq 0$ and 0 otherwise. Figure 1 illustrates the working process of LBP on a 3 * 3 neighborhood.

From Fig. 1c, a P-bit binary number is produced, and equivalent decimal representation is replaced as central pixel value. Bilinear interpolation is used whenever any neighboring pixel does not fall exactly on a pixel position. The histogram of the LBP operator applied image is further used as a texture descriptor. Figure 2 shows the LBP image and histogram representation.

Very minor muscular movement causes to vary the facial expressions, and LBP concentrates on these minor movements using labeled micropatterns which are distributed throughout the face region. Furthermore, facial expression variations are naturally characterized by areas like corners of mouth, inner corner of eyes, chin, and supplementary subregions being slightly significant.

3 Methodology

The proposed work consists of four key steps as given below.

Step 1: Conquering the noise and illuminated issues by using preprocessing.
Step 2: Obtain LBP, ULBP, and SNULBP features.
Step 3: Assess histograms of LBP, ULBP, E-SNULBP, and ULBP U E-SNULBP.
Step 4: Effective facial expression recognition by using a distance function.

3.1 Preprocessing

Gamma correction and Difference of Gaussian (DoG) filtering [1, 13] are the preprocessing methods used to address the effects of lighting variations and local shadowing in this paper. Gamma is the most significant characteristic of image which defines the pixel gray-level value and its real luminance association. Without gamma, shades captured by digital camera would not come into view as they appear to human eyes. Gamma correction (GC) is a nonlinear gray-level transformation, and new gray level I, is defined in Eq. (2)

$$\text{New grey level } (I) = I^{\gamma} \text{ for } \gamma > 0$$
$$\log(I) \text{ for } \gamma = 1$$
$$\text{Where } \gamma \in \{0, 1\}. \quad (2)$$

From, Eq. (2), the gray-level range of the pixels in dark or shadowed regions is enhanced and compresses the dynamic range at bright regions. The limitation of Gamma correction occasionally fails in removing shadowing effects, i.e., the influence of overall intensity gradients. To eliminate this problem on gamma-corrected image, DoG is used. DoG is a very successful and accurate grayscale image enhancement algorithm. To conquer the illumination and local shadowing problems, the present paper utilized DoG feature.

3.2 Uniform and Nonuniform Local Binary Patterns

An LBP is treated as uniform if it contains at most one 0–1 and one 1–0 transition in a circular manner. For example, 11111111 (zero transitions) and 11000011 (two transitions) are uniform, whereas 11001100 (four transitions) 10101100 (six transitions), and 01010101 (eight transitions) are not uniform. For $P = 8$, there are 58 uniform local binary patterns (ULBP) and 198 nonuniform local binary patterns (NULBP). Few researchers [12, 16–18] considered only ULBPs for classification and recognition because of the subsequent reasons—(a) 80–85% of the texture images hold

Table 1 Proportions (%) of ULBP values on sample images from the JAFFE database

Facial expression image	$P = 8, R = 1$	$P = 16, R = 2$
KA.FE2.46	62.04	52.84
KM.HA1.4	52.17	45.24
KM.HA4.7	48.72	35.81
YM.FE2.68	54.72	42.04

only ULBPs. (b) ULBPs are treated as the primary properties of texture image. (c) Treating all 198 NULBPs as one miscellaneous set will reduce a lot of dimensionality from 256 to 59 (58 + 1) without losing the texture content. (d) The background information is preserved by considering only the pixels with uniform patterns. Much of the pixels around the eyes (especially the eyebrows), mouth, and the nose are uniform patterns.

For some facial expressions, we have observed that the leading patterns are not always uniform. Table 1: shows the percentage (%) of uniform patterns on sample images of the JAFFE database with different values of P and R [12, 15]. By considering all the nonuniform patterns under a miscellaneous set causes or may lose helpful micropattern structural features in the facial expression images.

From Table 1, it implies that the dominant patterns are not mainly the ULBPs. In order to describe the leading patterns contained in the facial expression images, the conventional LBP is extended by considering the new set of patterns which also have some NULBPs. The researchers [12, 19, 20] considered a part or a small number of NULBPs along with ULBPs and proved that this combination yielded a better progress than by considering only ULBPs.

3.3 Extended Significant Nonuniform Local Binary Pattern

The major difficulty is choosing the subset from NULBPs to get better performance and to diminish large dimensionality problem. To achieve this, we have derived and used extended significant nonuniform local binary pattern (E-SNULBP) [19]. The E-SNULBP is a subset of NULBP, in which transition pattern is defined as two or more 1's to two or more 0's and vice versa is not true. The transitions are measured in a circular manner. No ULBP will have such transition pattern.

The derived E-SNULBPs are stable, since the transitions considered are from two or more consecutive 0's to two or more consecutive 1's only, instead of zero to one or vice versa. For efficient facial expression recognition, the present paper combined the derived E-SNULBPs with ULBP using union operation only. Ninety different LBPs are formed out of 256 by union operation between E-SNULBP and ULBP (E-SNULBP U ULBP).

The ULBP codes like 24, 15, 64, and 243 do not fall into E-SNULBP, since their binary representations are having the transitions from 00 to 11 and also 11 to 00.

For example, 24 equivalent binary representations are 000110000; it is having the transition from 00 to 11 and also 11 to 00. The ULBP code 64 also does not fall into E-SNULBP, since the binary value of 64 is 01000000 which does not have a transition from 00 to 11 at all.

The NULBP codes like 9, 51, and 87 do not fall into E-SNULBP, since their binary representations are not having the transitions from 11 to 00 at all. For example, 9 equivalent binary representation is 00001001; it is not having the transition from 11 to 00. The NULBP code 51 also does not fall into E-SNULBP, since the binary representation of 35 is 00110011 which is having transitions from 11 to 00 and also 00 to 11 in circular manner. The NULBP codes like 23, 97, 151, and 188 fall into the category of E-SNULBP patterns.

4 Results and Discussions

The performance of facial expression recognition is evaluated on popular database JAFFE [14] which has 213 facial expression images with seven varieties of expression of ten Japanese women. Each person has three to four facial expressions with varieties of facial expressions, including anger, disgust, fear, happy, sadness, surprise, and neutral. Figure 3 shows seven expressions of a person which are selected from JAFFE database.

For competent facial expression recognition the histograms of ULBP, ULBP U E-SNULBP with various (P, R) on each individual facial expression image and placed in training database. $P = 8, 16$ and $R = 1, 2, 3, 4$ are considered for experimentation purpose. In the comparable way, the above histograms are evaluated for test facial expression image and the facial expression recognition is evaluated based on chi-square distance technique as given in Eq. 3.

$$R(d, t) = \min\left(\sum_{i=1}^{n}\left((d_i - t_i)^2/(d_i + t_i)\right)/2\right) \quad (3)$$

where d_i represents the sum of histograms of all E-SNULBP U ULBPs of training database image i and t represents the sum of histograms of all E-SNULBP U ULBPs of the test facial expression image. $R(d, t)$ represents that the database image d is nearer to test image t. Hence, the recognized image for t is d. The same procedure is adopted for the ULBP, E-SNULBP, and E-SNULBP U ULBP, and the average percentage of recognition rate for each expression type with and without preprocessing is shown in the tables. The present paper evaluated the histograms of the above three texture features without and with preprocessing approach on the JAFFE database. Tables 2, 3, and 4 show the facial expression recognition rates of angry, disgust, and fear, respectively, without preprocessing.

Tables 5, 6, 7, and 8 show the facial expression recognition rates of happy, neutral, sad, and surprise, respectively, without preprocessing.

Table 2 Angry facial expression recognition rate without preprocessing

(P, R)	ULBP	E-SNULBP	E-SNULBP U ULBP
(8, 1)	65.56	20.56	68.82
(8, 2)	66.67	32.78	72.04
(8, 3)	69.89	40.67	76.48
(8, 4)	67.67	52.98	75.26
(16, 1)	76.56	15.67	80.82
(16, 2)	76.56	27.57	81.62
(16, 3)	69.89	29.78	75.71
(16, 4)	71.71	31.75	80.82
Average	70.56	31.47	76.45

Table 3 Disgust facial expression recognition rate without preprocessing

(P, R)	ULBP	E-SNULBP	E-SNULBP U ULBP
(8, 1)	66.57	19.42	71.09
(8, 2)	67.45	33.09	71.23
(8, 3)	66.56	38.38	73.67
(8, 4)	64.85	47.42	72.49
(16, 1)	74.53	17.59	78.01
(16, 2)	74.53	28.08	78.19
(16, 3)	70.15	29.51	75.90
(16, 4)	71.2	32.59	81.01
Average	69.48	30.76	75.20

Table 4 Fear facial expression recognition rate without preprocessing

(P, R)	ULBP	E-SNULBP	E-SNULBP U ULBP
(8, 1)	67.75	19.39	71.04
(8, 2)	68.25	25.41	71.36
(8, 3)	69.46	30.14	72.72
(8, 4)	67.67	51.76	71.60
(16, 1)	68.56	17.43	73.04
(16, 2)	70.45	25.44	75.02
(16, 3)	69.89	30.32	75.03
(16, 4)	70.77	31.73	76.05
Average	69.10	28.95	73.23

Table 5 Happy facial expression recognition rate without preprocessing

(P, R)	ULBP	E-SNULBP	E-SNULBP U ULBP
(8, 1)	67.94	24.08	73.33
(8, 2)	69.05	36.30	76.55
(8, 3)	72.27	44.19	80.99
(8, 4)	70.05	56.50	79.77
(16, 1)	78.94	19.19	85.33
(16, 2)	78.94	31.09	86.13
(16, 3)	72.27	33.30	80.22
(16, 4)	74.09	35.27	85.33
Average	72.94	34.99	80.96

Table 6 Neutral facial expression recognition rate without preprocessing

(P, R)	ULBP	E-SNULBP	E-SNULBP U ULBP
(8, 1)	66.51	17.95	70.47
(8, 2)	67.01	23.97	70.79
(8, 3)	68.22	28.70	72.15
(8, 4)	66.43	50.32	71.03
(16, 1)	67.32	15.99	72.47
(16, 2)	69.21	24.00	74.45
(16, 3)	68.65	28.88	74.46
(16, 4)	69.53	30.29	75.48
Average	67.86	27.51	72.66

Table 7 Sad facial expression recognition rate without preprocessing

(P, R)	ULBP	E-SNULBP	E-SNULBP U ULBP
(8, 1)	65.38	17.02	68.67
(8, 2)	65.88	23.04	68.99
(8, 3)	67.09	27.77	70.35
(8, 4)	65.30	49.39	69.23
(16, 1)	66.19	15.06	70.67
(16, 2)	68.08	23.07	72.65
(16, 3)	67.52	27.95	72.66
(16, 4)	68.40	29.36	73.68
Average	66.73	26.58	70.86

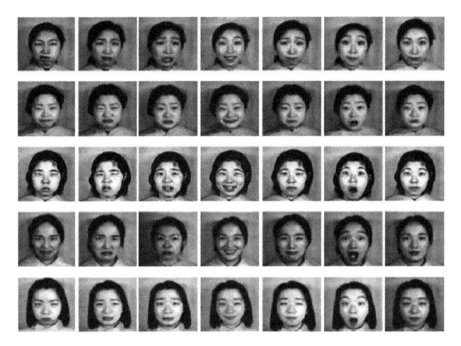

Fig. 3 Sample facial expression images from JAFFE dataset

Table 8 Surprise facial expression recognition rate without preprocessing

(P, R)	ULBP	E-SNULBP	E-SNULBP U ULBP
(8, 1)	67.45	22.45	70.71
(8, 2)	68.56	34.67	73.93
(8, 3)	71.78	42.56	78.37
(8, 4)	69.56	54.87	77.15
(16, 1)	78.45	17.56	82.71
(16, 2)	78.45	29.46	83.51
(16, 3)	71.78	31.67	77.60
(16, 4)	73.60	33.64	82.71
Average	72.45	33.36	78.34

Tables 9, 10, 11, 12, 13, 14, and 15 show the preprocessing results of seven facial expressions recognition rates.

From Tables 2, 3, 4, 5, 6, 7, 8, 9, 10, 11, 12, 13, 14, and 15, we can clearly observe that the facial expression recognition rate is high for (P_2, R) when compared to (P_1, R) where $P_2 > P_1$; i.e., the considered neighborhood points are more, for the same radius. This is evident for all facial expressions. The reason for this is the number of NULBP's will increase as we increase the number of neighboring points for the same radius, treating them as miscellaneous will reduce overall facial expression

Table 9 Angry facial expression recognition rate with preprocessing

(P, R)	ULBP	E-SNULBP	E-SNULBP U ULBP
(8, 1)	87.57	22.24	92.32
(8, 2)	87.57	32.98	92.62
(8, 3)	89.27	43.15	92.83
(8, 4)	90.40	57.27	93.40
(16, 1)	92.66	13.77	93.96
(16, 2)	92.53	28.46	94.12
(16, 3)	94.92	32.98	96.22
(16, 4)	94.35	33.54	95.66
Average	91.16	33.05	93.89

Table 10 Disgust facial expression recognition rate with preprocessing

(P, R)	ULBP	E-SNULBP	E-SNULBP U ULBP
(8, 1)	85.00	18.74	90.14
(8, 2)	86.25	31.24	90.34
(8, 3)	95.00	43.74	92.14
(8, 4)	91.25	59.99	93.24
(16, 1)	88.75	19.43	95.04
(16, 2)	95.00	21.42	96.25
(16, 3)	97.50	26.24	96.75
(16, 4)	92.50	49.24	97.44
Average	91.41	33.76	93.92

Table 11 Fear facial expression recognition rate with preprocessing

(P, R)	ULBP	E-SNULBP	E-SNULBP U ULBP
(8, 1)	80.97	16.78	82.65
(8, 2)	82.08	28.87	84.13
(8, 3)	83.19	41.11	86.85
(8, 4)	82.08	52.33	87.88
(16, 1)	86.53	14.56	87.88
(16, 2)	86.53	25.45	88.85
(16, 3)	83.19	28.94	88.64
(16, 4)	84.31	32.33	89.97
Average	83.61	30.05	87.11

Table 12 Happy facial expression recognition rate with preprocessing

(P, R)	ULBP	E-SNULBP	E-SNULBP U ULBP
(8, 1)	92.54	31.96	93.50
(8, 2)	92.54	37.33	94.25
(8, 3)	93.38	42.42	94.75
(8, 4)	93.95	49.48	95.45
(16, 1)	95.08	27.73	95.25
(16, 2)	94.52	35.07	95.65
(16, 3)	96.21	37.33	96.45
(16, 4)	95.93	37.61	96.75
Average	94.27	37.37	95.26

Table 13 Neutral facial expression recognition rate with preprocessing

(P, R)	ULBP	E-SNULBP	E-SNULBP U ULBP
(8, 1)	91.25	19.15	93.63
(8, 2)	91.88	31.13	94.45
(8, 3)	96.25	42.35	96.75
(8, 4)	94.38	51.55	95.45
(16, 1)	93.13	25.42	96.88
(16, 2)	96.25	24.54	97.25
(16, 3)	97.50	34.25	97.50
(16, 4)	95.00	42.56	98.83
Average	94.45	33.87	96.34

Table 14 Sad facial expression recognition rate with preprocessing

(P, R)	ULBP	E-SNULBP	E-SNULBP U ULBP
(8, 1)	84.90	19.57	89.65
(8, 2)	84.90	30.31	89.95
(8, 3)	86.60	40.48	90.16
(8, 4)	87.73	54.60	90.73
(16, 1)	89.99	11.10	91.29
(16, 2)	89.86	25.79	91.45
(16, 3)	92.25	30.31	93.55
(16, 4)	91.68	30.87	92.99
Average	88.49	30.38	91.22

Table 15 Surprise facial expression recognition rate with preprocessing

(P, R)	ULBP	SNULBP	SNULBP U ULBP
(8, 1)	82.86	20.07	83.86
(8, 2)	83.97	23.41	85.08
(8, 3)	85.08	34.52	86.14
(8, 4)	83.97	36.74	86.54
(16, 1)	88.42	23.41	90.64
(16, 2)	88.42	27.34	91.53
(16, 3)	85.08	30.83	87.54
(16, 4)	86.20	34.22	86.44
Average	85.50	28.82	87.22

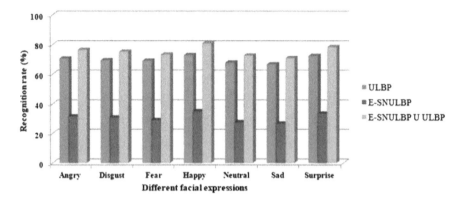

Fig. 4 Comparison of different facial expressions recognition rate without preprocessing

recognition rate. That's why one needs to consider E-SNULBPs to increase face recognition rate, and also by looking at E-SNULBP column in all tables, it is clearly evident that facial expression recognition rate is increasing gradually by increasing R. This is because as we increase R, the LBP contains more number of NULBPs. Therefore, one should consider the proposed E-SNULBPs for an accurate facial expression recognition, as R increases. Figures 4 and 5 show the average facial expression recognition rates without and with preprocessing.

From Figs. 4 and 5, we can understand that angry, disgust, happy and neutral expressions can be recognized with high accuracy (above 92%), but fear, sad, and surprise with moderate recognition rate (80–92%) with preprocessing. The facial expression recognition rates of the derived E-SNULBP U ULBP are higher than ULBP alone in both cases without and with preprocessing. This clearly reflects the advantages of the derived E-SNULBP and the importance of considering the nonuniform local binary patterns for facial expression recognition.

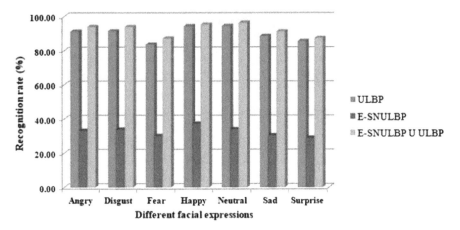

Fig. 5 Comparison of different facial expressions recognition rate with preprocessing

5 Conclusions

A facial expression recognition method is proposed which uses robust and simple LBP. The present work observed, the strong reason for considering NULBP's, because as the increase in values of P or R or both the number of NULBPs increases abnormally. If one treats such a huge number of patterns as miscellaneous, then definitely some image content will be lost and this degrades the overall performance. To overcome this and to deal with dimensionality, the present paper derived E-SNULBPs. The E-SNULBPs are stable because they considered the transition pattern is defined as two or more zeros to two or more ones and vice versa is not true. The results clearly indicate the proposed E-SNULBP U ULBP has shown high performance when compared to ULBP alone for all facial expressions. This clearly indicates that the significance of the proposed E-SNULBP will improve the overall facial expression recognition rate. Experimental results evidently show that facial expression recognition rate is improved by preprocessing. The proposed method can be extended to identify facial expressions recognition in real-time situations with video sequences.

References

1. Guo, G., & Dyer, C. (2005). Learning from examples in the small sample case: Face expression recognition. *IEEE Transactions on Systems, Man, and Cybernetics, Part B: Cybernetics, 35*(3), 477–488.
2. Zhang, Q., Liu, Z., Quo, G., Terzopoulos, D., & Shum, H. Y. (2006). Geometry-driven photorealistic facial expression synthesis. *IEEE Transactions on Visualization and Computer Graphics, 12*(1), 48–60.

3. Feng, X. Facial expression recognition based on local binary patterns and coarse-to-fine classification. In *Proceedings of the Fourth International Conference on Computer and Information Technology (CIT'04)*.
4. Shan, C., Gong, S., & McOwan, P. W. (2009). Facial expression recognition based on local binary patterns: A comprehensive study. *Image and Vision Computing, 27*, 803–816.
5. Huang, D., Shan, C., Ardabilian, M., Wang, Y., & Chen, L. (2011). Local binary patterns and its application to facial image analysis: A survey. *IEEE Transaction on Systems, Man and Cybernetics-Part C: Applications and Reviews, 41*(6), 765–781.
6. Ye, J. F., Zhan, Y. Z., & Song, S. L. (2004). Facial expression features extraction based on Gabor wavelet transformation. *IEEE International Conference on System, Man and Cybernetics, 3*, 2215–2219.
7. Kaur, M., Vashisht, R., & Neeru, N. (2010). Recognition of facial expressions with principal component analysis and singular value decomposition. *International Journal of Computer Applications, 9*(12), 36–40.
8. Feng, X., Pietikainen, M., & Hadid, A. (2005). Facial expression recognition with local binary patterns and linear programming. *Pattern Recognition and Image Analysis, 15*(2), 546–548.
9. Fellenz, W., Taylor, J., Tsapatsoulis, N., & Kollias, S. (1999). Comparing template-based, feature based and supervised classification of facial expression from static images. *Computational Intelligence and Applications, 19*, 9.
10. Ahonen, T., Hadid, A., & Pietikinen, M. (2004). Face recognition with local binary patterns. In *ECCV* (pp. 469–481).
11. Ojala, T., Pietikainen, M., & Maenpaa, T. (2002). Multiresolution gray-scale and rotation invariant texture classification with local binary patterns. *IEEE Transactions on Pattern Analysis and Machine Intelligence, 24*(7), 971–987.
12. Liao, S., Fan, W., Chung, A. C. S., & Yeung, D.-Y. (2006). Facial expression recognition using advanced local binary patterns, Tsallis entropies and global appearance features. In *ICIP* (pp. 665–668).
13. Tan X., & Triggs B. (2007). Enhanced local texture feature sets for face recognition under difficult lighting conditions. In *Proceedings of the 2007 IEEE International Workshop on Analysis and Modeling of Faces and Gestures (AMFG'07), Rio de Janeiro, Brazil* (pp. 168–182).
14. Lyons, M. J., Akamatsu, S., Kamachi, M., Gyoba, J., & Budynek, J. (1998). The Japanese female facial expression (JAFFE) database. http://www.kasrl.org/jaffe.html.
15. Vijaya Kumar, V., Srinivasa Reddy, K., & Krishna, V. V. (2015). Face recognition using prominent LBP model. *International Journal of Applied Engineering Research (IJAER), 10*(2), 4373–4384.
16. Fathi, A., & Naghsh-Nilchi, A. R. (2012). Noise tolerant local binary pattern operator for efficient texture analysis. *Pattern Recognition Letters, 33*(9), 1093–1100.
17. Liu, L., Zhao, L., Long, Y., Kuang, G., & Fieguth, P. (2012). Extended local binary patterns for texture classification. *Image and Vision Computing, 30*(2), 86–99.
18. Nguyen, H. V., Bai, L., Shen, L. (2009). Local gabor binary pattern whitened pca: A novel approach for face recognition from single image per person. In *Advances in biometrics* (pp. 269–278). Berlin, Heidelberg, Springer.
19. Srinivasa Reddy, K., Venkata Krishna, V., & Vijaya Kumar, V. (2016). A method for facial recognition based on local features. *International Journal of Mathematics and Computation (IJMC), 27*(03), 98–112.
20. Lowe, D. G. (2004). Distinctive image features from scale-invariant key points. *International Journal of Computer Vision, 60*(2), 91–110.

CPSIA information can be obtained
at www.ICGtesting.com
Printed in the USA
LVHW081005220919
631862LV00006B/698/P